Microbiomics

Dimensions, Applications, and Translational Implications of Human and Environmental Microbiome Research

Translational and Applied Genomics Series

Microbiomics

Dimensions, Applications, and Translational Implications of Human and Environmental Microbiome Research

Edited by

Manousos E. Kambouris

The Golden Helix Foundation, London, United Kingdom

Aristea Velegraki

Mycology Research Laboratory and UOA/HCPF culture collection, Department of Microbiology, School of Medicine, National and Kapodistrian University of Athens, Athens, Greece

ELSEVIER

ACADEMIC PRESS

An imprint of Elsevier

Academic Press is an imprint of Elsevier
125 London Wall, London EC2Y 5AS, United Kingdom
525 B Street, Suite 1650, San Diego, CA 92101, United States
50 Hampshire Street, 5th Floor, Cambridge, MA 02139, United States
The Boulevard, Langford Lane, Kidlington, Oxford OX5 1GB, United Kingdom

Notices
Knowledge and best practice in this field are constantly changing. As new research and experience broaden our understanding, changes in research methods, professional practices, or medical treatment may become necessary.

Practitioners and researchers must always rely on their own experience and knowledge in evaluating and using any information, methods, compounds, or experiments described herein. In using such information or methods they should be mindful of their own safety and the safety of others, including parties for whom they have a professional responsibility.

To the fullest extent of the law, neither the Publisher nor the authors, contributors, or editors, assume any liability for any injury and/or damage to persons or property as a matter of products liability, negligence or otherwise, or from any use or operation of any methods, products, instructions, or ideas contained in the material herein.

British Library Cataloguing-in-Publication Data
A catalogue record for this book is available from the British Library

Library of Congress Cataloging-in-Publication Data
A catalog record for this book is available from the Library of Congress

ISBN: 978-0-12-816664-2

For Information on all Academic Press publications
visit our website at https://www.elsevier.com/books-and-journals

Publisher: Andre Gerhard Wolff
Acquisitions Editor: Peter B. Linsley
Editorial Project Manager: Kristi Anderson
Production Project Manager: Stalin Viswanathan
Cover Designer: Miles Hitchen

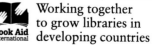

Typeset by MPS Limited, Chennai, India

Contents

List of Contributors

Cleo G. Anastassopoulou
Department of Biology, University of Patras, University Campus Rio, Patras, Greece

Michael Arabatzis
First Department of Dermatology-Venereology, Medical School, Aristotle University of Thessaloniki, Thessaloniki, Greece

Ilias Boltsis
Department of Cell Biology, Erasmus Medical Centre, Rotterdam, The Netherlands

Paraskevi Bouki
Laboratory of Food Microbiology and Biotechnology, Department of Food Science and Nutrition, University of Thessaly, Karditsa, Greece

Constantinos A. Chassomeris
Department of Pharmacy, University of Patras, Patras, Greece

Patrícia Pimentel de Barros
Department of Biosciences and Oral Diagnosis, Institute of Science and Technology, São Paulo State University (UNESP), São José dos Campos, São Paulo, Brazil

Georgios Gaitanis
Department of Skin and Venereal Diseases, Faculty of Medical Sciences, School of Medicine, University of Ioannina, Ioannina, Greece

Rose Ghemrawi
College of Pharmacy, Al Ain University, Abu Dhabi, United Arab Emirates

Ioannis Giavasis
Laboratory of Food Microbiology and Biotechnology, Department of Food Science and Nutrition, University of Thessaly, Karditsa, Greece; General Department, University of Thessaly, Karditsa, Greece

Dimitris Glistras
Department of History and Archaeology, National and Kapodistrian University of Athens, Athens, Greece

Konstantinos Grivas
Hellenic Military Academy, Kitsi, Greece

Juliana Campos Junqueira
Department of Biosciences and Oral Diagnosis, Institute of Science and Technology, São Paulo State University (UNESP), São José dos Campos, São Paulo, Brazil

Manousos E. Kambouris
The Golden Helix Foundation, London, United Kingdom

Maria Kantzanou
Department of Hygiene, Epidemiology & Medical Statistics, National Retrovirus Reference Center, School of Medicine, National and Kapodistrian University of Athens, Athens, Greece

Kristel Klaassen
Institute of Molecular Genetics and Genetic Engineering, University of Belgrade, Belgrade, Serbia

Martin Laurence
Shipshaw Labs, Montreal, QC, Canada

Yiannis N. Manoussopoulos
Laboratory of Virology, Plant Protection Division of Patras, ELGO-Demeter, NEO & Amerikis, Patras, Greece

Chrysanthi Mitsagga
Laboratory of Food Microbiology and Biotechnology, Department of Food Science and Nutrition, University of Thessaly, Karditsa, Greece

Eleftherios Mylonakis
Infectious Diseases Division, Alpert Medical School & Brown University, Providence, RI, United States

Basilis Papathanasiou
Department of Turkish and Contemporary Asian Studies, National and Kapodistrian University of Athens, Athens, Greece

George P. Patrinos
Department of Pharmacy, University of Patras School of Health Sciences, Patras, Greece; Department of Pathology, College of Medicine and Health Sciences, United Arab Emirates University, Al-Ain, Abu Dhabi, United Arabic Emirates; Faculty of Medicine and Health Sciences, Department of Pathology, Bioinformatics Unit, Erasmus University Medical Center, Rotterdam, The Netherlands

Sonja Pavlovic
Institute of Molecular Genetics and Genetic Engineering, University of Belgrade, Belgrade, Serbia

Felipe de Camargo Ribeiro
Department of Biosciences and Oral Diagnosis, Institute of Science and Technology, São Paulo State University (UNESP), São José dos Campos, São Paulo, Brazil

Rodnei Dennis Rossoni
Department of Biosciences and Oral Diagnosis, Institute of Science and Technology, São Paulo State University (UNESP), São José dos Campos, São Paulo, Brazil

Stavroula Siamoglou
Department of Pharmacy, University of Patras, Patras, Greece

Mohammad Al Sorkhy
College of Pharmacy, Al Ain University, Abu Dhabi, United Arab Emirates

Biljana Stankovic
Institute of Molecular Genetics and Genetic Engineering, University of Belgrade, Belgrade, Serbia

Maja Stojiljkovic
Institute of Molecular Genetics and Genetic Engineering, University of Belgrade, Belgrade, Serbia

Aristea Velegraki
Mycology Research Laboratory and UOA/HCPF culture collection, Department of Microbiology, School of Medicine, National and Kapodistrian University of Athens, Athens, Greece

Loukia Zerva
Department of Pathophysiology, School of Medicine, National and Kapodistrian University of Athens, Athens, Greece

Branka Zukic
Institute of Molecular Genetics and Genetic Engineering, University of Belgrade, Belgrade, Serbia

CHAPTER 1

INTRODUCTION: THE MICROBIOME AS A CONCEPT: VOGUE OR NECESSITY?

Manousos E. Kambouris[1] and Aristea Velegraki[2]

[1]The Golden Helix Foundation, London, United Kingdom [2]Mycology Research Laboratory, Department of Microbiology, School of Medicine, National and Kapodistrian University of Athens, Athens, Greece

The notion of "microbiome" has no negative or threatening sound on principle. Both its standard and its projected forms (with a multitude of new microbiota) are applicable and important in bioindustry, bioremediation, biotechnology, in environmental applications, and perhaps even in energy. But the aspect of pathogenicity is undeniably the most important, and the one able to threaten our species with extinction. Thus this aspect is always preferentially studied in terms of resources of any kind.

The universe of OMICS is now a tridimensional one; the first, original dimension, coined before the turn of the millennium, referred to collectivities within a cell: genomics, proteomics, transcriptomics, metabolomics, interactomics, and their special subsets such as pharmaco/toxico/immunogenomics. The second dimension emerged in the late 2000s, and referred to collectivities or interactions of multiple cells or organisms. It was the time of infectiomics, microbiomics, and the different levels and categories of the latter. A third iteration appeared in the 2010s, centered on the methods of study and not the organisms or their constituents: metagenomics, radiomics, and culturomics.

The -ome/-omic sciences are really, at least in part, something of a vogue. The use of the suffixes has grown out of all proportion and necessity, to retouch sectors, make studies more relevant and projects more promising, even in cases that are accurately and correctly described by previous, pre-Omics terminology. But in Microbiology, the Microbiomic Culture is a true necessity, as it encompasses the multidimensional and holistic view of an integrative picture, where biotic (microbiota and macrobiota) and abiotic elements interact in multiple levels, which are digested in much deeper zoom than previously done by ecological studies through the concept of niches, cycles, and networks.

The concept of *Microbiome* is not entirely novel: the term contains the word "microbe" almost intact, with the suffix "-ome," a latinicized version of the Greek suffix "-ωμα" denoting entirety, sum, or collectivity. Thus the Microbiome is the sum of all microbes sharing a common denominator, usually a common location/environment within a defined timeframe, as initially proposed in a largely forgotten stroke of foresight by Whipps et al. (1988), pg. 176: "A convenient ecological framework in which to examine biocontrol systems is that of the microbiome. This may be defined as a characteristic microbial community occupying a reasonably well defined habitat which has

Microbiomics. DOI: https://doi.org/10.1016/B978-0-12-816664-2.00001-3

distinct physio-chemical properties. The term thus not only refers to the microorganisms involved but also encompasses their theatre of activity."

There has been a tendency to use the term Microbiome to denote the sum of microbial genomes, as proposed by Hooper and Gordon (2001): "The Nobel laureate Joshua Lederberg has suggested using the term 'microbiome' to describe the collective genome of our indigenous microbes (microflora), the idea being that a comprehensive genetic view of *Homo sapiens* as a life-form should include the genes in our microbiome...." This context, apart from being hopelessly restrictive as it focuses solely in colonizing microbiota and, even worse, colonizers exclusively of *Homo sapiens*, and unimaginatively unidisciplinary, since it includes only a genetic/genomic aspect, is also both unclear, in its qualitative dimension, and inaccurate. Inaccurate, because there is not any letter in the word "microbiome" to attest to the genetic constituent of "genome," as the suffix "-ome" is about collectivity and entirety. And unclear, as it does not unequivocally describe, even qualitatively, the entirety of microbial genomes, as it does not define the genomic unit within a taxon horizon: genomes of different subgenera and, even more prominently, of different subspecies is unclear whether they are scored as one or as multiple entities.

In its organism-centered use the term Microbiome supplants a very well-established concept, the *microbial flora* (Tancrede, 1992). The latter term denoted possible heterogeneity and a dynamic, multi-level structure expressed by multiple and different interactions, mainly with the "environment" but also among the different species and niches; the concept of "microbiome" though, infers to much higher, more diverse and more impactful interactions. In addition, it remedies the conceptual pitfall of *flora*, which by definition denotes kingdom Plantae, or, in a functional, food-chain aspect, photosynthetic organisms that can be expanded to include autotrophs/producers in general. On the contrary, many if not most members of microbial floras are heterotrophs (consumers and decomposers).

Another similar term has been the *Microbial Community*, denoting microbes coexisting in space and time (Escobar-Zepeda et al., 2015). The term and concept were quite handy, lacking only the pertaining notion of collectivity and conceptual unity needed to describe the constituting microbiota as one entity, differentiated from and possibly opposed to other entities participating actively in the definition of their environment or coexisting in the same spatiotemporal window. Furthermore, the unity or common denominator which defines a certain Microbiome does not have to be a 3D location, possibly augmented by a temporal dimension, although it usually is. But the microbiome might be a functional or other sum of microbial populations.

Moreover, "microbiome" pertains to two aspects previously underappreciated and not appropriately covered by the notion of "microbial flora." The first such aspect is the motion of microbiota. Microbial flora echoes of motionlessness in spatial terms, while many microbiota are endowed with active motion/motility. Actually the concept of microbiome unifies both microbial flora and any functional concept of microbial fauna. Even in the lower microbiomatic levels, that is bacteriomes and viromes, the existence of actual predators or consumers, as are the predatory bacteria (Sockett, 2009) and the virophages (Bekliz et al., 2016; Katzourakis and Aswad, 2014), is analogous rather to faunal than to floral attributes and reminiscent of the more evolved, eukaryotic hyperparasites (Parratt and Laine, 2016).

The second aspect is that the *microbial flora* as a term and concept projects no notion whatsoever of the intense genetic processes inherent among microbiotes. The Plantae, which constitute the regular *flora* (or macroflora), are restricted to mispollination in this respect, with misfertilized plants producing sterile or dysfunctional fruit or ploidal variations. In stark contrast, microbiota, and especially *Prokarya*, present vigorous genetic mobility and exchange, a fact so prominent that the term "microbiome" is often erroneously used to denote the sum of genomes of the microbes, and not the microbiota proper, of a certain environment (Hooper and Gordon, 2001).

The concept of Microbiome is suitable for current and future needs to analyze and describe the explosive changes expected in microbiology due to biotechnology endeavors. It is the correct approach to investigate and comprehend intermingled and interacting microbial populations, as it does not imply the stable, dynamic but balanced condition of the microflora, where disturbances and changes are a diversion from the normality. The Microbiome may be understood as an instantaneous OR continuous entity, which incorporates the phenomena of nonlinear genetic exchange. It allows conceptual flexibility and room to accommodate (1) novel, engineered, or fully artificial microbiota (Hutchison et al., 2016; Smith et al., 2003; Malyshev et al., 2014), tentatively called "metapathogens" and "neopathogens" in their pathogenic capacities and depending on the degree of engineering (Kambouris et al., 2018); (2) the microbiota as appendages or symbiotic exo-organs of macroorganisms (Hooper and Goron, 2001), thus constituting (sub)microbiomes and individual microbiomes; (3) our expanded concept of microbiota, with the viruses being counted as proper lifeforms of acellular nature (Pearson, 2008) and an open door to extend such status to viroids; and (4) the genomic interactions implicating any nucleic acid form, from the single, possibly decaying DNA strands used in transformation to the elaborate mechanisms of transposition and transduction, which should be included, when in extracellular phase, in the "exogenome" (Kambouris et al., 2018).

The latter should be viewed as a collective, prospective pool of genetic information, which is potentially translatable depending on the receiver. The mind goes to prokaryotes, which incorporate any naked DNA by transformation, without any concern over its origin or sequence/meaning. Such randomly incorporated DNAs might introduce completely new protein threads, which may prime equal revolutionary events rather than evolutionary ones, by accumulating amino acid changes (individual or "en block") in existing proteins or adding/eliminating protein families by transferable elements.

When considering eukaryotes, naked DNAs are used for catabolism and lateral transfer intercellularly is of questionable applicability. But transduction by viruses is another issue altogether. Gene disruption and specialized transduction are well-attested events, but insertion in an exomic sequence, especially if the viral function is inert (pseudovirus) or becomes so (either randomly or by cellular defense mechanisms), equally produces a totally novel protein. The event introduces within an open reading frame a sequence possibly read in a different one and definitely having different length and embedded regulation, and this holds true for both eukaryotic and prokaryotic host genomes.

The alleged "democratization" in different scientific and technological areas provides prospects but also hides dangers. The "self-crisping" (Ireland, 2017) of interested individuals toward therapy or deranged concepts of superpowers is an upsetting and even worrisome fact, mostly in social terms, but not alarming or unnerving. On the contrary, self-acclaimed microbiologists and biotechnologists may have tremendous biological, not social-only, impact. In future, in each block or neighborhood, a genomics-capable microbiological workshop, mostly illegitimate, may be run to produce amenities which today are covered by very different products (from drugs to food and weapons or powerful chemicals); it also may not. But the problem is that it DOES may happen.

The inherent problems in research conduct, the current difficulty in performing second-generation sequencing, in terms of platform availability but also due to the exceptional needs in computational power (Escobar-Zepeda et al., 2015), might have been a safety feature, which will be removed by the oncoming third generation. Thus the methodological and cognitive tools to respond to a reality where the emergence of a new microbe happens many times a day in multiple, dispersed localities must be in place, even if this bleak prospect is considered improbable or unlikely. After all, the virtual humanity and the Internet of Things were considered scienceless fiction during the lifetime of today's mainstay scientists.

REFERENCES

Bekliz, M., Colson, P., La Scola, B., 2016. The expanding family of virophages. Viruses 8 (11), 317–331.

Escobar-Zepeda, A., Vera-Ponce de León, A., Sanchez-Flores, A., 2015. The road to metagenomics: from microbiology to DNA sequencing technologies and bioinformatics. Front. Genet. 6, 348.

Hooper, L.V., Gordon, J.I., 2001. Commensal host-bacterial relationships in the gut. Science 292, 1115–1118.

Hutchison, C.A., Chuang, R.Y., Noskov, V.N., Assad-Garcia, N., Deerinck, T.J., Ellisman, M.H., et al., 2016. Design and synthesis of a minimal bacterial genome. Science 351, 6253.

Ireland, T., December 24, 2017. I want to help humans genetically modify themselves. The Guardian. Available from: <https://www.theguardian.com/science/2017/dec/24/josiah-zayner-diy-gene-editing-therapy-crispr-interview> (accessed 15.11.18.).

Kambouris, M.E., Gaitanis, G., Manousopoulos, Y., Arabatzis, M., Kantzanou, M., Kostis, K., et al., 2018. Humanome versus microbiome: games of dominance and pan-biosurveillance in the Omics universe. OMICS-JIB 22 (8), 528–538.

Katzourakis, A., Aswad, A., 2014. The origins of giant viruses, virophages and their relatives in host genomes. BMC Biol. 12, 51–54.

Malyshev, D.A., Dhami, K., Lavergne, T., Chen, T., Dai, N., Foster, J.M., et al., 2014. A semi-synthetic organism with an expanded genetic alphabet. Nature 509, 385–388.

Parratt, S.R., Laine, A.-L., 2016. The role of hyperparasitism in microbial pathogen ecology and evolution. ISME J. 10 (8), 1815–1822.

Pearson, H., 2008. Virophage suggests viruses are alive. Nature 454, 677.

Smith, H.O., Hutchison, C.A., Pfannkoch, C., Venter, J.C., 2003. Generating a synthetic genome by whole genome assembly: phiX174 bacteriophage from synthetic oligonucleotides. Proc. Natl. Acad. Sci. U.S.A. 100 (26), 15440–15445.

Sockett, R.E., 2009. Predatory lifestyle of *Bdellovibrio bacteriovorus*. Annu. Rev. Microbiol. 63, 523–539.

Tancrede, C., 1992. Role of human microflora in health and disease. Eur. J. Clin. Microbiol. Infect. Dis. 11 (11), 1012–1015.

Whipps, J.M., Lewis, K., Cooke, R.C., 1988. Mycoparasitism and plant disease control. In: Burge, N.M., (Ed.), Fungi in Biological Control Systems. Manchester University Press, pp. 161–178.

PART

CLASSES AND KINDS OF MICROBIOMES

I

CHAPTER

2 BACTERIOME AND ARCHAEOME: THE CORE FAMILY UNDER THE MICROBIOMIC ROOF

George P. Patrinos[1,2,3], Loukia Zerva[4], Michael Arabatzis[5], Ioannis Giavasis[6] and Manousos E. Kambouris[7]

[1]*Department of Pharmacy, University of Patras School of Health Sciences, Patras, Greece* [2]*Department of Pathology, College of Medicine and Health Sciences, United Arab Emirates University, Al-Ain, Abu Dhabi, United Arabic Emirates* [3]*Faculty of Medicine and Health Sciences, Department of Pathology, Bioinformatics Unit, Erasmus University Medical Center, Rotterdam, The Netherlands* [4]*Department of Pathophysiology, School of Medicine, National and Kapodistrian University of Athens, Athens, Greece* [5]*First Department of Dermatology-Venereology, Medical School, Aristotle University of Thessaloniki, Thessaloniki, Greece* [6]*General Department, University of Thessaly, Karditsa, Greece* [7]*The Golden Helix Foundation, London, United Kingdom*

INTRODUCTION

The notion of bacteriome is focal in current studies of microbiome at both cognitive and research levels. Still, the use of the term is not proportional to the scientific weight of the relevant studies and the knowledge gathered, while there are frictions regarding the specific content of this particular biome. A brief historic account may assist in deconvoluting current misapprehension on the nature and use of the term.

Although PubMed paper returns are not an uncontested source of information, especially when literature search is limited to inclusion of keywords in titles, they may be considered as an indicative metric for significance and are thus preferred in this chapter to more meticulous approaches, including, but not restricted to, content browsing for terminology. There is the added advantage that the use of specific keywords usually implies a declaration or a mature cognitive environment for reviewers to accept and not edit the title. The first returns of PubMed featuring "bacteriome" in titles appear relatively early (Wang and Cheung, 1998) but are quite sparse. One may compare with the first uses of "microbiome" in PubMed titles, introducing the ecocellular notion of this term (Ordovas and Mooser, 2006; Friedrich, 2008), which described the collectivity of microbes in a spatiotemporal entity (microbe + -ome, a Greek suffix implying a collective entity, as in genome, etc.). This was much to the dismay of microbial geneticists who revived the term by giving it a clearly genomic spin: microbe + genomics = microbiomics (Gill et al., 2006). The comparison of microbiome entries shows that PubMed title returns lag behind browsing other, more expansive resources (see Chapter 1, for works coining the two uses of the term "microbiome"). It also shows that, by 2006, the term was used in both its genomic-centered and its environmental/cellular contexts (Gill et al., 2006; Ordovas and Mooser, 2006).

Microbiomics. DOI: https://doi.org/10.1016/B978-0-12-816664-2.00002-5

Still, numbers and metrics do sometimes lie: the notion of bacteriome appearing in late 1990s in PubMed had nothing to do with the early notion of microbiome—much less with the late, genome-centric context of the latter. It was used, as a term, to describe a chimeric organ where bacterial symbionts were residing within adapted host tissues, the adaptation and the resulting functionality defining the interface as an organ (Chen et al., 1999; Wang and Cheung, 1998). The term kept appearing in this context until 2013, when a study used it to describe the sum of bacteria in a spatiotemporal entity (Diaz et al., 2013), soon to be followed by a relevant definition (Oever and Netea, 2014; Probst et al., 2014). As mentioned in Chapter 1, the whole concept of microbiome had been structured on bacteria and was a spawn of the Human Genome Project (HGP). The massive investment to accomplish HGP became in essence a quest for "new problems to solve," or rather the case of a series of solutions seeking the right problem(s). There were two distinct possibilities: sequencing additional organisms, which raised issues of choice and processing of respective samples, which may require different handling principles than human ones, or elaborating on the human subject at large. The latter had some interesting prospects: apart from easier funding, it organized both genomics and the Omic sciences into an integrated system of study, which was to revolutionize the core of thinking in bioscience. Human physiologists and geneticists agreed that the genomic content of the human cell could not account for all metabolites encountered in the human organism (Turnbaugh et al., 2007; Mammen and Sethi, 2016; de Oliveira et al., 2017) and, most probably, could not support human life as we know it—in environmentally sustainable conditions, and not in laboratory-like, fully controlled environments. To have a spherical, all-aspect picture of human physiology, the genes controlling and encoding these metabolites had to be detected, sequenced, and, most importantly, annotated to their native genomes. The task of elucidating the genomic dimension of the human microbiome was really grandiose, as some figures were quite impressive: the number of microbial cells and the number of carried genes were astounding, the latter more or less equaling the ones of the human genome (Gill et al., 2006; Sender et al., 2016). Given that many microbes were considered fastidious, a genomic approach was in order, without intervention of the culturing phase that changes population representativeness both qualitatively and quantitatively (Lagier et al., 2012). Thus metagenomics were born, first introduced in PubMed titles in 2003 (Schloss and Handelsman, 2003), although the epithet "metagenomic" had appeared a year earlier (Gillespie et al., 2002). This was squarely to the point; a molecular biology—or rather genomic—approach to map the diversity of microbial genes encountered in the human body. But this procedure meant practically that the output of such analysis was to be *metagenomes*; there were no clearly defined and purified cells (cultured or not, irrelevant), the actual genome(s) of which were going to be explored. It would be a genomic soup, starting and ending with nucleic acids, initially DNA. It is obvious that the entity thus emerging, the microbiome, *was indeed* a genomic one. It did not really have to be defined as such. Whatever the definition of microbiome, the microbiome emerging by the said process was to be compiled from genomic elements, the sequences of which were recorded, aggregated, and annotated to create in silico microbial (meta) genomes attributable to *supposed* microbiota. To this point, a very thorny issue arose: the number of projected microbial genes, impressive at the very least as mentioned earlier (Gill et al., 2006; Sender et al., 2016), was not referring to *actual genes*, but to *gene types*. If there was a 1000-cell population of a certain microbe (or microorganism), its unique genes were counted in the estimates one time each, not 1000 times each.

This approach creates some issues with gene variants but is straightforward as a scheme. Still, there are a number of genes (and bacterial genes do not contain introns), which are expected to be identical, or at least very similar across many bacterial genera. Many enzymes participating in DNA processes were expected to be very common and highly conserved. Similarly, within the Gram groups, many genes coding for cell wall ingredients and enzymes for their synthesis were expected to be highly conserved among different taxa, but this conservation would be reversely proportional to the order of the compared taxa. Thus the gene population was estimated very differently in cases where different taxonomic levels—usually genera or species—were taken as the basis for declaring two bacteria as different within the microbiome. And that without even starting to take into consideration viruses and fungi, the latter making the definition of "difference" an impossible task, due to anamorphism and teleomorphism considerations in taxonomy, and hence in identification.

DNA EXCHANGE

The concept of a single prokaryotic biome, or rather a biome based on the concept of the two empires (Gupta, 1998), was founded securely, among other factors, on the practice of horizontal gene transfer (HGT; also known as lateral gene transfer - LGT), which allows an incorporation of novel genetic blueprints without procreation—and thus an adaptation of the recipient cells/organisms themselves and not of their spawn. Gene transfer actively changes the genomic content, thus perplexing phylogeny and identification in prokaryotic organisms and is in stark contrast to the *main* eukaryotic evolutionary principle of reshuffling existing genes and gene variants into allelic combinations through recombination. The change of genetic environment, which is brought about by moving genes and gene clusters within a chromosome, between different chromosomes and also by the chromosomes proper in different genomes usually—although by no means always—affects progeny and operates by altering the regulatory environment and the downstream combination of products or /and feedback signaling. Thus the division of a putative prokaryotic biome where Archaea and Bacteria coexist and interact (Egert et al., 2017) to two defined ones, bacteriome and archaeome, practically corresponds to the three-domain system (Woese et al., 1990) but takes into account this most archetypical genomic process of prokaryota, the HGT (Fig. 2.1). Although autonomous (nonviral-mediated) HGT occurs within and even across the two prokaryote biomes (Nelson et al., 1999; Fuchsman et al., 2017), the Archaea enact a number of proprietary such processes that are particular to them, such as DNA exchange through vesicles and through cell fusion

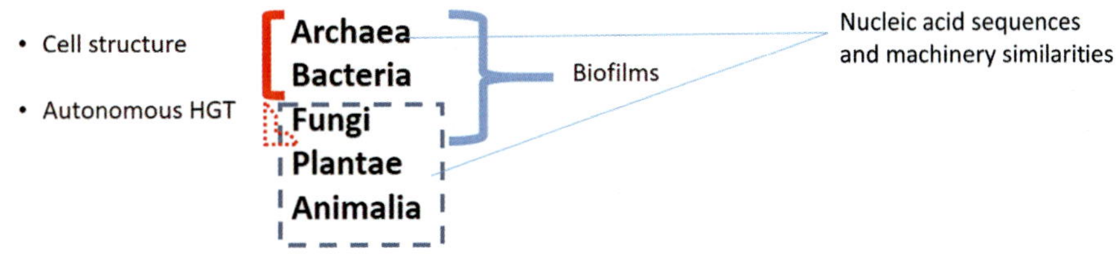

FIGURE 2.1

Common grounds across domains. Archaea and bacteria share three major similarities; Archaea and Eukarya one and two are found in all three domains.

(Wagner et al., 2017). Viral transduction is common in both biomes and although viruses usually infect either Bacteria or Archaea, there are some that do infect strains of both (Fuchsman et al., 2017), thus allowing for interdomain transduction. This fact, incidentally, argues in favor of using the term "archaeophage" rather than terms containing the "virus" theme, as heatedly proposed (Abedon and Murray, 2013) and despite the fact of virally mediated HGT (transduction) across domains (Boto, 2010).

THE PIONEERS, THEIR VISION AND THEIR MEANS

The objectives of the Human Microbiome Project (HMP), of the Metagenomics of the Human Intestinal Tract (MetaHIT) and the abovementioned realities regarding microbial genomic content and its peculiarities actually triggered a heated discussion on the relative merits of two very different approaches: the shotgun approach (literally "metagenomic whole genome shotgun sequencing") and the consensus gene metagenomics (Jovel et al., 2016). For certain, metagenomic approaches had been sought to tackle the issue of fastidious microbiota. The consensus gene—or, rather, sequences—approach offered some advantages: although the shotgun approach was the only way to efficiently catalogue and analyze the genomic content of human or any other microbiome, translating such data to microbiota present, meant that metagenomes had to be compiled. This sounds like a difficult puzzle to begin with, but it gets even worse in practice, as common sequences, both coding and regulatory ones—not to mention supposed "junk" sequences—are rather difficult to segregate from the genomic pool in order to determine to how many species/taxa they correspond, not to mention the endeavor to identify said taxa/species, a number of which are expected to be novel as they escaped isolation and characterization due to their inherent fastidiousness.

On the other hand, consensus sequences may not reveal genes or any genomic information, but they do provide a qualitative list of concerned biota, a fact making the compilation of sequences to metagenomes and their annotation to actual species feasible at a later stage. They represent a far more robust and informative approach to delineate mixed samples as output sequences variability permits not only identification of unknown contacts most probably corresponding to unique taxa, but also phylogenetic categorization of novel taxa. By producing far less sequence noise, the approach is clearly at its best when striving to determine the content and possibly the abundance of species/taxa in a sample, and not the genomic content of each member of the community or of the community as a whole, neither in qualitative nor in quantitative terms. The most suitable approach in this case was, inescapably, the combination of the two methods (Turnbaugh et al., 2007).

Thus both HMP and MetaHIT may have been initiated explicitly to discover the genes (Human Microbiome Project Consortium, 2012) and, if possible, the (meta)genomes implicated in human physiology and responsible for the synthesis of many metabolites, but the explosive microbiomic research was oriented toward cataloging microbial presence in various habitats, and thus pushed for the optimization of the consensus gene metagenomic—and metatranscriptomic—approach (Chapter 7).

Consensus metagenomics, as occasionally referred to for brevity, had a very important drawback: such use and approaches were conditional on the very consensus sequences. If these did not exist, or were not universal, the respective genomes were not analyzed; they were actually passed over, as nonexistent. This uneasy detail led to successive segregation of various biomes from the microbiome, although it was a known fact that actually when speaking of microbiome, it had been the bacteria or, in some cases, the prokaryotes as a whole, that were actually discussed or studied (Ghannoum et al., 2010). Thus viruses were the first to be detached from the microbiome, as the virome (Anderson et al., 2003), since they possessed no consensus sequences. In addition, the massive percentage of RNA viruses allowed a reverse-transcriptase-primed procedure, which was more akin to metatranscriptomics. Thus at least such viruses, amounting to four out of seven Baltimore classes, were treated in a standardized fashion across all sample types, which was different to that of entities containing DNA genomes (see Chapter 4). After all the above taken into consideration, it remains unexplained why a 7-year time lapse was required for the first appearance of the word "virome" in a title in PubMed (Coetzee et al., 2010).

Fungi followed suit with the term "mycobiome" appearing to describe the "fungal microbiome" (Ghannoum et al., 2010); the timing and the alternative descriptive term insinuate that the already mentioned paper introducing it might have been influenced in terms of terminology by the almost concurrent emergence of Virome in titles (Coetzee et al., 2010). Possessing consensus sequences in the vicinity of 18S rDNA, patently different than those of the Prokarya (in the environs of 16S rDNA), fungi were easy to discriminate and analyze with consensus metagenomics. Originally disregarded in terms of significance and research resources allocated to their study, they are now considered highly important, and not only due to their interaction with bacteria (see Chapter 3). They were underrepresented in any given environment: their cell number is a fraction of that of Prokarya (Lai et al., 2019; Wisecaver et al., 2014)—a very small fraction, to the tune of 0.1% (Qin et al., 2010; Seed, 2015)—and their genomic content is lower than that of the prokarya, albeit not proportionately to their cell number. Eukarya have a much higher genomic content due to their size and the existence of a defined and organized nucleus with sequentially modular genomes (chromosomes) and intriguing gene sets encoding smart approaches in order to adapt to different environments in a spatiotemporal context (Wisecaver et al., 2014).

The emergence of the term "bacteriome" in the sense of a biome was belated, appearing for the first time in a paper title during 2014 (Probst et al., 2014), while the other two biomes (i.e., virome and mycobiome) had already been described by 2010. The respective paper not only introduces bacteriome but goes a step further and introduces archaeome as well, clearly distinguishing the two prokaryotic biomes. The domain theory of the phylogenetic tree of life was published a quarter-century earlier (Woese et al., 1990) and became very popular, as the term "archaeon" appears in the title of 2080 Pubmed-registered papers ever since. As a result, it remains odd that the concept of domains needed such a long time to be "officially" acknowledged in the context of biomes. Moreover, archaeome still shows an absolute minimum of popularity: there exist only two PubMed returns for paper titles containing this term, from 2014 to early summer of 2019.

Notably, the number of papers with the term "bacteriome" in their title remains extremely low as well. There are only 44 returns for the biomic sense of the bacteriome, while the field of microbiomics flourishes and bacterial populations remain the main focus of microbiomic research (Seed, 2015; Lai et al., 2019; Cui et al., 2013). Microbiologists, and not only the ones involved in the clinical setting, clearly feel that the term "microbe" and any of its spinoffs suit bacteria better than anything else.

The most striking issue, in this widespread reluctance to recognize or refer to bacteriome and archaeome as such, is that the concept of domains is a purely genomic one. It came around by

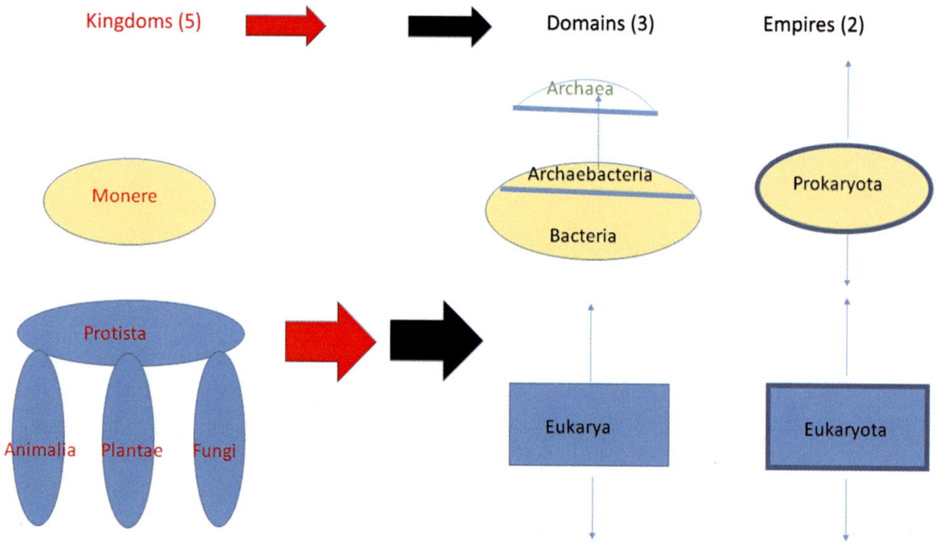

FIGURE 2.2

The virtual mechanics of the kingdom/domain/empire system for understanding and viewing the tree of life.

taking the Archaebacteria out of the Kingdom of Monera, to assign them to a status higher than any other taxon, equal to the rest of Prokarya (which were recast as Bacteria) and Eukarya (which aggregate all kingdoms of eukaryotes), and finally by changing their name from Archaebacteria to Archaea (Woese et al., 1990). This massive repositioning and reshuffling (Fig. 2.2) was the result of stringent comparisons of rDNA sequences, which showed a disproportional difference to respective sequences of other Prokarya, as well as the result of sanctioning of DNA differences as the yardstick for phylogenetic relations.

DIVERSITY

Initially the human microbiome was supposed to differ among individuals due to illness and also due to a number of factors, such as geographical locale, diet, lifestyle, age, and climate (Wilantho et al., 2017; Lloyd-Price et al., 2016; Christensen and Brüggemann, 2014; O'Toole and Jeffery, 2018; Prohic et al., 2016); but later on, many other issues were implicated as well. Body sites (biocompartments) of the same individual may be colonized by wildly dissimilar microbiota per biocompartment as a baseline setup of diversity in health (Ma et al., 2018; Perez Perez et al., 2016; Rogers et al., 2016; Ding and Schloss, 2014). As mentioned earlier, aging is another factor contributing to this diversity. Turning points of human life such as birth, puberty, adulthood, and old age represent temporal highlights (Desai and Landay, 2018; Chu et al., 2017) characterized by intense biochemical and microenvironmental changes within the host, which affect standard microbial symbionts (Wilantho et al., 2017). Notably, microbiomic diversity by biocompartments and differential responses/effects of their respective microbiomes to various challenges are also illustrated in plants (Liu et al., 2017).

Although the distinction between Archaea and Bacteria, and thus between respective biomes, is very popular and widely accepted, it has to be underlined that it is *not* unchallenged. Their differentiation is based on nucleic acid sequences and relevant mechanisms (Woese et al., 1990). Still, this distinction remains a genomic approach, or at the very best a genomic/post-genomic one. In terms of cellular biology the facts mentioned earlier constitute an important set of differences but not the most important one, especially in the context of the wildly diversified prokaryotic cells. Archaea and Bacteria have more in common than not: transformation and conjugation, although mediated by different effectors, are instances of direct HGT and occur in both these groups (whatever their name and status) and in these two groups only (García-Aljaro et al., 2017; Stingl and Koraimann, 2017), creating a network of lineages and sequences rather than a phylogenetic tree (Toussaint and Chandler, 2012). Interestingly, transkingdom conjugation seems to set the tune for reconsidering (Lacroix and Citovsky, 2016) the exclusivity of autonomous HGT outside the Eukarya domain/empire.

HGTs are nuclear (*sensu lato*) and genetic, but not genomic, events, implicating cellular mechanisms exclusive to prokaryotic organisms. Transduction, an indirect HGT event taking place through sequential viral infections, is a complex nuclear and cellular event (Touchon et al., 2017). This process is applicable in all three domains (Touchon et al., 2017; Hashemi et al., 2018), but there are viral strains that infect both Bacteria and Archaea (Fuchsman et al., 2017), a cross-domain flexibility in host specificity and selection not observed with viruses infecting Eukarya: eukaryote-specific viruses infect neither Bacteria nor Achaea. In cellular terms, Bacteria and Archaea are even closer, which resulted in their being brigaded together as Prokaryota (or Monera) since genomic differences were not considered paramount in principle (although at the time, some of them had not been identified). Instead, they were viewed as significant variations within an extremely diverse kingdom. Cell structure, size, and basic physiology were considered much more important, a view giving precedence to "hardware" over "software" in modern parlance. The observation that Archaea and Bacteria may very well form polymicrobial biofilms (Probst et al., 2014), being thus perfectly capable of cooperating in a far more integrated manner, similar (or precedent) to multicellular structures/tissues, has been downplayed due to the fact that fungi and bacteria can also participate in the same biofilm as elaborated in Chapter 3 (Ghannoum, 2016; Hoarau et al., 2016). Still, the latter is a transkingdom event, whereas the domain system makes the former case of far greater impact, since it escalates to a transdomain integration, or, actually, a chimera.

Even more perplexing has been a couple of uneasy observations. The first was that the domain Eukarya was comprised of organisms which had undergone two (plants), or one (fungi and animals) symbiotic events (Dolezal et al., 2005), an evolutionary distance of considerably greater magnitude and significance compared to some chemical differences in lipids and some DNA sequences and *material* (Woese et al., 1990). The second was the observation that prokaryotic cells could sponsor either one cell membrane (monoderms), which is the case with all Gram-positive Bacteria and Archaea, or, among Gram-negative Bacteria, a very intriguing two-membrane envelope (diderms). This difference in cell structure is far more prominent than its consequences in cell motility, cell wall thickness and permeability, drug and pest resistance, and sporulation efficiency (Gupta, 2011). As a conclusion, the actual, genome-centric opinion is that bacteriome and archaeome are two different and distinct biomes within a microbiome, but the case that the archaeome is actually a part of the bacteriome is not as thoroughly discarded as the sponsors of the genome-centric opinion would like to hold.

HABITATS, SETTINGS, AND FORMATS

Studies on the bacteriome, either as such or in its guise as microbiome, highlighted two basic issues. The first is the massive interactions among its members, which are cooperating, competing, or indifferent to each other. The actual status of interactions may change by the slightest alteration in any one out of a large number of factors relating to the environment. Such are temperature or living in an abiotic environment, the existence of host(s) or factors directly related with the biota, like age, fitness, population density, and gene transfer (Deshpande et al., 2018; Izhar and Ben-Ami, 2015; Egert et al., 2017; Miller and Bassler, 2001). The second issue refers to the collective interactions of bacteriome's members with the host or the abiotic habitat (Foster et al., 2017; Gould et al., 2018; Marrero et al., 2015). These two issues are actually one and the same if taken into consideration in a more relaxed perspective and from the point of view of the bacteria—an approach definitely heretical a decade or so earlier (Turnbaugh et al., 2007; Ley et al., 2007). Actually, although biomedicine simply detested the idea, it is true that for *any* metabolically active microbe a biocompartment of a human, animal or plant multicellular organism is simply a potentially toxic but rich in nutrients environment (de Oliveira et al., 2017; Foster et al., 2017). An environment characterized also by potentially destructive and abrupt changes in temperature and chemical composition, as well as fraught with enemies and predators (including, without being restricted to, natural killer cells and phagocytes respectively).

This unified view of possible interactions with both biotic and abiotic elements implied in its turn a number of interesting projections: to start with, the bacteriome could have repercussions extending much further than its actual position. Microbiomes hosted in gut may affect the cardiovascular system or the brain (Liu et al., 2019; Ordovas and Mooser, 2006). In addition, a microbiome could be used positively to shape different environments according to needs and wishes; the acidification of fruit juice to alcoholic beverages (brewing) is an example, and the use of bacteria for biomining (Banerjee et al., 2017; Donati et al., 2016; Marrero et al., 2015) or for bioremediation (Verma and Sharma, 2017) are additional ones. Furthermore, by extending the latter thought, it is possible to use communal relationships so as not only to postpone or avert adverse effects caused by (members of) the bacteriome (Tanaka and Itoh, 2019)—thus limiting it to neutrality as a best case scenario—but also to actually mobilize it for supporting or restoring good health (Huang et al., 2019).

Within a bacteriome, multiple interactions can be dissected to very few baseline events. The first, most probably in importance if not in occurrence, is antagonism, sometimes causing unilateral extinction of one of the antagonists. The antagonistic setup, where one or more bacterial strains forestall the development or impede the survival of usually one homogenous population, or sometimes more than one (Johnson and Foster, 2018) whether similar or widely different to each other, offers considerable leverage for shaping bacteriomes at will, usually through either the production of specialized antimicrobial substances, such as, but not restricted to, antibiotics; or extreme competition for resources such as nutrients (Egert et al., 2017).

The other baseline interaction is cooperation (Coyte et al., 2015). In that case, (some of) the constituent microbiota enhance the development, fitness and/or survival of others, with different degrees of reciprocity. The more advanced cooperative condition, synergy, implies a mutually, or rather omnilaterally beneficial set of interactions, for all (minimum, two) participant microbiotes

(Troselj et al., 2018; Miller and Bassler, 2001). Different ranges and scopes accommodate cognitively the unilaterally neutral (commensalistic) versus the mutually beneficial (mutualistic) interaction, although such terms usually describe symbiotic relations which are directional, with a symbiont and a host, not mere interactive coexistence formats, which are better described by the term "syntrophy." The latter may expand to more complicated association networks than the rather straightforward host–symbiont pairs (Dimijian, 2000).

Similar relationships are observed among different biomes of microbiota. The truth is that viruses are practically the ultimate predators and different strains plague prokaryota (Fuchsman et al., 2017), and eukaryota (Li et al., 2010). However, the bacteriome demonstrates a more complex relationship with the mycobiome, from joint polymicrobial biofilms (Probst et al., 2014) through bipartite serial metabolic reactions as in the production of acetate from carbohydrates and all the way to the production of antibiotics (Mishra et al., 2017; Petatán-Sagahón et al., 2011). Symbiosis is a different breed of interactions and concerns the relationship, on a comparative beneficial basis, between one or many microbiota and their host. If a whole microbiome is investigated, the host can be considered an all-biotic environment, usually extremely homogenous (single-host, in terms of individuals), but the idea can be expanded to consider multihost setups, usually limited in a spatiotemporal and/or host-type context. In other words, identical or similar hosts being colonized by one or by the same community of symbionts within the same space, whatever its definition, or within a time unity, whatever its length or its period. In the context of prokarya, such interactions may be beneficial (mutualism), neutral (commensalism), or negative (parasitism) for the host, while they are, at their basic setup, beneficial for the symbiont. Parasitic events usually lead to infections of multicellular hosts, mostly animals, plants, and occasionally fungi (especially mushrooms). Mutualistic events are highlighted by bacteriomes in insects, benign gut bacteria in various animals (Egert et al., 2017; Ipci et al., 2017) or cyanolichens consisting of cyanobacterium and fungus (Henskens et al., 2012). Nonharmful symbionts, commensals *sensu lato*, are endophytic and epiphytic bacteria residing in/on host plants (Balsanelli et al., 2019), a great number of animal gut-residing microbiota and skin-colonizing microbiota (Christensen and Brüggemann, 2014; Egert et al., 2017). The dynamic balance in commensalism may be upset by the relative dominance (survival or overgrowth) of a specific symbiont (Egert et al., 2017; Weiss and Hennet, 2017), by the arrival and propagation of a virulent strain (Ipci et al., 2017; Weiss and Hennet, 2017) or due to a drastic alteration of host immunocompetence (de Oliveira et al., 2017), be it immunocompromization (Egert et al., 2017) or derailment to allergy or autoimmunity (de Oliveira et al., 2017; Ipci et al., 2017). All these factors derail symbioses to dysbioses, initiate the pathophysiology of infection, and, possibly cause, as a consequence, disease. Although the latter may be of other, noninfectious etiology, dysbiosis may still participate in the process (Weiss and Hennet, 2017). Many bacteria show an effort, or at least a tendency to evolve into multicellular organisms (Meysman, 2018; Risgaard-Petersen et al., 2015), which is a step further from colonial organization. Different to common knowledge, a bacterial colony will not expand indefinitely even if substrate is available; under the quorum sensing concept, there are different suppression signals, including, but not restricted to, catabolism by-products of toxic effect (Chacón et al., 2018; Miller and Bassler, 2001). The notion of continuous cultures includes many more intrinsic factors subject to rebooting by the presence of the dynamic, influx–efflux processes than the simple supplementation with additional quantities of nutrients. Among them are the efflux of biomass to keep cells as young as possible and—the most vital—stirring. The latter not only distributes evenly nutrients and toxins (preventing the latter from reaching toxic levels locally) but also impairs a spatial arrangement by age, as happens on solid media.

In the multicellular context of bacteria, the biofilm is a focal entity, as it bears the characteristics of a tissue, due to the existence of a most bioactive extracellular matrix and structurally and functionally differentiated (sessile) cells. A further step is represented by multispecies biofilms (Gabrilska and Rumbaugh, 2015), where the participants differ in the extent and function of their participation but also in the timing of settling within the spatial and functional context of the biofilm. All the above introduce the concept of chimeric tissues, if not organs, in a prokaryotic context. Chimeras mostly concern eukaryotes, with different numbers and structures of chromosomes; the simplistic structure of the prokaryotic genome, where ploidy is attestable only in detached plasmids and other, basically nonessential, facultative episomes, makes the concept of prokaryotic "chimeras" controversial.

The transformation of planktonic cells, which do live in a congested and densely populated shared environment, to a sessile form includes morphological as well as biochemical alterations to a communal standard so as to live, function, and, most importantly, communicate and cooperate within a biofilm. Biofilm formation is observed in both Bacteria and Archaea (Fig. 2.1). The latter do form biofilms of their own, unispecies or multispecies ones (van Wolferen et al., 2018), they also incorporate bacteria or are themselves incorporated in bacterial biofilms, thus producing neither mere multispecies nor transkingdom ones (Ghannoum, 2016; Hager and Ghannoum, 2017; Gabrilska and Rumbaugh, 2015), but transdomain biofilms (Probst et al., 2014). The five-kingdom and the two-empire systems would provoke no frictions and argumentation in cases of such observations, as Bacteria and Archaea, under different names, are considered akin; but within the three-domain system, such coexistence and synergy crosses the ultimate phylogenetic border and sounds, seems or reads impressive.

ENVIRONMENTAL ADAPTABILITY, MONITORING, AND ENGINEERING

There are two basic cases in the bacteriome-to-habitat two-way relationships. The first refers to *association* and the second to *causality*. The former is basically a quest for biomarkers, seeking to use microbiomic parameters, such as presence and representativeness, for prognostic, diagnostic, and possibly agnostic applications, especially in hosted bacteriomes and secondarily to other, abiotic habitats. There are many studies indicating that changes in bacteriomes are associated to certain health conditions. Even with unestablished causality, the detection of specific bacteriomic formats, representing individual or collective biomarkers, could lead to early recognition of pathophysiology, prior to evolution into disease, thus allowing prevention, prophylaxis or, at the very least, administration of suitable treatment at the earliest possible (Egert et al., 2017; Al Khodor et al., 2017; Aragón et al., 2018).

The associations where a specific bacteriome (in both qualitative and quantitative terms) causes, to any degree, an altered health status (not necessarily adverse) and *vice versa*, that is, a specific health status promotes certain bacteriomic formats and supports preferentially certain biota, are characterized as direct associations. If the causal polarity of a specific case of direct association remains elusive (Rivas et al., 2016), any relevant applications (clinical, environmental, or technological/industrial) are bound to be somewhat hampered. Indications for prognosis or diagnosis may still be provided by indirect associations, where there is no causality, or where partial causality can be established, in connection with a variety of other, extrinsic factors (Egert et al., 2017). In such cases though, the assistance of bioinformatic tools will be unavoidable and compulsory, in order to enact an approach similar to mining genomic biomarkers. It is a plain fact that bacteriome screening is achievable either through metagenomics or through culturomics or both (Dubourg et al., 2013; Kambouris et al., 2018), since metabolomes are more difficult to deconvolute and associate to unknown organisms or mixed metagenomes. For routine uses in clinical or public health samples, culturomics is slower and less discriminative, but more robust and, conditionally, more sensitive than metagenomic/metatranscriptomic approaches can be (see Chapter 7 and Chapter 8).

The established causality is actually a much more robust subcase of direct association. In such events the stakes are higher, as there is a prospect for intervention (Reid et al., 2017; Aragón et al., 2018; Chen et al., 2017; Mao et al., 2018; Pevsner-Fischer et al., 2017). If a given bacteriome, in terms of locality, quantitative and qualitative composition, actually causes or exacerbates an adverse condition, shaping the bacteriome into a stable alternative format may remove the cause and thus initiate alleviation of symptoms and health restoration. Successful such cases are the Fecal Microbiome Transplants (FMT) to treat recurring *Clostridium difficile*-associated diarrhea (Moayyedi et al., 2017) and irritable bowel syndrome (Huang et al., 2019). In addition, causality seems to be established between gut bacteriome and behavior patterns (Johnson and Foster, 2018), heart conditions (Tang et al., 2019), or hypertension (Pevsner-Fischer et al., 2017) as well as between lung microbiome and lung cancer (Mao et al., 2018) or urinary tract microbiome and urologic disorders (Aragón et al., 2018). Still, it must be stressed that interactions may be anatomically remote, occurring over considerable distances through the use of chemical effectors, as are neurotransmitters in behavioral changes caused by gut bacteria (Johnson and Foster, 2018; Liu et al., 2019; O'Toole and Jeffery, 2018; Reid et al., 2017) or other means of signaling through appropriate databuses and networks (Whalen et al., 2019; Meysman, 2018; Miller and Bassler, 2001).

For the entire field of microbiomics, at the cognitive level the dichotomy between their biomedical and nonmedical dimensions (the latter contain, but are not limited to environmental ones, as industry is also a valid field) has been partially erased and the event has been celebrated (Ley et al., 2007). When the applicability of the microbiomics, though, is taken into consideration, and especially for translational research, such dichotomy still holds value. Nonmedical contexts are more diverse in the intended effects and allow increased flexibility in shaping the approaches and effectors, as the diverse environments/habitats are abiotic and far more robust than the relatively fragile physiologies of living hosts. Furthermore, the coexistence of two separate sets of genes in a holobiont, the genome of the host on one hand and the accumulative set of microbial genes on the other, the latter originating from multiple genomes with possible overlaps, introduces a number of challenges that do not befall in environmental contexts, nor in industrial ones. In addition, medical contexts have to consider not only host fitness and survival, but also the host's direct, multidimensional, and irregular—or rather "adapting"—participation in the diverse networks of interactions, which shape the qualitative, quantitative, and spatial—if not spatiotemporal- structure of a microbiome (Coyte et al., 2015; Egert et al., 2017; Foster et al., 2008).

The most basic case of indirect interaction is when a bacterium does not affect the host directly (modifier) but decisively affects another one that does so (effector), either in a benign fashion, as occurs with the already mentioned FMT (Moayyedi et al., 2017; Huang et al., 2019) or in a malign fashion (Bäumler and Sperandio, 2016), as illustrated in Fig. 2.3. This is the cardinal difference between the microbiological studies of previous ages and the microbiomic studies as they come of age: the recognition of said concept of compound mediators, formed by the combination of effector(s) and modifier(s); followed by the allocation of attention and resources to both, which exemplifies the modern, networked-causality concept, instead of concentrating on effectors only, as dictated by the original, sole-causality idea.

The direct, cooperative interrelation of two bacteria of the same bacteriome may be serial or parallel. In the first case, one metabolizes a by-product of the other, irrespective of anabolism or catabolism. Both steps may either be of the same metabolic type (anabolic or catabolic) or different, thus presenting four distinct cases: anabolic–anabolic, catabolic–catabolic, anabolic–catabolic, and catabolic–anabolic, although catabolic first steps are expected to be much more common, so as to produce metabolite readily available for a next step overtaken by another strain. The production of vinegar by *Acetobacter* spp. and *Gluconobacter* spp., through alcohol produced by *Saccharomyces cerevisiae* from sugars, is a most common example (Alauzet et al., 2010). Although two different biomes are implicated, this is the best example for bipartite serial metabolic reactions and a just as good example of cross-domain microbiomics interactions. A bacteriome-only example is *Clostridium tyrobutyricum* metabolizing to butyrate the lactate produced by *Bacillus lacticus* in dairy products (Detman et al., 2019) and a very similar sequence in human gut (Fig. 2.4A): Bacteria, such as—indicatively but not exclusively—*Bifidobacterium adolescentis*

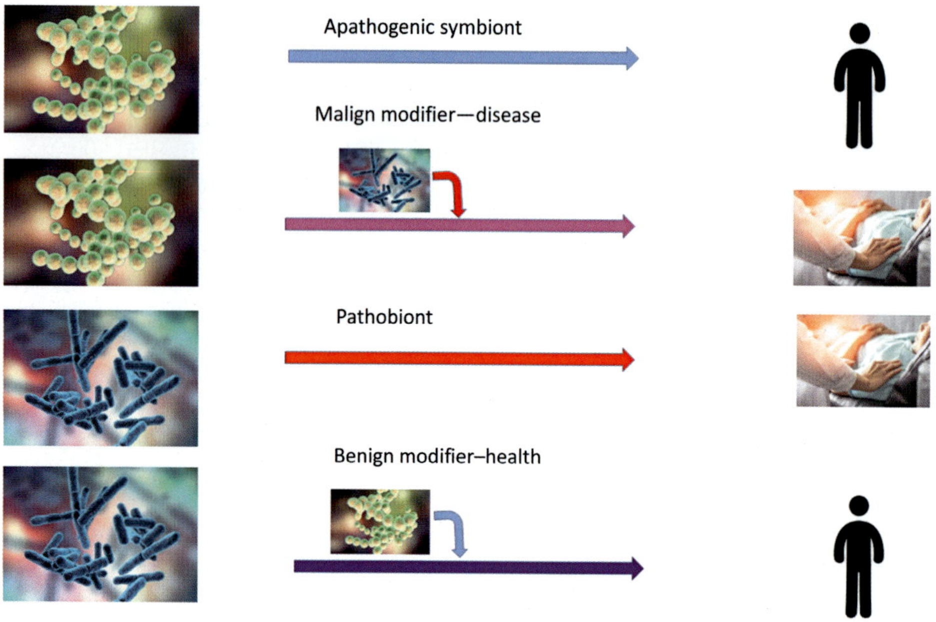

FIGURE 2.3

The role played by modifiers: through the action of benign modifiers, pathobionts fail to provoke disease, while apathogenic symbionts may contribute to dysbiosis or ailment by interference of malign modifiers.

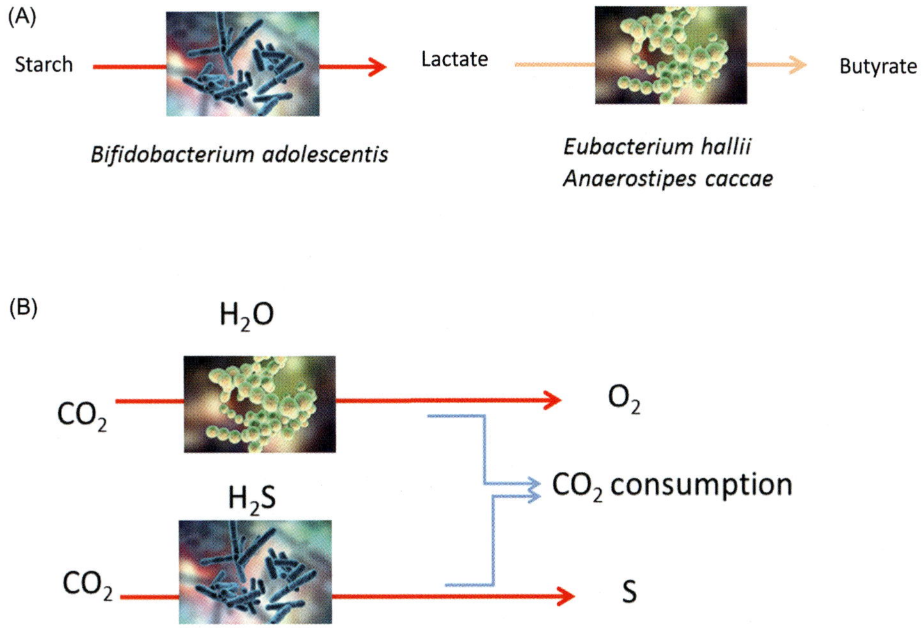

FIGURE 2.4

The direct cooperative interrelation of two bacteria of the same bacteriome may be serial (A) or parallel (B). (A) *Bifidobacterium adolescentis* catabolizes starch to lactate, which is taken up by other bacteria, such as *Eubacterium hallii* and *Anaerostipes caccae* and further metabolized to butyrate. (B) CO_2 consumption occurs independently by different photoautotrophs, such as oxygenic Cyanobacteria and anoxygenic photosynthetic sulfur bacteria.

catabolize starch to lactate, which is taken by other bacteria, such as *Eubacterium hallii* and *Anaerostipes caccae* and further metabolized to butyrate (Duncan et al., 2004). For more than two steps the combinations increase exponentially, accounting for the universally observed diversity of microbiomes and the respective ecosystems.

In parallel formats (Fig. 2.4B), two bacteria, or, more expansively, two biota present in the same ecosystem, perform similar functions (catabolic *or* anabolic) to different substrates, which lead to the same end result. For example, the decomposition of different macromolecules from different saprophytes to recycle inorganic ingredients is one of the most well-known such catabolic cases; the same goes, in an anabolic setup, with CO_2 consumption by different (classes of) photoautotrophs, such as oxygenic Cyanobacteria and anoxygenic photosynthetic bacteria (Ludwig et al., 2006). With more bacteria participating, more possibilities emerge, with higher degrees of complexity.

Cases within the same microbiome, with opposite effects being sought for, are very common in food industry. In many instances, certain bacteria need to thrive in order to promote processing of foodstuff, as in fermentations, while other, spoilage bacteria have to be suppressed so as not to degrade or destroy the end product. For example, the vigorous growth of lactobacilli and other lactic acid bacteria (LAB), which produce, among other compounds, lactic acid and bacteriocins during cheese fermentation, is instrumental so as to inhibit the growth of counterproductive strains. Such are spoilage bacteria, like coliforms (e.g., *Enterobacter* and *Klebsiella*) and *Pseudomonas*; pathogenic bacteria, such as *Escherichia coli* and *Listeria monocytogenes*, especially in the case of naturally fermented cheese made from nonpasteurized (fresh) milk, are important players too (Kamal et al., 2018; Gemechu, 2015; Florou-Paneri et al., 2013; Ortolani et al., 2010). Similarly, the prevalence of LAB during meat fermentation, especially lactobacilli and pediococci, safeguards fermented salami and prosciutto from food pathogens, such as *L. monocytogenes* (Woraprayote et al., 2016; Rattanachaikunsopon and Phumkhachorn, 2010). Also, many LAB, such as *Lactobacillus rhamnosus* and *Lactobacillus plantarum*, as well as Propionibacteria (which produce propionic acid, a very effective antifungal agent) are natural inhibitors of several spoilage fungi, including both yeasts and molds, in many fermented or nonfermented foods, in which the bioprotective bacteria are either naturally present, or can be added for shelf-life extension (Fernandez et al., 2017; Abbaszadeh et al., 2015; Cheong et al., 2014; Delavenne et al., 2013). Such fine-tuning may be applicable to environmental engineering and even to therapeutics.

THE (NEAR-TERM) WAY AHEAD

The case of differential, and especially opposite effects in bacterial OTUs (operational taxonomic units, meaning the taxa of the level considered in each case/study) of the same bacteriome will undoubtedly witness the need to differentially regulate probiotics and pharmabiotics on one hand and colonizers/pathogens/facultative and/or opportunistic pathogens on the other (Bäumler and Sperandio, 2016). Such coexistence may appear in a spatiotemporal unity, in one and the same hosted bacteriome, or, in more than one, possibly—but not necessarily—communicating biocompartments within the same host. Different biocompartments may present the clinician with different situations at any given time. In one biocompartment, dysbiosis or infection may be in progress, while, simultaneously, other biocompartments may show perfectly functional symbioses. As a matter of fact, this differential picture is the quintessence of the development of local ailments and infections. Anatomical distance may promote the isolation of diverse hosted microbiomes within the host, thus inhibiting, up to a point, surging dysbioses from occurring regularly. However, many clinical conditions are now linked to distant symbiont alterations affecting host physiology at remote sites in regard to symbiont's own habitat. Hence, local dysbioses linked to remote adverse effects to the host are attributable to alterations of symbionts' communal characteristics, regardless of anatomic distance or length of time required for disease establishment (Ipci et al., 2017; Rivas et al., 2016).

In these cases, biphasic pharmacogenomics might become the hottest issue in managing and maintaining mutually beneficial host–microbiome symbiosis by proactive intervention through antimicrobial regimen administration to avert dysbiotic derailing. Locally applied, locally acting pharmaceutical agents, basically but not exclusively antibiotics (see Chapter 14), are prone to gain preference and prominence against drugs administered per os or parenterally, with the added advantage of reduced adverse effects among the former.

Biphasic pharmacogenomics is likely to be a catalyst in precision medicine as far as infectious diseases are concerned. Its prospective use, especially in volume, does pose some interesting issues:

a cardinal issue will be what kind of metagenomic methodology is to be followed, as suggested in Chapter 7. Another issue, just as thorny, regards the needed algorithms to assign the signs +/− to the three different pharmacogenomic parameters (namely, **safety** for the host, **susceptibility** of pathogen(s) and **susceptibility** of benign microbiota), and especially to the two last ones, which refer to characters of the microbiome (Fig. 2.5). The latter are to be described by a most diverse range of biomarkers in a case-by-case basis due to the multitude of implicated microbial taxa. Subsequently, the tripartite weighting of the analysis, instead of the bipartite in conventional pharmacogenomics (Fig. 2.5), will be one of the trickiest issues for translational research and decision-making.

Optimization, standardization, and cost- and time-sensitive processing guidelines usually followed in routine genomic analyses would suggest a one-vial assay for both genomic phases in biphasic pharmacogenomics. It is more than obvious that the shotgun approach is the only applicable solution, possibly followed by generic priming or by unprimed sequence reading. This is so because consensus gene metagenomics reveals perhaps the present biota but these originate from one or two biomes only, as viruses do not contain consensus sequences and bacteria and fungi, the most usual cellular pathogens, have very different such sets. Furthermore, neither host pharmacogenomes nor microbiomics susceptibility/resistance sequences can be detected by consensus gene metagenomics. Generic priming and unprimed formats, on the other hand, at least in theory analyzes the whole metagenome of the holobiont and answers the abovementioned specific queries. Of course, there are problems with misallocation of amplification resources (primers, enzyme, and dNTPs), as the full genome of the host is not informative. What is needed is the ability to analyze pharmacoloci (biomarkers implicated with response to drugs and positioned anywhere in the genome, not only on coding sequences) and in no great depth. Thus other, more selective methods, such as multiplex amplification may reemerge within the current methodology stream, so as to

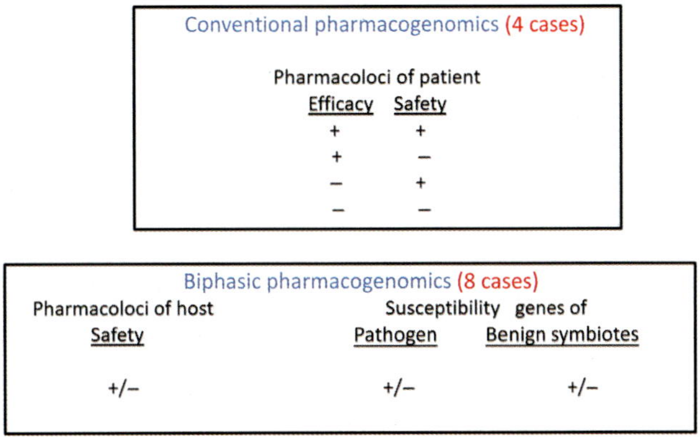

Conventional pharmacogenomics (4 cases)
Pharmacoloci of patient
Efficacy Safety
+ +
+ −
− +
− −
Biphasic pharmacogenomics (8 cases)
Pharmacoloci of host
Safety
Susceptibility genes of
Pathogen
Benign symbiotes
+/−
+/−
+/−

FIGURE 2.5

The factors examined per proposed or available drug in conventional (upper panel) and biphasic pharmacogenomics (lower panel). The latter result in an 8-case combination scheme, where in principle only one, the "triple + " is clinically acceptable for subscription. The efficacy loci for antimicrobial compositions reside with the microbial genomes.

exploit current investments in training, know-how, instrumentation, and logistics. The fact that this approach will be primarily genomic, whatever its exact nature, is perhaps the only safe bet.

REFERENCES

Abbaszadeh, S., Tavakoli, R., Sharifzadeh, A., Shokri, H., 2015. Lactic acid bacteria as functional probiotic isolates for inhibiting the growth of *Aspergillus flavus*, *A. parasiticus*, *A. niger* and *Penicillium chrysogenum*. J. Mycol. Med. 25, 263–267.

Abedon, S.T., Murray, K.L., 2013. Archaeal viruses, not archaeal phages: an archaeological dig. Archaea 2013, 251245.

Al Khodor, S., Reichert, B., Shatat, I.F., 2017. The microbiome and blood pressure: can microbes regulate our blood pressure? Front. Pediatr. 5, 138.

Alauzet, C., Teyssier, C., Jumas-Bilak, E., Gouby, A., Chiron, R., Rabaud, C., et al., 2010. Gluconobacter as well as *Asaia* species, newly emerging opportunistic human pathogens among acetic acid bacteria. J. Clin. Microbiol. 48, 3935.

Anderson, N.G., Gerin, J.L., Anderson, N.L., 2003. Global screening for human viral pathogens. Emerg. Infect. Dis. 9, 768–773.

Aragón, I.M., Herrera-Imbroda, B., Queipo-Ortuño, M.I., Castillo, E., Del Moral, J.S.-G., Gómez-Millán, J., et al., 2018. The urinary tract microbiome in health and disease. Eur. Urol. Focus 4, 128–138.

Balsanelli, E., Pankievicz, V.C., Baura, V.A., de Oliveira Pedrosa, F., de Souza, E.M., 2019. A new strategy for the selection of epiphytic and endophytic bacteria for enhanced plant performance. Methods Mol. Biol. (Clifton, N.J.). 1991, 247–256.

Banerjee, I., Burrell, B., Reed, C., West, A.C., Banta, S., 2017. Metals and minerals as a biotechnology feedstock: engineering biomining microbiology for bioenergy applications. Curr. Opin. Biotechnol. 45, 144–155.

Bäumler, A.J., Sperandio, V., 2016. Interactions between the microbiota and pathogenic bacteria in the gut. Nature 535, 85–93.

Boto, L., 2010. Horizontal gene transfer in evolution: facts and challenges. Proc. R. Soc. B: Biol. Sci 277, 819–827.

Chacón, J.M., Möbius, W., Harcombe, W.R., 2018. The spatial and metabolic basis of colony size variation. ISME J. 12, 669–680.

Chen, X., Li, S., Aksoy, S., 1999. Concordant evolution of a symbiont with its host insect species: molecular phylogeny of genus *Glossina* and its bacteriome-associated endosymbiont, *Wigglesworthia glossinidia*. J. Mol. Evol. 48, 49–58.

Chen, J., Domingue, J.C., Sears, C.L., 2017. Microbiota dysbiosis in select human cancers: evidence of association and causality. Semin. Immunol. 32, 25–34.

Cheong, E.Y., Sandhu, A., Jayabalan, J., Le, T.T.K., Nhiep, N.T., Ho, H.T.M., et al., 2014. Isolation of lactic acid bacteria with antifungal activity against the common cheese spoilage mould *Penicillium commune* and their potential as biopreservatives in cheese. Food Control. 46, 91–97.

Christensen, G.J.M., Brüggemann, H., 2014. Bacterial skin commensals and their role as host guardians. Benef. Microbes 5, 201–215.

Chu, D.M., Ma, J., Prince, A.L., Antony, K.M., Seferovic, M.D., Aagaard, K.M., 2017. Maturation of the infant microbiome community structure and function across multiple body sites and in relation to mode of delivery. Nat. Med. 23, 314–326.

Coetzee, B., Freeborough, M.-J., Maree, H.J., Celton, J.-M., Rees, D.J.G., Burger, J.T., 2010. Deep sequencing analysis of viruses infecting grapevines: virome of a vineyard. Virology 400, 157–163.

Coyte, K.Z., Schluter, J., Foster, K.R., 2015. The ecology of the microbiome: networks, competition, and stability. Science 350 (80-.), 663–666.

Cui, L., Morris, A., Ghedin, E., 2013. The human mycobiome in health and disease. Genome Med. 5, 63.

de Oliveira, G.L.V., Leite, A.Z., Higuchi, B.S., Gonzaga, M.I., Mariano, V.S., 2017. Intestinal dysbiosis and probiotic applications in autoimmune diseases. Immunology 152, 1–12.

Delavenne, E., Ismail, R., Pawtowski, A., Mounier, J., Barbier, G., Le Blay, G., 2013. Assessment of lactobacilli strains as yogurt bioprotective cultures. Food Control. 30, 206–213.

Desai, S.N., Landay, A.L., 2018. HIV and aging. Curr. Opin. HIV AIDS 13, 22–27.

Deshpande, N.P., Riordan, S.M., Castaño-Rodríguez, N., Wilkins, M.R., Kaakoush, N.O., 2018. Signatures within the esophageal microbiome are associated with host genetics, age, and disease. Microbiome 6, 227.

Detman, A., Mielecki, D., Chojnacka, A., Salamon, A., Błaszczyk, M.K., Sikora, A., 2019. Cell factories converting lactate and acetate to butyrate: *Clostridium butyricum* and microbial communities from dark fermentation bioreactors. Microb. Cell. Fact. 18, 36.

Diaz, P.I., Hong, B.-Y., Frias-Lopez, J., Dupuy, A.K., Angeloni, M., Abusleme, L., et al., 2013. Transplantation-associated long-term immunosuppression promotes oral colonization by potentially opportunistic pathogens without impacting other members of the salivary bacteriome. Clin. Vaccine Immunol. 20, 920–930.

Dimijian, G.G., 2000. Evolving together: the biology of symbiosis, part 1. Proc. (Bayl. Univ. Med. Cent.) 13, 217–226.

Ding, T., Schloss, P.D., 2014. Dynamics and associations of microbial community types across the human body. Nature 509, 357–360.

Dolezal, P., Smid, O., Rada, P., Zubacova, Z., Bursac, D., Sutak, R., et al., 2005. Giardia mitosomes and trichomonad hydrogenosomes share a common mode of protein targeting. Proc. Natl. Acad. Sci. U.S.A. 102, 10924–10929.

Donati, E.R., Castro, C., Urbieta, M.S., 2016. Thermophilic microorganisms in biomining. World J. Microbiol. Biotechnol. 32, 179.

Dubourg, G., Lagier, J.C., Armougom, F., Robert, C., Hamad, I., Brouqui, P., et al., 2013. The proof of concept that culturomics can be superior to metagenomics to study atypical stool samples. Eur. J. Clin. Microbiol. Infect. Dis. 32, 1099.

Duncan, S.H., Louis, P., Flint, H.J., 2004. Lactate-utilizing bacteria, isolated from human feces, that produce butyrate as a major fermentation product. Appl. Environ. Microbiol. 70, 5810–5817.

Egert, M., Simmering, R., Riedel, C., 2017. The association of the skin microbiota with health, immunity, and disease. Clin. Pharmacol. Ther. 102, 62–69.

Fernandez, B., Vimont, A., Desfossés-Foucault, É., Daga, M., Arora, G., Fliss, I., 2017. Antifungal activity of lactic and propionic acid bacteria and their potential as protective culture in cottage cheese. Food Control 78, 350–356.

Florou-Paneri, P., Christaki, E., Bonos, E., 2013. Lactic acid bacteria as source of functional ingredients. In: Lactic Acid Bacteria-R & D for Food, Health and Livestock Purposes. IntechOpen.

Foster, J.A., Krone, S.M., Forney, L.J., 2008. Application of ecological network theory to the human microbiome. Interdiscip. Perspect. Infect. Dis. 2008, 839501.

Foster, K.R., Schluter, J., Coyte, K.Z., Rakoff-Nahoum, S., 2017. The evolution of the host microbiome as an ecosystem on a leash. Nature 548, 43–51.

Friedrich, M.J., 2008. Microbiome project seeks to understand human body's microscopic residents. JAMA 300, 777.

Fuchsman, C.A., Collins, R.E., Rocap, G., Brazelton, W.J., 2017. Effect of the environment on horizontal gene transfer between bacteria and archaea. PeerJ 5, e3865.

Gabrilska, R.A., Rumbaugh, K.P., 2015. Biofilm models of polymicrobial infection. Future Microbiol. 10, 1997–2015.

García-Aljaro, C., Ballesté, E., Muniesa, M., 2017. Beyond the canonical strategies of horizontal gene transfer in prokaryotes. Curr. Opin. Microbiol. 38, 95–105.

Gemechu, T., 2015. Review on lactic acid bacteria function in milk fermentation and preservation. Afr. J. Food Sci. 9, 170–175.

Ghannoum, M., 2016. Cooperative evolutionary strategy between the bacteriome and mycobiome. mBio 7, 10–13.

Ghannoum, M.A., Jurevic, R.J., Mukherjee, P.K., Cui, F., Sikaroodi, M., Naqvi, A., et al., 2010. Characterization of the oral fungal microbiome (mycobiome) in healthy individuals. PLoS Pathog. 6, e1000713.

Gill, S.R., Pop, M., Deboy, R.T., Eckburg, P.B., Turnbaugh, P.J., Samuel, B.S., et al., 2006. Metagenomic analysis of the human distal gut microbiome. Science 312, 1355–1359.

Gillespie, D.E., Brady, S.F., Bettermann, A.D., Cianciotto, N.P., Liles, M.R., Rondon, M.R., et al., 2002. Isolation of antibiotics turbomycin a and B from a metagenomic library of soil microbial DNA. Appl. Environ. Microbiol. 68, 4301–4306.

Gould, A.L., Zhang, V., Lamberti, L., Jones, E.W., Obadia, B., Korasidis, N., et al., 2018. Microbiome interactions shape host fitness. Proc. Natl. Acad. Sci. U.S.A. 115, E11951–E11960.

Gupta, R.S., 1998. Life's third domain (Archaea): an established fact or an endangered paradigm? Theor. Popul. Biol., 54. pp. 91–104.

Gupta, R.S., 2011. Origin of diderm (Gram-negative) bacteria: antibiotic selection pressure rather than endosymbiosis likely led to the evolution of bacterial cells with two membranes. Antonie Van Leeuwenhoek 100, 171–182.

Hager, C.L., Ghannoum, M.A., 2017. The mycobiome: role in health and disease, and as a potential probiotic target in gastrointestinal disease. Dig. Liver Dis. 49, 1171–1176.

Hashemi, H., Condurat, A.-L., Stroh-Dege, A., Weiss, N., Geiss, C., Pilet, J., et al., 2018. Mutations in the non-structural protein-coding sequence of protoparvovirus H-1PV enhance the fitness of the virus and show key benefits regarding the transduction efficiency of derived vectors. Viruses 10, 150.

Henskens, F.L., Green, T.G.A., Wilkins, A., 2012. Cyanolichens can have both cyanobacteria and green algae in a common layer as major contributors to photosynthesis. Ann. Bot. 110, 555–563.

Hoarau, G., Mukherjee, P.K., Gower-Rousseau, C., Hager, C., Chandra, J., Retuerto, M.A., et al., 2016. Bacteriome and mycobiome interactions underscore microbial dysbiosis in familial Crohn's disease. mBio 7, 10.1128/mBio.01250-16.

Huang, H.L., Chen, H.T., Luo, Q.L., Xu, H.M., He, J., Li, Y.Q., et al., 2019. Relief of irritable bowel syndrome by fecal microbiota transplantation is associated with changes in diversity and composition of the gut microbiota. J. Dig. Dis. 1751-2980.12756.

Human Microbiome Project Consortium, 2012. A framework for human microbiome research. Nature 486, 215–221.

Ipci, K., Altıntoprak, N., Muluk, N.B., Senturk, M., Cingi, C., 2017. The possible mechanisms of the human microbiome in allergic diseases. Eur. Arch. Oto-Rhino-Laryngol. 274, 617–626.

Izhar, R., Ben-Ami, F., 2015. Host age modulates parasite infectivity, virulence and reproduction. J. Anim. Ecol. 84, 1018–1028.

Johnson, K.V.-A., Foster, K.R., 2018. Why does the microbiome affect behaviour? Nat. Rev. Microbiol. 16, 647–655.

Jovel, J., Patterson, J., Wang, W., Hotte, N., O'Keefe, S., Mitchel, T., et al., 2016. Characterization of the gut microbiome using 16S or shotgun metagenomics. Front. Microbiol. 7, 459.

Kamal, R.M., Alnakip, M.E., El Aal, S.F.A., Bayoumi, M.A., 2018. Bio-controlling capability of probiotic strain *Lactobacillus rhamnosus* against some common foodborne pathogens in yoghurt. Int. Dairy J. 85, 1–7.

Kambouris, M.E., Pavlidis, C., Skoufas, E., Arabatzis, M., Kantzanou, M., Velegraki, A., et al., 2018. Culturomics: a new kid on the block of OMICS to enable personalized medicine. OMICS 22, 108–118.

Lacroix, B., Citovsky, V., 2016. Transfer of DNA from bacteria to eukaryotes. mBio 7, 10.1128/mBio.00863-16.
Lagier, J.-C., Armougom, F., Million, M., Hugon, P., Pagnier, I., Robert, C., et al., 2012. Microbial culturomics: paradigm shift in the human gut microbiome study. Clin. Microbiol. Infect. 18, 1185–1193.
Lai, G.C., Tan, T.G., Pavelka, N., 2019. The mammalian mycobiome: a complex system in a dynamic relationship with the host. Wiley Interdiscip. Rev. Syst. Biol. Med. 11, e1438.
Ley, R.E., Knight, R., Gordon, J.I., 2007. The human microbiome: eliminating the biomedical/environmental dichotomy in microbial ecology. Environ. Microbiol. 9, 3–4.
Li, L., Victoria, J.G., Wang, C., Jones, M., Fellers, G.M., Kunz, T.H., et al., 2010. Bat guano virome: predominance of dietary viruses from insects and plants plus novel mammalian viruses. J. Virol. 84, 6955–6965.
Liu, H., Carvalhais, L.C., Schenk, P.M., Dennis, P.G., 2017. Effects of jasmonic acid signalling on the wheat microbiome differ between body sites. Sci. Rep. 7, 41766.
Liu, Z., Yuan, T., Dai, X., Shi, L., Liu, X., 2019. Intermittent fasting alleviates diabetes-induced cognitive decline via gut microbiota-metabolites-brain axis (OR32-04-19). Curr. Dev. Nutr. 3.
Lloyd-Price, J., Abu-Ali, G., Huttenhower, C., 2016. The healthy human microbiome. Genome Med. 8, 51.
Ludwig, M., Schulz-Friedrich, R., Appel, J., 2006. Occurrence of hydrogenases in cyanobacteria and anoxygenic photosynthetic bacteria: implications for the phylogenetic origin of cyanobacterial and algal hydrogenases. J. Mol. Evol. 63, 758–768.
Ma, Z., Li, L., Li, W., 2018. Assessing and interpreting the within-body biogeography of human microbiome diversity. Front. Microbiol. 9, 1619.
Mammen, M.J., Sethi, S., 2016. COPD and the microbiome. Respirology 21, 590–599.
Mao, Q., Jiang, F., Yin, R., Wang, J., Xia, W., Dong, G., et al., 2018. Interplay between the lung microbiome and lung cancer. Cancer Lett. 415, 40–48.
Marrero, J., Coto, O., Goldmann, S., Graupner, T., Schippers, A., 2015. Recovery of nickel and cobalt from laterite tailings by reductive dissolution under aerobic conditions using *Acidithiobacillus* species. Environ. Sci. Technol. 49, 6674–6682.
Meysman, F.J.R., 2018. Cable bacteria take a new breath using long-distance electricity. Trends Microbiol. 26, 411–422.
Miller, M.B., Bassler, B.L., 2001. Quorum sensing in bacteria. Annu. Rev. Microbiol. 55, 165–199.
Mishra, V.K., Passari, A.K., Chandra, P., Leo, V.V., Kumar, B., Uthandi, S., et al., 2017. Determination and production of antimicrobial compounds by *Aspergillus clavatonanicus* strain MJ31, an endophytic fungus from *Mirabilis jalapa* L. using UPLC-ESI-MS/MS and TD-GC-MS analysis. PLoS One 12, e0186234.
Moayyedi, P., Yuan, Y., Baharith, H., Ford, A.C., 2017. Faecal microbiota transplantation for *Clostridium difficile*-associated diarrhoea: a systematic review of randomised controlled trials. Med. J. Aust. 207, 166–172.
Nelson, K.E., Clayton, R.A., Gill, S.R., Gwinn, M.L., Dodson, R.J., Haft, D.H., et al., 1999. Evidence for lateral gene transfer between Archaea and Bacteria from genome sequence of *Thermotoga maritima*. Nature 399, 323–329.
O'Toole, P.W., Jeffery, I.B., 2018. Microbiome–health interactions in older people. Cell. Mol. Life Sci. 75, 119–128.
Oever, J., ten, Netea, M.G., 2014. The bacteriome-mycobiome interaction and antifungal host defense. Eur. J. Immunol. 44, 3182–3191.
Ordovas, J.M., Mooser, V., 2006. Metagenomics: the role of the microbiome in cardiovascular diseases. Curr. Opin. Lipidol. 17, 157–161.
Ortolani, M.B.T., Yamazi, A.K., Moraes, P.M., Viçosa, G.N., Nero, L.A., 2010. Microbiological quality and safety of raw milk and soft cheese and detection of autochthonous lactic acid bacteria with antagonistic activity against *Listeria monocytogenes*, *Salmonella* spp., and *Staphylococcus aureus*. Foodb. Path. Dis. 7, 175–180.

Perez Perez, G.I., Gao, Z., Jourdain, R., Ramirez, J., Gany, F., Clavaud, C., et al., 2016. Body site is a more determinant factor than human population diversity in the healthy skin microbiome. PLoS One 11, e0151990.

Petatán-Sagahón, I., Anducho-Reyes, M.A., Silva-Rojas, H.V., Arana-Cuenca, A., Tellez-Jurado, A., Cárdenas-Álvarez, I.O., et al., 2011. Isolation of bacteria with antifungal activity against the phytopathogenic fungi *Stenocarpella maydis* and *Stenocarpella macrospora*. Int. J. Mol. Sci. 12, 5522–5537.

Pevsner-Fischer, M., Blacher, E., Tatirovsky, E., Ben-Dov, I.Z., Elinav, E., 2017. The gut microbiome and hypertension. Curr. Opin. Nephrol. Hypertens. 26, 1–8.

Probst, A.J., Birarda, G., Holman, H.-Y.N., DeSantis, T.Z., Wanner, G., Andersen, G.L., et al., 2014. Coupling genetic and chemical microbiome profiling reveals heterogeneity of archaeome and bacteriome in subsurface biofilms that are dominated by the same archaeal species. PLoS One 9, e99801.

Prohic, A., Jovovic Sadikovic, T., Krupalija-Fazlic, M., Kuskunovic-Vlahovljak, S., 2016. *Malassezia* species in healthy skin and in dermatological conditions. Int. J. Dermatol. 55, 494–504.

Qin, J., Li, R., Raes, J., Arumugam, M., Burgdorf, K.S., Manichanh, C., et al., 2010. A human gut microbial gene catalogue established by metagenomic sequencing. Nature 464, 59–65.

Rattanachaikunsopon, P., Phumkhachorn, P., 2010. Lactic acid bacteria: their antimicrobial compounds and their uses in food production. Ann. Biol. Res. 1, 218–228.

Reid, G., Abrahamsson, T., Bailey, M., Bindels, L.B., Bubnov, R., Ganguli, K., et al., 2017. How do probiotics and prebiotics function at distant sites? Benef. Microbes 8, 521–533.

Risgaard-Petersen, N., Kristiansen, M., Frederiksen, R.B., Dittmer, A.L., Bjerg, J.T., Trojan, D., et al., 2015. Cable bacteria in freshwater sediments. Appl. Environ. Microbiol. 81, 6003–6011.

Rivas, M.N., Crother, T.R., Arditi, M., 2016. The microbiome in asthma. Curr. Opin. Pediatr. 28, 764–771.

Rogers, M.B., Firek, B., Shi, M., Yeh, A., Brower-Sinning, R., Aveson, V., et al., 2016. Disruption of the microbiota across multiple body sites in critically ill children. Microbiome 4, 66.

Schloss, P.D., Handelsman, J., 2003. Biotechnological prospects from metagenomics. Curr. Opin. Biotechnol. 14, 303–310.

Seed, P.C., 2015. The human mycobiome. Cold Spring Harb. Perspect. Med. 5, 103–122.

Sender, R., Fuchs, S., Milo, R., 2016. Revised estimates for the number of human and bacteria cells in the body. PLoS Biol. 14, e1002533.

Stingl, K., Koraimann, G., 2017. Prokaryotic information games: how and when to take up and secrete DNA. Curr. Top. Microbiol. Immunol. 413, 61–92.

Tanaka, M., Itoh, H., 2019. Hypertension as a metabolic disorder and the novel role of the gut. Curr. Hypertens. Rep. 21, 63.

Tang, W.H.W., Li, D.Y., Hazen, S.L., 2019. Dietary metabolism, the gut microbiome, and heart failure. Nat. Rev. Cardiol. 16, 137–154.

Touchon, M., Moura de Sousa, J.A., Rocha, E.P., 2017. Embracing the enemy: the diversification of microbial gene repertoires by phage-mediated horizontal gene transfer. Curr. Opin. Microbiol. 38, 66–73.

Toussaint, A., Chandler, M., 2012. Prokaryote genome fluidity: toward a system approach of the mobilome. Methods Mol. Biol. (Clifton, N.J.). 804, 57–80.

Troselj, V., Cao, P., Wall, D., 2018. Cell-cell recognition and social networking in bacteria. Environ. Microbiol. 20, 923–933.

Turnbaugh, P.J., Ley, R.E., Hamady, M., Fraser-Liggett, C.M., Knight, R., Gordon, J.I., 2007. The human microbiome project. Nature 449, 804–810.

van Wolferen, M., Orell, A., Albers, S.-V., 2018. Archaeal biofilm formation. Nat. Rev. Microbiol. 16, 699–713.

Verma, N., Sharma, R., 2017. Bioremediation of toxic heavy metals: a patent review. Recent Pat. Biotechnol. 11, 171–187.

Wagner, A., Whitaker, R.J., Krause, D.J., Heilers, J.-H., van Wolferen, M., van der Does, C., et al., 2017. Mechanisms of gene flow in archaea. Nat. Rev. Microbiol. 15, 492–501.

Wang, J.B., Cheung, W.W., 1998. Multiple bacteroids in the bacteriome of the lantern bug *Pyrops candelaria* Linn. (Homoptera: Fulgoridae). Parasitol. Res. 84, 741–745.

Weiss, G.A., Hennet, T., 2017. Mechanisms and consequences of intestinal dysbiosis. Cell. Mol. Life Sci. 74, 2959–2977.

Whalen, K.E., Becker, J.W., Schrecengost, A.M., Gao, Y., Giannetti, N., Harvey, E.L., 2019. Bacterial alkyl-quinolone signaling contributes to structuring microbial communities in the ocean. Microbiome 7, 93.

Wilantho, A., Deekaew, P., Srisuttiyakorn, C., Tongsima, S., Somboonna, N., 2017. Diversity of bacterial communities on the facial skin of different age-group Thai males. PeerJ 5, e4084.

Wisecaver, J.H., Slot, J.C., Rokas, A., 2014. The evolution of fungal metabolic pathways. PLoS Genet. 10, e1004816.

Woese, C.R., Kandler, O., Wheelis, M.L., 1990. Towards a natural system of organisms: proposal for the domains Archaea, Bacteria, and Eucarya. Proc. Natl. Acad. Sci. U.S.A. 87, 4576–4579.

Woraprayote, W., Malila, Y., Sorapukdee, S., Swetwiwathana, A., Benjakul, S., Visessanguan, W., 2016. Bacteriocins from lactic acid bacteria and their applications in meat and meat products. Meat Sci. 120, 118–132.

CHAPTER

MYC(ET)OBIOME: THE BIG UNCLE IN THE FAMILY

3

Manousos E. Kambouris[1] and Aristea Velegraki[2]

[1]*The Golden Helix Foundation, London, United Kingdom* [2]*Mycology Research Laboratory, Department of Microbiology, School of Medicine, National and Kapodistrian University of Athens, Athens, Greece*

INTRODUCTION

The mycobiome constitutes one of the vertical divisions of the microbiome (Stefanaki et al., 2017), and it has been coined since 2010 (Ghannoum et al., 2010), precisely as a part of the microbiome in a human biocompartment (Dworecka-Kaszak et al., 2016). It is considered an upper iteration, as it comprises eukaryotes, and many of them not microbial in size and multicellular in structure. Although originally used in an ambiguous manner, so as to allow loose interpretation, as an entity of organisms or of their genomes (Ghannoum et al., 2010), it is currently used to describe the sum of mycobiota in a spatiotemporal unit plus the interactions in which they are involved (Cui et al., 2013).

EMERGENCE AND ESTABLISHMENT

One favored metric in assessing the acceptance and the interest invested in any new concept or entity is its temporal population curve in PubMed; for mycobiome, the procedure has been repeatedly implemented and published, although rather abstractedly (Cui et al., 2013; Limon et al., 2017). It is especially revealing that PubMed returns containing the term "mycobiome" in general (implying the keywords) are 550% more than the ones containing the same term in the title by 12/2018 compared to less than 200% till 2017 (Fig. 3.1). This metric currently presents an exponential growth of doubling its value yearly from 2010 to 2018 with a lull in 2016; at the same timeline the returns containing the term "mycobiome" within the title presented a lull in 2017 and 2018, after a less steep but steady increase between 2010 and 2016. The obvious interpretation is that the term and concept are used descriptively and perhaps cognitively within current research schemes but without constituting a vogue and thus having a real impact on the conceptual phase of research definition and planning.

Interestingly, the mycobiome was conceptualized in 2010 as previously noted and cited, 11 years after the bacteriome first appeared in bibliography—although in a different context. Still, it was the mycobiome that made a lasting impression. It registers 96 total returns in titles, all between

Microbiomics. DOI: https://doi.org/10.1016/B978-0-12-816664-2.00003-7

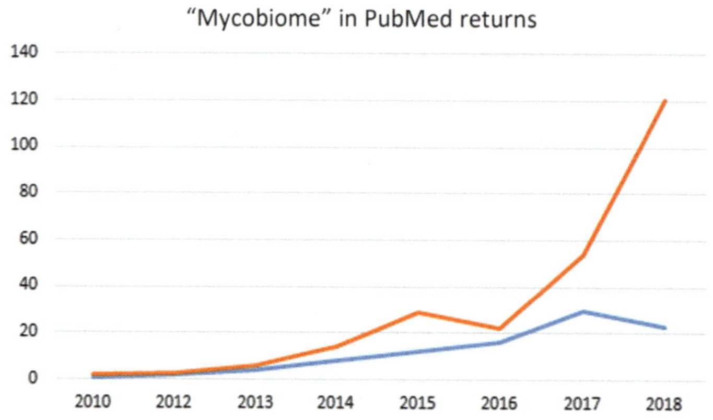

FIGURE 3.1

Chart of the returns containing the word "mycobiome" from PubMed. Returns from keywords are in steep increase and depicted in red and returns with the term in the title are in slight decline and depicted in blue.

2010 and 2018 compared to 58 for the bacteriome, of which some should be rejected due to the difference in cognitive context (see Chapter 2).

DEFINITION AND IDENTITY

The mycobiome as is known today, namely, the collectivity that is encompassing the fungal biota (Cui et al., 2013) emerged in successive iterations. The first was the virtual absence of fungi/"Mycobiota" (Lai et al., 2018; Enaud et al., 2018) from the Human Microbiome Project (NIH HMP Working Group et al., 2009) and from the Metagenomics of the Human Intestinal Tract (Qin et al., 2010). Both projects, with the latter clearly stating it, were based on genome-oriented methods; the former simply followed a vogue where "microbiome" meant a sum of spatially and temporally related microbial genomes (Hooper and Gordon, 2001). At the same time, it furthered a point of view characteristic of human (or rather medical) microbiologists that equals "microbe" with "bacterium" plus or minus the Archaea (Dworecka-Kaszak et al., 2016; Gundogdu and Nalbantoglu, 2017).

From a standpoint of biological orthodoxy the inclusion of the mycobiome in the microbiome, which is an all-encompassing concept simply containing all categories of microbiota (also referred to as subordinate biomes) and their interactions within the same spatiotemporal entity (Whipps et al., 1988), is controversial: a large number of fungi are facultative unicellular organisms and may develop multicellular forms, such as mycelia, which are entities of relatively low differentiation, and possibly of predetermined growth. The macrobiotic option, with highly visible and occasionally massive multicellular formations (mycelia of *Amanita*) is in some fungi a matter of environmental triggering, as with thermal dimorphism (Gauthier, 2015), making the relationship between mycobiome and microbiome a partial overlap rather than a full subordination/inclusion.

Table 3.1 Horizontal Categorization and Vertical Iterations of Microbiomes

Subordinate Biomes	Hosted Biocompartments	Environmental Locations
Virome, bacteriome, mycobiome	Skin (armpit, scalp, navel, back, hand, foot)	Soil
	GIT (mouth, stomach, esophagus, small and large intestines)	Freshwater
	Ear	Seawater
	Respiratory tract (nasal cavity, pharynx, larynx, trachea, lung)	Sewage
	Leave	Urban structures
	Root	Pollution site
	Stem	Cave/chasm

The main concept in mycobiome, as in other biomes, does not just refer to its constituents but rather to the interactions amongst them on one hand (Tipton et al., 2018) and between them and their net, integrated function on the other—should there be one such function, which is controversial. This fact—or aspect—is perplexed by the innate nature of mycobiota—a new, embellished way to say "fungi"—and their niches in respective environments (see Table 3.1). The similarities between fungi and bacteria result in vivid exchanges and interaction (Tipton et al., 2018; Enaud et al 2018; Hager and Ghannoum, 2017; Seed, 2014), which make a clear differentiation between mycobiome and bacteriome extremely difficult within the same spatiotemporal parameters, and, in reality, implausible. The two entities are entangled in so many cases, contexts, and so much depth that efforts toward deconvolution affect the accurate understanding of the wider, integrated picture. In both appendage and environmental microbiomes the integral picture, both in dynamic and static—snapshot—terms is the result of a network of interactions among cells and their products and moieties belonging to both biomes (Table 3.1), and, usually, to others as well (Hager and Ghannoum, 2017). Hence, the extraction and study of only one kind/kingdom leads to untenable, misguiding, and practically artificial conclusions (Tipton et al., 2018).

In addition, the presence of both fungi and bacteria within the same—"polymicrobial"—biofilms (Ghannoum, 2016; Hoarau et al., 2016; Hager and Ghannoum, 2017) shows vividly the challenges in trying to dissect different descriptive subordinate biomes of a microbiome (Fig. 3.2). The single-strain biofilm by definition is a good approximation of a multicellular organism of low differentiation; the matrix is a functional and possibly structural approximation of the basal membranes or of the extracellular matrices of tissues in multicellular organisms, while the differentiated form of microbial cells embedded in a matrix (*sessile* cells) can be considered a clear adaptation of individual, *planktonic* cells to communal standards (Rollet et al., 2009). Thus a polymicrobial biofilm entails a much higher degree of integration and interface than that described by the inventory-type list of the members of a microbiome (Ghannoum, 2016; Hoarau et al., 2016; Hager and Ghannoum, 2017).

The decision to practically isolate one subordinate biome from a functional microbiome is based on technical rather than biological reasons. Mycobiota are heterotrophs, basically saprobiotes/decomposers, as are most bacteria; they are not gifted with motion, as are not many bacteria, but

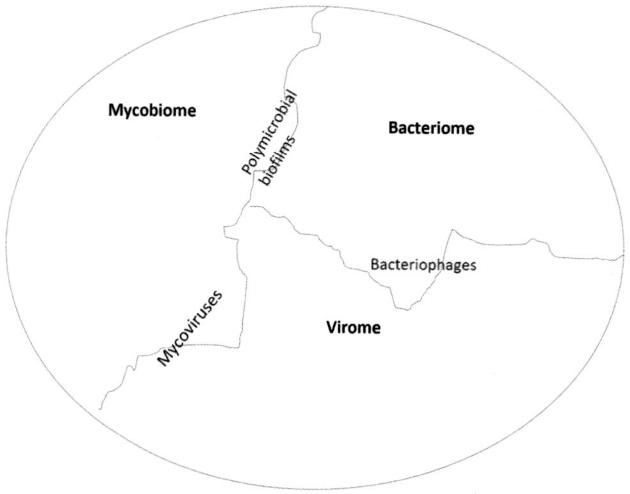

FIGURE 3.2

Interfaces among the three main subordinate biomes within a microbiome. Symbiotic forms and phenomena (sensu lato) including the, but not being restricted to, polymicrobial biofilms and viral transfection make separate studies inaccurate.

they are eukaryotes and thus possess considerable differing attributes: a much more evolved biochemical machinery, from codification to posttranslational and enzymatic processes, so as to afford individual adaptability to environtomic changes *sensu lato* (Hekim and Özdemir, 2017); much more energy-efficient, labor-distributing, and sizeable cells; a higher ability to have cells coordinate, interface, and create interdependencies leading to effective cooperation and formation of multicellular structures initially and, ultimately, to the formation of multicellular organisms—which bear greater biochemical similarities with other multicellular organisms than with bacteria sharing the same habitats.

These differences have practical but also biological consequences: the former is exemplified by the inability to use the same detection and classification techniques. As mentioned before, the two major genomic-oriented projects aiming to analyze microbiomes (human appendage microbiomes, to be exact) focused on bacteria and excluded fungi—practically the studies where of bacteriomic and not of microbiomic context (Cui et al., 2013). For detecting fungi, and then for classifying them, the internal transcribed spacer (ITS)-1/2 sequences are the established method, which also allows barcoding in some cases (Irinyi et al., 2015; Schoch et al., 2012; Dworecka-Kaszak et al., 2016; Thomas et al., 2014), while other taxa must be referred to tubulin or other genes to complete barcoding down to species level (Frisvad and Samson, 2004; Schoch et al., 2012; Seifert, 2009).

Bacteria, on the other hand, are identifiable by 16S rDNA sequences, and the respective protocols differ from the mycological ones, more or less. The "more or less" makes multiplexing least favorable in most microbiology labs that usually possess low-level PCR and molecular biology know-how. In effect, the distinction between bacteriome and mycobiome is largely artificial in ecosystemic terms and is due to the diversity of scientific background, availability of laboratory protocols and infrastructures tasked to studying them (Lai et al., 2018) rather than to any intrinsic functional reason.

Last but not least, the virome is easier to define but just as difficult to extract from an integrated, distinct microbiome, due to the existence of virophages and mycoviruses (Fig. 3.2) that keep it tangled with the other two subordinate biomes (Ghabrial et al., 2015).

Practicalities notwithstanding, there is a very sound biological dimension that sets the mycobiota apart from other microbiota. The basic cellular functions, known as "housekeeping" in genetic slang of previous generations, are very similar among animal and fungal cells and, slightly less, plant cells as well. This simple fact is focal in every targeted manipulation approach, especially in the use of biocides and, most notably, antibiotics (*sensu lato*, including antibacterial and antifungal substances causing necrosis or stasis, no matter the source or origin). The different biochemical machineries present a natural barrier to retasking classes of antibiotics, although this may change in the near future (Liu et al., 2019); and also—allegedly—preclude the development of cross-resistance between the kingdoms of bacteria and fungi.

The antifungals—with few exceptions—demonstrate intrinsic toxicity to other eukaryotes. The toxic effect of amphotericin B to human kidneys, due to cross-reactivity with human cholesterol instead of its main target, the fungal ergosterol, is a paradigm of such ill effects (Kamiński, 2014). There are metabolomics exclusivities that allow highly specific targeting, such as targeting the fungal cell wall pathways by echinocandins (Perlin, 2011), but such examples are scarce. As a result, the dispersion of antifungals in any environment may cause both acute and accumulating ill-effects not only to populations of benign mycobiota but also to other in-place or in-chain/network eukaryotes, especially of the animal kingdom.

MYCOBIOME: STATUS, CATEGORIES, AND ESSENCE

The idea of treating the human microbiome (and these of animals and possibly of plants as well) as an organ since it performs a number of necessary and occasionally exclusive functions (Limon et al., 2017) and accounts, in terms of mass, to the human liver (Hawrelak and Myers, 2004; Rowland et al., 2018) is undeniably novel. The previous concept was to consider the *microflora*, a more descriptive and less function- and interaction-defined entity, as an adaptable environmental factor that was fragile, unstable, and variable to the extreme. If the microbiome is to be considered an organ, the mycobiome should rather be viewed as one of its tissues. The problem is that this organ concept is not well-defined in structure (especially in terms of integration), nature (especially in terms of interdependence), locality/site, and individuality.

If the microbiome is considered as an organ, the obvious differentiation of prokarya, mycobiota, and viruses as homogenous subordinate biomes within the same microbiome clearly deserves a "tissue"-level status. The different subordinate biomes, of course, do not share the same genome—actually not even the different bits of any single biome (i.e. Operational Taxonomic Units (OTUs) of the same Kingdom) do so amongst themselves. This inconvenient little detail is usually explained away by an "eagle's eye" approach, where the entirety of the cell populations is seen from afar as a community with their functional interactions being considered normal, although somewhat lax, homeostasis.

There are, though, some pitfalls: in the simplest of cases, when speaking of one human, or animal (Kearns et al., 2017; Dworecka-Kaszak et al., 2016), or even plant (Schlegel et al., 2018)—although the latter to a lesser degree—they are considered individual hosts. It is a valid query to ask, then, whether there is, at any given time, only one hosted microbiome (Human Microbiome Project Consortium, 2012) dispersed in different biocompartments of each individual host, constituting a two-part holobiont (Dworecka-Kaszak et al., 2016). The alternative is to conceptualize a series of appendage microbiomes (and their subordinate biomes) defined by spatial terms/biocompartments of their hosting (Table 3.1) and possibly interacting in different degrees (Cui et al., 2013) but in essence discrete and possibly different from each other, constituting a holobiont of proteic nature. Said holobiont may shed, reshuffle, and replace the microbiotic *constituents* (plural) either to adapt or as a result of imposed challenges and stimuli (Dworecka-Kaszak et al., 2016). It is true that the oral cavity and the lung or the gut microbiomes may well communicate and interact (Seed, 2014; Dworecka-Kaszak et al., 2016). There is respective evidence aplenty: a disturbance of the gut mycobiome by *Candida albicans* impacts allergic pulmonary disease induced by *Aspergillus fumigatus* of the lung mycobiome (Cui et al., 2013; Enaud et al., 2018; Limon et al., 2017), while the fungal distribution in different mycobiomes changes proportionately to the distance (Cui et al., 2013). Significant differences in distribution are observed among distant body sites, whereas similar patterns of distribution were found in mycobiomes from nearby sites; for example, *Cladosporium*, *Aspergillus*, and *Penicillium* dominate other fungal genera in both oral and nasal cavities (Cui et al., 2013). The same goes in the case of vaginal and respective near-skin mycobiomes.

But the opposite is true as well. The gut microbiome and that of the left armpit are not readily communicating, nor interacting, at least not continually and directly. They include different microbiota both qualitatively and quantitatively; present different collective metagenomes; and they perform, enact, or undergo different functions. Thus it is perhaps more correct to speak of human *microbiomes* (plural) and submicrobiomes/subordinate biomes within them (Table 3.1). The former implies spatiotemporal differentiation, the latter taxonomical differentiation on a kingdom basis.

In this line of thought the functionality is important. Microbiomes and subordinate biomes are by definition either "plain"/environmental or "hosted" (Table 3.1). The latter, when engaged in a complex and bilateral set of interactions that feature a mutualistic intercourse with the host in terms of anatomical sites/biocompartments should be considered as "appendages." Still, even if some microbiota of the gut microbiome are essential for the assimilation of nutrients, the production of vitamins, and other necessary metabolites (Hawrelak and Myers, 2004; Oever and Netea, 2014; Seed, 2014; Lai et al., 2018; Limon et al., 2017) and their demise causes real health issues at distant sites, beyond local dysbiosis (Hawrelak and Myers, 2004; Cui et al., 2013; Lai et al., 2018; Enaud et al., 2018), there is no indication that this is the case for *all* human- (or animal-) hosted microbiomes.

Thus the microbiome of a single host (i.e., a single human) is not unitary but rather a selection of site-specific ones with differing in constitution, variability, and rate of communication, which makes the whole sound less and less as an organ. The wide, almost all-encompassing distribution of microbial populations is not so much a problem for apprehending them as a collective organ, much like the skin. But once the microbiota of a microbiome lack the common denominator of an identical genome, some other common elements should be detected instead. The functionality is *not* a good choice: it is not universal in terms of site and microbial diversity. Even if microbiomes, where selected microbiota do perform a vital function, *could* be considered "organs" (Hawrelak and Myers, 2004), there is still a cognitive gap in their definition as such. They cannot be defined as external organs, because the gut microbiome, the most functional, is not easy to be considered "external;" the gut environment interfaces between the outside world and the human internal environment (Hawrelak and Myers, 2004). Furthermore, other microbiomes are present in sites previously considered sterile—at least for healthy individuals—as is the lower respiratory tract and the lung proper (Seed, 2014; Dworecka-Kaszak et al., 2016; Culibrk et al., 2016). The latter is found sparsely but steadily colonized by environmental organisms such as *Aspergillus* and *Cladosporium* (a fact changing in immunosuppression, where the fungal loads are higher), although the detected nucleic acids may originate from transient loads, cleared by the immune system and reacquired through inhalation rather than from colonizers (Limon et al., 2017; Seed, 2014; Cui et al., 2013; Delhaes et al., 2012). Thus the definition of internal and external environment becomes more complicated than it used to be.

A cognitively more sound approach would be to revert to terms such as "exo" or "xeno-organ" of a superorganism (Gundogdu and Nalbantoglu, 2017), which is still more rigid and integrated than the symbiotic concept of holobiont (Dworecka-Kaszak et al., 2016) but less so than the idea of the conventional organ. The concept of the superorganism and its xeno-organs is intended to demonstrate not only the genomic and ontogenetic internal diversity of this organ but also the expected and accepted degree of differentiation among individual hosts and between host and organ. The difference among individuals of the same species, which are usually composed of the same cell types, as, for example, two humans of the same sex, is mostly observable at a molecular level, relevant to individuality rather than basic function; the compatibility in major histocompatibility complex is the best example. But the microbiomes of two persons might be *very* different at the cellular level; different microbiotes, at the very least in terms of species if not of higher taxa, may perform the same vital function (homologues) and occupy a similar niche (Ghannoum et al., 2010).

To this uneasy observation, one must add the fact that some microbiomes do not perform any positive function. The maturation or/and modulation of the immune system (Schlegel et al., 2018; Dworecka-Kaszak et al., 2016; Seed, 2014; Hawrelak and Myers, 2004) and the containment of malign species (Limon et al., 2017; Oever and Netea, 2014; Hawrelak and Myers, 2004) of obligate or opportunistic parasitic nature and in some (but not in all) cases pathogenic (pathobionts—Lai et al., 2018), are *not* functions attributable to an organ, external or otherwise, but events developing within the context of a heterogeneous and complex community which interacts with its environment—or rather environtome—and presents processes of social nature amongst its members. In some cases the local microbiome seems simply to abstain from any hostile manifestation rather than contribute to the welfare of the macroorganism, human or other (the healthy lung microbiome being such an example), let alone to perform vital and indispensable functions.

At the same time the sum of the different microbial genomes was projected as a collective supergenome which creates, in translational terms, a big picture; a picture which is practically reflected in the growth and health of the host (Rowland et al., 2018; Enaud et al., 2018) or, in an environmental context, in the shaping of the character of the respective microenvironment (Schlegel et al., 2018).

Such relationships point rather to a more complicated than previously thought but basically similar aspect of symbiosis and community, than to any highly integrated (deep) physiological incorporation to the like of chimeric organs. Consequently, individualized entities of one host and the sum of its symbiotic microbes, termed holobiont (Dworecka-Kaszak et al., 2016), must be apperceived, while the term hologenome engulfs the sum of the genomic sequences: DNA–RNA, chromosomal, episomatic or exonuclear, and viral. The sum of the symbiotic microbiota is the microbiome, perceived by any adjective, which can be dissected to appendage microbiomes by topology (oral, gut, skin microbiomes) and to subordinate biomes by rough microbial taxonomy (virome, bacteriome, mycetobiome).

A concept of obligatory mutualism might be more appropriate for the hardcore cases where interdependence seems to be established, leaving commensalism and other forms to describe less beneficial and necessary microbiomes in various biocompartments. Once the holobiont is in harmony (commensal or mutualistic events), the condition is termed normobiosis. The concept of dysbiosis lately becomes more and more used to interpret adverse or even harmful events and conditions occurring after qualitative or quantitative derailment—for whatever reason—of a balanced normobiosis, such as the aggressive attitude of hitherto commensal fungi in AIDS patients (Enaud et al., 2018; Hager and Ghannoum, 2017; Limon et al., 2017; Dworecka-Kaszak et al., 2016; Hoarau et al., 2016; Hawrelak and Myers, 2004).

Actually, there seems to be an intrinsic relationship between mycobiome, coined in 2010, and appearing ever since in 96 titles until the end of 2018, and dysbiosis—in terms of cognition and conceptual interest. The latter is included in 848 English-speaking paper titles returned by PubMed, of which two before 2000s, 47 between 2000 and 2010, and the rest 799 between 2010 and the end of 2018.

On top of the above, the term *microbiome* and its subordinate biomes are used in an environmental context as well, as already mentioned. In such a context, there is no room for assignment of the substance/status of organ to microbiomes, even of organs celebrated by the use of exo-, xeno-, meta-, or whatever fancy prefix. The subordinate biomes are easily and clearly definable as the collectivity of their members *and* of the networks of interactions implicating them. Such interactions belong to three tiers but are not ordered in this manner, as they develop in networks, not in chains, pairs, or pathways. The first tier is the interactions among the biotes of the subordinate biome (Kearns et al., 2017); the second is the interactions among biotes of different subordinate biomes of the same microbiome (Oever and Netea, 2014); and the third comprises the interactions between the whole microbiome and the environment (Seed, 2014). Should a host be present (thus defining "appendage biomes"), the host *is* the environment or a part of it. The concept of holobiome (Dworecka-Kaszak et al., 2016) was advanced exactly to describe the sum of appendage microbiomes and their host, as opposed to unhosted, environmental microbiomes. Thus appendage microbiomes are subject to two environments: the immediate one which is afforded by the host and the wider one where the holobiont is placed and affected, a condition favoring the concept of environtome. There is no conceptual interaction of a subordinate biome with the host or the environment; the interactions refer to the entire microbiome.

MYCOBIOME: STRUCTURE AND COMPOSITION

Regarding hosted mycobiomes, the current understanding is (1) that they demonstrate considerable diversity, as they include a fair number of species, but a number of taxa constitute a core which shows consistency (Lai et al., 2018); (2) that they differ not only among different biocompartments of the same host but also among the same biocompartments of different hosts of the same species, thus being practically personalized and indicative of their host (Dworecka-Kaszak et al., 2016; Cui et al., 2013; Hoarau et al., 2016; Hager and Ghannoum, 2017); and (3) that they also differ in their entirety—or regarding homologous biocompartments—among different host species as humans versus other mammals (Lai et al., 2018).

Moreover, individual mycobiomes are mostly steady in temporal terms once ontogenesis and development of the host are complete but may restructure if environtomic factors, host-defined (diet, use of antifungals), host-specific (age, sex) and host-independent (environment), change. Such changes might be random, spontaneous, or imposed and may alter important characteristics of a whole microbiome, the respective mycobiome included. In hosted microbiomes this might disrupt a beneficial or uneventful normobiosis to a dysbiosis which may promote, cause, or evolve to disease of any kind, severity, or etiology (Dworecka-Kaszak et al., 2016; Tang et al., 2015; Seed, 2014; Wu et al., 1998; Kearns et al., 2017; Limon et al 2017; Lai et al., 2018; Hall and Noverr, 2017; Hawrelak and Myers, 2004; Enaud et al., 2018; Cui et al., 2013). The interactions which define a mycobiome are of three main categories (Lai et al., 2018) as already mentioned:

1. The interactions among mycobiota, which can usually be reduced to paired interactions (Mar Rodríguez et al., 2015; Limon et al., 2017). Such events are usually antagonistic (Kearns et al., 2017). These interactions typically refer to one species trying to occupy, expand, and retain a niche against others, thus initiating killer and recession events (see Fig. 3.3). The latter might

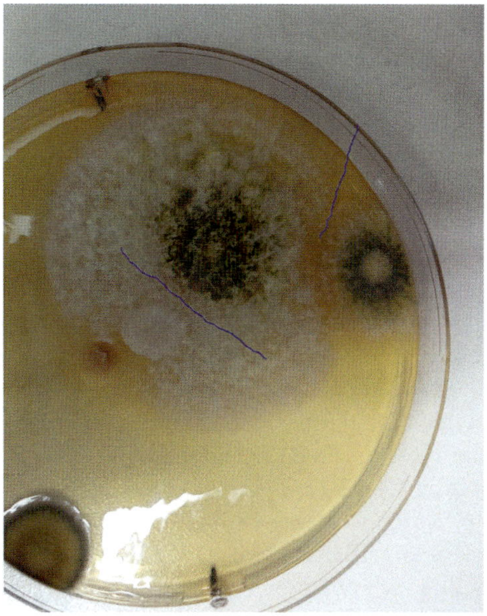

FIGURE 3.3

Antagonistic phenomena in microbiomes/mycobiomes. The *aggressive* microbiote presents a typical radial growth despite the presence of other species' colonies. The *recessive* microbiote shows a "negative" growth vector in comparison to its expected growth as it concesses ground to the *aggressive* one. The interface between *aggressive* and *recessive* mycelia/colonies is indicated by lines (*light blue*).

be defined as the development of a colony or mycelium against the actual growth orientation due to contact or proximity with other mycelia or colonies. An exemplary such case is *Candida* pathobionts usually being kept at bay in the oral cavity, most probably by commensal mycobiota such as *Pichia* (Mukherjee et al., 2014; Seed, 2014; Limon et al., 2017). Mutually beneficial events are rather rare, but there are exceptions, as is the combined metabolic process resulting in the production of novel, advantageous functional macromolecules (Thota et al., 2017; Stierle et al., 2017), or the formation of intrakingdom multispecies biofilms (Costa-Orlandi et al., 2017).

2. The interactions between the mycobiome and other microbiota/biomes of the respective microbiome (Tipton et al., 2018; Sovran et al., 2018; Enaud et al., 2018; Ghannoum, 2016; Cui et al., 2013; Haq et al., 2014; Mar Rodríguez et al., 2015) are the most important, due to the comparatively low abundance of mycobiota in a microbiome compared to bacteria and viruses (Lai et al., 2018; Seed, 2014). Despite relatively low representation, mycobiota do exist and affect, sometimes disproportionally to their representation metrics, both hosted and environmental biomes. The interaction between a mycobiome and a bacteriome may be cooperative or competitive (Cui et al., 2013; Seed, 2014; Limon et al., 2017; Hager and Ghannoum, 2017; Lai et al., 2018), usually both. These opposing interactions occur in different degrees and coordinates and thus have different impact, depending to a significant degree, but not totally, on the different interacting pairs or groups of said biomes in each individual case.

   The case of cooperation is exemplified by *Mycobacterium tuberculosis* superinfection occurring in lung aspergillosis cases (Darling, 1976); also by bacterial peptidoglycan-derived moieties that are detected in human tissues and induce hyphal growth in *C. albicans* (Xu et al., 2008) and by the observation that *Pseudomonas aeruginosa* and *Escherichia coli* increase the virulence of *C. albicans* (Neely et al., 1986). Moreover, the development of polymicrobial biofilms populated by both mycobiota and bacteria is a typical finding (Wargo and Hogan, 2006). A handy example of the latter are biofilms formed (1) by *C. albicans* and *Staphylococcus epidermis*, which demonstrate resistance to both fluconazole and vancomycin (Lai et al., 2018; Adam et al., 2002); (2) by *C. albicans* and *Citrobacter freundii* or *Trichosporon asahii* and *Staphylococcus simulans* in human/animal wound sites, where they delay or forestall the healing process (Ghannoum, 2016); (3) by *Candida tropicalis* and *E. coli* and/or *Serratia marcescens*. The latter are instrumental in gastrointestinal tract (GIT) dysbioses are causative of, or at least associated with, irritable bowel disease (Hager and Ghannoum, 2017; Limon et al., 2017; Hoarau et al., 2016; Ghannoum, 2016).

   For the case of competition the most renowned examples are (1) the adverse effect of *Penicillium* to neighboring bacteria (Fleming, n.d.), but exemplary are also the cases: (2) of *Candida* and *P. aeruginosa* in Cystic Fibrosis, where the bacterium suppresses the mycobiote (Kerr, 1994); and (3) of *P. aeruginosa and Staphylococccus aureus*, which inhibit *A. fumigatus* biofilm formation (Mowat et al., 2010; Ramírez Granillo et al., 2015).

   Still, the associations, although observed may be incidental; causality cannot be established despite intense theorizing (Cui et al., 2013; Lai et al., 2018). Irrespective of the competitive or cooperative nature of the interaction between hosted mycobiome and bacteriome, the end results include an altered host immune response to pathogens, different mycobiome susceptibility, and decreased bacteriome response to medical therapy (Cui et al., 2013; Ghannoum, 2016).

3. The interactions between the mycobiota and the environtome, either the host or the environment—the former being a specific case of the latter with more interactive dynamics (Cui et al., 2013; Haq et al., 2014; Mar Rodríguez et al., 2015; Dworecka-Kaszak et al., 2016; Culibrk et al., 2016; Babikova et al., 2013). Whether these interactions; in the case of hosted mycobiomes, are ubiquitous or tissue-specific is open to further research.

Interestingly, the diversity of sampled mycobiomes tends to be more representative of individuals and biocompartments rather than of health status (Hager and Ghannoum, 2017; Hoarau et al., 2016; Dworecka-Kaszak et al., 2016; Cui et al., 2013), as their constitution depends heavily not only on immunological parameters but also on environtomic ones as well. As part of the latter, the microecologic/metabolic parameters tend to be affected by environmental conditions (including weather/seasonality), dietary and life style patterns of the host, possibly its sex and age and, of course, exposure to antifungal agents, especially medical, veterinary, or agricultural formulations (Lai et al., 2018; Gundogdu and Nalbantoglu, 2017). As some of these factors might change over time, the mycobiome may fluctuate accordingly, restructuring the holobiome. The skin mycobiome, for example, is demonstrated to change from its initial state to an adult type within 30 days after birth (Cui et al., 2013) with the diverse original mycobiome slowly being replaced by steady pathobiotic *Malassezia* populations, better attuned to the predominantly fatty excretions of the adult skin (Limon et al., 2017); still the skin of feet shows a distinct character (Dworecka-Kaszak et al., 2016). Bolstering the individuality factor mentioned above, a genomic dimension was added, which linked genomic variants directly to microbiomic composition (Blekhman et al., 2015; Enaud et al., 2018).

In the mycorrhizal context the soil may protect the interacting bodies from extreme variability even in terms of temperature, thus providing a steady and friendly environment for the typical exchange—organic carbon for nitrogen and metal-containing metabolites (Babikova et al., 2013; Song et al., 2010).

STUDYING THE MYCOBIOME

The microbiomic era was catapulted to prominence due to the sequencing revolution; although the microbiome as a term had been coined earlier, and more accurately and correctly (see Chapter 1), it came to prominence due to the development of deep sequencing capabilities that allowed characterization of fastidious and uncultivable microbiota and latent viruses without suitable hosts in functional proximity (Cui et al., 2013; Ghannoum et al., 2010; Hager and Ghannoum, 2017; Limon et al., 2017; Enaud et al., 2018).

The problem with all DNA-based methods, though, is that the obtained results are an indication of a different organism, not a proof. The definition of "different" is, of course, crucial; still, DNA methodologies were always plighted by artifacts and false-positives. Full sequencing options supposedly tackle this issue, but they actually do not *produce* genomic data and causally linked identification, classification, and phylogeny results; they rather *assemble* genomic sequence data from molecular reactions (*reads*) by fit algorithms, much like a puzzle (Thomas et al., 2014). The genomes thus assembled (*metagenomes*) constitute *possible* and even *probable* (in terms of annotation to existing genomes submitted in databases) but not *actual* ones. As the number of metagenomes in a sample is usually larger than the number of genome carriers detected by culture-based approaches in the same sample (Dworecka-Kaszak et al., 2016; Ghannoum et al., 2010), which is one of the advantages of metagenomics methods, there is not a solid proof that the discovered or detected metagenomes do indeed originate from the respective—elusive—organisms, especially when the latter are fastidious. The detection of so many novel organisms without any other kind of confirmation would have been discarded as low specificity in many applications of biosciences and had been a major issue that promoted, especially in Mycology, the application of polyphasic identification (Frisvad and Samson, 2004). This was the case for years on end during the 1990s for the PCR protocols used for molecular diagnosis in medicine. But in the context of microbiomic studies, it is explained away as simply a case of superior sensitivity.

As a result, the current exclusive use of the concept of Microbiome, and of mycobiome as well, was introduced by the metagenomics revolution and was tailored to the latter in concept and in methodology. Still, as genomes do not provide, by any interpretation, the full potential and traits of the respective organisms, the exclusive use of metagenomic approaches in complex, mixed samples bears an inherent risk of misinterpreting data and misannotating sequences. This is especially so whenever quantitative discrepancies are expected among different populations of a microbiome.

Initially samples were collected, DNA was extracted, and shotgun procedures for cloning into vectors were used followed by sequencing (Thomas et al., 2014; Dworecka-Kaszak et al., 2016). This process had been actually molecular microbiomics rather than metagenomics, as there was a culture step (the propagation of the vectors), albeit not of the studied organisms, in order to acquire increased, meaningful quantities of sequences. Practically, the procedure was a nascent genomic approach to microbiomics, as whole microbiomes where analyzed, meaning at the time populations of microbes (Thomas et al., 2014; Seed, 2014; Dworecka-Kaszak et al., 2016).

The key approach that introduced metagenomics has been the next-generation sequencing (NGS), which allowed individualized sequencing and readout of inherently pooled amplicons produced by generic priming of common, but diversified sequences flanked by highly conserved primer targets. There had been lots of commotion in the first half of 2010s over the choice of the target, with rDNA being favored and from which the ITS regions were finally downselected (Cui et al., 2013).

Still there was a problem: sequences not matched to database submissions were considered novel, characterized as "unknown," which they truly were, and were considered as proof that a high proportion of the mycobiome was uncultivable (Ghannoum et al., 2010). This might have not been so; in more than one century of mycological research, there have been identified, named, and renamed many fungal taxa, and it is not certain that ITS sequences had been sequenced and submitted for all of them, especially since there had been no real consensus for the most suitable genes/sequences to be used universally as markers. Thus some sequences appear in metagenomics-only research, they are not annotated in any database to any fungus and an uncultivable contact is registered. Even when metagenomics and cultures (as in some culturomics publications) are used in tandem or in parallel, there are challenges in the annotation of the sequence to the strain/colony, and even more challenges if a strain is not grown and leaves one (or some tenths) of the sequences "orphan". These sequences are *not* indication, much less proof, of uncultivable species, but, in many cases, species and strains failing to grow in culture *then* and *there*, a common problem with surveillance cultures in medical mycology. Blood cultures rarely turn positive for fungi and usually do so when it is too late for proper action. Cultures are much less robust than a good molecular protocol due to factors relating to the health and well-being of the cell, and failure to grow is not necessarily attributable to uncultivable/fastidious status of a strain.

An obvious choice would have been to fund a massive worldwide project for the ITS sequencing of all strains isolated and kept stored in laboratories and collections, an approach, which, by pooling and tagging, might have been much less expensive than sounded. The retirement and actual obsolescence of the Roche 454 chemistry, though, which was standardized for bacterial 16S and for fungal ITS analysis, left some uncertainty on what the next standard technical approach for metagenomic analysis was going to be (Tang et al., 2015). The truly genomic option, on the other hand, which is to proceed rather to whole genome sequencing as the method of choice, allows much richer informational context, unprecedented depth of knowledge, and prospect of discovery (genetic, translational, and other), a great sales pitch for such commodities as their price drops and their robustness increases and enjoys some additional, less important "pros."

Still, the issue of cognitive infrastructure, with databases to compare the retrieved sequences so as to annotate assembled genomes, is not resolved (Nilsson et al., 2019; Tang et al., 2015). Many mistakes in the systematic characterization of submitted sequences make necessary either a worldwide editing effort, which presupposes a fixed nomenclature not amenable to the vogue and whims of the taxonomists every 5–10 years; or a novel start of whole genome sequencing of well-characterized reference/type strains of centralized collections and depositories.

A method to annotate genomes to cells and confirm bioinformatics output (Tipton et al., 2018) is currently sorely missing for the correct and balanced qualitative analysis of the mycobiome in a sample. Culturomics (see Chapter 8) seem to fill the bill nicely, rather by complementing metagenomics, than by being an alternative to it (Lagier et al., 2012). It is interesting that culturomics, conceived to isolate and study cells detected through metagenomics, were found to detect, occasionally, microbiota undetectable by metagenomics (Dubourg et al., 2013); in these cases the results are definite: organisms are grown and characterized thoroughly despite the fact that they are missing from the reports of metagenomic analyses. This may be due to assembly failures and/or errors, to inadequate database resources coupled to a metagenomic software pipeline or to amplification and/or sequencing bias, exemplified by, but not restricted to, the abovementioned numerical disproportion among different populations (Thomas et al., 2014; Dworecka-Kaszak et al 2016; Nilsson et al., 2019; Anslan et al., 2018; Tang et al., 2015).

It is possible that new, extremely sensitive microscopy amenities (Liu et al., 2018; Xu et al., 2017), capable of observing and tracking single cells *in vivo*, may prove capable of providing data indicating intercellular processes and allow annotation of metagenomes to cells, thus resulting in a revolution similar to the original unification of morphological and physiological observations during the early decades of the 20th century.

The truth be told, exotic technologies are bound to assist greatly, but mycobiome research can still progress within novel iterations of existing modalities. A "-biomic" study can be either focused to detecting one or more entities/taxa of interest (*detection*); or to dissecting and revealing the present and constituent entities/organisms, in qualitative but possibly in quantitative terms as well (*analysis*); or to locating and possibly identifying or at least characterizing novel/unknown entities (*discovery*). The former application, *detection*, has been the object of conventional and molecular microbiology for decades, in all diagnostic and environmental/biotechnological contexts. Being a selective approach, detection needs a differentiation element as well as an element of purely detective function, which are usually—but not always—one and the same.

To implement *analysis*, however, originally required the pooling of *detection* protocols, as many as possible, either in aliquoted or in pooled formats. Wide-spectrum techniques, from cultures and staining dyes to molecular targets and probes, constituted the backbone of such research which, however, had an all-significant limitation: only *known* entities (a condition including also the notion of *expected*) could be analyzed. In this particular case the deep sequencing revolution afforded an agnostic application hitherto unavailable even to the most evolved and smart spinoffs of classic molecular protocols.

Finally, *discovery* demands agnostic approaches as well. Although it reads as the most meddlesome case, for centuries microorganisms were discovered and practically the path to a discovery passes through analysis. A good analytical concept with adequate discriminatory power sorts the known entities and leaves the unknown ones to be examined. As one may imagine, the novel technologies and methods which afforded exceptional analysis have worked wonders in terms of discovery (Lagier et al., 2012).

MICROSCOPY

All the above in a mycobiomic context translate to the continuous usability and usefulness of microscopy for all three modes of mycobiomes' study (El-Kirat-Chatel et al., 2017; Guo et al., 2017; Culibrk et al., 2016; Leveau et al., 2018). Although currently in steep decline, this is not due to a methodological problem but rather to one of vogue and specifications. Microscopy remains difficult, hard, and demands patience, persistence, and insistence. In Mycology the size of the fungal cell (Limon et al., 2017) and the variability of fungal structures allowed less laborious and more permissive techniques with dry lenses and up to 400 × magnification, which result in excellent discrimination and sensitivity in environmental samples and moderate in appendage samples, as in many cases a host is not the most suitable substrate for a symbiont to develop elaborate morphology (Vyzantiadis et al., 2012).

Microscopy is one quintessential asset for medical diagnosis even now, and in environmental contexts the identification through conidial morphology (Saccardo, 1882–1906) is still used more universally than expected. A number of dyes and the modern iterations of microscopy, which allow *in vivo* observation of subcellular entities, may allow matching genomic variations and taxonomic iterations to functional and physiological differentiations (Culibrk et al., 2016; Xu et al., 2017).

CULTURE AND CULTUROMICS

The advent of culturomics refreshed the second pillar of conventional mycology. With the exception of fastidious organisms, which cannot be cultured *yet*—a very important word—as culture is conditional to the creation of the correct set of conditions (Ghannoum, 2016), fungi are laborious and time-consuming to cultivate, but rewarding. Their morphology is patently diverse and fungal cultures are very informative even under the most rudimentary macroscopic examination (Huseyin et al., 2017). Standardizations and exhaustive observation, as proposed by culturomics (Kambouris et al., 2018) is bound to multiply the use of cultures, as it may lead to (1) an increase in the discriminatory power, conditional on graphs and continuous observation replace empiric practices; (2) an exhaustive postgenomic study as it furnishes live samples where translational and genomic data will be fused to create Big Data; and (3) more inclusive growth outputs; since more combinations will become available to match the preferences of currently fastidious organisms. In addition, by conceivably pairing culturomics with flow cytometry, a really detailed segregation of cellular populations is possible in order both to culture-develop as much as possible of said distinct populations and also to determine true fastidious fractions in mycobiomes, which might have been grossly overestimated by metagenomic-only approaches.

IMMUNOASSAYS

Immunoassays have been a fast and sensitive methodology in medical mycology and still allow excellent detection performance with very advantageous operational parameters. In purely medical formats, antibodies might be sought in tests, thus increasing their sensitivity manifold while keeping excellent specificity; of course, there are issues in quantification and temporal relevance, as even after the clearance of the antigen antibodies are still present in considerable titer; the specifics of the titer depend on the individual patient, among other parameters, making quantification uneasy (Azar and Hage, 2017).

Detecting antigens with monoclonal antibodies is more orthodox in a wide range of assays and tests. Monoclonal antibodies, primarily targeting cell wall moieties, have worked miracles on the issue, offering an interesting combination of excellent specificity (conditional to taxonomy details), satisfactory sensitivity, and practical universality of the application in almost any context (Vidal and Boulware, 2015). The high degree of cross-reactions (there is a finite number of cell wall moieties and many fungal taxa share each of them) and the need for a specific, defined target turned immunoassays unsuitable for analysis and discovery, respectively. Coupling immunoassays to culturomics is likely to increase the formers' detection spectrum but possibilities to become mainstay or even mainstream analyses and/or discovery methods are remote.

METAGENOMICS

The fungal genome is a decent substrate for metagenomic analysis. If background issues, mostly associated with the sample type and origin are worked out, there is a trio of (conditional) properties that make such analyses relatively straightforward: first, the extremely robust fungal cell wall that may pose challenges in bacteria-oriented extraction protocols (Limon et al., 2017; Seed, 2014; Huseyin et al., 2017), but allows enrichment steps, where less robust cells such as human/animal or bacterial can be ruptured and their nucleic acids disposed of *before* cracking the fungal cells and extracting their nucleic acids (Thomas et al., 2014).

Second, as eukaryotes the mycobiota sport poly-A tails in mRNA that allow metagenomic analysis based on transcriptome profiling or, more correctly, metatranscriptomic analysis (Seed, 2014; Dworecka-Kaszak et al., 2016). And third, contrary to viruses but similar to bacteria and other eukaryotes, fungi contain multiple copies of rDNA sequences that present sequence differentiation similar and in cases proportional to phylogenetic divergence and thus extremely accurate detection, analysis, and even discovery can be pursued if new-generation sequencing (Heather and Chain, 2016; Seed, 2014; Cui et al., 2013; Dworecka-kaszak et al., 2016; Huseyin et al., 2017) and perhaps other, agnostic-tailored genotyping methods are applied.

Both poly-A/transcriptome fishing and targeted rDNA sequencing and alignment are methods incompatible with whole genome or even whole exome analysis, and without the latter approaches, mycobiomic study will be an imprecise modality as there are a series of nuclear phenomena in fungi that need accurate genome mapping in order to be resolved. The ploidal status is one such example, since meiosis occurs only in teleomorphic fungi and to different phases of their life cycle in different taxa. It may not be a defining parameter for identifying a taxon, but one has to remember that the interactions are key elements of mycobiomes, setting them apart from *flora* concepts and thus should be considered in depth when studying the former. The morphic status (anamorph–teleomorph), each with different taxonomy lines and algorithms, further complicates the annotation of sequences to an organism and the scanning of libraries to perform detection and analysis (Tang et al., 2015).

In addition, some of the characteristics of mycobiota may be counterproductive for metagenomic studies. For starters, their cell typically is perhaps 100-fold the size of a typical bacterial one, making microscopy easy, but their abundance in a microbiome is two orders of magnitude lower than that of bacteria (Enaud et al., 2018; Lai et al., 2018; Limon et al., 2017; Dworecka-Kaszak et al., 2016; Seed, 2014; Cui et al., 2013). In an extracted DNA environment, which would further the difference in favor of bacterial genomes as the fungal cell is a tougher nut to crack, fungal genomic targets would be lost amongst bacterial sequences and thus hardly accessible to the multitude of the amplification reaction factors, further limiting the yield. Moreover, the fungal rDNA is much more polymorphic than the bacterial one, presenting differentiations in both size and sequence (Tang et al., 2015; Limon et al., 2017), thus perplexing alignment and comparison algorithms of bioinformatics pipelines. The comparatively low quality of fungal genome databases and entries (Nilsson et al., 2019; Limon et al., 2017; Tang et al., 2015; Seed, 2014; Cui et al., 2013), combined with the abovementioned wet laboratory complications, sap the reliability of the metagenomic approaches. The issue worsens given the low concurrence output for the sequence dataset of the same samples by different pipelines (Nilsson et al., 2019; Anslan et al., 2018), insinuating computational biases of unknown etiology and impact.

Although the study of viruses is disproportionally based on nucleic acids, in the end, there was a decision to be made: whether the microbiome and its subordinate biomes should be defined by genomic criteria or on an organism/cell/particle basis (Dworecka-Kaszak et al., 2016). This question has been quintessential in the context of a 20-year old—or even older—problem: whether Koch's postulates should be adhered to, even with a fair degree of revamping, as their original form and function had been made obsolete once molecular diagnostics became standard. As the metagenomic revolution on one hand and the ever-present issue of uncultivable and fastidious fungi on the other (Dworecka-Kaszak et al., 2016; Ghannoum et al., 2010) were attracting ever more attention, due to economies of scale and scope (the former) and due to the prospect of high profitability by biotechnological exploitation (the latter), the microbiome in general and the mycobiome along with it were initially poised to become translational entities (Hooper and Gordon, 2001; Gundogdu and Nalbantoglu, 2017; Ghannoum et al., 2010). In this concept, there has been an innate antithesis of some importance: the genomic dimension of microbiome was projected in terms of genes, sequences, and genomes, but still the majority of microbiome genomics are simply massively parallel analyses by next generation sequencing (NGS) of singular descriptive consensus sequences amplified by PCR—and this holds true for both bacteria and fungi. In other words an extremely small part of the microbial genomes was analyzed to reliably annotate sequences to existing or discovered/predicted microbial taxa, which, however, are based on nontranslational, cell-centered criteria (Thomas et al., 2014; Dworecka-Kaszak et al., 2016). This approach *sensu stricto* is *not* metagenomics, despite being used and pitched as such, but phylogenetic analysis reconditioned with NGS. To call an approach metagenomic, genomes must be analyzed in some extent, historically with shotgun metagenomics (Thomas et al., 2014; Seed, 2014; Dworecka-Kaszak et al., 2016; Limon et al., 2017) but in the near future quite possibly through third-generation sequencing approaches, which do not demand fracture of the DNA as they produce extremely long sequencing reads (Lee et al., 2016).

SELECT MYCOBIOMIC RESEARCH: SOME WORKING EXAMPLES

As already mentioned, the mycobiome is an artificial iteration of the microbiome in any relevant spatiotemporal unit. It is discrete in terms of analyzing techniques and approaches, as it requires specific molecular protocols and specialized database infrastructure. Issues in target selection for barcoding and/or metagenomic analysis (Cui et al., 2013) point to whole genome sequencing and shotgun approaches as the way ahead. Moreover, as a result, the applicable software pipelines are also different and adapted to the particularities of the fungal genome (Cui et al., 2013; Anslan et al., 2018).

Mycobiomes that have been analyzed and described include a collection of hosted ones from human biocompartments (Seed, 2014; Cui et al., 2013), from other living hosts such as plant or animal tissue (Schlegel et al., 2018; Kearns et al., 2017; Dworecka-Kaszak et al., 2016; Anslan et al., 2018) and from environmental samples such as soil (Anslan et al., 2018) or urban/industrial site environment (Green, 2018; Hamdy et al., 2017); moreover, *inpromptu* mycobiomes of economic importance have begun being analyzed (Sha et al., 2018)—a tendency expected to skyrocket in the near future.

HUMAN MYCOBIOMES

In terms of human appendage mycobiomes, the main focus has steadily been for the human gut, but other biocompartments have attracted attention, including oral and nasal cavities, vagina, skin, and respiratory tract and lung (Cui et al., 2013; Seed, 2014; Dworecka-Kaszak et al., 2016; Limon et al., 2017; Enaud et al., 2018; Lai et al., 2018). The common denominators in all these sites are the pathobiont genera *Candida* and *Aspergillus* (Lai et al., 2018), the former being at times detected and identified as a predominant mycobiote in vaginal samples of individuals who never experienced vaginal candidiasis (Seed, 2014; Limon et al., 2017).

Dysbiosis usually—but not necessarily—settles with the increase of specific fungal load, as in oral candidiasis in HIV patients (Limon et al., 2017), while the diversity of the mycobiota allegedly decreases. The correlation between the diversity of the entire mycobiome and disease progression is less clearly defined though, with the best example being irritable bowel disease where the correlation is controversial (Lai et al., 2018).

Regarding other syndromes, fungal diversity increases with disease progression in atopic dermatitis, inflammatory bowel disease, cirrhosis, or chronic hepatitis B. But in other cases, fungal diversity decreases with disease progression, as with respiratory mycobiome in cystic fibrosis and nasal mycobiome in allergies (Cui et al., 2013). And finally, in some syndromes, the mycobiome shows no signs of alteration with disease progression, as observed in human psoriasis (Cui et al., 2013; Lai et al., 2018).

Of course, there are some issues remaining, especially regarding causality: there is no solid evidence whether the mycobiome alterations cause changes in health/disease status and disease severity, or the latter trigger the alterations of the mycobiome, which should then be considered effects, not causes, of changes in health status. Similarly, the possibility that the changes of the mycobiome may be secondary, caused by primary changes of the bacteriome, remains appealing and possible but unproven (Cui et al., 2013; Lai et al., 2018).

In the skin mycobiome, though, which presents multiple immunomodulatory and immunoregulatory functions (Lai et al., 2018), the genus *Malassezia* reigns supreme in mammals at large (Lai et al., 2018) and produces antimicrobial and anti-inflammatory moieties as is the azelaic acid (Schulte et al., 2015); and modulators for both immune and regenerative/homeostatic responses (Gaitanis et al., 2008; Esser and Rannug, 2015; Vlachos et al., 2012; Lai et al., 2018). Most important dysbiotic events include, without being restricted to, *Malassezia*-induced folliculitis, pityriasis versicolor (Velegraki et al., 2015) and most probably seborrheic and atopic dermatitis (Prohic et al., 2016). The superficial wound mycobiome, though, seems bereft of *Malassezia* and is populated mostly by *Cladosporium* and *Candida* instead (Limon et al., 2017).

The dysbiotic events of mycobiomic etiology in the lower airways, when concerning *Pneumocystis*, are rather common single-agent pathologic events of allergic or infectious nature, since this agent is *not* a benign colonizer (Calderón et al., 2007; Lai et al., 2018). In cystic fibrosis though, the fungal diversity is lower in patients—where *Candida* reigns supreme in the lower respiratory tract—than in healthy people (Cui et al., 2013; Limon et al., 2017).

The GIT, starting from the mouth and ending at the anus and with a length of 30 ft/9 m (Limon et al., 2017), hosts at least two major discrete mycobiomata (actually microbiomata): one in the gut proper and the other at the oral cavity. The latter was the first appendage human mycobiome to be studied by metagenomic approaches; it was found colonized by numerous fungal species (85 fungal genera detected) in diverse combinations that create significant individual variability in healthy individuals, each of whom hosted ~15 genera. These included expected (*Candida*) and rather unexpected mycobiota (*Aspergillus*, *Fusarium*, and *Cryptococcus*) while one-third of the detected metagenomes belonged to supposedly uncultivable fungi (Limon et al., 2017; Hager and Ghannoum, 2017; Seed, 2014; Ghannoum et al., 2010).

The oral mycobiome may be affected quantitatively and qualitatively by several individual qualities of the host, not all of them biological (Seed, 2014; Ghannoum et al., 2010). It may extend either toward the gut; oral candidiasis extending to the stomach and intestines was described very long ago (Hager and Ghannoum, 2017). Or toward the respiratory tract, *Candida* being prominent in its upper part, most probably migrating from the oral cavity and the nasopharynx and in stark contrast to the lower respiratory tract mycobiome (Dworecka-Kaszak et al., 2016; Limon et al., 2017). Throughout the GIT semisecluded, practically isolated positions develop separate niches and host diverse microbiomata with different community structures (Limon et al., 2017).

The gut is colonized by as many as 66 fungal genera (Limon et al., 2017; Seed, 2014) or as few as 10, the latter dominated by *C. tropicalis* (65%), which was the only fungal species with significantly increased abundance in Crohn's disease (CD) patients relative to their non-CD healthy relatives and controls (Hager and Ghannoum, 2017; Hoarau et al., 2016). A balanced bacteriome acts suppressively or at least restrictively to fungal presence, possibly by stimulating production of host antifungal peptides, among other possible mechanisms. The abundance of said peptides may be affected differentially in the short term by current diet (Limon et al., 2017). The other way around, *Saccharomyces boulardii*, closely related to *Saccharomyces cerevisiae*, shows promise for the prevention and mitigation of antibiotic-associated diarrhea, including diarrhea caused by *Clostridium difficile* (Seed, 2014; Dworecka-Kaszak et al., 2016; Enaud et al., 2018). Commensal and/or mutualistic fungi most probably compete with pathogenic organisms by beneficially modifying the intestinal function while also inhibiting inflammatory reactions in bowel diseases (Dworecka-Kaszak et al., 2016)—a pathway leading to tolerance for themselves and to shirking inflammation-instigated episodes for the host. The types of fungal species residing in the gut correlate with the quantitative distribution of the bacterial taxa and not with their mere presence, so the culprit(s) for the aggravation of irritable bowel disease (IBD) symptoms should be sought for in their midst, and *not* in the respective bacteriome (Hager and Ghannoum, 2017).

REMOTE EFFECTS, COMMUNICATION, AND CONTROL FUNCTIONS OF MYCOBIOMES

The remote effects, usually adverse, caused by the reshuffling of an appendage mycobiome (or microbiome, for that matter) to distant anatomic locations, have been cardinal pillars for the conceptualization of the microbial biomes. Appendage mycobiomes are not relevant exclusively in human/medicinal settings (mycorrhizae being a perfect example of the opposite), but their most celebrated applications are related to this field. A cross-talk exists between opportunistic fungi and their hosts by means of the metabolites of the mycobiota. The immune system does not ignore commensal or opportunistic fungi but recognizes them through pattern-recognition receptors (PRRs) and reacts gradually, allowing a range of response intensities that result in dynamic though fragile balancing.

When PRR initiate low or null response, equaling immunotolerance, normobiosis between symbionts and host ensues. Different PRR signals trigger more intense responses, which result in dysbiosis either directly, by inducing pro-inflammatory pathways, or indirectly, by initiating secretion of antimicrobial/antifungal agents that deplete the symbiont communities and allow resistant strains to grow unchallenged, in competition-free environments, to malign status by exceeding the threshold of specific population metrics. The latter include, indicatively but not exclusively, abundance, concentration, and expression-metabolomics profiles (Dworecka-Kaszak et al., 2016).

More explicitly, the normobiosis, the state of neutral or beneficial balance (accounting for commensal or mutualistic symbioses respectively) may be perturbed not only as mentioned above, but if any from a range of factors conserving the fragile balances becomes un- or dysregulated. Such events may cause communal frictions within the appendage microbiome, leading to local dysbiosis, a state of affairs where newcomers and/or deregulated old commensals either turn aggressive to the host to improve their own vitality (parasitism) or cause ill effects even without securing any advantage for themselves (amensalism).

Translocation into the bloodstream and subsequent circulation of fungal molecules, such as RNA, DNA, or peptidoglycans, may initiate systemic immune responses and lead to disease remote from the initial site of fungal infection (Cui et al., 2013). This mechanism though is little different from bloodstream-borne sepsis or toxicity caused by the migration of infectious cells, standard in systemic septic syndromes and of toxins in systemic toxic ones, respectively. Much more complex and evolved pathomechanisms are indicative of the microbiomic concepts: the association of genomic variability to microbiomic profiles and, at the same time, to noninfectious pathophenotypes, especially of autoimmune nature (Blekhman et al., 2015), indicates an external link in genotype–pathophenotype pathway which seems to be causal and fitting well with the observation of parasitic microbiota meddling with the actual expression of the host genome (Culibrk et al., 2016).

GUT–BRAIN AXIS AND THE MYCOBIOME FACTOR

Fungi participate in gut–brain axis (GBA) interactions through neuro-immuno-endocrine mediators (Enaud et al., 2018). The mechanism, as understood to this day, suggests that, on one hand, gut mycobiome interacts with cells of the immune system that produce neuroactive moieties traveling to the brain. Thus it has been proposed that immunological pathways are focal in this, indirect, interaction and are mediated by cytokines produced in the gut, which reach the brain via the bloodstream and then cross the blood–brain barrier to modulate stimulation of specific brain areas, particularly the hypothalamus.

On the other hand, the quintessence of mycobiome/microbiome—GBA interaction lies with direct interaction. The neuroactive and phsychotropic molecules produced by gut mycobiota/microbiota themselves travel—mainly but perhaps not entirely—by bloodstream to the brain, to inflict short or long-term behavioral alterations (Tang et al., 2015; Dworecka-Kaszak et al., 2016). Psychiatric and neurological disorders are associated with this interaction, but once more causality remains to be determined (Enaud et al., 2018). Such syndromes may be due to the microbial behavior or may affect it: the host activities' pattern signals endocrine secretions—hunger, cravings—which alter the functions of gut cells and thus affect the mycobiome/microbiome. There is an innate similarity in this direct and dynamic relationship with the principle of remote toxic effects caused by infectious agents, which manifest in distant organs, not always sharing physical continuity with the site of the infection.

MYCORRHIZAL DATABUSES

An environmentally far-reaching function of mycobiomes is the expansion of the mutualistic formation of mycorrhizae, where the fungal symbiont extends mycelial hyphae to considerable distances so as they become plugged to neighboring mycorrhizae. The modularity of the concept, in terms of whether mycorrhizae differing in host plant, in mycobiont or in both, can plug into such a network, is unclear, as the original strategy of the fungus must have been to propagate the symbiosis to as many plants in an area as possible.

The underground hyphal network transmits fiber optics–like chemical signals among plants, thus bypassing environmental jamming and background noise that is inherent in the airwaves through which conventional signaling functions by means of volatile chemical moieties. The signaling is performing an early warning function, allowing the community to initiate as fast as possible defense mechanisms so as to limit any threat to the population (Babikova et al., 2013; Song et al., 2010); threats which may develop and reach existential levels otherwise.

REFERENCES

Adam, B., Baillie, G.S., Douglas, L.J., 2002. Mixed species biofilms of *Candida albicans* and *Staphylococcus epidermidis*. J. Med. Microbiol. 51, 344–349.

Anslan, S., Nilsson, R.H., Wurzbacher, C., Baldrian, P., Tedersoo, L., Bahram, M., 2018. Great differences in performance and outcome of high-throughput sequencing data analysis platforms for fungal metabarcoding. MycoKeys 39, 29–40.

Azar, M.M., Hage, C.A., 2017. Laboratory diagnostics for histoplasmosis. J. Clin. Microbiol. 55, 1612–1620.

Babikova, Z., Gilbert, L., Bruce, T.J.A., Birkett, M., Caulfield, J.C., Woodcock, C., et al., 2013. Underground signals carried through common mycelial networks warn neighbouring plants of aphid attack. Ecol. Lett. 16, 835–843.

Blekhman, R., Goodrich, J.K., Huang, K., Sun, Q., Bukowski, R., Bell, J.T., et al., 2015. Host genetic variation impacts microbiome composition across human body sites. Genome. Biol. 16, 191.

Calderón, E.J., Rivero, L., Respaldiza, N., Morilla, R., Montes-Cano, M.A., Friaza, V., et al., 2007. Systemic inflammation in patients with chronic obstructive pulmonary disease who are colonized with *Pneumocystis jiroveci*. Clin. Infect. Dis. 45, e17–e19.

Costa-Orlandi, C.B., Sardi, J.C.O., Pitangui, N.S., de Oliveira, H.C., Scorzoni, L., Galeane, M.C., et al., 2017. Fungal biofilms and polymicrobial diseases. J. Fungi (Basel, Switzerland) 3, 22.

Cui, L., Morris, A., Ghedin, E., 2013. The human mycobiome in health and disease. Genome Med. 5, 63.

Culibrk, L., Croft, C.A., Tebbutt, S.J., 2016. Systems biology approaches for host-fungal interactions: an expanding multi-omics frontier. OMICS 20, 127–138.

Darling, W.M., 1976. Co-cultivation of mycobacteria and fungus. Lancet (London, England) 2, 740.

Delhaes, L., Monchy, S., Fréalle, E., Hubans, C., Salleron, J., Leroy, S., et al., 2012. The airway microbiota in cystic fibrosis: a complex fungal and bacterial community—implications for therapeutic management. PLoS One 7, e36313.

Dubourg, G., Lagier, J.C., Armougom, F., Robert, C., Hamad, I., Brouqui, P., et al., 2013. The proof of concept that culturomics can be superior to metagenomics to study atypical stool samples. Eur. J. Clin. Microbiol. Infect. Dis. 32, 1099.

Dworecka-Kaszak, B., Dąbrowska, I., Kaszak, I., 2016. The mycobiome – a friendly cross-talk between fungal colonizers and their host. Ann. Parasitol. 62, 175–184.

El-Kirat-Chatel, S., Puymege, A., Duong, T.H., Van Overtvelt, P., Bressy, C., Belec, L., et al., 2017. Phenotypic heterogeneity in attachment of marine bacteria toward antifouling copolymers unraveled by AFM. Front. Microbiol. 8, 1399.

Enaud, R., Vandenborght, L.-E., Coron, N., Bazin, T., Prevel, R., Schaeverbeke, T., et al., 2018. The mycobiome: a neglected component in the microbiota-gut-brain axis. Microorganisms 6, 22.

Esser, C., Rannug, A., 2015. The aryl hydrocarbon receptor in barrier organ physiology, immunology, and toxicology. Pharmacol. Rev. 67, 259–279.

Fleming, A., n.d. Classics in infectious diseases: on the antibacterial action of cultures of a *Penicillium*, with special reference to their use in the isolation of *B. influenzae* by Alexander Fleming, Reprinted from the British Journal of Experimental Pathology 10:226–236, 1929. Rev. Infect. Dis. 2, 129–139.

Frisvad, J., Samson, R.A., 2004. Polyphasic taxonomy of *Penicillium* subgenus *Penicillium*. A guide to identification of food and air-borne terverticillate penicillia and their mycotoxins. Stud. Mycol. 49, 1–174.

Gaitanis, G., Magiatis, P., Stathopoulou, K., Bassukas, I.D., Alexopoulos, E.C., Velegraki, A., et al., 2008. AhR ligands, malassezin, and indolo [3,2-b]carbazole are selectively produced by *Malassezia furfur* strains isolated from seborrheic dermatitis. J. Invest. Dermatol. 128, 1620–1625.

Gauthier, G.M., 2015. Dimorphism in fungal pathogens of mammals, plants, and insects. PLoS Pathog. 11, e1004608.

Ghabrial, S.A., Castón, J.R., Jiang, D., Nibert, M.L., Suzuki, N., 2015. 50-Plus years of fungal viruses. Virology 479–480, 356–368.

Ghannoum, M., 2016. Cooperative evolutionary strategy between the bacteriome and mycobiome. mBio 7, 10–13.

Ghannoum, M.A., Jurevic, R.J., Mukherjee, P.K., Cui, F., Sikaroodi, M., Naqvi, A., et al., 2010. Characterization of the oral fungal microbiome (mycobiome) in healthy individuals. PLoS Pathog. 6, e1000713.

Green, B.J., 2018. Emerging insights into the occupational mycobiome. Curr. Allergy Asthma Rep. 18, 62.

Gundogdu, A., Nalbantoglu, U., 2017. Human genome-microbiome interaction: metagenomics frontiers for the aetiopathology of autoimmune diseases. Microb. Genomics 3, e000112.

Guo, F., Li, S., Caglar, M.U., Mao, Z., Liu, W., Woodman, A., et al., 2017. Single-cell virology: on-chip investigation of viral infection dynamics. Cell Rep. 21, 1692–1704.

Hager, C.L., Ghannoum, M.A., 2017. The mycobiome: role in health and disease, and as a potential probiotic target in gastrointestinal disease. Dig. Liver Dis. 49, 1171–1176.

Hall, R.A., Noverr, M.C., 2017. Fungal interactions with the human host: exploring the spectrum of symbiosis. Curr. Opin. Microbiol. 40, 58–64.

Hamdy, A.M., El-massry, M., Kashef, M.T., Amin, M.A., Aziz, R.K., 2017. Toward the drug factory microbiome: microbial community variations in antibiotic-producing clean rooms. OMICS 22, 133–144. omi.2017.0091.

Haq, I.U., Zhang, M., Yang, P., van Elsas, J.D., 2014. The interactions of bacteria with fungi in soil: emerging concepts. Adv. Appl. Microbiol. 89, 185–215.

Hawrelak, J.A., Myers, S.P., 2004. The causes of intestinal dysbiosis: a review. Altern. Med. Rev. 9, 180–197.

Heather, J.M., Chain, B., 2016. The sequence of sequencers: the history of sequencing DNA. Genomics 107, 1–8.

Hekim, N., Özdemir, V., 2017. A general theory for "post" systems biology: iatromics and the environtome. OMICS 21, 359–360.

Hoarau, G., Mukherjee, P.K., Gower-Rousseau, C., Hager, C., Chandra, J., Retuerto, M.A., et al., 2016. Bacteriome and mycobiome interactions underscore microbial dysbiosis in familial Crohn's disease. mBio 7. Available from: https://doi.org/10.1128/mBio.01250-16.

Hooper, L.V., Gordon, J.I., 2001. Commensal host-bacterial relationships in the gut. Science 292, 1115–1118.

Human Microbiome Project Consortium, 2012. Structure, function and diversity of the healthy human microbiome. Nature 486, 207–214.

Huseyin, C.E., Rubio, R.C., O'Sullivan, O., Cotter, P.D., Scanlan, P.D., 2017. The fungal frontier: a comparative analysis of methods used in the study of the human gut mycobiome. Front. Microbiol. 8, 1–15.

Irinyi, L., Serena, C., Garcia-Hermoso, D., Arabatzis, M., Desnos-Ollivier, M., Vu, D., et al., 2015. International Society of Human and Animal Mycology (ISHAM)-ITS reference DNA barcoding database--the quality controlled standard tool for routine identification of human and animal pathogenic fungi. Med. Mycol. 53, 313–337.

Kambouris, M.E., Pavlidis, C., Skoufas, E., Arabatzis, M., Kantzanou, M., Velegraki, A., et al., 2018. Culturomics: a new kid on the block of OMICS to enable personalized medicine. OMICS 22, 108–118.

Kamiński, D.M., 2014. Recent progress in the study of the interactions of amphotericin B with cholesterol and ergosterol in lipid environments. Eur. Biophys. J. 43, 453–467.

Kearns, P.J., Fischer, S., Fernández-Beaskoetxea, S., Gabor, C.R., Bosch, J., Bowen, J.L., et al., 2017. Fight fungi with fungi: antifungal properties of the amphibian mycobiome. Front. Microbiol. 8, 1–12.

Kerr, J., 1994. Inhibition of fungal growth by *Pseudomonas aeruginosa* and *Pseudomonas cepacia* isolated from patients with cystic fibrosis. J. Infect. 28, 305–310.

Lagier, J.-C., Armougom, F., Million, M., Hugon, P., Pagnier, I., Robert, C., et al., 2012. Microbial culturomics: paradigm shift in the human gut microbiome study. Clin. Microbiol. Infect. 18, 1185–1193.

Lai, G.C., Tan, T.G., Pavelka, N., 2018. The mammalian mycobiome: a complex system in a dynamic relationship with the host. Wiley Interdiscip. Rev. Syst. Biol. Med. 11, e1438.

Lee, H., Gurtowski, J., Yoo, S., Nattestad, M., Marcus, S., Goodwin, S., et al., 2016. Third-generation sequencing and the future of genomics bioRxiv:048603.

Leveau, J.H.J., Hellweger, F.L., Kreft, J.-U., Prats, C., Zhang, W., 2018. Editorial: The individual microbe: single-cell analysis and agent-based modelling. Front. Microbiol. 9, 2825.

Limon, J.J., Skalski, J.H., Underhill, D.M., 2017. Commensal fungi in health and disease. Cell Host Microbe 22, 156–165.

Liu, T.-L., Upadhyayula, S., Milkie, D.E., Singh, V., Wang, K., Swinburne, I.A., et al., 2018. Observing the cell in its native state: imaging subcellular dynamics in multicellular organisms. Science 360. Available from: https://doi.org/10.1126/science.aaq1392.

Liu, Y., Wang, W., Yan, H., Wang, D., Zhang, M., Sun, S., 2019. Anti-Candida activity of existing antibiotics and their derivatives when used alone or in combination with antifungals. Future Microbiol. 14, 899–915.

Mar Rodríguez, M., Pérez, D., Javier Chaves, F., Esteve, E., Marin-Garcia, P., Xifra, G., et al., 2015. Obesity changes the human gut mycobiome. Sci. Rep. 5, 14600.

Mowat, E., Rajendran, R., Williams, C., McCulloch, E., Jones, B., Lang, S., et al., 2010. *Pseudomonas aeruginosa* and their small diffusible extracellular molecules inhibit *Aspergillus fumigatus* biofilm formation. FEMS Microbiol. Lett. 313, 96–102.

Mukherjee, P.K., Chandra, J., Retuerto, M., Sikaroodi, M., Brown, R.E., Jurevic, R., et al., 2014. Oral mycobiome analysis of HIV-infected patients: identification of *Pichia* as an antagonist of opportunistic fungi. PLoS Pathog. 10, e1003996.

Neely, A.N., Law, E.J., Holder, I.A., 1986. Increased susceptibility to lethal Candida infections in burned mice preinfected with *Pseudomonas aeruginosa* or pretreated with proteolytic enzymes. Infect Immun. 52, 200–204.

NIH HMP Working Group, Peterson, J., Garges, S., Giovanni, M., McInnes, P., Wang, L., et al., 2009. The NIH Human Microbiome Project. Genome Res. 19, 2317–2323.

Nilsson, R.H., Anslan, S., Bahram, M., Wurzbacher, C., Baldrian, P., Tedersoo, L., 2019. Mycobiome diversity: high-throughput sequencing and identification of fungi. Nat. Rev. Microbiol. 17, 95–109.

Oever, J.T., Netea, M.G., 2014. The bacteriome-mycobiome interaction and antifungal host defense. Eur. J. Immunol. 44, 3182–3191.

Perlin, D.S., 2011. Current perspectives on echinocandin class drugs. Future. Microbiol. 6, 441–457.

Prohic, A., Jovovic Sadikovic, T., Krupalija-Fazlic, M., Kuskunovic-Vlahovljak, S., 2016. *Malassezia* species in healthy skin and in dermatological conditions. Int. J. Dermatol. 55, 494–504.

Qin, J., Li, R., Raes, J., Arumugam, M., Burgdorf, K.S., Manichanh, C., et al., 2010. A human gut microbial gene catalogue established by metagenomic sequencing. Nature 464, 59–65.

Ramírez Granillo, A., Canales, M.G.M., Espíndola, M.E.S., Martínez Rivera, M.A., de Lucio, V.M.B., Tovar, A.V.R., 2015. Antibiosis interaction of *Staphylococccus aureus* on *Aspergillus fumigatus* assessed in vitro by mixed biofilm formation. BMC Microbiol. 15, 33.

Rollet, C., Gal, L., Guzzo, J., 2009. Biofilm-detached cells, a transition from a sessile to a planktonic phenotype: a comparative study of adhesion and physiological characteristics in *Pseudomonas aeruginosa*. FEMS Microbiol. Lett. 290, 135–142.

Rowland, I., Gibson, G., Heinken, A., Scott, K., Swann, J., Thiele, I., et al., 2018. Gut microbiota functions: metabolism of nutrients and other food components. Eur. J. Nutr. 57, 1–24.

Saccardo, P.A., 1882–1906. Sylloge Fungorum Omnium Hucusque Cognitorum. Saccardo P.A., Pavia.

Schlegel, M., Queloz, V., Sieber, T.N., 2018. The endophytic mycobiome of European ash and Sycamore maple leaves - geographic patterns, host specificity and influence of ash dieback. Front. Microbiol. 9, 1–20.

Schoch, C.L., Seifert, K.A., Huhndorf, S., Robert, V., Spouge, J.L., Levesque, C.A., et al., 2012. Nuclear ribosomal internal transcribed spacer (ITS) region as a universal DNA barcode marker for Fungi. Proc. Natl. Acad. Sci. U.S.A. 109, 6241–6246.

Schulte, B.C., Wu, W., Rosen, T., 2015. Azelaic acid: evidence-based update on mechanism of action and clinical application. J. Drugs Dermatol. 14, 964–968.

Seed, P.C., 2014. The human mycobiome. Cold Spring Harb. Perspect. Med. 5, 103–122.

Seifert, K.A., 2009. Progress towards DNA barcoding of fungi. Mol. Ecol. Resour 9 (Suppl. s1), 83–89.

Sha, S.P., Suryavanshi, M.V., Jani, K., Sharma, A., Shouche, Y., Tamang, J.P., 2018. Diversity of yeasts and molds by culture-dependent and culture-independent methods for mycobiome surveillance of traditionally prepared dried starters for the production of Indian alcoholic beverages. Front. Microbiol. 9, 2237.

Song, Y.Y., Zeng, R., Sen, Xu, J.F., Li, J., Shen, X., Yihdego, W.G., 2010. Interplant communication of tomato plants through underground common mycorrhizal networks. PLoS One 5, e13324.

Sovran, B., Planchais, J., Jegou, S., Straube, M., Lamas, B., Natividad, J.M., et al., 2018. Enterobacteriaceae are essential for the modulation of colitis severity by fungi. Microbiome 6, 152.

Stefanaki, C., Peppa, M., Mastorakos, G., Chrousos, G.P., 2017. Examining the gut bacteriome, virome, and mycobiome in glucose metabolism disorders: are we on the right track? Metabolism 73, 52–66.

Stierle, A.A., Stierle, D.B., Decato, D., Priestley, N.D., Alverson, J.B., Hoody, J., et al., 2017. The Berkeleylactones, antibiotic macrolides from fungal coculture. J. Nat. Prod. 80, 1150–1160.

Tang, J., Iliev, I.D., Brown, J., Underhill, D.M., Funari, V.A., 2015. Mycobiome: approaches to analysis of intestinal fungi. J. Immunol. Methods 421, 112–121.

Thomas, T., Gilbert, J., Meyer, F., 2014. Metagenomics: a guide from sampling to data analysis. Role Bioinf. Agric. 2, 357–383.

Thota, S.P., Badiya, P.K., Yerram, S., Vadlani, P.V., Pandey, M., Golakoti, N.R., et al., 2017. Macro-micro fungal cultures synergy for innovative cellulase enzymes production and biomass structural analyses. Renew. Energy 103, 766–773.

Tipton, L., Müller, C.L., Kurtz, Z.D., Huang, L., Kleerup, E., Morris, A., et al., 2018. Fungi stabilize connectivity in the lung and skin microbial ecosystems. Microbiome 6, 1–14.

Velegraki, A., Cafarchia, C., Gaitanis, G., Iatta, R., Boekhout, T., 2015. *Malassezia* infections in humans and animals: pathophysiology, detection, and treatment. PLoS Pathog. 11, e1004523.

Vidal, J.E., Boulware, D.R., 2015. Lateral flow assay for cryptococcal antigen: an important advance to improve the continuum of HIV care and reduce cryptococcal meningitis-related mortality. Rev. Inst. Med. Trop. Sao Paulo 57 (Suppl. 19), 38–45.

Vlachos, C., Schulte, B.M., Magiatis, P., Adema, G.J., Gaitanis, G., 2012. Malassezia-derived indoles activate the aryl hydrocarbon receptor and inhibit Toll-like receptor-induced maturation in monocyte-derived dendritic cells. Br. J. Dermatol. 167, 496–505.

Vyzantiadis, T.-A.A., Johnson, E.M., Kibbler, C.C., 2012. From the patient to the clinical mycology laboratory: how can we optimise microscopy and culture methods for mould identification? J. Clin. Pathol. 65, 475–483.

Wargo, M.J., Hogan, D.A., 2006. Fungal-bacterial interactions: a mixed bag of mingling microbes. Curr. Opin. Microbiol. 9, 359–364.

Whipps, J.M., Lewis, K., Cooke, R.C., 1988. Mycoparasitism and plant disease control. In: Burge, M.N. (Ed.), Fungi in Biological Control Systems. Manchester University Press, Manchester, pp. 161–187.

Xu, X.-L., Lee, R.T.H., Fang, H.-M., Wang, Y.-M., Li, R., Zou, H., et al., 2008. Bacterial peptidoglycan triggers *Candida albicans* hyphal growth by directly activating the adenylyl cyclase Cyr1p. Cell Host Microbe 4, 28–39.

Xu, J., Ma, B., Su, X., Huang, S., Xu, X., Zhou, X., et al., 2017. Emerging trends for microbiome analysis: from single-cell functional imaging to microbiome big data. Engineering 3, 66–70.

Wu, C.T., Li, Z.L., Xiong, D.X., 1998. Relationship between enteric microecologic dysbiosis and bacterial translocation in acute necrotizing pancreatitis. World J. Gastroenterol. 4, 242–245.

CHAPTER

VIROME: THE PRODIGIOUS LITTLE COUSIN OF THE FAMILY

4

Yiannis N. Manoussopoulos[1,*] and Cleo G. Anastassopoulou[2,*]

[1]*Laboratory of Virology, Plant Protection Division of Patras, ELGO-Demeter, NEO & Amerikis, Patras, Greece*

[2]*Department of Biology, University of Patras, University Campus Rio, Patras, Greece*

INTRODUCTION

The implementation of the Human Genome Project at the dawn of the century ushered in a new era in biology and medicine. The concerted attempts of the scientific community to read the genetic code hiding parsimoniously in the four alternating nitrogenous bases of the DNA molecule accelerated the development of novel sequencing methodologies and computational platforms. Sanger's "dideoxy method" of reading DNA sequences was replaced by the much cheaper and faster next-generation sequencing which matured into long-read sequencing (LRS). LRS technologies enable the study of genomes and transcriptomes at an unprecedented resolution while allowing for sequencing of RNA molecules without prior reverse transcription (van Dijk et al., 2018). These recent advances in sequencing technologies that helped us read —but not yet analyze, interpret, and fully comprehend— the genetic code of life, also rendered possible the identification of microorganisms directly from their genomes or genomic signatures at their ecological niches (metagenomics), without the need for the laborious procedures of isolation, culturing, and morphological characterization of individual species. Microbiome research thus emerged as an innovative area of science dealing with the exploration and exploitation of the ever dynamically interacting communities of microbes associated with all living things and the environment of various types, including soils and the oceans.

The totality of microorganisms that reside in a strictly defined habitat at a given time span comprises the microbiome. The term refers either to the collective genomes of the microorganisms in an environmental milieu (that can also be referred to as the metagenome of the microbiota) or the microorganisms themselves. Typical microbiomic components include prokaryotic (bacteria and archaea) and eukaryotic species (protozoa and fungi) as well as viruses. Mutualism, commensalism, and parasitism are the principal types of symbiotic associations governing the continuum of interactions among microbiota and their components. These complex interactions play critical roles in the shaping and functionality of microbiomic communities, ultimately determining each community's equilibrium state with the microbiomic ecological niche. Viruses are the least understood elements among microbiomic communities. Up to a few

*Both authors contributed equally to this work.

Microbiomics. DOI: https://doi.org/10.1016/B978-0-12-816664-2.00004-9

years ago, they were plainly considered as agents of disease, if not death. These fascinating molecular wizards still provoke fear to the uninformed, but the notion that is gaining momentum as evidence accumulates and our understanding increases is that viruses can play essential roles in biosphere maintenance, biotic adaptation, and human health.

Although methodologically challenging, particularly due to the absence of a universal marker gene for viruses, progress in viral metagenomics (viromics, metaviromics, or the "metaomics" of viruses) over the last few years has boosted virology to exciting new dimensions that also warrant the evolution of Koch's postulates for establishing causality in infectious diseases (Antonelli and Cutler, 2016; Byrd and Segre, 2016). The great potential of advances in viromics for the mitigation of new disease outbreaks becomes immediately apparent with the paradigm of the discovery of Bombali (BOMV), the first Ebola virus to be detected from sequencing the virome of bats rather than infected individuals (Goldstein et al., 2018). BOMV has been shown to infect human cells in culture, while its pathogenicity in the field remains unknown; still, its discovery was monumental given that the provision of glimpses into emerging viral pathogens could allow for the timely, or even proactive preparedness of appropriately designed therapeutics or prophylactic strategies before the spillover to humans. Virome research could also provide answers to a wide range of profound questions about the virosphere, the entire "greater virus world" (Iranzo et al., 2016), beyond the continued search for new viruses as potential sources of pathogens in diverse marine and terrestrial environments and extreme habitats such as the arctic. These questions revolve around two main axes: the role of viruses as environmental drivers of communities' structural homeostasis and ecological processes, and their role as gatekeepers of human (and possibly animal and plant) health.

In this chapter, we focus on the virome. We begin by decrypting the current rather perplexing relevant terminology and continue with a detailed description of the various viral components of the microbiome that collectively form the virome. Challenges that complicate the characterization of these components compared to other microbiomic entities are also presented. Next, we document the viruses and viral sequences that have been catalogued to date (at the time of writing this chapter in August 2019) by the International Committee on Taxonomy of Viruses (ICTV) (Lefkowitz et al., 2018) by genome type, according to the Baltimore classification system. We then proceed with an ambitious effort to map the interactions among the most frequent double-stranded DNA (dsDNA) viruses and their hosts using a cutting-edge approach, network analysis, which, to the best of our knowledge, has never been attempted before. We conclude by discussing the implications of the observed interfaces for living organisms —and humans in particular— as hosts, and the environment. Despite their miniscule dimensions for the most part, the great importance of the viral components of the microbiome within, on, and all around us justifies the title of this chapter.

THE VIRAL COMPONENTS OF THE MICROBIOME

The combination of the words "virus" and "genome" resulted in the blend "virome." As the viral component of the microbiome, the virome includes eukaryotic viruses, bacteriophages, or simply "phages" (bacteria-infecting viruses), viruses that infect all other hosts, including other viruses —“virophages” (Mougari et al., 2019)— as well as viral elements integrated into the host genome not only of animals, including humans, higher primates, and other vertebrates, but also of other species such as plants, for instance (Diop et al., 2018; Harper et al., 2002). The virome therefore refers to the collection of nucleic acids, both RNA and DNA, that make up the viral community associated with a particular ecosystem or holobiont. We are just beginning to apprehend the multifaceted importance of viromes in such roles as the nutrient and energy cycling of the biosphere (Wegley et al., 2007), development of immunity (Barr et al., 2013), and host fitness improvement with acquisition of control both at the transcription and translation and posttranslation levels. In fact, the oceanic virome is an almost unlimited source of naturally bioengineered genes (Sharon et al., 2011).

THE ENVIRONMENTAL VIROME

Obtained from seawater samples collected off the California coast, the first uncultured viral genome was sequenced in 2002 (Breitbart et al., 2002). More than 65% of the sequences, most of which were viral in origin and belonged to all of the major families of dsDNA tailed phages —and to algal viruses to a lesser extent— were novel, demonstrating how little of the marine viral biodiversity was —and still is today— known, despite of its enormous influence on global biogeochemical cycles and ecological processes (Fuhrman, 1999). Analyses have since expanded to many other environments, such as marine sediments (Breitbart et al., 2004), soil (Fierer et al., 2007), and even air (Rosario et al., 2018); and hosts, such as animals, including insect vectors (Nanfack Minkeu and Vernick, 2018; Vandegrift and Kapoor, 2019) and humans as well as to the microenvironments/biocompartments within them (Ramírez-Martínez et al., 2018; Zárate et al., 2017). The role of the virome in health and disease is under intense investigation (Cadwell, 2015; Virgin, 2014). Both endogenous and exogenous factors may modify the virome profile that has been suggested to be host-specific (Lim et al., 2015). Studies on chickens, pigs, and humans have shown that the virome composition may indeed be affected by such factors as drugs, diet, and infections (Rampelli et al., 2017; Sachsenröder et al., 2014; Shah et al., 2016).

THE PLANT VIROME

Plants are crucial participants of the ecosystem. Several plant viruses have been recognized as agents of specific and often severe diseases, especially in the economically important crop species compared to the less studied wild plants. Ecogenomic applications in recent years disclosed a plethora of viruses in plant communities bearing no or a distant association to known plant viruses (Roossinck, 2010); the role of the plant virome, that is, the viral component of the phytobiome, still remains unknown (Roossinck, 2019). Moreover, recent metagenomic analyses in cultivated plant species with viral disease symptoms unveiled many previously unknown viruses (Gutiérrez Sánchez et al., 2016; Jo et al., 2017, 2018; Li et al., 2018). Thus, in contrast to the up until recently prevailing view, plant viruses appear to be common among plant diversities with specific types of interactions with hosts, which in many cases are beneficial for either the plant, the virus, or both (Roossinck, 2019).

THE HUMAN VIROME

The collection of viruses, including eukaryotic and prokaryotic viral species (bacteriophages), with which we coexist comprises the human virome (Zárate et al., 2017). Genomic fossils of past retroviral infections, human endogenous retroviruses (HERVs), and elements thereof account for about 8% of our genome (Lander et al., 2001). We are just beginning to appreciate the beneficial roles of these symbiotic viruses that coevolved with humans, although HERVs occasionally may have deleterious consequences for the host (Weiss, 2016). Meanwhile, the distinct viral communities harbored in each anatomical part of the human body (or "biocompartment") define "local" viromes, for example the notorious gut virome, which interact with other components of the microbiome, as do the bacteria in the gut (Handley, 2016). Prokaryotic viruses affect bacterial community structure and function, thereby influencing human health (Waldor and Mekalanos, 1996). The effects of eukaryotic viruses on human health are clearly more substantial, although they can range from mild, self-limited acute, or chronic infections to those with serious or even fatal consequences, as in acquired immunodeficiency syndrome (AIDS) (Wylie et al., 2012).

Each person carries a unique virome at a metastable equilibrium point with the rest of the microbial communities of the normal flora (Williams, 2013). Such continuous interactions lead to perpetual adaptations of the human virome, a condition reminiscent of the "Red Queen's hypothesis" (van Valen, 1973), according to which (in Lewis Carroll's words in *Through the Looking-Glass*) "it takes all the running you can do to keep in the same place" (Clarke et al., 1994). Adaptation to and improved fitness in response to selective pressures on existing or changing environments through natural selection are facilitated by the great genetic diversity within a virus quasispecies population, particularly for RNA viruses, so long as population sizes are large (Andino and Domingo, 2015). Factors such as lifestyle, age, geographic location, and even the season of the year can affect an individual's exposure to viruses; preexisting immunity and both viral and human genetics also determine one's susceptibility to viral diseases (Delwart, 2013).

Advances in high-throughput, deep-sequencing technology have driven progress in human virome analysis, yielding novel viruses that may be important pathogens and associating others with clinical phenotypes (Haynes and Rohwer, 2010). Diseases of unknown etiology are often thought to be of viral origin, but proving causality is difficult (Foxman and Iwasaki, 2011). As we aim to understand how microbial communities influence human health and disease, analysis of the human virome is critical and ultimately it may affect the treatment of patients with a variety of clinical syndromes. The way forward to precision medicine and individualized healthcare will have to take into consideration this least appreciated component of the microbiome (Balasopoulou et al., 2016).

METHODOLOGICAL CHALLENGES ASSOCIATED WITH VIROME STUDIES

Virome studies for virus discovery and virome analysis entail the generation of viral metagenomic sequence libraries from different biomes via the application of wet lab and heuristic bioinformatic methods. A sample could be processed to isolate viral nucleic acids suitable for high-throughput sequencing in about one week (Thurber et al., 2009). Depending on the source of samples and type of viral particles, each step of the process has to be individualized, with no single technique affording a generic approach. The general procedure is as follows: viral particles are first concentrated and separated from cellular components using a combination of filtration and density centrifugation methods, followed by enzymatic treatment to eliminate contaminating free nucleic acids. Viral nucleic acids are then extracted, amplified, purified, and finally sequenced and analyzed using metagenomic methods (Kumar et al., 2017). In an alternative approach, recent computational methods use directly metagenomic assembled sequences to discover viruses from microbiomic samples in about half a day (Paez-Espino et al., 2017).

Virus genomes consist of either DNA or RNA; thus to obtain the full virome, sequencing protocols for both types of nucleic acids have to be implemented. Given the typically small amount of viral genetic material present, in clinical specimens in particular, sample enrichment for viral particles is often necessary to increase the number of viral reads obtained during sequencing. The lack of universally conserved genes or even sequences, analogous to the 16S rRNA gene in bacteria or the internal transcribed spacer (ITS) in fungi, poses a particular challenge for the identification of viruses and subviral agents compared to other microbes (Krishnamurthy and Wang, 2017). Moreover, viral diversity remains mostly uncharacterized, as the so-called dark matter (Waldron, 2015), thereby complicating further the identification and assembly of viral sequences into metagenomes of the different viruses that form the virome. Three factors have been reported to contribute to the existence of viral dark matter: the divergence and length of virus sequences, the limitations of alignment-based classification, and the limited representation of viruses in reference sequence databases (Krishnamurthy and Wang, 2017).

Moreover, viral sequence databases are biased toward human viruses. As a result, many viruses that show no sequence similarity with those reported in the databases remain unknown. Owing to their compact genome organization, fast mutation rate, and low evolutionary conservation, RNA viruses pose additional problems to many standard bioinformatics analyses (Marz et al., 2014). It should be stressed that virus populations do not consist of a single member with a defined nucleic acid sequence. Instead, they are dynamic distributions of nonidentical but genetically related members called a "quasispecies" (Andino and Domingo, 2015). Hence, the genome sequences of viruses cluster around an average sequence, but every single genome is very likely different from that consensus. Advances in sequencing technologies expand our capacity to examine the composition of mutant spectra of viral quasispecies in infected cells and host organisms. Despite the methodological challenges, intense research efforts continue, utilizing viral metagenomics to discover novel viruses and characterize environmental and human viromes and to understand the mechanisms by which the virome as a whole contributes to the states of health and disease.

Deep sequencing of nucleic acids is indeed a very powerful method that has accelerated the pace of virus discovery; however, the enhanced sensitivity of the method in conjunction with the absence of the need to design specific primers to preamplify target sequences, render it prone to detect contaminants just as easily. The story of the NIH-CQV/PHV-1 (Xu et al., 2013) proves precisely the point: two nearly identical strains of a DNA virus, at the interface between the *Parvoviridae* and *Circoviridae*, supposed to account for non-(A-E) hepatitis cases based on findings in Chinese and US patients, proved to be the result of laboratory reagent contaminants and not *bona fide* infectious agents of humans (Naccache et al., 2013). The $\sim$99% nucleotide and amino acid identity shared by the geographically, temporally, and clinically diverse strains should have raised suspicion of viral contamination. Indeed, the origin of the virus was eventually traced to diatoms-contaminated silica-binding spin columns used for nucleic acid extraction.

Another type of contamination could be inert viral sequences from reagents or other sources; methods relying on the sequencing of newly synthesized RNAs, an unequivocal sign of the presence of a transcriptionally active virus, as reported very recently (Cheval et al., 2019), could alleviate this problem. An extensive list of novel contaminating viral sequences that can be used for the evaluation of viral findings in future virome and metagenome studies is now made available (Asplund et al., 2019). In any case, great care should be taken while conducting virome studies; and rigorous controls for each step of the process are needed. Disclosure and sharing of resulting sequence data, as in the above example, should also help minimize errors. Minimum information standards for reporting sequences of uncultivated virus' (meta)genomes were recently published (Roux et al., 2018).

THE HOST–VIRUS INTERACTOME

NETWORK ANALYSIS: A ROADMAP TO EXPLORE HOST–VIRUS INTERACTIONS

At first glance, adding the effects of each component in the simple equation shown (4.1) appeared to suffice to describe the interactions between the environmental and plant and human viromes and the respective hosts:

$$\text{The host-virus interactome} = \text{Hosts} + \text{environmental virome} + \text{plant virome} + \text{human virome} \quad (4.1)$$

However, the simplicity of the otherwise reasonable Eq. (4.1) is deceitful since describing the ever dynamically interacting viral components with their respective hosts and other elements of the microbiome, each exhibiting diverse spatiotemporal patterns in response to diverse genetic and epigenetic factors, appeared to be a scientifically unattainable goal. Outside the relatively narrow, tight groups, the applicability of traditional tree model and tree-based methods of phylogenetics and phylogenomics, representing vertical descent from a last common ancestor, seemed inadequate for probing the intricate evolutionary relationships of the virosphere (Iranzo et al., 2016). The two main characteristics of viral evolution, high rates of horizontal gene transfer (HGT) and fast sequence divergence, in addition to the lack of universal genes or even sequences shared by all viruses, challenge the tree-like interpretation of the virus world (Iranzo et al., 2017). Network analysis that considers introgressive descent events may provide the answer to this perplexing scientific puzzle. In contrast to the tree representation that stresses how entities evolve *ex unibus plurum*, that is expansively, the network representation, which uses graphs to analyze microbial complexity and evolution, emphasizes how entities evolve *ex pluribus unum*, that is convergingly (Corel et al., 2016).

Recent revolutionary advances in *in silico* genomic analyses were accompanied by notable progress in the development of mathematical explorative concepts and tools for Big Data analysis, including data mining and network research (Corel et al., 2016). Networks provide a unique explorative tool for determining interfaces among interacting species within the holobiont and for understanding complex interactions by locating and interpreting structural and functional patterns in Big Data. A network consists of vertices (sing. vertex) and edges. Vertices (or nodes), represented as points, are the interacting items and edges (or links), represented as lines, are the connection types. Networks consisting of two classes of nodes, such as the host–virus interactome described in this work, are known as bipartite networks. Edges in these networks run only between nodes of different kinds. A bipartite network illustrates the intricate web of connections among the items of a particular system. The relative importance of nodes in networks is evaluated by calculating the so-called centralities. The most common centrality is the degree, which indicates the number of node connections. Further information on network structure and functionality can be found in Newman (2018). As metagenomics data accumulate, host–virus networks may assist in the interpretation of the former by probing the interactions of viruses with their hosts, unraveling thus the dynamics of viral communities; and, just as importantly, by tracing the pathways of viral evolution. Comparative analysis of microbial and bacteriophage genomes using evolutionary networks resulted in the identification of preferred routes and patterns of HGT (Jachiet et al., 2014; Popa and Dagan, 2011; Popa et al., 2011).

PROSPECTING THE CONTINUUM OF INTERACTIONS WITHIN THE VIROSPHERE

Viruses cannot replicate outside their hosts; therefore viruses can be considered as symbionts with their hosts. Symbiosis, originally defined as "two dissimilar entities living in an intimate association," in fact encompasses several different kinds of relationships, including parasitism (where one partner benefits at the expense of the other), commensalism (where one partner benefits and the other is unaffected), and mutualism (where all parties benefit, thus leading to increased fitness, as measured by increased reproduction). Furthermore, the symbiotic relationship can vary with the environment or other circumstances and often it cannot be strictly classified as belonging to solely one of these three categories (Roossinck, 2011). The phenomena of satellite viruses (Kassanis and Nixon, 1961) and virophages (Mougari et al., 2019) seem to be closely related; multipartitism (Lucía-Sanz and Manrubia, 2017), if considered as something totally alien to a version of breaking down a genome to chromosomes, may join the previous two in introducing new iterations in types of symbioses and evolution via recombination. Between the viral constituents the procedures of satellite viruses are commensal, but toward their karyotic host, they may be more offensive, rather parasitic. The change of balance —or a less balanced balance— between the viral constituents may lead to virophagy, definitely a parasitic event between the two viruses, but the host may benefit from such proceedings and the host phenotype may be mutualistic or mildly parasitic. On the other hand, multipartitism allows a much neater, modular approach to recombination, testing different mutated/evolved genomic parts in various combinations with nonevolved parts, without tainting a whole genome and bringing about its effective termination or even its commitment to a limited range of evolutionary events.

Examples of viruses in mutualistic symbiotic relationships with their hosts include the following: (1) symbiogenic viruses which due to their long, in evolutionary time, association became part of the host forming a new host species, while probably also leading to the speciation of viruses themselves (Roossinck, 2005). (2) Virus infections could lead to milder disease forms caused by other virus species, strains or other pathogens, as in the case of human immunodeficiency virus type 1 (HIV-1)-infected individuals who progress to full-blown AIDS more slowly if they are coinfected with human pegivirus (also called GBV-C virus or hepatitis G virus) (Heringlake et al., 1998; Tillmann et al., 2001), typically a common nonpathogenic virus that may nonetheless be associated with certain idiopathic forms of fulminant hepatitis (Anastassopoulou, 2002). (3) Viruses that destroy competitors and are thereby useful to their hosts, as in the cases of phages that can kill competing bacteria that are not lysogenic for the virus (Bossi et al., 2003; Lehnherr et al., 1993). (4) Viruses that assist in the adaptation of their hosts to extreme environmental changes. Such is the case of the three-way mutualistic symbiosis of *Curvularia thermal tolerance virus* (CThTV) with the fungal endophyte *Curvularia protuberate* and the latter's plant host, the panic grass *Dichanthelium lanuginosum*. This symbiotic scheme allows both cellular elements to survive harsh geothermal soils (Márquez et al., 2007), a trait that may be increasingly important as the Earth's climate changes. (5) Viruses interacting tortuously with a multiplicity of species including transducing fitness genes amongst their host taxa (Chen et al., 2018; Haaber et al., 2016; Roossinck, 2011).

Ever since its origin, virus–host coevolution has been prominent in the evolution of life as implied by the ubiquity of viruses in the extant biosphere (Forterre and Prangishvili, 2009, 2013; Koonin and Dolja, 2013). In the course of evolution, viruses and their components are believed to have been repeatedly recruited (exapted) for host functions that typically involve defense from other viruses or cellular competitors or transfer of nucleic acids between cells, or storage functions (Koonin and Krupovic, 2018). Exaptation is a term conceptualized by Gould and Vrba (1982) to denote recruitment and repurposing of a biological entity for a new function unrelated to the originally intended one. Different depths, with respect to what remains of the virus genome, have been reached by the exaptation of viral genomes, individual genes, or smaller fragments for host functions (Koonin and Krupovic, 2018). Viruses have indeed played a major part in the evolution of life on earth, being responsible for such major evolutionary leaps, as the emergence of placental mammals from egg-laying species after retrovirus infection (Harris, 1991). Thus, it is obvious that the relationships between viruses and hosts extend beyond the proverbial arms race (Carpentier and Geballe, 2016).

Viruses are infectious obligate intracellular parasites, consisting, in their simplest form, of a proteinaceous capsid that is protecting the virus genetic material, which may be either DNA or RNA, double stranded (ds) or single stranded (ss), and of either positive or negative polarity. By convention, mRNA is defined as a positive (+) strand because it is coding (as opposed to complementary); that means it contains information that can be immediately translated by cellular ribosomes. Negative polarity indicates complementarity to the positive strands, but no compatibility with ribosomes as is. Within an appropriate cell (that defines the virus tropism) the viral genome orchestrates the synthesis of viral components by cellular systems. Newly synthesized components are *de novo* self-assembled, forming progeny virus particles, termed virions, which will continue to propagate the viral genetic material to the next host (cell or organism), initiating a new infection cycle (Flint et al., 2015).

Viruses constitute the most abundant microbiomic entities on the planet (Rohwer and Thurber, 2009; Suttle, 2005, 2007). They are omnipresent and infect almost all cellular life forms (Koonin and Dolja, 2013), and even other viruses, in the case of virophages. Interdependent for replication, satellite viruses infecting the same host cell, were discovered back in 1961 (Kassanis and Nixon, 1961). Presumably true infections of (giant) viruses by another group of viruses that were named "virophages" were first described in 2008 (Fischer, 2011; La Scola et al., 2008). Giant viruses, with larger virions and genomes encoding similar if not greater numbers of predicted genes compared to some bacteria (Aherfi et al., 2016; Colson et al., 2019), were discovered in 2003. *Acanthamoeba polyphaga mimivirus* (APMV) was the first giant virus ever discovered (La Scola et al., 2003). Due to their sharing of many characteristics with intracellular microbes, these giants challenge the definition of a virus (Colson et al., 2017); still they are only cultured by inoculating on amoebae and being devoid of ribosomal DNA, they have been largely neglected in microbiome and virome investigations.

Although the viruses of the virosphere exhibit striking differences with respect to their size, virion structure, genome architecture, and strategies for propagation (Baltimore, 1971; Koonin, 1991; Koonin and Dolja, 2014; Suttle, 2007), they all have to synthesize mRNA that is decoded by the host's translational machinery. The pathways from viral genomes to mRNA are used to define specific virus classes in the Baltimore classification system, based on the nature (DNA or RNA) and polarity [(+) or (−)] of their genomes, which define their "survival" strategies; DNA genomes allow for in-cell dormancy (lysogeny) in a nucleic-acid-only stage ("provirus"/"prophage" if integrated in host DNA, "episome" if remaining as a circular DNA molecule separate from the host DNA molecules) which allows active cellular infection (lytic cycle), virion production, and propagation of whole virus progeny only if optimal conditions ensue (Howard-Varona et al., 2017). RNA genomes allow faster mutagenesis and evolution (Elena and Sanjuán, 2005) and forego the prospect for the adaptability inherent in in-cell dormancy for a more diverse expression pathway, possibly away from the well-guarded and curated nucleus (Kobiler et al., 2012) and in cases applicable very fast and in relatively isolated —or, on the contrary, multiple and ubiquitous— areas of the cytoplasm, thus shirking or overwhelming the cellular defense mechanisms, respectively.

Viruses with dsDNA or ssDNA (Baltimore Groups I and II, respectively) transcribe their genome directly to mRNA either by virally encoded enzymes, thus foregoing any need to enter the nucleus and the respective survivability complications, as is the case with poxviruses (Kobiler et al., 2012), or by using the transcriptional machinery of the cell (Kobiler et al., 2012); dsRNA and ssRNA(−) viruses (Baltimore Groups III and V, respectively) initiate transcription through an RNA-dependent RNA polymerase (RdRp) packaged into the virus particle. All ssRNA(+) viruses (Baltimore Group IV), except retroviruses, are directly translated in the host ribosomes. There are two additional strategies typified by retroviruses [ssRNA(+)] and pararetroviruses (gapped dsDNA). Retroviruses (e.g., HIV) reverse transcribe their ssRNA(+) initially into ssDNA(-) by a reverse transcriptase (RT) enzyme carried within the virion and fuse the produced dsDNA into the host genome by a virally encoded integrase molecule. Gapped dsDNA viruses, as those of the families *Caulimoviridae* (e.g., *Cauliflower mosaic virus*) and *Hepadnaviridae* (e.g., *Hepatitis B virus*), first have their genome filled by a host DNA polymerase and then transcribe their genes to mRNAs via ssRNA(−) intermediates. These viruses encode for an RT to produce cDNA and dsDNA from ssRNA(+).

The viruses that have been cataloged to date by the ICTV (Lefkowitz et al., 2018) according to the type of their genome are presented in Table 4.1. Viruses-harboring dsDNA genomes were found to dominate the virosphere (2312/5846 or ~40%). Of those, most target bacteria (1331/2312), and the rest vertebrate (688/2312) or, to a lesser extent, invertebrate (201/2312) species. Viruses of Group IV with ssRNA(+) were the second more frequently encountered entities (1376/5846 or ~23.5%), with the vast majority of them targeting plants (846/1376).

THE DOUBLE-STRANDED DNA VIRUS–HOST INTERACTOME

In Fig. 4.1 the host–virus interaction network of all available dsDNA viruses is presented; it is an integral part of the whole Virome network that includes all types of virus–host interactions (Manoussopoulos and Anastassopoulou, unpublished data). To construct the network, all relevant information was retrieved from the NCBI database (NCBI Resource Coordinators, 2018).

Table 4.1 Type of Viral Genomes Related to Infected Organisms, Summarized From the International Committee on Taxonomy of Viruses, Virus Metadata Repository (version June 1, 2019; MSL34)

	Group I	Group II	Group III	Group IV	Group V	Group VI	Group VII	
Organisms	**dsDNA**	**ssDNA**	**dsRNA**	**ssRNA (+)**	**ssRNA (−)**	**ssRNA-RT**	**dsDNA-RT**	**Subtotals**
Algae	34	5	1	9	0	2	0	51
Archaea	48	1	0	0	0	0	0	49
Bacteria	1331	54	7	4	0	0	0	1396
Fungi	1	1	104	33	1	11	0	151
Invertebrates	201	43	29	105	161	34	0	573
Plants	0	848	45	846	22	26	81	1868
Protozoa	9	0	9	1	1	1	0	21
Vertebrates	688	305	41	378	219	70	36	1737
Total	2312	1257	236	1376	404	144	117	5846

ds, *Double stranded;* RT, *reverse transcriptase;* ss, *single stranded.*

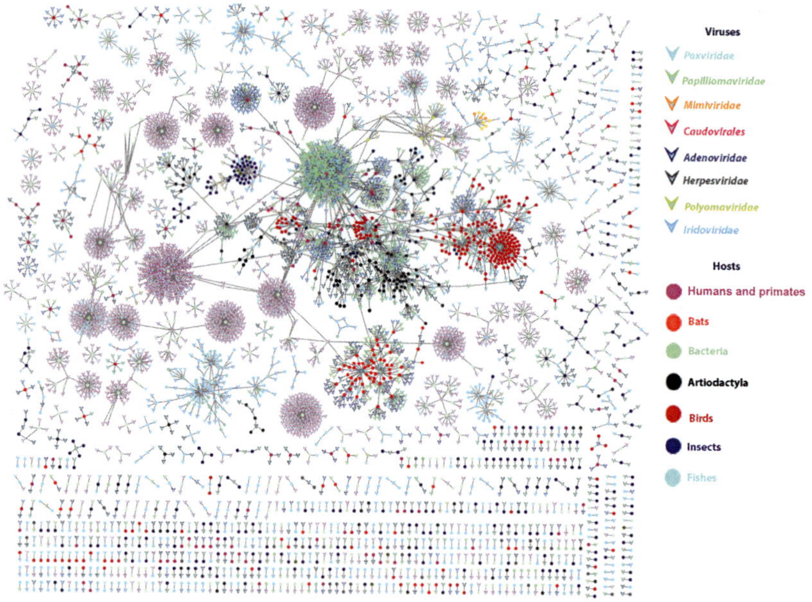

FIGURE 4.1

The bipartite network of dsDNA virus–host interactions. Reversed triangles and filled dots represent virus and host species, respectively. Families of the most common virus species and groups of hosts are indicated by colors. The periphery of the network consists of unlinked components, the size of which increases toward the center. The central part of the network consists of large overconnected submodules. Fig. 4.2 zooms into this central part. *ds*, Double stranded.

However, the exploitable part was less than 10% and most of it needed further curation and validation (e.g., cleaning of duplicates and removal of self-loops). Unclean and incomplete (dirty) data may well be the biggest problem in Big Data exploration. Such problems may be alleviated by appropriate regulatory initiatives and, perhaps more importantly, through exercising caution when contributing data. For the purposes of this work the collected information on host–virus interactions was considered a random sample of the whole dsDNA virus–host population, fixed in time. No further optimization or sample weighting was applied. All virus–host associations reported here are at the species level except where otherwise stated. Graphs were constructed and analyzed using the Cytoscape software (v.3.7.1) (Shannon et al., 2003).

The resulting network (Fig. 4.1) consisted of 6552 nodes and 5896 links. The network was highly fragmented, having 876 unlinked fragments of varying size. Fragment size is defined by the number of interacting nodes that each fragment possesses. Table 4.2 shows the structural parameters of this fragmented component of the dsDNA virus–host network according to fragment size that ranged from 2 (the minimum) to 217 interacting nodes. Most fragments of the network (577/876) were of the most basic form (size 2), meaning that they had only two interacting nodes. Fragments of larger sizes (i.e., with more numerous interacting nodes) were sparse. For example, only one fragment was of size 217 (interacting items), with an equal number of edges (links). Nevertheless, as fragment size increases, presumably over the course of evolutionary time, the complexity of the system also increases, while the occurrence of large fragments and the number of involved virus families decreases. This type of fragmented, but ordered, network organization is suggestive of a tendency toward lower entropy of the system, most likely driven by evolutionary events.

The central and largest part of the dsDNA virus–host network (Fig. 4.2) involved 2297 vertices and 2469 edges. It consisted of several components, each of which comprised a single (e.g., *Escherichia coli*), several (e.g. *Bacillus* sp.), or many interacting species and also showed a high level of fragmentation. Most of its modular components had at least one overconnected node (i.e., a hub) that provided direct or indirect connections to other network components. Perhaps not surprisingly, the largest module of this subnetwork, which was further separated into smaller submodules, was that of mammals. The overwhelming majority of dsDNA viruses associated with mammalian species were papillomaviruses, followed by adenoviruses, herpesviruses, and poxviruses.

HUMANS, APES, AND MONKEYS

The mammals' module included the following primate hosts (denoted by *pink dots* in Fig. 4.2): humans (*Homo sapiens*), apes (*Pan troglodytes* and *Gorilla gorilla*), and monkeys (*Macaca* sp. and *Saguinus* sp., *inter alia*). Human hosts were the most heavily connected node of the mammalian module, possibly reflecting the anthropocentric prominence of sequence databases that are biased toward human viruses. The primate submodule was found to be joined to the second largest mammalian submodule involving bovid and deer species of the families *Bovidae* and *Cervidae*, respectively, via adenovirus and herpesvirus links (Table 4.3). Furthermore, two poxvirus species, *Orf virus* and *Camelpox virus*, directly linked *Bos taurus* (cows) or *Camelus dromedarius* (camels) to humans, whereas *B. taurus* and humans were indirectly connected by *Orf virus* via camels.

Table 4.2 Structural Parameters of the Fragmented Component of the Double-Stranded DNA Virus–Host Network, According to Fragment Size (i.e., Determined by the Number of Interacting Nodes)

Fragment Size	Occurrences	Nodes	Edges	Number of Virus Families
2	577	1154	577	24
3	136	408	272	16
4	39	156	116	16
5	32	160	129	15
6	15	90	75	14
7	15	105	91	10
8	8	64	56	8
9	6	54	48	8
10	3	30	27	5
11	6	66	61	7
12	4	48	46	5
13	2	26	25	4
14	2	28	26	3
16	2	32	30	8
17	2	34	36	5
18	4	72	68	3
19	1	19	18	3
24	4	96	92	4
25	1	25	24	3
26	2	52	52	6
29	1	29	30	1
31	1	31	30	4
35	1	35	41	2
40	1	40	40	3
46	1	46	45	3
50	1	50	49	4
54	1	54	56	3
62	1	62	61	2
98	1	98	97	1
119	1	119	127	3
124	1	124	126	4
134	1	134	146	4
147	1	147	146	2
167	1	167	168	3
217	1	217	217	3

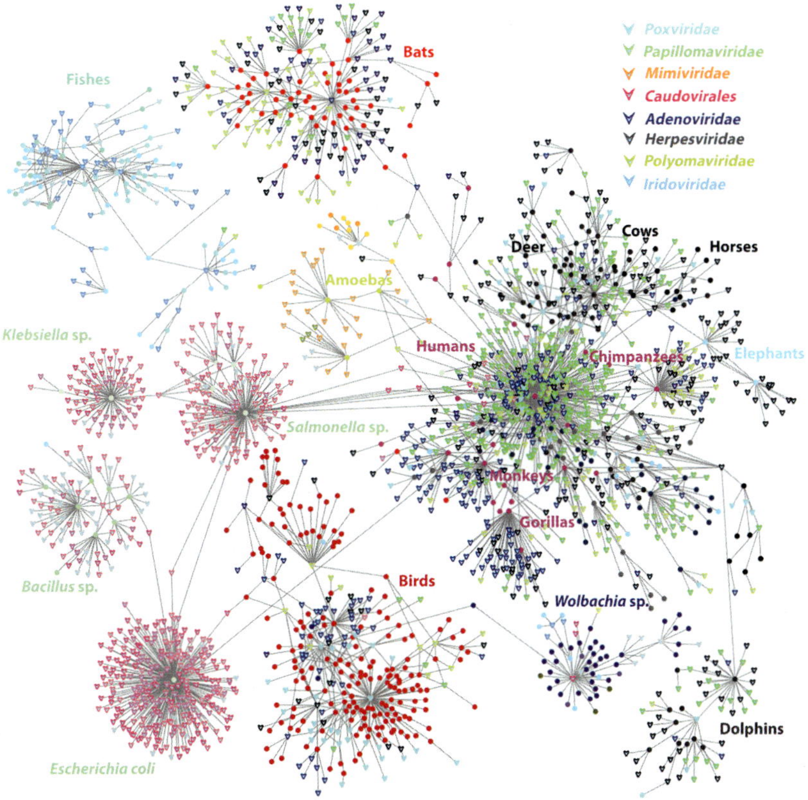

FIGURE 4.2

The most densely interconnected section of the dsDNA virus–host interactome as a bipartite network. Nodes corresponding to viruses are denoted by inverted triangles and nodes corresponding to hosts are denoted by dots. Links represent interactive connections between the most common viruses and their hosts. The modular structure of the network is highlighted by coloring virus nodes according to the submodule to which they belong. The topology of some major color-coded viral families and hosts is indicated. *ds*, Double stranded.

DOLPHINS

In a smaller submodule of aquatic mammalians where dolphins predominated (denoted by *black dots* in Fig. 4.2), associations with herpesviruses, papillomaviruses, and poxviruses were uncovered. Interestingly, this submodule was linked to the main mammals' module via *Human alphaherpesvirus 1* (*Herpes simplex virus 1*) which infects both humans and bottlenose dolphins (*Tursiops truncatus*). This type of association may have surfaced under the shifting environmental conditions in dolphins' niches as the human population increases; in the Canary Islands, different cetacean populations can be found close to the coast, where they are exposed to increasingly greater amounts of human microorganisms through urban sewage (Esperón et al., 2008). The same species is also infected by *Cetacean poxvirus 1*, connecting it to the rest of dolphin species and seals.

Table 4.3 Some of the Species and Their Links Involved in Submodular Connections in the Double-Stranded DNA Virus–Host Network

Connected Participants	Host Species	Virus Link
Humans–cows–camels	*Homo sapiens*	*Orf virus*
	Bos taurus	*Orf virus*
	Camelus dromedarius	*Orf virus*
	C. dromedarius	*Camelpox virus*
	H. sapiens	*Camelpox virus*
Humans–chimpanzees	*Pan troglodytes*	*Human adenovirus 5*
	H. sapiens	*Human adenovirus 5*
	P. troglodytes	*Human betaherpesvirus 6B*
	H. sapiens	*Human betaherpesvirus 6B*
Humans–gorillas	*H. sapiens*	*Human mastadenovirus B*
	Gorilla gorilla	*Human mastadenovirus B*
Humans–monkeys	*Macaca mulatta*	*Macaca mulatta polyomavirus 1*
	H. sapiens	*Macaca mulatta polyomavirus 1*
Humans–amoebas	*H. sapiens*	*Acanthamoeba polyphaga mimivirus*
	Acanthamoeba sp.	*Acanthamoeba polyphaga mimivirus*
Bacillus–Escherichia	*Bacillus thuringiensis*	*Escherichia virus T4*
	Escherichia coli	*Escherichia virus T4*
Escherichia–Salmonella	*E. coli*	*Escherichia virus KP26*
	Salmonella enterica	*Escherichia virus KP26*
Humans–dolphins	*Tursiops truncatus*	*Human alphaherpesvirus 1*
	H. sapiens	*Human alphaherpesvirus 1*

BATS

Another submodule of the network was that of bats (denoted by *orange dots* in Fig. 4.2), which was associated mostly with poxviruses, papillomaviruses, adenoviruses, herpesviruses, and polyomaviruses. *Bat mastadenovirus* and *Bat polyomavirus* (the species nomenclature here follows the one in the database *verbatim*) were the viruses with the most connections, found in the majority of bat species. Notably, the bat-interacting component seems to be directly connected with the submodules of humans and other primates through a common link provided by one species of the genus *Mastadenovirus* found both in the Cape serotine bat (*Neoromicia capensis*), a species of vesper bat occurring in sub-Saharan Africa, and in humans. Unfortunately, no further details on the virus species are provided in the database, so this link is restricted at the genus level.

BIRDS

Bird species (denoted by *reddish brown dots* in Fig. 4.2) were mainly associated with poxviruses, herpesviruses, and adenoviruses, forming another cluster of interactions. Not surprisingly, since chickens are of the most common and widespread domestic animals, the most connected node of the birds' submodule was *Gallus gallus domesticus*, found to be infected by 33 viral species. Species of the genus *Avipoxvirus* infect more than 87 bird species. Another highly spread bird virus, infecting more than 23 species, was *Fowlpox virus*. This virus has also been found in the zoophilic mosquito *Coquillettidia crassipes*, which is related to the highly prevalent *Wolbachia* bacterial species (Yeo et al., 2019) as well as to *Wolbachia phage WO* (Chauvatcharin et al., 2006); thus a link is revealed between birds and insects infected with *Wolbachia* sp. *Wolbachia* is a genus of Gram-negative bacteria that infect arthropods, including a wide range of insects and certain nematodes. *Wolbachia phage WO* has been found in association with over 38 insect species of different orders, apparently linked to *Wolbachia* and *Rickettsia*.

AMOEBAS

An interesting small submodule of the dsDNA virus–host network is the one formed by amoebas (denoted by *olive green dots* in Fig. 4.2) and their associated, recently discovered nucleocytoplasmic large DNA viruses (NCLDV), an order of viruses that contain the *Megavirales* or giant viruses. The gigantism of these viruses that include mimiviruses and marseilleviruses endows them with unique properties structurally and functionally: genome size range typical of Bacteria and Archaea and genes that encode components of the translation system (which is the signature cellular molecular machinery), respectively (Koonin and Yutin, 2018). This submodule was directly connected to the primates' module via a common infection route of *Acanthamoeba* species and humans, or of the Rhesus macaque (*Macaca mulatta*) with *Human polyomavirus 1*. Another direct connection was through a common infection of *Acanthamoeba polyphaga* and humans by APMV. Amoebal mimiviruses have been linked to pneumonia in humans, while marseilleviruses have been mostly described in asymptomatic persons and in a lymph node adenitis case (Colson et al., 2016).

FISHES

Fishes (denoted by *turquoise dots* in Fig. 4.2) form another cluster that is disconnected from the rest of the network, but it is shown for comparison purposes. They were mainly associated with virus species of the family *Iridoviridae*. Although amphibia, invertebrates, and insects are natural hosts of iridoviruses, no direct or indirect connection could be made with them.

BACTERIA

As expected, bacteria (denoted by *light green dots* in Fig. 4.2) were exclusively associated with phage viruses, mainly of the order *Caudovirales*. *E. coli*, *Salmonella enterica*, and *Klebsiella pneumoniae*, three important human pathogens, had degrees of 281, 116, and 67, respectively, all representing phage associations. Apart from the high number of associated phages, these bacteria were linked to each other by one or two connections. Thus, *E. coli* and *S. enterica* were joined by *Escherichia virus KP26*. In turn, *Salmonella* was linked to humans via phages SEN1, SEN4, SEN5, SEN 8, SEN 22, SEN 34, obviously isolated from human *Salmonella* infections. Moreover, six bacteria of the genus *Bacillus*, namely *Bacillus cereus*, *B. subtilis*, *B. thuringiensis*, *B. megaterium, B. pumilus, B. anthracis* were associated with 40, 18, 9, 8, 5, and 4 phages, respectively. These species were connected by infection with the same phage species, although some links were at the genus level due to a lack of specific information in the database.

FUTURE PERSPECTIVES, ASPECTS, AND PROSPECTS

Viruses constitute the most abundant microbiomic entities on the planet. They are present in all ecosystems, from the poles to deep oceans, and infect almost all cellular life forms, from Archaea to multicellular eukaryotes and, perhaps astonishingly, even other viruses. Although until recently viruses were considered life-threatening pathogens and some are indeed, new evidence proves the majority of them to be beneficial, if not to living organisms individually, definitely to living systems; since they regulate the biosphere conditions, they exert evolutionary pressure that results in increased fitness within and among species and by their function to enact genomic exchanges among organisms, promoting evolution. In the course of evolution, viruses and their components are believed to have been repeatedly deputized to implement host functions.

Technological advances, particularly in sequencing formats, together with the explosive increase of computational resources are expected to contribute significantly to our apprehension and awareness of viruses in the near future. Much information has been accumulated in big databases, and new computational tools have been developed to properly exploit them; still, such data are highly biased. The research has always been focused on viruses relevant to high-impact cellular species, with humans being by far the most important and most highly prioritized. Thus, less intense research resulting in much fewer detected viruses and submitted sequences and hence lower analytical depth for viruses of other species or nonhuman-affecting niches and habitats restrict the available interactions and produce falsely weighted results. The situation is exacerbated by the fact that viruses do no share consensus sequences, thus existing detection protocols based on PCR or hybridization of nucleic acids in general cannot be used to detect unknown viral entities, especially apathogenic ones lurking in extreme habitats.

The constructed network of dsDNA virus–host interactions serves as a roadmap for the exploration of the whole virome and for understanding —and, much more importantly, interpreting— the intricate relationships among viruses themselves and their hosts in the successive iterations, which are to emerge as massive metagenomics, metatranscriptomics and, in effect, metaviromics data will start pouring in.

REFERENCES

Aherfi, S., Colson, P., La Scola, B., Raoult, D., 2016. Giant viruses of amoebas: an update. Front. Microbiol. 7, 349.

Anastassopoulou, C., 2002. Fulminant hepatic failure in a pediatric patient with active GB virus C (GBV-C)/ hepatitis G virus (HGV) infection. Hepatol. Res. 23 (2), 85–89.

Andino, R., Domingo, E., 2015. Viral quasispecies. Virology 479–480, 46–51.

Antonelli, G., Cutler, S., 2016. Evolution of the Koch postulates: towards a 21st-century understanding of microbial infection. Clin. Microbiol. Infect. 22 (7), 583–584.

Asplund, M., Kjartansdóttir, K.R., Mollerup, S., Vinner, L., Fridholm, H., Herrera, J.A.R., et al., 2019. Contaminating viral sequences in high-throughput sequencing viromics: a linkage study of 700 sequencing libraries. Clin. Microbiol. Infect. 25, 1277–1285.

Balasopoulou, A., Patrinos, G.P., Katsila, T., 2016. Pharmacometabolomics informs viromics toward precision medicine. Front. Pharmacol. 7, 411.

Baltimore, D., 1971. Expression of animal virus genomes. Bacteriol. Rev. 35 (3), 235–241.

Barr, J.J., Auro, R., Furlan, M., Whiteson, K.L., Erb, M.L., Pogliano, J., et al., 2013. Bacteriophage adhering to mucus provide a non-host-derived immunity. Proc. Natl. Acad. Sci. U.S.A. 110 (26), 10771–10776.

Bossi, L., Fuentes, J.A., Mora, G., Figueroa-Bossi, N., 2003. Prophage contribution to bacterial population dynamics. J. Bacteriol. 185 (21), 6467–6471.

Breitbart, M., Salamon, P., Andresen, B., Mahaffy, J.M., Segall, A.M., Mead, D., et al., 2002. Genomic analysis of uncultured marine viral communities. Proc. Natl. Acad. Sci. U.S.A. 99 (22), 14250–14255.

Breitbart, M., Miyake, J.H., Rohwer, F., 2004. Global distribution of nearly identical phage-encoded DNA sequences. FEMS Microbiol. Lett. 236 (2), 249–256.

Byrd, A.L., Segre, J.A., 2016. Infectious disease. Adapting Koch's postulates. Science (New York, N.Y.) 351 (6270), 224–226.

Cadwell, K., 2015. The virome in host health and disease. Immunity 42 (5), 805–813.

Carpentier, K.S., Geballe, A.P., 2016. An evolutionary view of the arms race between protein kinase R and large DNA viruses. J. Virol. 90 (7), 3280–3283.

Chauvatcharin, N., Ahantarig, A., Baimai, V., Kittayapong, P., 2006. Bacteriophage WO-B and Wolbachia in natural mosquito hosts: infection incidence, transmission mode and relative density. Mol. Ecol. 15 (9), 2451–2461.

Chen, J., Quiles-Puchalt, N., Chiang, Y.N., Bacigalupe, R., Fillol-Salom, A., Chee, M.S.J., et al., 2018. Genome hypermobility by lateral transduction. Science (New York, N.Y.) 362 (6411), 207–212.

Cheval, J., Muth, E., Gonzalez, G., Coulpier, M., Beurdeley, P., Cruveiller, S., et al., 2019. Adventitious virus detection in cells by high-throughput sequencing of newly synthesized RNAs: unambiguous differentiation of cell infection from carryover of viral nucleic acids. mSphere 4 (3).

Clarke, D.K., Duarte, E.A., Elena, S.F., Moya, A., Domingo, E., Holland, J., 1994. The red queen reigns in the kingdom of RNA viruses. Proc. Natl. Acad. Sci. U.S.A. 91 (11), 4821–4824.

Colson, P., Aherfi, S., La Scola, B., Raoult, D., 2016. The role of giant viruses of amoebas in humans. Curr. Opin. Microbiol. 31, 199–208.

Colson, P., La Scola, B., Levasseur, A., Caetano-Anollés, G., Raoult, D., 2017. Mimivirus: leading the way in the discovery of giant viruses of amoebae. Nat. Rev. Microbiol. 15 (4), 243–254.

Colson, P., Ominami, Y., Hisada, A., La Scola, B., Raoult, D., 2019. Giant mimiviruses escape many canonical criteria of the virus definition. Clin. Microbiol. Infect. 25 (2), 147–154.

Corel, E., Lopez, P., Méheust, R., Bapteste, E., 2016. Network-thinking: graphs to analyze microbial complexity and evolution. Trends Microbiol. 24 (3), 224–237.

Delwart, E., 2013. A roadmap to the human virome. PLoS Pathog. 9 (2), e1003146.

Diop, S.I., Geering, A.D.W., Alfama-Depauw, F., Loaec, M., Teycheney, P.-Y., Maumus, F., 2018. Tracheophyte genomes keep track of the deep evolution of the Caulimoviridae. Sci. Rep. 8.

Elena, S.F., Sanjuán, R., 2005. Adaptive value of high mutation rates of RNA viruses: separating causes from consequences. J. Virol. 79 (18), 11555–11558.

Esperón, F., Fernández, A., Sánchez-Vizcaíno, J.M., 2008. Herpes simplex-like infection in a bottlenose dolphin stranded in the Canary Islands. Dis. Aquat. Org. 81 (1), 73–76.

Fierer, N., Breitbart, M., Nulton, J., Salamon, P., Lozupone, C., Jones, R., et al., 2007. Metagenomic and small-subunit rRNA analyses reveal the genetic diversity of bacteria, archaea, fungi, and viruses in soil. Appl. Environ. Microbiol. 73 (21), 7059–7066.

Fischer, M.G., 2011. Sputnik and Mavirus: more than just satellite viruses. Nat. Rev. Microbiol. 10 (1), 78. author reply 78.

Flint, S.J., Racaniello, V.R., Skalka, A.M., Rall, G.F., Enquist, L.W., 2015. Principles of Virology, fourth ed. ASM Press, Washington, DC.

Forterre, P., Prangishvili, D., 2009. The great billion-year war between ribosome- and capsid-encoding organisms (cells and viruses) as the major source of evolutionary novelties. Ann. N. Y. Acad. Sci. 1178, 65–77.

Forterre, P., Prangishvili, D., 2013. The major role of viruses in cellular evolution: facts and hypotheses. Curr. Opin. Virol. 3 (5), 558–565.

Foxman, E.F., Iwasaki, A., 2011. Genome-virome interactions: examining the role of common viral infections in complex disease. Nat. Rev. Microbiol. 9 (4), 254–264.

Fuhrman, J.A., 1999. Marine viruses and their biogeochemical and ecological effects. Nature 399 (6736), 541–548.

Goldstein, T., Anthony, S.J., Gbakima, A., Bird, B.H., Bangura, J., Tremeau-Bravard, A., et al., 2018. The discovery of Bombali virus adds further support for bats as hosts of ebolaviruses. Nat. Microbiol. 3 (10), 1084–1089.

Gould, S.J., Vrba, E.S., 1982. Exaptation—a missing term in the science of form. Paleobiology 8 (1), 4–15.

Gutiérrez Sánchez, P.A., Jaramillo Mesa, H., Marin Montoya, M., 2016. Next generation sequence analysis of the forage peanut (*Arachis pintoi*) virome. Rev. Fac. Nac. Agron. 69 (2), 7881–7891.

Haaber, J., Leisner, J.J., Cohn, M.T., Catalan-Moreno, A., Nielsen, J.B., Westh, H., et al., 2016. Bacterial viruses enable their host to acquire antibiotic resistance genes from neighbouring cells. Nat. Commun. 7, 13333.

Handley, S.A., 2016. The virome: a missing component of biological interaction networks in health and disease. Genome Med. 8 (1), 32.

Harper, G., Hull, R., Lockhart, B., Olszewski, N., 2002. Viral sequences integrated into plant genomes. Annu. Rev. Phytopathol. 40, 119–136.

Harris, J.R., 1991. The evolution of placental mammals. FEBS Lett. 295 (1–3), 3–4.

Haynes, M., Rohwer, F., 2010. The human virome. In: first ed. Nelson, K.E. (Ed.), Metagenomics of the Human Body, vol. 102. Springer, New York, pp. 63–77.

Heringlake, S., Ockenga, J., Tillmann, H.L., Trautwein, C., Meissner, D., Stoll, M., et al., 1998. GB virus C/hepatitis G virus infection: a favorable prognostic factor in human immunodeficiency virus—infected patients? J. Infect. Dis. 177 (6), 1723–1726.

Howard-Varona, C., Hargreaves, K.R., Abedon, S.T., Sullivan, M.B., 2017. Lysogeny in nature: mechanisms, impact and ecology of temperate phages. ISME J. 11 (7), 1511–1520.

Iranzo, J., Krupovic, M., Koonin, E.V., 2016. The double-stranded DNA virosphere as a modular hierarchical network of gene sharing. mBio 7 (4), pii: e00978-16.

Iranzo, J., Krupovic, M., Koonin, E.V., 2017. A network perspective on the virus world. Commun. Integr. Biol. 10 (2), e1296614.

Jachiet, P.-A., Colson, P., Lopez, P., Bapteste, E., 2014. Extensive gene remodeling in the viral world: new evidence for nongradual evolution in the mobilome network. Genome Biol. Evol. 6 (9), 2195–2205.

Jo, Y., Choi, H., Kim, S.-M., Kim, S.-L., Lee, B.C., Cho, W.K., 2017. The pepper virome: natural co-infection of diverse viruses and their quasispecies. BMC Genomics 18 (1), 453.

Jo, Y., Lian, S., Chu, H., Cho, J.K., Yoo, S.-H., Choi, H., et al., 2018. Peach RNA viromes in six different peach cultivars. Sci. Rep. 8 (1), 1844.

Kassanis, B., Nixon, H.L., 1961. Activation of one tobacco necrosis virus by another. J. Gen. Microbiol. 25, 459–471.

Kobiler, O., Drayman, N., Butin-Israeli, V., Oppenheim, A., 2012. Virus strategies for passing the nuclear envelope barrier. Nucleus (Austin, TX) 3 (6), 526–539.

Koonin, E.V., 1991. Genome replication/expression strategies of positive-strand RNA viruses: a simple version of a combinatorial classification and prediction of new strategies. Virus Genes 5 (3), 273–281.

Koonin, E.V., Dolja, V.V., 2013. A virocentric perspective on the evolution of life. Curr. Opin. Virol. 3 (5), 546–557.

Koonin, E.V., Dolja, V.V., 2014. Virus world as an evolutionary network of viruses and capsidless selfish elements. Microbiol. Mol. Biol. Rev. 78 (2), 278–303.

Koonin, E.V., Krupovic, M., 2018. The depths of virus exaptation. Curr. Opin. Virol. 31, 1–8.

Koonin, E.V., Yutin, N., 2018. Multiple evolutionary origins of giant viruses. F1000Research 7, 10.12688/f1000research.16248.1.

Krishnamurthy, S.R., Wang, D., 2017. Origins and challenges of viral dark matter. Virus Res. 239, 136–142.

Kumar, A., Murthy, S., Kapoor, A., 2017. Evolution of selective-sequencing approaches for virus discovery and virome analysis. Virus Res. 239, 172–179.

La Scola, B., Audic, S., Robert, C., Jungang, L., Lamballerie, X., de, Drancourt, M., et al., 2003. A giant virus in amoebae. Science (New York, N.Y.) 299 (5615), 2033.

La Scola, B., Desnues, C., Pagnier, I., Robert, C., Barrassi, L., Fournous, G., et al., 2008. The virophage as a unique parasite of the giant mimivirus. Nature 455 (7209), 100–104.

Lander, E.S., Linton, L.M., Birren, B., Nusbaum, C., Zody, M.C., Baldwin, J., et al., 2001. Initial sequencing and analysis of the human genome. Nature 409 (6822), 860–921.

Lefkowitz, E.J., Dempsey, D.M., Hendrickson, R.C., Orton, R.J., Siddell, S.G., Smith, D.B., 2018. Virus taxonomy: the database of the International Committee on Taxonomy of Viruses (ICTV). Nucleic Acids Res. 46 (D1), D708–D717.

Lehnherr, H., Maguin, E., Jafri, S., Yarmolinsky, M.B., 1993. Plasmid addiction genes of bacteriophage P1: doc, which causes cell death on curing of prophage, and phd, which prevents host death when prophage is retained. J. Mol. Biol. 233 (3), 414–428.

Li, Y., Jia, A., Qiao, Y., Xiang, J., Zhang, Y., Wang, W., 2018. Virome analysis of lily plants reveals a new potyvirus. Arch. Virol. 163 (4), 1079–1082.

Lim, E.S., Zhou, Y., Zhao, G., Bauer, I.K., Droit, L., Ndao, I.M., et al., 2015. Early life dynamics of the human gut virome and bacterial microbiome in infants. Nat. Med. 21 (10), 1228–1234.

Lucía-Sanz, A., Manrubia, S., 2017. Multipartite viruses: adaptive trick or evolutionary treat? NPJ Syst. Biol. Appl. 3 (1), 34.

Márquez, L.M., Redman, R.S., Rodriguez, R.J., Roossinck, M.J., 2007. A virus in a fungus in a plant: three-way symbiosis required for thermal tolerance. Science (New York, N.Y.) 315 (5811), 513–515.

Marz, M., Beerenwinkel, N., Drosten, C., Fricke, M., Frishman, D., Hofacker, I.L., et al., 2014. Challenges in RNA virus bioinformatics. Bioinformatics (Oxford, Engl.) 30 (13), 1793–1799.

Mougari, S., Sahmi-Bounsiar, D., Levasseur, A., Colson, P., La Scola, B., 2019. Virophages of giant viruses: an update at eleven. Viruses 11 (8), 10.3390/v11080733.

Naccache, S.N., Greninger, A.L., Lee, D., Coffey, L.L., Phan, T., Rein-Weston, A., et al., 2013. The perils of pathogen discovery: origin of a novel parvovirus-like hybrid genome traced to nucleic acid extraction spin columns. J. Virol. 87 (22), 11966–11977.

Nanfack Minkeu, F., Vernick, K.D., 2018. A systematic review of the natural virome of anopheles mosquitoes. Viruses 10 (5), 10.3390/v10050222.

NCBI Resource Coordinators, 2018. Database resources of the National Center for Biotechnology Information. Nucleic Acids Res. 46 (D1), D8–D13.

Newman, M., 2018. Networks, second ed. Oxford University Press, 793 pp.

Paez-Espino, D., Pavlopoulos, G.A., Ivanova, N.N., Kyrpides, N.C., 2017. Nontargeted virus sequence discovery pipeline and virus clustering for metagenomic data. Nat. Protoc. 12 (8), 1673–1682.

Popa, O., Dagan, T., 2011. Trends and barriers to lateral gene transfer in prokaryotes. Curr. Opin. Microbiol. 14 (5), 615–623.

Popa, O., Hazkani-Covo, E., Landan, G., Martin, W., Dagan, T., 2011. Directed networks reveal genomic barriers and DNA repair bypasses to lateral gene transfer among prokaryotes. Genome Res. 21 (4), 599–609.

Ramírez-Martínez, L.A., Loza-Rubio, E., Mosqueda, J., González-Garay, M.L., García-Espinosa, G., 2018. Fecal virome composition of migratory wild duck species. PLoS One 13 (11), e0206970.

Rampelli, S., Turroni, S., Schnorr, S.L., Soverini, M., Quercia, S., Barone, M., et al., 2017. Characterization of the human DNA gut virome across populations with different subsistence strategies and geographical origin. Environ. Microbiol. 19 (11), 4728–4735.

Rohwer, F., Thurber, R.V., 2009. Viruses manipulate the marine environment. Nature 459, 207–212.

Roossinck, M.J., 2005. Symbiosis versus competition in plant virus evolution. Nat. Rev. Microbiol. 3 (12), 917–924.

Roossinck, M.J., 2010. Lifestyles of plant viruses. Philos. Trans. R. Soc. B: Biol. Sci. 365 (1548), 1899–1905.

Roossinck, M.J., 2011. The good viruses: viral mutualistic symbioses. Nat. Rev. Microbiol. 9 (2), 99–108.

Roossinck, M.J., 2019. Viruses in the phytobiome. Curr. Opin. Virol. 37, 72–76.

Rosario, K., Fierer, N., Miller, S., Luongo, J., Breitbart, M., 2018. Diversity of DNA and RNA viruses in indoor air as assessed via metagenomic sequencing. Environ. Sci. Technol. 52 (3), 1014–1027.

Roux, S., Adriaenssens, E.M., Dutilh, B.E., Koonin, E.V., Kropinski, A.M., Krupovic, M., et al., 2018. Minimum information about an uncultivated virus genome (MIUViG). Nat. Biotechnol. 37, 29. EP -.

Sachsenröder, J., Twardziok, S.O., Scheuch, M., Johne, R., 2014. The general composition of the faecal virome of pigs depends on age, but not on feeding with a probiotic bacterium. PLoS One 9 (2), e88888.

Shah, J.D., Desai, P.T., Zhang, Y., Scharber, S.K., Baller, J., Xing, Z.S., et al., 2016. Development of the intestinal RNA virus community of healthy broiler chickens. PLoS One 11 (2), e0150094.

Shannon, P., Markiel, A., Ozier, O., Baliga, N.S., Wang, J.T., Ramage, D., et al., 2003. Cytoscape: a software environment for integrated models of biomolecular interaction networks. Genome Res. 13 (11), 2498–2504.

Sharon, I., Battchikova, N., Aro, E.-M., Giglione, C., Meinnel, T., Glaser, F., et al., 2011. Comparative metagenomics of microbial traits within oceanic viral communities. ISME J. 5 (7), 1178–1190.

Suttle, C.A., 2005. Viruses in the sea. Nature 437 (7057), 356–361.

Suttle, C.A., 2007. Marine viruses—major players in the global ecosystem. Nat. Rev. Microbiol. 5 (10), 801–812.

Thurber, R.V., Haynes, M., Breitbart, M., Wegley, L., Rohwer, F., 2009. Laboratory procedures to generate viral metagenomes. Nat. Protoc. 4 (4), 470–483.

Tillmann, H.L., Heiken, H., Knapik-Botor, A., Heringlake, S., Ockenga, J., Wilber, J.C., et al., 2001. Infection with GB virus C and reduced mortality among HIV-infected patients. N. Engl. J. Med. 345 (10), 715–724.

Vandegrift, K.J., Kapoor, A., 2019. The ecology of new constituents of the tick virome and their relevance to public health. Viruses 11 (6), 10.3390/v11060529.

van Dijk, E.L., Jaszczyszyn, Y., Naquin, D., Thermes, C., 2018. The third revolution in sequencing technology. Trends Genet.: TIG 34 (9), 666–681.

van Valen, L., 1973. A new evolutionary law. Evolut. Theor. 1, 1–30.

Virgin, H.W., 2014. The virome in mammalian physiology and disease. Cell 157 (1), 142–150.

Waldor, M.K., Mekalanos, J.J., 1996. Lysogenic conversion by a filamentous phage encoding cholera toxin. Blood 272 (5270), 1910–1914.

Waldron, D., 2015. Microbial ecology: sorting out viral dark matter. Nat. Rev. Microbiol. 13 (9), 526–527.

Wegley, L., Edwards, R., Rodriguez-Brito, B., Liu, H., Rohwer, F., 2007. Metagenomic analysis of the microbial community associated with the coral *Porites astreoides*. Environ. Microbiol. 9 (11), 2707–2719.

Weiss, R.A., 2016. Human endogenous retroviruses: friend or foe? APMIS 124 (1–2), 4–10.

Williams, S.C.P., 2013. The other microbiome. Proc. Natl. Acad. Sci. U.S.A. 110 (8), 2682–2684.

Wylie, K.M., Weinstock, G.M., Storch, G.A., 2012. Emerging view of the human virome. Transl. Res. 160 (4), 283–290.

Xu, B., Zhi, N., Hu, G., Wan, Z., Zheng, X., Liu, X., et al., 2013. Hybrid DNA virus in Chinese patients with seronegative hepatitis discovered by deep sequencing. Proc. Natl. Acad. Sci. U.S.A. 110 (25), 10264–10269.

Yeo, G., Wang, Y., Chong, S.M., Humaidi, M., Lim, X.F., Mailepessov, D., et al., 2019. Characterization of *Fowlpox virus* in chickens and bird-biting mosquitoes: a molecular approach to investigating *Avipoxvirus* transmission. J. Gen. Virol. 100 (5), 838–850.

Zárate, S., Taboada, B., Yocupicio-Monroy, M., Arias, C.F., 2017. Human virome. Arch. Med. Res. 48 (8), 701–716.

PART

THE STUDY OF MICROBIOTA AND MICROBIOMES

II

CHAPTER

IDENTIFYING MICROBIOTA: GENOMIC, MASS-SPECTROMETRIC, AND SERODIAGNOSTIC APPROACHES

5

Aristea Velegraki[1] and Loukia Zerva[2]

[1]Mycology Research Laboratory, Department of Microbiology, School of Medicine, National and Kapodistrian University of Athens, Athens, Greece [2]Department of Pathophysiology, School of Medicine, National and Kapodistrian University of Athens, Athens, Greece

INTRODUCTION—THE ROMANTIC PAST

The discovery of the microbial world is directly attributable to microscopy. The progress was incremental: initially discovering and detecting, then observing followed by studying and, last but not least, categorizing. The exhilaration of early days was short-lived, as it became obvious that the optical properties of microbes were ill-suited for observation—as if they possessed innate properties for avoiding observation and staying clandestine under the veil of their miniscule size.

Coloring using dyes proved to be a valuable tool for studying microbes, as their semitransparent, white-gray microscopic appearance was very common, making detection—not to speak of differentiation—a laborious issue. Shapes and morphology initially allowed a crude categorization, which led to taxonomy. But it became apparent that morphology had a limited discriminatory power, or else very similar appearing microbes demonstrated very different properties.

The fixation of cells allowed more accurate description and observation, as well as filing. The advent of dyes increased discrimination manifold, allowing differentiation and improving detection even in high-clutter backgrounds. The problem was still that the dyes were usually toxic and thus necrosing the cells. Consequently, even if said, dyes had in theory differential visual effects associated with the physiology of the cells, the respective processes could not be observed. The—partial—solution to this problem has been the single-field applications, which allow continuous tracking, thus making fixation unnecessary. Video technology furthered this approach considerably, allowing visual tracking and recording, for better and more thorough observation. Still, it took literally centuries to come to that.

The second issue was that morphology, although extremely descriptive (Escobar-Zepeda et al., 2015), could not give any clue as to some vital qualities of the microbes. Tropisms, preferences, metabolism, adaptability, and a long list of biological properties were neither discernible from nor associable to morphological features, and this worsened when realizing that microbial qualities may

Microbiomics. DOI: https://doi.org/10.1016/B978-0-12-816664-2.00005-0

change due to environmental circumstances and also during ontogeny. One and the same microbe may transform and look different under the microscope.

Cultures provided a valuable tool to alleviate such issues. By enabling isolation from a sample, they permitted microbial multiplication to a homogenous population the biological properties of which could be subsequently defined by multiple assays. They could even stimulate the expression of some of these properties by substrate modification, and finally, they allowed observation over time and in diverse environmental contexts, thus elucidating the morphological variability of transformational events. Their most invaluable characteristic was the ability to provide massive amounts of identical cells for further testing after growth under specified conditions. Still, they could not cover all possible conditions describing the transformational potential of a microbe, much less of many microbes, and even less their functional profiles. Moreover, some microbes were not cultivable under testing conditions, especially those living in demanding environments. Being morphologically similar to other species or hidden into their habitats, they remain still unknown.

THE MODERN PEDIGREE

The main volume of resources in microbiology has been earmarked for the detection and classification of known microbes in their natural milieu where they demonstrate a detrimental effect–medicine being one but by no means the only field of such interest. A vast effort has been undertaken in order to detect them specifically and sensitively (in very low population numbers) under conditions of significant homologous (microbial) and heterologous (environmental, sensu lato) background. Possibly the best tool in this effort has been the bioengineered (once more sensu lato) antibodies, or parts of them, manufactured for the detection of a specific 3-D entity (epitope) of a given molecule or formation (antigen). This approach selects specifically for antigen-bearing cells within high clutter, aggregates them into clusters, occasionally insoluble and thus easier to isolate. But, more importantly and widely applied, it is possible to label these antibodies with a marker molecule, be it a chromo-enzyme (Dubois-Dalcq et al., 1977), a radioactive (Yalow, 1980; Towbin et al., 1992) or fluorophore (Coons and Kaplan, 1950) moiety, or any other conceivable signal- or beacon-generating entity. Occasionally, such a staining system may be one part of a very complex chain of effectors, as in sandwich ELISA (Maiolini et al., 1978; Adams and Clark, 1977) or indirect immunofluorescence assays so as to reveal antigen presence and position.

As each antibody detects a single epitope, an antigen may be detected by a number of different antibodies targeting different epitopes on it, if the latter are produced by animal immunization (polyclonal antibodies). The monoclonal antibodies (Köhler and Milstein, 1975) were a quantum leap allowing the production of molecularly identical antibodies against the same epitope in large amounts and over long and repeated production runs over time. This progress reduced serological cross-reactions drastically, and deterministic and probabilistic models may determine the probability, or rather the possibility of two identical 3-D epitopes occurring on different antigens within the same sample, clinical or environmental. Serial steps of purification result in a diminished probability of cross-reaction to heterologous antigens.

As morphology, physiology/biochemistry and even the serological signature of a microbe are being subject to change due to environmental and genetic factors, while structural similarities to counterparts of the environment and other microbes are frequently observed, a more constant but just as variable and discriminatory signature was sought for the definitive recognition/identification of microbial cells. The genetic material, mainly DNA, proved to be a very good candidate. Most of it, the core DNA (positioned in the nucleus of eukaryotes or forming the nucleoid in bacteria) is very stable irrespective of form, function and condition of the cell or environmental parameters. The discriminatory potential of DNA between closely related cells allowed such detail in categorizing and then characterizing them, that it became the defining factor in taxonomy. Thus molecular biology provided the ultimate diagnostic and analytical approach, which could reveal the presence of a microbe, identify it by comparison of signature sequences to recorded ones and even project functional aspects by detecting the presence of defining or contributing genes.

METAMODERNISM: THE CHANGING ENVIRONMENT

The initial reaction to the discovery of the tiny, living particles was their christening as *microbes* (Fig. 5.1); a clear projection of the understanding of their nature but also of the awkwardness associated with them in regard to their identities and position within the Creation—what was later to be considered as *niche*. The term *microorganisms*, defining clearly their fully autonomous nature and suggesting the inter- and intracellular organization and physiology, was undertaken by life sciences readily, as accurately and properly descriptive while BioMedicine insisted on the term *microbe*, as microbiology was but a tiny part of its scope—a word readily adopted by the nonspecialist audiences. The quantum leap has been their recasting as *microbiota*, a word clearly implying their functional position within collective entities in which they are interacting as units either in a collaborative way, or in an antagonistic one—the particulars of which may change over time and due to any number of conditions. The *biota* were part of respective-*biomes*, namely of the microbiome, further subdivided to bacteriome, archaeome (Probst et al., 2014), mycobiome (Ghannoum et al., 2010), virome (McDaniel et al., 2008) and are a direct result of the omics revolution, combined with brand new, albeit already established (at the time) views of microbiology, such as the DNA-sanctioned division of prokaryotes to Prokarya and Archaea (Woese et al., 1990).

The microbial sciences have since taken a different course, enabled by technological and methodological progress, but sanctioned primarily by the new concepts of *-biota/-biomes*. The much more integrated, communal (instead of collective in the near past) aspect of the microbiota actually triggered an interest in the individual cells of a population, and that not only in a sociological sense, but also resulting from viewing microbial populations as striving to form multicellular contexts, in many cases of a chimeric setup (Solé and Valverde, 2013).

The coming decades will witness an exponential increase of workload in terms of analysis and especially identification of microbiota. The spontaneous emergence of new microbiota, the engineering of even more, the growth of biotechnology as a more sustainable processing form for the satisfaction of many needs, all these fields and many more (Kambouris et al., 2018a,b) require monitoring of implicated microbiota (Chapter 15), where applicable. An example to the differentiation and specialization of such applications are microbial inocula implanted to clear organic leftovers in hardware used in the processing industry. But to a far greater extent than monitoring, the awareness, identification and containment of contaminant microbial populations or even single cells

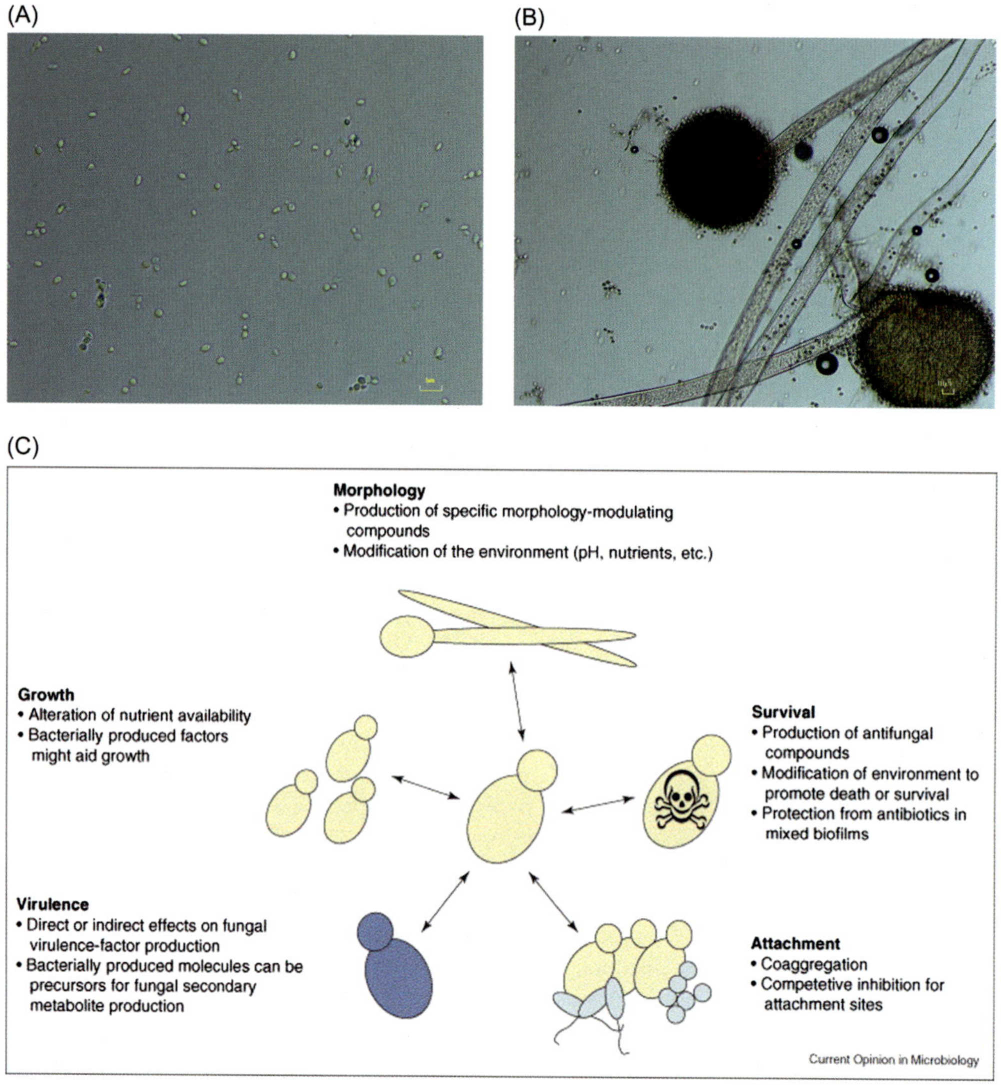

FIGURE 5.1 The conceptual and natural pathway from *Microbes* to *Microbiota*.

(A) *Microbes*: small, diverse, unseen by naked eye particles showing characteristics of lifeforms (yeast cells of *Candida parapsilosis*)...; (B) *Microorganisms*: ... which may bear individual and social organization of different levels and kinds (stacks, conidial heads and conidia of conidiophores of *Aspergillus niger*)...; and (C) *microbiota*: ...and associate in diverse ways with similar and dissimilar lifeforms, to shape environment into a niche.

(C) From Wargo, M.J., Hogan, D.A., 2006. Fungal-bacterial interactions: a mixed bag of mingling microbes. Curr. Opin. Microbiol. 9, 359–364; with permission.

is required. The latter may multiply to form a population which infringes on or infiltrates beneficial microbiomes, be they pure or multistrain/multispecies ones.

The above context, combined with the increased volume of exploitations of microbiota for every plausible application, may result in a demand for the aforementioned services which projected microbiological and microbiomic assets may never satisfy. To address this inadequacy, automation, and interventionism, coupled to decentralization and networking seem to be the operational mix offering the best prospects. Accordingly, operational options, the RAPID device of Idaho Technologies at 22 kg total weight being a prime example (Christensen et al., 2006) while they were previously reserved for very few, highly privileged users, nowadays they can be acquired by modest operators for very reasonable prices. The 2MoBiL field polymerase chain reaction (PCR) system (Siamoglou and Patrinos, 2019) is a very good example: developed initially in a biodefense context with strict protocols rigidly developed and optimized for less than two dozen microbial agents, it has evolved to an affordable portable instrument, with enough flexibility to incorporate interfaces so as to be deployed and operated in austere environments and lugged to local or central headquarters for reporting and support. Even more to the task, portable sequencers with a very compact footprint for presequencing processing of samples and after-sequencing data mining and exploration may revolutionize any application based on genomic analysis.

These newcomers were preceded by a previous generation of infrastructure-independent, point-of-care diagnostics, based on enzymes but primarily on serology (Piarroux et al., 2019), much alike pregnancy tests. Even before that, field microscopes had been introduced (Goldman and Sawyer, 1959) to assist decontamination and eradication campaigns during a golden age of public health initiatives.

Additional requirements in the microbiomics context relate to the ability to watch events in vivo, and to track processes, either in real time or by using recording media, which allows manipulation of speed to fast-forward or to pause, according to requirements. The in vivo tracking and observation were a prerogative of cultures and remained limited to visual or at most stereoscopic microscopy, with little dye enhancement and very limited magnification, unsuitable for monitoring intracellular processes. Serology can assist microscopy by marking cells, without degrading them, making it possible to keep track of them and thus dissect population-scale events and processes. Moreover, ontogenetic pathways in nonlinear development may be chartered, while intermingled development of mixed species populations in given (micro)environments may be resolved. And, of course, the precise sequence of events in known intercellular and host–symbiont interactions may be crucial for the development of a multimicrobial infection or of the incurrence of self-infections and infections either complicated or contained or even resolved due to resident microbiomes in any kind of host.

The expansion of the microbiome in space and influence over the previous conceptual "microbial communities" will result in an increased need for tackling unknown microbes in an agnostic setting (Kambouris et al., 2018b), which means without a hypothesis on their affiliation so as to use specific reagents and approaches. Up to—a most relative—"now" the focus was to detect specific entities, usually at the species but occasionally the genus or, alternatively, the variety/strain levels in a range of contexts and samples, such as clinical samples, stored goods, processing or processed foodstuff, beverages, fodder, raw materials, degrading artifacts, and environmental samples including urban/built environment. Detection and identification were in cases separate, but interdependent events and sometimes polyphasic approaches (Schoch et al., 2012; Prakash et al., 2007) were adopted. Their overlapping data generation was able to provide confirmation for each other and cover identification gaps present in each particular phase. Still, barcoding, the standard and proven approach of identifying and classifying isolates with one, standard technique or a series of techniques remained the Holy Grail of Microbiology (Schoch et al., 2012). This was so in the Golden Days of Microscopy (Sogin et al., 1972), when microscopy, assisted by Identification Keys, Guides and Manuals, reigned supreme and is still so as Genomics take over, replacing microscopes with sequencers as well as keys, guides, and manuals with databases (Popescu and Cao, 2018).

The new requirements call for descriptive examination and analysis, meaning identification of all present microbiota and possibly exogenomes (Kambouris et al., 2018a) within a sample, as the potential for detailed identification of OTUs (operational taxonomic units) under suitable conditions seems inherent in different sample types. The paradigm shift of microbiology, through Metagenomics, is the leap from *barcoding* an isolate (Schoch et al., 2012; Sogin et al., 1972) to *metabarcoding* of a population (Elbrecht and Leese, 2015; Popescu and Cao, 2018). This need for detecting and identifying all present microbiota and then mapping them on interaction networks is the only way to truly dissect and describe a microbiome and is agnostic to its core, thus necessitating specific experimental setups and data confirmation and exploitation steps. Such process constitutes the ultimate challenge, in terms of the first quarter of 21st century and is supposed to overcome current bottlenecks in the attempt to revolutionize the field, not only through its capabilities and inherent opportunities but—mainly—by its sheer volume of application and level of lateral integration.

That is not to say that the current concept of selective (as opposed to descriptive) analysis will be redundant. Far from that, high volumes of samples with emerging and novel microbiota will be needed for the actual, routine, and productive applications of biotechnological and diagnostic microbiology, as opposed to research microbiology that will be tuned to descriptive efforts as just mentioned. The just mentioned facts and prospects directly imply that a large part, if not the larger, of the projected analyses will be of the selective brand.

All the above, both qualitative and quantitative terms (the former implying performing the analysis and the latter referring to processing the volume of analyses requested), require very high rates of data processing, ideally using improved, more efficient algorithms. The volume of data produced by the new generation of methodology applied in a massive context will reach the Big Data level fast and easily, and the need to fuse data and compare entries will increase technical requirements exponentially. It is a fact, not a projection, that the volume of streaming data will be overwhelming if high distribution rate, merging, fusion and automation of processing are not adopted, and preferably from the onset of the development of an application or technology. Interfacing will be a primary requirement in any analysis or mining tool, being as vital as specificity, sensitivity, discrimination, and other front-end metrics have been proven to be during the last two decades. The Internet-of-Things as a part of machine-learning efforts will be the only way to seamlessly integrate complementary techniques in variable contexts and high volumes of queries, despite the uneasiness of many scientists and even more other stakeholders whenever the touch and feeling of raw data becomes inaccessible or rather incoherent and/or irrelevant and unfathomable for human operators (Özdemir, 2018; Özdemir and Hekim, 2018).

METAMODERNISM: THE METHODS

Microbiomics studies are prone to become multiphasic, using more than one applicable front-end technology in any context, be it in vivo, ex vivo, in situ, in vitro (https://www.novoprolabs.com/

Table 5.1 Characteristic Properties of Methodologies for Studying Microbiomes and Microbiota

Methodology	Properties
Mass spectrometry	Analysis of qualitative and quantitative molecular content
Microscopy	Delineation of form, color, shape, and localization/departmentalization
Culture	Study of selected developmental sequences/processes with their distinct physiological and morphological features
Immunoassay	Specific recognition of small-molecule-sized 3-D entities
Nuclear acid analysis	Definition of genetic programming of morphology, physiology, qualitative macromolecular context and developmental pathways (growth-adaptation)

Table 5.2 Methodologies for Studying Microbiomes and Microbiota and Characteristics of Their Applicability

Methodology	Agnostic	Time-Sensitivity (Sample Alteration-Degradation)	Portability	Low TAT
Mass spectrometry	Yes	Yes	Yes	Yes
Microscopy	Yes	Yes	Yes	Yes
Culture	No	Conditional	No	No
Immunoassay	No	Yes	Yes	Yes
Nucleic acid analysis	Yes	No	Yes	Yes

TAT, *Turnaround time.*

support/articles/what-is-the-difference-between-ex-vivo-in-vivo-and-in-vitro-201802261300.html) in order to acquire data for back-end processing (bioinformatics/in silico). The whole concept becomes complicated, as different combinations are needed for different occasions (Tables 5.1 and 5.2). Whether all needed expertise, equipment, and infrastructure can be concentrated in a single or in different laboratory units, conceivably spaced quite apart, which will be required to cooperate on an everyday basis, and some of them having to content with virtual samples or metasamples (files of data describing qualities and properties in the absence of actual samples), all these represent just some of the several issues that must be tackled. Additional such issues include, without being restricted to, defining and describing which metasamples are to be acceptable for further analysis and which lack the relevant specifications and conditions rendering them acceptable.

A good example stems from the need to resolve infections. The usual approach is to try to obtain the pathogen involved by sampling. Quite often this translates into the occurrence of a few pathogenic cells among a great number of host cells, a fact easily resulting in false-negative results. The specificity of serological approaches allows labeling, enrichment, and collection of such rare cells while at the same time identifies—possibly with some kinetic help (Bass et al., 2017)—molecular entities, such as toxins, which do not contain DNA/RNA and are very hard to extract and enrich from the site of their in vivo secretion. Since cells of microbiota in an infectious disease context (and even more if under therapy) neither present standard metabolomic profile and physiology (thus negating a structure-based detection approach), nor retain their typical morphology, the consistency of molecular approaches, especially genomic analysis and/or fingerprinting/barcoding is warranted (Escobar-Zepeda et al., 2015). Cultures, or even culturomics (Lagier et al., 2012; Kambouris et al., 2018c), described in detail later (Chapter 8), are very slow for such an application, but complex therapeutic regimens, especially if not entirely based on conventional drugs, but including electroceuticals (Chapter 14) and/or biopharmaceuticals, can only be tested in culture (Kambouris et al., 2018c).

At least for the time being, readouts of a genome, although in the near future extractable from single cells, do not reveal the sequence of cellular events. They allow a very accurate projection of virulence factors, from antibiotic and other resistance to toxin, slime, and capsule production; thus toxicogenomics and other translational genomics do apply, but the quantum leap is expected to be Biphasic Pharmacogenomics, an in situ, combined application which will explore and examine two very different genomic entities and draw conclusions by a series of associations between them. An indicative such association is, of course, with the current patient environment (indicated but not definitively determined through his genome) for establishing the projected safety profiles of candidate drugs and, at the same time, pathogen susceptibility, to create the projected efficacy profiles for the implicated strain, thus compiling a panel of candidate drugs. This approach partially negates or, at least, lowers the priority of pathogen identification by directly assessing its vulnerabilities (instead of inferring them by virtue of identification) and thus drastically decreasing the time period of empirical administration of antimicrobials.

In any case, for disease awareness (and modeling) in situ, tracking of single cells is a must to assign metagenomes to observable entities, especially if there are a number of candidate carriers to be assigned to candidate signatures. A combination of culture- and structure-based methods may reveal an identity where consensus metagenomics cannot, and may furnish functional data, which shotgun metagenomics may, but also may not, be able to produce (Escobar-Zepeda et al., 2015). Once more in situ microscopy may reveal the interactions among microbial and host cells, the processes and formations caused by their initial and their prolonged contact and the—possibly focal—roles of noncausal cells and entities (Feng et al., 2018). The latter include, without being restricted to, inflammatory responses, foreign objects and bodies and/or malignancies, the presence of which may be complicating baseline infection events. All the above are not easily discernible, nor mechanistically associated with infection, its development and prognosis. Similarly, if the concept of pathogenicity and host entity are deleted, the capabilities, processes, and output which emerged in the previous example can be readily applied to environmental issues focusing on ecosystem shaping and, at a later stage, engineering (Solé and Valverde, 2013).

As is obvious, an image, and even more a series of images and streaming videos, demonstrates much more clearly the spatiotemporal sequence of events than indirect metabolome/physiology observation and analysis. This is even more so if accurate, high-fidelity simulations can be created and visually observed and tracked. At this point microfluidics may enter the fray, allowing a high-fidelity simulation by challenging low populations or single cells by serial stimuli of selective order, intensity, and level of concurrence (Yarmush and King, 2009; Hong et al., 2017) to map responses and adaptability and finally define susceptibilities and vulnerabilities.

MASS SPECTROMETRY

Mass spectrometry (MS) produces an extremely detailed list, in quantitative and qualitative terms, of the macromolecular makeover of an isolated organism or of any other sample. It does not reveal, nor is affected by, the exact distribution of such molecules in spatial terms, that is their location, interactions, and bonds. Needing a rigid common denominator of sample preparation and testing conditions to normalize ontogenetic/developmental differentiations in the macromolecular context, it is at its best in fanning out details between similar and/or closely related taxa, or detecting and discovering cryptic species—converging anamorphic species in mycology is a good such example. All this is achieved fast and affordably and, in some cases, in portable formats (Prieto et al., 2002).

On the other hand, virulent strains and lineages of usually harmless species are less easy to detect and discriminate, as there are many pathogenicity-defining factors, the expression of which may be conditional. Thus, in principle, genomic scanning and analysis seem preferable, despite the ability of MS systems to detect single-base substitutions and other subtle alterations, once they are informed by other approaches on the whereabouts in chemical and cellular terms and significance of the said alterations.

The method presents difficulties with mixed samples or not well normalized pure samples. It is really plagued with even more challenging agnostic applications; cutting edge methods, as MALDI-TOF and electrospray ionization are able to accomplish an objective molecular scan from an environmental or multispecies-containing sample (Wieser et al., 2012). This scan though records the molecular context of the sample and does so with adequate resolution. But the molecules are grouped-and that not always exclusively-in multiple patterns and thus both form and define different cells and organisms; such datasets are limited in number and variety and occasionally proprietary. Big Data might be able to correctly segregate readouts of multispecies samples, but these metataxa will always be a feat of mix and match rather than of accurate discrimination. Furthermore, the readouts are conditional to existing standard curves. Thus, by definition, the methods cannot tackle an unknown agent by projecting a novel curve pattern over similar datasets, since phylogenetic diversion is not always proportional to molecular variability, both qualitatively and quantitatively (Emami et al., 2016; Ferreira et al., 2011). Still, once a pure sample is read and recorded and the pattern is informed by any other method to a taxon, next encounters will be automatically identified and previous ones hopefully resolved and attributed. Such retrospective attribution might imply a continuous reconsideration of recorded data that poses a serious problem of keeping databases and recorded entries current and valid. Even more challenging, and much more important, is the task to continually, or even occasionally review and confirm/validate or amend exported resources, as in publications and guidelines.

IMMUNOASSAYS

Cross-reaction is an inherent feature of immunoassays (IAs) (Schroeder and Cavacini, 2010; Vojdani and Cooper, 2004; Vojdani, 2015), and is expected to exacerbate as more microbiota emerge. Most tests, based on immunoagglutination (Kamiya et al., 1986) or immunolabeling (Bass et al., 2017) show impressive specificity. Sensitivity in many formats, such as, but not restricted to, sandwich ELISA (Adams and Clark, 1977) is satisfactory, at the very least.

Although conventional antibody-based methods cannot tackle novel agents, artificial or natural, should pure samples of said novel agents be available, an epitopic profile may be compiled by antibody microarrays containing thousands of paratopic loci. In such a case the collective signal may compensate for the perceived lack of specificity of the individual ones, providing a better alternative to differential detection of known pathogens than current formats with mutually exclusive and highly specific loci. By fusing the binary signals of the reacting array loci, collective epitopic imaging may be achievable. Thus typing becomes possible even for novel agents. The relation between phylogenetic distance and immunoprofile is neither proportional nor causal, and so in silico methods cannot correct any raw data to mine relationships in a reproducible way, in order to extend applicability from typing to identification/classification.

On the other hand, IAs are uniquely suited to field operations and point-of-care applications. Many IA-based tests neither require powered instrumentation nor other infrastructure to work, nor refrigeration, or other active conservation method; thus they are independent of electricity. Furthermore, IAs are adequately specific in detecting target epitopes in the high clutter of degraded, mixed, and impure samples (Bartholomew et al., 2017; Morel et al., 2012) and are thus able to definitively identify known pathogens under such conditions (Ramage et al., 2016).

Obviously, the conventional use of purified antigens in order to develop monoclonal and polyclonal antibodies (Vojdani, 2015) still is and will remain a valid approach for microorganism detection and identification, especially if drastic curtailing of the time required for antibody production can be achieved. Even for novel microbiota, the approach makes sense since said microbiota are likely to persist in the environtome practically indefinitely; for their fast diagnostics and effective therapeutics the commercial development of antibody reagents seems a sound investment.

Fast-track immunology is prone to tackle emerging pathogens as mentioned above, under the assumption that they express temporally stable epitopes, which do not degrade and remain unaltered over time or over successive generations. This format is a medium-term solution as it takes a minimum of 1–2 months to develop monoclonal antibodies de novo (Ouisse et al., 2017). In addition, double that time is required for assay deployment and distribution. Furthermore, IAs offer a cardinal advantage: they are likely to detect biotoxins (Velegraki-Abel, 1986) and presumably other acellular agents undetectable by nucleic acid amplification tests (NAATs), such as prions and possibly viroids (Petrovick et al., 2007; McCutcheon et al., 2014).

An outbreak caused by unheard-of pathogens cannot be dealt with serology diagnostics, as antibody development and production is a relatively slow process. Still, with the possible exception of malevolent, perpetrated events (see Chapter 15), the prospect for using serodiagnostics for the prompt detection of novel pathogens should perhaps be considered in a case-by-case basis. This prospect arises due to the possibility of cross-reactions which, if over a given threshold of binding strength and specificity, may allow retasking of an antibody against a novel strain or, in a less dramatic development, the use of a given antibody against novel species phylogenetically close to known ones.

GENOMICS

The term *Genomics* has been introduced since the turn of the century; still, its predecessor, the "molecular" detection/identification of pathogens—actually the nucleic acid analysis (NAA)—had already been established as a valid, if not advantageous approach once Molecular Biology came of age, and the more so as NAATs were coming into vogue as the most handy and multitasking versions of NAA. Their merits were the detection and identification of facultatively reproducing bioagents, even noncultivable/fastidious ones, even in extremely low concentrations, and all that is achievable in a short turnaround time (TAT) compared to culture-based methods. Furthermore, the methodology shows high specificity, and imperviousness to phenotypic shifts and antigenic variation (Escobar-Zepeda et al., 2015; Velegraki et al., 1999a,b).

Detection by nucleic acid signature is straightforward, but simple detection—or the lack of it—does not establish the presence—or absence—of a microbe, as at least two and usually multiple genomes coexist in clinical and environmental samples. Thus analyzing a nucleic acid sequence to achieve classification or identification presupposes the attribution of a number of known/existing similar/analogous sequences to different organisms so as to allow comparisons (Doggett et al., 2016). Consequently, molecular identification of microbiota and diagnosis of infections, required microbiote sequences extracted from pure cultures to be used as targets for oligonucleotide PCR primers and probes designed accordingly, and cross-tested with abundant genomes (i.e., of hosts) so as to establish specificity. Further noncompetitive or competitive testing including clinically or phylogenetically relevant taxa—depending on the application—determined specificity and discriminatory power (Kambouris et al., 1999). Once new modalities, exemplified by next/new generation sequencing (NGS), started mass-producing sequencing data from diverse organisms that were accessible in public depositories, the previous process was simplified and implemented in silico instead of massive testing of diverse isolates (Kambouris, 2009; Wang et al., 2005).

NAATs are characterized by the capability to detect/identify dormant and fastidious agents, exemplified by viruses and latent forms of cellular microbiota (spores, conidia, and cysts). They thus became the preferred CIDT (culture-independent diagnostic tests) (Doggett et al., 2016). Overtime, the informational context of NAATs increased in depth and scope. An important field of improvement has been the ability to quantify results, initially in relative terms within a batch of samples or compared to a control run, and later on in absolute terms with the use of proper standards and processing. The main—but not exclusive—vector of this improvement has been RT-PCR (real-time PCR) (Doggett et al., 2016; Sapsford et al., 2008; Kambouris, 2010).

Another such area of improvement refers to the TAT as already mentioned, which was drastically shortened due to high throughput. It was a result attributable not so much to decreasing the implementation times of the actual assays, as to the establishment of reliable, massively parallel formats, which analyze large numbers of samples simultaneously. This was brought about initially by multiplexing, which is the similar but not identical coprocessing of multiple analytes present within the same sample, either in mixed or in pooled samples (Deshpande et al., 2016). Initially, multiplexing had been a prerogative of PCR (Dubois-Dalcq et al., 1977; Lin et al., 1996) and then of microarrays (Pastinen et al., 2000), to be promptly superseded by a combination of the two (Wang et al., 2005; Song et al., 2005), a development that actually spawned the term "genomics" in wet lab context. The ultimate multiplex format is the NGS, in both *consensus gene* and *shotgun* metagenomics and metatranscriptomics (Escobar-Zepeda et al., 2015), as described in Chapter 7. The performance of successive iterations of platforms and chemistries, combined with a drastic drop of cost, allows resolving multimicrobial samples and microbiomes down to species level, whether striving to describe the biodiversity of soil or other environmental samples, or deciphering an appendage microbiome colonizing or parasitizing human, animal, or plant hosts (Popescu and Cao, 2018; Wargo and Hogan, 2006).

Unfortunately quantification and high throughput were never convincingly combined to an integrated approach. Microarrays were able to tackle simultaneously four-digit singular queries (Wang et al., 2005; Doggett et al., 2016), although redundancies were unavoidable, to increase reliability. At the same time, they could be used in a semiquantative way rather easily and with proper controls they aspired to quantification as well (Taniguchi et al., 2001). However, they proved too cumbersome and went speedily out of fashion within a decade, despite offering genetic discretion, a cardinal advantage in social and legal terms. Any number of loci can be interrogated, but the actual interrogation is limited to relevant ones, thus negating exposure of the sum of the genomic background of the individual, the privacy of which is otherwise compromised and raises multiple legal and ethical issues (Kambouris, 2016).

Novel and modified molecular techniques have been developed to improve reliability and enhance sensitivity and applicability, based on innovative chemistries for synthesis, hybridization, and labeling on one hand and improved incubation and detection hardware on the other. Still, they are being introduced into the routine of diagnostic laboratories slowly and hesitantly. This is due to the domination of PCR over the extensive and diverse range of new molecular techniques and methodologies that discourages moving to new approaches and due to the requirements in terms of expensive dedicated instrumentation, and of support and logistics. The latter include diverse support including, but not restricted to, maintenance, stockpiling, and managing inventories of consumables and reagents as well as expert personnel. The need for standardization, if not of techniques, at least of operating procedures and data presentation, curation, and recording to enable interlaboratory exchanges, comparison, cooperation, and interaction, further complicates the choice and applicability of these new and promising, yet of unproven value, methods (Kyriazis et al., 2014).

A spin-off of NAATs, the isothermal assays are characterized by chemistries, which are easier to miniaturize and to dissociate from the supply of electrical power and other infrastructure, thus being more suitable for portable, and/or instrumentation-free assays. This implies a much easier pitch as point-of-care tests for affordable in situ processing of samples and subjects (Euler et al., 2013; Doggett et al., 2016; Mayboroda et al., 2016; Rohrman and Richards-Kortum, 2012).

MICROSCOPY

Microscopy remains a quintessential parameter for medical diagnosis even now, and in environmental contexts the identification of fungi through conidial morphology (Saccardo, 1882–1906) is still widely used. Although currently in steep decline, at least relative to its Golden Days, this is not due to a methodological problem but rather to one of vogue and specifications.

Quite the opposite: a number of improvements in microscopy proper (dark/bright field, phase contrast, polarization, different spin-offs of fluorescent microscopy) and of labeling methods (antibody-conjugated fluorophore, immunofluorescence, which improves functional and spatial accuracy) and dyes have increased its usefulness and applicability with even more dedicated formats. Thus molecule/organelle specific dyes (as is cotton blue) were supplemented by dye systems that stain differentially cell components and/or cells (i.e., Gram stain) and in some cases may confer etiologic diagnosis of infection (as does the Ziehl–Neelsen stain for mycobacterial infections). The capability of some dyes to actually necrose processed cells may not only infringe with continuous tracking and in some cases with fidelity but also enhances safety during sample processing, especially when performed under pressure, as it practically sterilizes them in real time (Vyzantiadis et al., 2012).

On the other hand, phase contrast microscopy and laser scanning confocal microscopy allow observation of living cells, the latter with up to 200 nm 2-D resolution in very thin layers which can be reconstructed to produce a 3-D aspect.

Microscopy remains difficult to master, physically hard to perform for any length of time and demands patience, persistence, and insistence. In contemporary context, it still presents spin-offs based on magnification and optical means (visual, electronic, and probed) as well as on optical pathway (transmission and reflection/scanning). Reflection/scanning modalities output images that display true or computational 3-D context with detail. Transmission modalities produce internal and external aspects of 3-D entities either in compressed projection or in resolved multilayer frames. Proper staining allows detection, tracking, and discrimination of observable entities by differential images in terms of position and content of an entity due to the affinity of the stain, be it direct or indirect (Avrameas, 1969).

A number of dyes and the modern spin-offs of visual microscopy, along with atomic force microscopy (Shi et al., 2012)—as opposed to electron microscopy—allow in vivo observation of cellular motion and processes and subcellular entities (Liu et al., 2018; Nishiyama and Arai, 2017) without necrosing or even degrading the cells. This has been achievable by, without being restricted to, phase contrast microscopy, which is a transmission microscopy amenity, and laser scanning confocal microscopy, which is a scanning amenity.

Capitalizing on this in vivo observation microscopy modality, the just as benign and much more effective current approaches are prone to jump-start the matching of genomic variations and taxonomic grades to functional and physiological differentiation (Culibrk et al., 2016; Xu et al., 2017). In some contexts, said approaches provide a dynamic, continuous track of the motion and form of developing and reacting entities, all these in real time considering the cell life. The above in a microbiomic context translate to the continuous usability and usefulness of microscopy (El-Kirat-Chatel et al., 2017; Guo et al., 2017; Culibrk et al., 2016; Leveau et al., 2018). On the contrary, among the other four methodologies, only culture provides a continuous track of developing living entities but does that in the context of a more or less interacting and interdependent aggregation of cells, in cases forming an organism, not of a single isolated cell.

In a different but parallel evolutionary line, superresolution microscopy (Dersch and Graumann, 2018; Spahn et al., 2017) and then nanoscopy (Spahn et al., 2017) emerged. They include microscopic amenities and formats that allow extremely high discrimination and magnification levels, all the way down to a single molecule, defining thus the single molecule localization microscopy (SMLM) (Spahn et al., 2017). Different fluorescence amenities (Dersch and Graumann, 2018) pushed down observation limits to macromolecule aggregations enabling both superresolution and SMLM.

Scanning probe microscopy (SPM), the continuous evolution of atomic force microscopy (the most renowned species of the SPM ilk), is not only able to provide images of surface in 3-D, much like scanning electron microscopy, but also to evaluate molecular and cellular interactions (as in ligand-receptor binding force), thus accelerating the fields of nanobiology and nanomedicine (Shi et al., 2012). Electron microscopy, using electrons instead of light beams, provided—and still does—unprecedented magnification and detail but has always been a category of its own, cumbersome, demanding and lacking flexibility.

FTIR (Fourier-transformed infra-red) microscopy increases the discriminatory power of the process and detects cells and tissues infected by fungi, bacteria, and viruses. Ideally it allows identification of microbes at species level (Erukhimovitch et al., 2007); it is fast, relatively straightforward, affordable, and massively applicable for the dynamic analyzing prescribed by microbiomics. Raman spectrometry in a microscopy format (Smith et al., 2016), on the other hand, is also very fast, and discriminates species of the same genus, an ability simply invaluable when mutated microbes (engineered or spontaneously evolved) must be detected and identified within similar populations of ancestral cells.

CONCLUSION: A PEEK OF THE FUTURE

Smart and extremely enhanced microscopic approaches allow nondestructive single cell tracking and observation of subcellular structures, events, and processes (Liu et al., 2018), thus directly substantiating processes and interactions along with localized physiology and genomic events. Genomics delineate irrespective of adaptive transformations the true identity of a cell, while revealing genetic circuitry and blueprints indicative of its physiology and ecology. Cultures provide the only means to properly confirm developmental and physiology wiring and the connection of all relevant signatures to morphology, while IAs and serology specifically identify structural patterns in high clutter and implement labeling, which, along with a dedicated detection methodology (microscopy or other imaging solutions) may locate, differentiate and even segregate or enrich cellular populations of interest (or even acellular ones, exemplified by but not restricted to viruses) within multispecies and multistrain samples. Last but not least, MS is able to differentiate morphologically and even genetically similar taxa by qualitative and quantitative comparison of their macromolecular context. Combinations of these methodologies in terms of priority, specific weighting of results, and correct choice of available assays and procedures/protocols may represent the proper course of action to address different needs. The basics of this approach have already been validated by the polyphasic-identification concept (Schoch et al., 2012; Arabatzis et al., 2011). Still, the feasibility and sustainability of thus processing massive numbers of samples may require imaginative logistics and administrative initiatives. Contrary to previous administrative guidelines, which insisted on the choice of a single method, the future points to the choice of combination methodology on a case-by-case basis, while a full range of techniques must be available, to call upon.

REFERENCES

Adams, A.N., Clark, M.F., 1977. Characteristics of the microplate method of enzyme-linked immunosorbent assay for the detection of plant viruses. J. Gen. Virol. 34, 475–483.

Arabatzis, M., Kambouris, M., Kyprianou, M., Chrysaki, A., Foustoukou, M., Kanellopoulou, M., et al., 2011. Polyphasic identification and susceptibility to seven antifungals of 102 *Aspergillus* isolates recovered from immunocompromised hosts in Greece. Antimicrob. Agents Chemother. 55, 3025–3030.

Avrameas, S., 1969. Indirect immunoenzyme techniques for the intracellular detection of antigens. Immunochemistry 6, 825–831.

Bartholomew, R.A., Ozanich, R.M., Arce, J.S., Engelmann, H.E., Heredia-Langner, A., Hofstad, B.A., et al., 2017. Evaluation of immunoassays and general biological indicator tests for field screening of *Bacillus anthracis* and ricin. Health Secur. 15, 81–96.

Bass, J.J., Wilkinson, D.J., Rankin, D., Phillips, B.E., Szewczyk, N.J., Smith, K., et al., 2017. An overview of technical considerations for Western blotting applications to physiological research. Scand. J. Med. Sci. Sports 27, 4–25.

Christensen, D.R., Hartman, L.J., Loveless, B.M., Frye, M.S., Shipley, M.A., Bridge, D.L., et al., 2006. Detection of biological threat agents by real-time PCR: comparison of assay performance on the R.A.P.I.D., the LightCycler, and the Smart Cycler platforms. Clin. Chem. 52, 141–145.

Coons, A.H., Kaplan, M.H., 1950. Localization of antigen in tissue cells; improvements in a method for the detection of antigen by means of fluorescent antibody. J. Exp. Med. 91, 1–13.

Culibrk, L., Croft, C.A., Tebbutt, S.J., 2016. Systems biology approaches for host-fungal interactions: an expanding multi-omics frontier. OMICS 20, 127–138.

Dersch, S., Graumann, P.L., 2018. The ultimate picture — the combination of live cell superresolution microscopy and single molecule tracking yields highest spatio-temporal resolution. Curr. Opin. Microbiol. 43, 55–61.

Deshpande, A., McMahon, B., Daughton, A.R., Abeyta, E.L., Hodge, D., Anderson, K., et al., 2016. Surveillance for emerging diseases with multiplexed point-of-care diagnostics. Health Secur. 14, 111–121.

Doggett, N.A., Mukundan, H., Lefkowitz, E.J., Slezak, T.R., Chain, P.S., Morse, S., et al., 2016. Culture-independent diagnostics for health security. Health Secur. 14, 122–142.

Dubois-Dalcq, M., McFarland, H., McFarlin, D., 1977. Protein A-peroxidase: a valuable tool for the localization of antigens. J. Histochem. Cytochem. 25, 1201–1206.

El-Kirat-Chatel, S., Puymege, A., Duong, T.H., Van Overtvelt, P., Bressy, C., Belec, L., et al., 2017. Phenotypic heterogeneity in attachment of marine bacteria toward antifouling copolymers unraveled by AFM. Front. Microbiol. 8, 1399.

Elbrecht, V., Leese, F., 2015. Can DNA-based ecosystem assessments quantify species abundance? Testing primer bias and biomass-sequence relationships with an innovative metabarcoding protocol. PLoS One 10, e0130324.

Emami, K., Nelson, A., Hack, E., Zhang, J., Green, D.H., Caldwell, G.S., et al., 2016. MALDI-TOF mass spectrometry discriminates known species and marine environmental isolates of *Pseudoalteromonas*. Front. Microbiol. 7, 104.

Erukhimovitch, V., Tsror Lahkim, L., Hazanovsky, M., Talyshinsky, M., Souprun, Y., Huleihel, M., 2007. Early and rapid detection of potato's fungal infection by Fourier transform infrared microscopy. Appl Spectrosc. 61, 1052–1056.

Escobar-Zepeda, A., Vera-Ponce de León, A., Sanchez-Flores, A., 2015. The road to metagenomics: from microbiology to DNA sequencing technologies and bioinformatics. Front. Genet. 6, 348.

Euler, M., Wang, Y., Heidenreich, D., Patel, P., Strohmeier, O., Hakenberg, S., et al., 2013. Development of a panel of recombinase polymerase amplification assays for detection of biothreat agents. J. Clin. Microbiol. 51, 1110–1117.

Feng, H., Wang, X., Xu, Z., Zhang, X., Gao, Y., 2018. Super-resolution fluorescence microscopy for single cell imaging. Adv. Exp. Med. Biol. 1068, 59–71.

Ferreira, L., Sánchez-Juanes, F., García-Fraile, P., Rivas, R., Mateos, P.F., Martínez-Molina, E., et al., 2011. MALDI-TOF mass spectrometry is a fast and reliable platform for identification and ecological studies of species from family Rhizobiaceae. PLoS One 6, e20223.

Ghannoum, M.A., Jurevic, R.J., Mukherjee, P.K., Cui, F., Sikaroodi, M., Naqvi, A., et al., 2010. Characterization of the oral fungal microbiome (mycobiome) in healthy individuals. PLoS Pathog. 6, e1000713.

Goldman, L., Sawyer, F., 1959. A simple portable dark-field microscope. AMA Arch. Derm. 79, 589–590.

Guo, F., Li, S., Caglar, M.U., Mao, Z., Liu, W., Woodman, A., et al., 2017. Single-cell virology: on-chip investigation of viral infection dynamics. Cell Rep. 21, 1692–1704.

Hong, H., Koom, W., Koh, W.-G., 2017. Cell microarray technologies for high-throughput cell-based biosensors. Sensors 17, 1293.

Kambouris, M.E., 2016. Population screening for hemoglobinopathy profiling: is the development of a microarray worthwhile? Hemoglobin 40, 240–246.

Kambouris, M.E., 2009. Staged oligonucleotide design, compilation and quality control procedures for multiple SNP genotyping by multiplex PCR and Single Base Extension microarray format. e-JSTech 4, 21–40.

Kambouris, M.E., 2010. Integrated real-time PCR formats: methodological analysis and comparison of two available industry options. e-JSTech 5, 33–40.

Kambouris, M.E., Reichard, U., Legakis, N.J., Velegraki, A., 1999. Sequences from the aspergillopepsin PEP gene of *Aspergillus fumigatus*: evidence on their use in selective PCR identification of I species in infected clinical samples. FEMS Immunol. Med. Microbiol. 25, 255–264.

Kambouris, M.E., Gaitanis, G., Manoussopoulos, Y., Arabatzis, M., Kantzanou, M., Kostis, G.D., et al., 2018a. Humanome versus microbiome: games of dominance and pan-biosurveillance in the Omics Universe. OMICS 22, 528–538.

Kambouris, M.E., Manoussopoulos, Y., Kantzanou, M., Velegraki, A., Gaitanis, G., Arabatzis, M., et al., 2018b. Rebooting bioresilience: a multi-OMICS approach to tackle global catastrophic biological risks and next-generation biothreats. OMICS 22, 35–51.

Kambouris, M.E., Pavlidis, C., Skoufas, E., Arabatzis, M., Kantzanou, M., Velegraki, A., et al., 2018c. Culturomics: a new kid on the block of OMICS to enable personalized medicine. OMICS 22, 108–118.

Kamiya, S., Nakamura, S., Yamakawa, K., Nishida, S., 1986. Evaluation of a commercially available latex immunoagglutination test kit for detection of *Clostridium difficile* D-1 toxin. Microbiol. Immunol. 30, 177–181.

Köhler, G., Milstein, C., 1975. Continuous cultures of fused cells secreting antibody of predefined specificity. Nature 256, 495–497.

Kyriazis, I.D., Kambouris, M.E., Poulas, K., Patrinos, G.P., 2014. Molecular techniques for the detection and characterization of microorganisms. Arch. Hell. Med. 31, 23–40 (in Greek).

Lagier, J.-C., Armougom, F., Million, M., Hugon, P., Pagnier, I., Robert, C., et al., 2012. Microbial culturomics: paradigm shift in the human gut microbiome study. Clin. Microbiol. Infect. 18, 1185–1193.

Leveau, J.H.J., Hellweger, F.L., Kreft, J.-U., Prats, C., Zhang, W., 2018. Editorial: The individual microbe: single-cell analysis and agent-based modelling. Front. Microbiol. 9, 2825.

Lin, Z., Cui, X., Li, H., 1996. Multiplex genotype determination at a large number of gene loci. Proc. Natl. Acad. Sci. U.S.A. 93, 2582–2587.

Liu, T.-L., Upadhyayula, S., Milkie, D.E., Singh, V., Wang, K., Swinburne, I.A., et al., 2018. Observing the cell in its native state: imaging subcellular dynamics in multicellular organisms. Science 80 (360), eaaq1392.

Maiolini, R., Ferrua, B., Quaranta, J.F., Pinoteau, A., Euller, L., Ziegler, G., et al., 1978. A sandwich method of enzyme-immunoassay. II. Quantification of rheumatoid factor. J. Immunol. Methods 20, 25–34.

Mayboroda, O., Gonzalez Benito, A., Sabaté del Rio, J., Svobodova, M., Julich, S., Tomaso, H., et al., 2016. Isothermal solid-phase amplification system for detection of *Yersinia pestis*. Anal. Bioanal. Chem. 408, 671–676.

McCutcheon, S., Langeveld, J.P., Tan, B.C., Gill, A.C., de Wolf, C., Martin, S., et al., 2014. Prion protein-specific antibodies that detect multiple TSE agents with high sensitivity. PLoS One 9, e91143.

McDaniel, L., Breitbart, M., Mobberley, J., Long, A., Haynes, M., Rohwer, F., et al., 2008. Metagenomic analysis of lysogeny in Tampa Bay: implications for prophage gene expression. PLoS One 3, e3263.

Morel, N., Volland, H., Dano, J., Lamourette, P., Sylvestre, P., Mock, M., et al., 2012. Fast and sensitive detection of *Bacillus anthracis* spores by immunoassay. Appl. Environ. Microbiol. 78, 6491–6498.

Nishiyama, M., Arai, Y., 2017. Tracking the movement of a single prokaryotic cell in extreme environmental conditions. Methods Mol. Biol. (Clifton, N.J.). 1593, 175–184.

Ouisse, L.-H., Gautreau-Rolland, L., Devilder, M.-C., Osborn, M., Moyon, M., Visentin, J., et al., 2017. Antigen-specific single B cell sorting and expression-cloning from immunoglobulin humanized rats: a rapid and versatile method for the generation of high affinity and discriminative human monoclonal antibodies. BMC Biotechnol. 17, 3.

Özdemir, V., 2018. The dark side of the moon: the Internet of Things, Industry 4.0, and the Quantified Planet. OMICS 22, 637–641.

Özdemir, V., Hekim, N., 2018. Birth of Industry 5.0: making sense of big data with artificial intelligence, "The Internet of Things" and next-generation technology policy. OMICS 22, 65–76.

Pastinen, T., Raitio, M., Lindroos, K., Tainola, P., Peltonen, L., Syvänen, A.C., 2000. A system for specific, high-throughput genotyping by allele-specific primer extension on microarrays. Genome Res. 10, 1031–1042.

Petrovick, M.S., Petrovick, M.S., Harper, J.D., Nargi, F.E., Schwoebel, E.D., Hennessy, M.C., et al., 2007. Rapid sensors for biological-agent identification. Lincoln Lab. J. 17, 63–84.

Piarroux, R.P., Romain, T., Martin, A., Vainqueur, D., Vitte, J., Lachaud, L., et al., 2019. Multicenter evaluation of a novel immunochromatographic test for anti-aspergillus IgG detection. Front. Cell. Infect. Microbiol. 9, 12.

Popescu, L., Cao, Z.-P., 2018. From microscopy to genomic approach in soil biodiversity assessment. Curr. Issues Mol. Biol. 27, 195–198.

Prakash, O., Verma, M., Sharma, P., Kumar, M., Kumari, K., Singh, A., et al., 2007. Polyphasic approach of bacterial classification - an overview of recent advances. Indian J. Microbiol. 47, 98–108.

Prieto, M.C., Kovtoun, V.V., Cotter, R.J., 2002. Miniaturized linear time-of-flight mass spectrometer with pulsed extraction. J. Mass. Spectrom. 37, 1158–1162.

Probst, A.J., Birarda, G., Holman, H.-Y.N., DeSantis, T.Z., Wanner, G., Andersen, G.L., et al., 2014. Coupling genetic and chemical microbiome profiling reveals heterogeneity of archaeome and bacteriome in subsurface biofilms that are dominated by the same archaeal species. PLoS One 9, e99801.

Ramage, J.G., Prentice, K.W., DePalma, L., Venkateswaran, K.S., Chivukula, S., Chapman, C., et al., 2016. Comprehensive laboratory evaluation of a highly specific lateral flow assay for the presumptive identification of *Bacillus anthracis* spores in suspicious white powders and environmental samples. Health Secur. 14, 351–365.

Rohrman, B.A., Richards-Kortum, R.R., 2012. A paper and plastic device for performing recombinase polymerase amplification of HIV DNA. Lab Chip 12, 3082–3088.

Saccardo, P.A., 1882–1906. Sylloge Fungorum Omnium Hucusque Cognitorum. Saccardo P.A., Pavia, Italy.

Sapsford, K.E., Bradburne, C., Delehanty, J.B., Medintz, I.L., 2008. Sensors for detecting biological agents. Mater. Today 11, 38–49.

Schoch, C.L., Seifert, K.A., Huhndorf, S., Robert, V., Spouge, J.L., Levesque, C.A., et al., 2012. Nuclear ribosomal internal transcribed spacer (ITS) region as a universal DNA barcode marker for Fungi. Proc. Natl. Acad. Sci. U.S.A. 109, 6241–6246.

Schroeder, H.W., Cavacini, L., 2010. Structure and function of immunoglobulins. J. Allergy Clin. Immunol. 125, S41–S52.

Shi, X., Zhang, X., Xia, T., Fang, X., 2012. Living cell study at the single-molecule and single-cell levels by atomic force microscopy. Nanomedicine 7, 1625–1637.

Siamoglou, S., Patrinos, G.P., 2019. Role of the Mobile Molecular Biology Laboratory (2MoBiL) in raising awareness of school students in the field of molecular genetics and personalized medicine. Per. Med. 1, 17–25.

Smith, R., Wright, K.L., Ashton, L., 2016. Raman spectroscopy: an evolving technique for live cell studies. Analyst 141, 3590–3600.

Sogin, S.J., Sogin, M.L., Woese, C.R., 1972. Phylogenetic measurement in prokaryotes by primary structural characterization. J. Mol. Evol. 1, 173–184.

Solé, R.V., Valverde, S., 2013. Before the endless forms: embodied model of transition from single cells to aggregates to ecosystem engineering. PLoS One 8, e59664.

Song, L., Ahn, S., Walt, D.R., 2005. Detecting biological warfare agents. Emerg. Infect. Dis. 11, 1629–1632.

Spahn, C., Glaesmann, M., Gao, Y., Foo, Y.H., Lampe, M., Kenney, L.J., et al., 2017. Sequential super-resolution imaging of bacterial regulatory proteins, the nucleoid and the cell membrane in single, fixed *E. coli* cells. Methods Mol. Biol. (Clifton, N.J.). 1624, 269–289.

Taniguchi, M., Miura, K., Iwao, H., Yamanaka, S., 2001. Quantitative assessment of DNA microarrays - comparison with northern blot analyses. Genomics 71, 34–39.

Towbin, H., Staehelin, T., Gordon, J., 1992. Electrophoretic transfer of proteins from polyacrylamide gels to nitrocellulose sheets: procedure and some applications. 1979. Biotechnology 24, 145–149.

Velegraki, A., Kambouris, M., Kostourou, A., Chalevelakis, G., Legakis, N.J., 1999a. Rapid extraction of fungal DNA from clinical samples for PCR amplification. Med. Mycol. 37, 69–73.

Velegraki, A., Kambouris, M.E., Skiniotis, G., Savala, M., Mitroussia-Ziouva, A., Legakis, N.J., 1999b. Identification of medically significant fungal genera by polymerase chain reaction followed by restriction enzyme analysis. FEMS. Immunol. Med. Microbiol. 23, 303–312.

Velegraki-Abel, A., 1986. Mycotoxins. Athens (in Greek).

Vojdani, A., 2015. Reaction of monoclonal and polyclonal antibodies made against infectious agents with various food antigens. J. Clin. Cell. Immunol. 06, 1–9.

Vojdani, A., Cooper, E.L., 2004. Identification of diseases that may be targets for complementary and alternative medicine (CAM). Adv. Exp. Med. Biol. 546, 75–104.

Vyzantiadis, T.-A.A., Johnson, E.M., Kibbler, C.C., 2012. From the patient to the clinical mycology laboratory: how can we optimise microscopy and culture methods for mould identification? J. Clin. Pathol. 65, 475–483.

Wang, H.-Y., Luo, M., Tereshchenko, I.V., Frikker, D.M., Cui, X., Li, J.Y., et al., 2005. A genotyping system capable of simultaneously analyzing >1000 single nucleotide polymorphisms in a haploid genome. Genome Res. 15, 276–283.

Wargo, M.J., Hogan, D.A., 2006. Fungal-bacterial interactions: a mixed bag of mingling microbes. Curr. Opin. Microbiol. 9, 359–364.

Wieser, A., Schneider, L., Jung, J., Schubert, S., 2012. MALDI-TOF MS in microbiological diagnostics - identification of microorganisms and beyond (mini review). Appl. Microbiol. Biotechnol. 93, 965–974.

Woese, C.R., Kandler, O., Wheelis, M.L., 1990. Towards a natural system of organisms: proposal for the domains Archaea, Bacteria, and Eucarya. Proc. Natl. Acad. Sci. U.S.A. 87, 4576–4579.

Xu, J., Ma, B., Su, X., Huang, S., Xu, X., Zhou, X., et al., 2017. Emerging trends for microbiome analysis: from single-cell functional imaging to microbiome big data. Engineering 3, 66–70.

Yalow, R.S., 1980. Radioimmunoassay. Annu. Rev. Biophys. Bioeng. 9, 327–345.

Yarmush, M.L., King, K.R., 2009. Living-cell microarrays. Annu. Rev. Biomed. Eng. 11, 235–257.

CHAPTER

PANMICROBIAL MICROARRAYS 6

Aristea Velegraki

Mycology Research Laboratory, Department of Microbiology, School of Medicine, National and Kapodistrian University of Athens, Athens, Greece

INTRODUCTION

The concept of panmicrobial arrays emerged as both a need and a promise. It was the promise of early omics era—especially, but not exclusively, of genomics—to resolve the deteriorating threat of infectious diseases in both spontaneous and perpetrated spin-offs (Malanoski et al., 2006; Palacios et al., 2007; Lin et al., 2006; Wang et al., 2006; Heller, 2002). It is no coincidence that most pedigrees of such arrays are more or less associated to institutions controlled by the US defense establishment (Palacios et al., 2007; Lin et al., 2006). The concept evolved in iterations. Initially the prefix "Pan-" (meaning "all encompassing" in Greek) had the notion of "all available" or, rather, "all sequenced", as known microbial sequences were the restrictive factor (Gardner et al., 2010).

The concept evolved to "all relevant" when sequences became more abundant, with the exact meaning of "relevant" being debatable. The one plane of relevance was the taxonomic and the other the functional, with notions of microecology. The taxonomic plane was easier to tackle (that is to design and operate suitable arrays), with microarrays specialized to viruses being the most common (Chou et al., 2006; Palacios et al., 2007; Chiu et al., 2007; Huguenin et al., 2012), but others for bacteria, fungi, and protozoa being developed as well (Hong et al., 2004; Wang et al., 2004; Sakai et al., 2014; Sato et al., 2010). The functional plane of relevance was more applicable in real-world scenarios, as different agents, possibly implicated in similar syndromes, as for example respiratory infections (Huguenin et al., 2012; Chiu et al., 2007; Wang et al., 2006; Lin et al., 2006), were assorted and arrayed.

A more literary use of "panmicrobial" implying detection and broad classification, not identification, of *all* microbes, was possible with more elaborate array structures and probe designs (Gardner et al., 2010; Palacios et al., 2007). Long probes targeting generic, conserved barcoding sequences with a tolerance so as to bind the analyte despite mismatches (Lin et al., 2006) allowed the capture of more or less distant relatives, ensuring the capture and detection of an agent, even if the strain belonged to a rare or hitherto unknown/unsequenced lower taxon (Sakai et al., 2014). Thus, a pattern, with signal detected from generic reporters targeting conserved regions and lack of signal from more specific identification reporters, could arise in cases where a novel unsequenced isolate of a given genus is present (Gardner et al., 2010).

Microbiomics. DOI: https://doi.org/10.1016/B978-0-12-816664-2.00006-2

INVENTION, DEFINITION, AND RATIONALE OF MICROARRAYS

The concept of microarrays was first projected in the early 1980s for immunoassay applications (Chang, 1983). In a way, the assays conducted in multiwell plates, such as ELISA, and the assimilation assays in strips of one or two rows of wells, had already introduced parallel processing of different kinds (the former of multiple samples and the latter of multiple biomarkers in one sample), which were furthermore coupled to an association between location of the testing agent (reporter) and applicable data readings.

The prefix "micro" of the "microarrays" is due to the microscopic size of the individual loci, rather than of the array itself. The former were of 10–500 μm in diameter, and around 100 μm apart (Lindroos et al., 2001), which could not be read by naked eye, contrary to some multiwell plates and assimilation tests. On the contrary, a microarray may be sized from 25 mm^2 or even less (Kambouris, 2016) to the surface of a whole glass slide, which is more than 10 cm^2 (Palacios et al., 2007). But the novel concept of (micro)arraying went some steps further: the sample is presented in a single aliquot, and a large number of singular assays ("features" or "loci"), each containing a homogenous population of "reporters" in high density, defines the complete assay. Thus a number of biomarkers can be interrogated simultaneously, coining the definition of "parallel" processing, and that number can be quite high, to six digits or even more, thus justifying the characterization "high throughput" (Liu et al., 2009; Heller, 2002). Parallel processing, contrarily to serial one, allows concomitant interrogation for all applicable queries and thus avoids the degradation of the sample either by time or by products and by-products of the previous tests. Compared to parallel, low-throughput assays, where different aliquots of the sample are used for different tests even if these are run in parallel (as in conventional assimilation tests), the novel format also enhances the specific sensitivity (detected moiety-to-volume ratio) as the sum of the sample is used for all inquiries.

In the microarrays proper the differential, simultaneous, and parallel capturing or otherwise signal-generating processes on the different loci are enabled by different strategies, which define the nature of the microarray: An external force may stir the sample aliquot so as to continuously circulate its analytes throughout the array in order to have them repeatedly presented to the reporters of each different fixed locus (passive arrays). Alternatively, electric fields from electrodes embedded in the microarray structure (active arrays) can be used to address the analytes (Heller, 2002; Shin, 2008; Patirupanusara and Jackrit Suthakorn, 2012). The term "proper" is meant to emphasize the fact that contrary to common wisdom, the term "microarray" is used loosely even to describe "solution" or "self-assembled" microarrays on top of the "planar/positional" ones (Hong et al., 2017).

PEDIGREE AND CATEGORIES OF MICROARRAYS

There are multiple iterations of the microarray technology for a vast number of applications, even beyond bioscience (see Table 6.1). The initial concept is a two-dimensional (2D) placement of unique positions each performing a single assay; from there on, the variability is astounding. Active or microelectronic microarray utilizes electric fields produced by electrodes embedded on

Table 6.1 Categories of Microarrays.

Categorization of Microarrays	Exemplary Categories
By reporter/interaction nature	Oligo, substrate–enzyme, Ab–Ag, aptamer, ligand–receiver, live cell physiology, histochemistry
By interrogation method	Radioactivity, enzymatic activity, illuminance, fluorescence, mass spectra analysis, nuclear resonance; binary vs differential signal; imaging vs scanning; instrumentation requirement vs instrumentation-free readout
By logical association among loci	Exclusive, inclusive, conclusive
By circulation strategy	Active, passive
By manufacture	Printed/spotted, synthesized, microfabricated
By substrate	Plate, slide, membrane
By throughput	High, medium, low
By redundancy	Self-array, multifeature, degenerate, redundant, stacked
By single assay concurrency and sequence	Parallel, sequential, serial, radial, phased, convergent–divergent
By phase	Fixed/planar, suspension/self-assembled, solid state, 3D, microfluidics

Classification of microarrays by different integral or functional attributes.

the microarray surface that allow rapid and controlled circulation of charged DNA/RNA or other charged/chargeable (electrophoresis) or polarizable (dielectrophoresis) analyte and enable faster hybridization (in mere minutes), compared to passive microarray that may require easily 16 h or more (Heller, 2002; Shin, 2008; Yarmush and King, 2009; Patirupanusara and Jackrit Suthakorn, 2012). The electric fields produced are by definition reconfigurable and allow both electrophoresis and dielectrophoresis, depending on the nature of the array and its targets (Heller, 2002; Shin, 2008; Yarmush and King, 2009; Patirupanusara and Jackrit Suthakorn, 2012). Except their application in transporting and addressing the target molecules through the array and to their reporters, the fields have more elaborate uses; reversing the electric field on the test site produces an electronic stringency effect, which can greatly improve binding specificity, especially in nucleic acid arrays, where highly specific hybridization is needed to detect point mutations and single nucleotide polymorphisms (SNPs) (Sosnowski et al., 2002; Heller, 2002).

Conventional active microarrays that utilize 2D electrodes have been used, but with limited success, especially in high-volume samples, where they display suboptimal trapping due to wash away phenomena over the imprinted loci (z-axis). In such cases, 3D carbon electrodes, fabricated using C-MEMS (Carbon MEMS) technology, which enable manipulation of large volumes, suitable for pathogen detection, come in useful (Shin, 2008), whereas more basic substrates are also applicable. Such are microfabricated polyacrylamide pads created on a slide for depositing reporter oligonucleotides (Zlatanova and Mirzabekov, 2001).

In the field of passive arrays, there are a number of options for promoting the circulation, addressing, and binding between reporter and analyte. From the heat hybridization chambers for oven placement and the slide moats, where diffusion is the sole factor—aided by the elevated temperature, which increases the Brownian motion—all the way to the planar/orbital shakers for slides and plates, to rotating multichamber assemblies and, to the most sophisticated, automated microarray hybridization stations that use micropumping or other features such as microfluidic principles to continuously circulate the sample on the array (Peytavi et al., 2005).

Except the addressing/circulation strategy, other iterations are based on the nature of the reporters, the geometry of the loci, the singularity or collectiveness of the array substrate or matrix, its composition, the logic/association among loci (if any) and their information context, and the interrogation/reporting method. Not all microarray types are suitable for mibrobiologic or microbiomic applications.

The major distinction among microarrays is their functional nature; this means the molecule/moiety used as reporter and the interaction used for capturing its analyte. The original microarray had printed antibodies to detect antigens, using the epitope–paratope interaction, but the widely used nucleic acid (or oligonucleotide) microarrays use hybridization to detect nucleic acid fragment analytes. Still, proteins are the most common moiety, either as reporters or as analytes in arrays based on receptor–ligand; antibody–antigen; enzyme–substrate; nucleic acid–protein interactions. In all the abovementioned cases, at least one of the pair reporter–analyte is protein or peptide. Other microarray formats include captured viable cells to test bioactive compounds and sources of energy; metabolites to elucidate phenotypic characters from cell samples; histological assortments of paraffin-embedded fine sections for massive, parallel preparations and treatment; chemical compounds for nonbioscience-related interaction research.

The positional arrangement of the microarray loci is a most complicated issue; the fixed arrays have a standardized layout, usually rectangular and eventually divided in subarrays. The introduction of subarrays decreases the overall density but was developed to improve the massive, unwieldy, and homogenous formats of the initial iterations of array technology. The division to subarrays improves on data extraction, especially in automated setups, as it facilitates all functions based on alignment; there are multiple points for the accurate application of the reading/imaging frames. In more practical terms the use of subarrays allows either articulated, logically compartmentalized arrays of high order (as in Fig. 6.1), or repetitive identical subarrays for statistical robustness.

The self-assembled, or suspension microarrays, do not have loci defined by two dimensions on a planar substrate (slide, plate, membrane, either of glass or of synthetic material). They rather use coated beads (metallic or, more often, synthetic) as features; the concept of "locus" with its notion of positional determinism is unwarranted in this case.

Each bead carries a homogenous population of reporters on its surface and is encoded to differ from the other beads, which bear different reporters and represent different features. The encoding is achieved by differential coloring or by surface markers different than the reporter biomarkers and constitutes the functional analog of the 2D location-based identification of the loci of planar arrays. Thus, after the interaction stage is complete and the reporters loaded, the beads are presented serially to a reading apparatus, possibly fiber optic in nature and/or in a flow cytometer context, and read in different channels for feature coding and reporter loadout; or they are aggregated and shot in both channels by wide-angle imagers, with the two images superimposed and then deconvoluted, thus assembling *ad hoc* microarray due to the serial nature of the readings (Walt, 2000; Heller, 2002; Hong et al., 2017).

Regarding the manufacture of a microarray, the reporter entities can either be readied beforehand and attached to the substrate any time after their synthesis is complete (spotted and printed

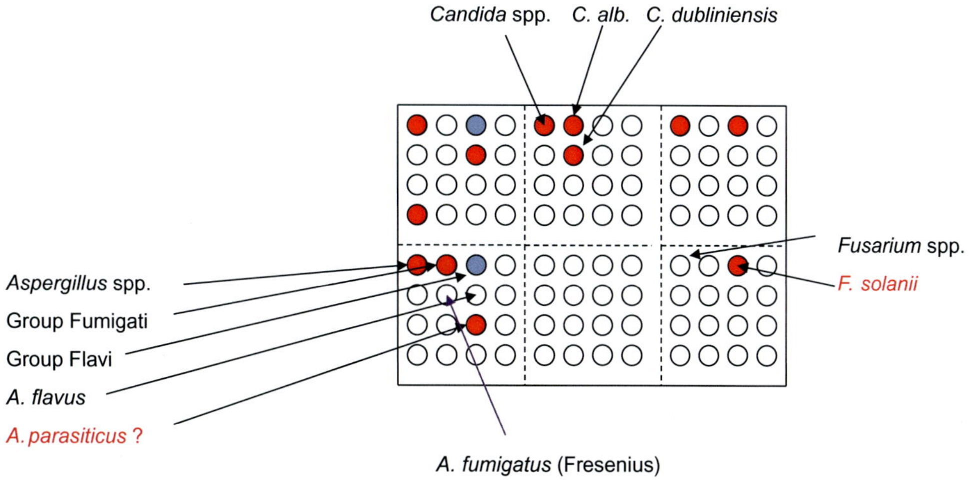

FIGURE 6.1

Positionally relevant subarrays produce the basic conclusive microarray setup. In such a format, any number of loci might be positive and the result is a function of the number *and* position of the positive loci. Shaded circles insinuate loci of low calling score, making high scores (*filled circles* coupled to red font) of subordinate loci inconsistent.

arrays) or be developed/differentiated on the substrate in a guided, paced manner (synthesized arrays). Beyond interactive synthesis, microfabrication allows more elaborate results and microfluidics are expected to revolutionize relevant research with dedicated, multifunctional self-contained assays (Heller, 2002; Yarmush and King, 2009; Patirupanusara and Jackrit Suthakorn, 2012; Hong et al., 2017). Different techniques and options are available for both the abovementioned featuring strategies, which offer different advantage–disadvantage mixtures. Massive production, in industrial scale, favors synthesis, while research applications with limited production runs are best served by spotting/printing.

To produce spotted arrays, simple pipetting microvolumes by hand on a marked glass slide to create dot blots or in a multiwell plate have given their place to robotic instruments (arrayers), which use robotic arms with one or more solid or hollow pins for deposition of solutions containing reporter molecules, exemplified by but not restricted to DNA fragments and oligonucleotides. The hollow pins dip into the solution of the reporter and load a steady quantity through surface tension and then touch with submicrometric precision a spot on the solid substrate (silicon, glass, or plastic) to deposit a droplet (Heller, 2002), the size of which depends on the surface tension and the material of the substrate. The substrate must have been previously treated and activated chemically so as to bind the spotted reporters. A series of thermal and chemical processes, after the whole array has been spotted, stabilizes the bonds. The use of uncoated plastic slides as substrate for immobilizing spotted oligonucleotide reporters with UV irradiation simplifies the procedure and drives costs down (Sakai et al., 2014). In printed microarrays, droplets are sprayed following the operating principle of inkjet printers, which simplifies the respective chemistry for correct loading, transportation, and spotting as the surface tension phenomena and their dependence on ambient environmental conditions play minimal role or are completely irrelevant (Heller, 2002; Palacios et al., 2007; Patirupanusara and Jackrit Suthakorn, 2012).

Synthesized arrays have their reporters assembled *in situ*, with successive rounds of masking–unmasking events during adapted photolithography protocols. Synthesis might refer to reporters only, or, in the case of active arrays, to the electrodes as well, which can be synthesized similarly, with completely different rounds of photolithography (Heller, 2002; Shin, 2008; Patirupanusara and Jackrit Suthakorn, 2012).

The notion of arrays in general and microarrays in particular does not necessarily point to actually different loci/features. It points to a multitude of them, and ordered in a specific way; but this multitude may—or may not be—identical, considering the reporters used. Although such single-feature arrays (self-arrays) are not easily coming to mind in bioscience, they are indispensable in electronics, and even in bioscience, they come out handy for accurate quantification of targets. But their most popular application is the cellular array, practically the inverted type of phenotype microarrays (discussed later on) for pharmacological and other physiological tests on any newly considered microorganism or cell type (Bochner, 2009; Yarmush and King, 2009; Hong et al., 2017). Of course, on top of the above, self-arrays are excellently suited to be used as affinity chromatography matrices to fish an analyte off a mixture by selectively binding it and then, after washing steps, unbinding it, thoroughly purified and potentially quantified (Palacios et al., 2007).

The usual array format, though, incorporates different loci in a more or less canonical geometry. If homogenous populations of receptors are placed in more than one locations, resulting in multiple representation of single features within the same array instead of unique ones, the array is *degenerate* as one locus associates with one receptor only, but this association is unilateral and not mutual/reciprocal (Diamandis, 2000; Wang et al., 2005; Lin et al., 2006).

The issue gets even more complicated with the actual array target: it is possible that the features are unique (one unique locus presents each reporter), but a number of such loci refer to a single target entity (Lin et al., 2006; Palacios et al., 2007; Liu et al., 2009; Gardner et al., 2010). In such case the array is *redundant*. Different degrees of extend and homogeneity can be expected in a redundant array, as the number of loci assigned to a single target (redundancy factor) differs among arrays and may not be standard even within the same array (Palacios et al., 2007; Liu et al., 2009; Gardner et al., 2010). Last, but not least, a redundant array may well be degenerate as well, the combination defining a *second degree of degeneration*.

The relations among different loci of a microarray do define the microarray proper and affect its utility and logistics. A nondegenerate, nonredundant microarray in its simplest application may produce one positive locus and the fact practically excludes all others from being positive; such an array is called *exclusive*. Identification arrays in particular fall in this category. Multiple positive loci mean multiple targets co-existing in the sample (Malanoski et al., 2006; Lin et al., 2006; Liu et al., 2009; Gardner et al., 2010). It is the simplest array format, short of the self-array, presenting few challenges in generation and interpretation of results.

The *differential* or *inclusive* array expects a positive signal to all loci, but the signal of each locus is differential, either quantitatively (different signal intensities among different loci) or qualitatively (signal may appear in different forms, as are colors or frequencies, irrespective of intensity). Such are the typing arrays, especially the genotyping ones (Wang et al., 2005).

The *conclusive* array is by far the most complex, comprehensive, and variable. In such a format, any number of loci might be positive and the result is a function of positive loci (Fig. 6.1), possibly

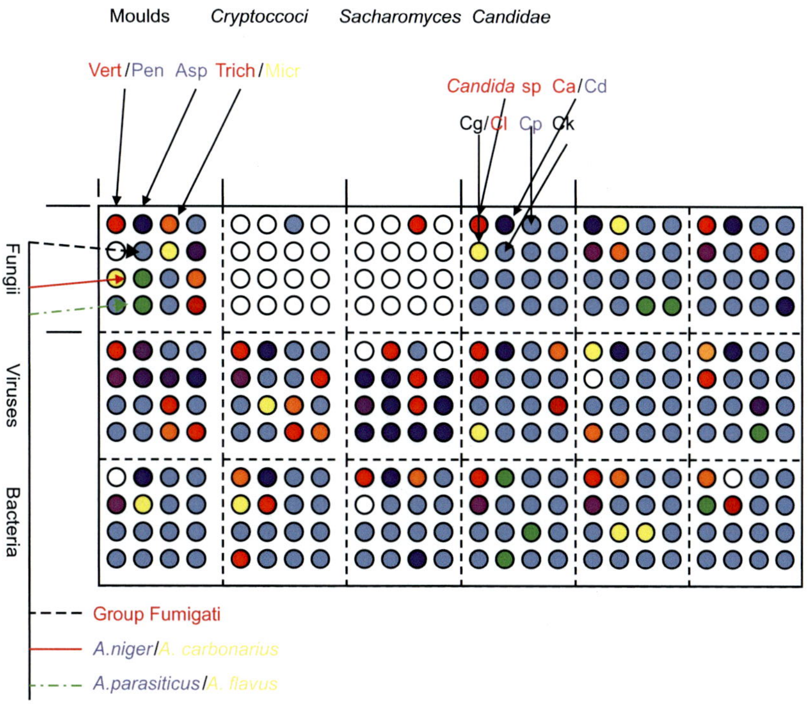

FIGURE 6.2

Higher order conclusive microarrays, by far the most complex, comprehensive, and variable, take into account the combination of digital and differential signal of the loci and their on-array locations, as there may be intrinsic relations among individual loci or among and within subarrays.

enhanced by their allelic status, which creates a differential signal and possibly taking into account the combination of their locations, as there may be intrinsic relations among individual loci or among and within subarrays (Chou et al., 2006; Gardner et al., 2010; Lin et al., 2006; Liu et al., 2009; Palacios et al., 2007; Malanoski et al., 2006; Wang et al., 2006) as shown in Fig. 6.2. Degenerate and redundant microarrays by definition belong to this category, but all conclusive arrays are not degenerate, nor redundant. Moreover, the scoring criteria differ between arrays with multiple identical loci, intended for robustness, and arrays with associated loci intended to extract more elaborate conclusions (Fig. 6.2).

COMPARISON TO THE STATE OF THE ART

New generation sequencing (NGS) provides a definite answer to any current or projected genomic query and detects insertions/deletions easily and positively with increased discrimination power, although its cost still is prohibitive for massive applications. It is not fast, but rather straightforward, as it depends on a small number of alternative solutions and a handful of implicated technologies. This allows better allocation of research funds, standardization, and cross-talk among different providers (Quail et al., 2012).

But the most important issue is that, both in whole genome or whole exome sequencing (WGS/WES) formats, it reveals the genetic content of an individual in unneeded detail, an act of extreme controversy liable to spark heated reactions within societies by sensitive fractions, cultures, and religions. Discrimination in professional and social environments, targeted promotional, extortional or lethal events, and other forms of unwanted intervention become more likely by the day, as collected data are never safeguarded up to specs, especially collective public data (Kambouris, 2016).

In addition, the discretion in sequencing will be exceptionally important for testing/screening proprietary and patented microorganisms (natural or engineered) for granting biosecurity/biosafety clearance before using them commercially and/or in industrial levels (Kambouris et al., 2018d).

In this context the microarray technology may offer a viable option in genotyping, as its routine use is affordable (Wang et al., 2005), discrete (only the stated loci of interest are revealed and genotyped), and adaptable. The latter characteristic refers to selective utilization of a standardized microarray. This approach includes all known loci of interest but reads preferentially the loci of priority for the target of each application and is thus software-defined, at least in principle. Examples include, without being restricted to, screening populations in different countries or areas, implementing geoepidemiology (Gkantouna et al., 2015), which is an important issue in microbial diseases and public health microbiomics.

Thus to retain or implement discretion, the full extent of information may neither be called nor extracted. The former approach is achievable by interrogating and saving only the loci of interest, an approach easier in implementation if interrogation by scanning is used. Alternatively, and concerning mostly the latter approach, data selectivity can be achieved even in interrogation through imaging, by operating at a deeper level; that is by omitting specific events in the amplification/analyte producing/addressing/binding chain of reactions for analytes captured to loci of low specific interest.

Arrays engage a galaxy of different methods and technologies (Table 6.1); to pursue and compare them so as to finally select just one format is a herculean feat. Their application goes much further than genotyping and genome-related applications, may these be direct or indirect. The microarrays as a broad methodology should have been funded to develop by themselves, as a technology culture more than as an approach, without near-term applicability in mind. A supported boom would result in a survival of the fittest and selection processes for different applications and uses in the scientific spectrum; a generic and somewhat rough example are the staring arrays of infrared sensors and the arrays of the last generations of radars (passive and active electronically scanned arrays).

TRADE-OFFS AND PROSPECTS

The microarrays as a method are cumbersome and laborious in development, possibly needing individualized conditions and laws for calling each locus. Still, in terms of methodology, they are flexible: once having invested in it, many similar and dissimilar problems may be tackled by changing loci (Heller, 2002; Liu et al., 2009; Gardner et al., 2010; Palacios et al., 2007) and by creating spin-off arrays (Palacios et al., 2007; Gardner et al., 2010). Furthermore, in terms of utilization, the moment a microarray has finished development, it becomes very cost-effective: its use is affordable, easy, and massive (Heller, 2002; Gardner et al., 2010; Palacios et al., 2007; Liu et al., 2009) while it entails very high throughput (Palacios et al., 2007; Gardner et al., 2010). But, especially in their first generation, the microarrays, which were not degenerate and one locus reported on one characteristic (Wang et al., 2005), have been proven mostly inflexible.

In addition, due to subarrays and 2D geometry, different functional zoom-ins become possible, along with repeatability and selective depth of processing/analysis (Gardner et al., 2010; Liu et al., 2009; Palacios et al., 2007). One approach, the degenerate microarray, promotes multiple prints of the same reporter in different positions of the same array (Wang et al., 2005) or in different subarrays (Yarmush and King, 2009); it can be used for acquiring statistically robust signal assessment but also to improve redundancy and thus may tackle local anomalies in circulation, binding, and signal production.

Another, more widespread, approach of redundant microarray is to use different reporters (each might be printed in multiple loci, of course) for the same target in order to increase certainty and/or interrogate multiple secondary markers so as to achieve identification (Palacios et al., 2007; Gardner et al., 2010; Liu et al., 2009). The tiling oligonucleotide microarrays, which can achieve actual (re)sequencing without electrophoresis (Malanoski et al., 2006; Lin et al., 2006; Wang et al., 2006), are a special case of the second approach.

Moreover, in qualitatively dissimilar applications, where different analytes are pursued, explored, and exploited, microarrays are more robust than sequencing as they can unshadow low-copy analytes by individualized scoring parameters for each locus; thus sensitivity becomes manageable and tunable.

On the other hand a microarray needs ample sample to work well. The "ample" is, understandably, rather subjective and depends on the case; still it may be defined as enough to saturate the printed reporters plus some spare, which may degrade or be lost during active or passive circulation of the sample particles among loci. Increased sample quantities would, in turn, increase cross-reactions, thus degrading accuracy; lower quantities would fail to make full use of the reporter potential.

Of course, this is a very optimistic scenario, especially with collected samples as opposed to produced ones. Thus the microarray needs either massive samples or amplification of the analyte or of the target bearing the analyte (Lindroos et al., 2001; Heller, 2002; Wang et al., 2005; Lin et al., 2006; Malanoski et al., 2006; Palacios et al., 2007; Kambouris, 2016). Many approaches are applicable for sample amplification and cultures are the definitive solution in some cases, although they tend to increase demands on resources and/or infrastructure (Palacios et al., 2007; Wang et al., 2005; Kambouris, 2016). Of course, in oligonucleotide arrays the favored approach is sample amplification through PCR spin-offs or isothermal technologies (Wang et al., 2005; Kambouris, 2016; Sakai et al., 2014), in order to keep biowaste and need for labor-intensive and infrastructure-demanding *in vivo* steps to a minimum.

METHODOLOGY

DEVELOPMENT AND OPTIMIZATION

Practically, the main challenge of an array is the optimization of different steps and standardization of the multiplex products of these steps and the reporter molecules of the loci.

The "*reverse staged pooling*" is a "bottom-up" approach. The idea here is that the results of two successive steps are compared and the inconsistencies detected are resolved in terms of nature and causality (Kambouris, 2016).

The first step is "simplex-all-the-way," where each analyte extracted, transformed or amplified, and tagged is captured on a single-feature, multidot microarray. Genotyping is compared to reference scores as is standard procedure (Jiang et al., 2001) to qualify the design of the oligonucleotides and to determine the gain and threshold values at the allele-calling step. This step is frequently disregarded, as it would necessitate as many simplex reactions as are the actual different features of the array.

The second step is a "multiplex capture" whence the pooled, labeled analytes are captured onto a proper multifeature microarray to detect possible crossover binding of analytes to heterologous array features. This step is crucial for the containment of false-positive binding. The third step is "multiplex labeling." The development of the analytes is implemented in a multiplex fashion and they are then presented on the multiplex array for capture. In this stage, misincorporation incidents are determined. The "multiplex labeling" step exists in cases where the production of the analyte mixture entails more than one steps, usually a first step for amplification and a second for labeling. The case is exemplified by, without being restricted to, PCR amplification and single-base extension labeling (Wang et al., 2005; Kambouris, 2016). If the production of the analyte mixture is achievable in one step, then the "multiplex labeling" step is omitted.

The final stage is "multiplex all-the-way," where the targets are pooled and mixed with the reagents, which will produce the analytes that will then be presented to the reporters of the microarray. This stage pinpoints the most severe and numerous problems, which are attributable to the multiplex production/amplification procedure proper. Reagents' cross-reacting or interacting events increase noise, deplete reagents,' change the chemical balance, and deoptimize in every conceivable way the environment of the reactions, resulting in low yield and extremely unfavorable signal-to-noise ratios.

For reproducibility testing, at least two identical arrays have to be processed simultaneously (duplicate experiment to test spontaneous reproducibility); additionally, at least one more repetition must be implemented with the same batches of reagents after some time (24–48 h) to establish test reproducibility over time, an attribute contributing to the robustness of the method (Unknown, 1994). This procedure should be repeated for every one of the abovementioned steps.

A single microarray format, once set, offers no flexibility; loci cannot be replaced or added without repeating the qualification procedure. Similar or even vastly different platforms that allow significant flexibility in adapting, enriching, and evolving their scope have been developed (Cremonesi et al., 2007), but they address lower numbers of loci, usually down to low two digit. On the contrary, in the presented paradigm, loci can be functionally deleted by simply not adding the respective PCR and/or SNE (single-nucleotide extension) primers in the reaction mix, as happens with gene panel assays; such a course is fully reversible in subsequent assays by adding the respective primers if they are needed. The microarray is plighted by expensive and laborious development for lower data context compared to high-efficiency WGS/WES. It simply bears a much lower cost per test that drops proportionately with the number of tests produced and performed as the research and development (R&D) costs, which consist the main liability, are dissipated.

TYPES OF LABELING SIGNAL

Labeling and calling chemistries were always at the heart of the array methodology. Radioactive dyes were the first labeling method for quite some time, but except the lurking dangers for the personnel, it also produced much waste, demanded radiosensitive materials of large logistic and environmental footprint (films) or respective equipment such as phosphoimagers (Lindroos et al., 2001), and the output signal was difficult to analyze and standardize. Overlapping signals were regularly seen when probing libraries using *in situ* hybridization, and signal footprint larger than the labeled locus was considered normal. Thus high density of labeled loci readable simultaneously was a challenge, if not unattainable.

Precision and standardization came with fluorescent dyes, which were more accurate in spatial terms than usual chromophores. Cy-3 and Cy-5 allowed very predictable and manageable signal characteristics, facilitating semiquantitative analysis of features both dense and canonically arranged. More importantly, they allowed qualitatively differential labeling, opening the way for two- and then multi-channel dying, compared to the single-channel dying afforded by radioactivity, which achieved digital, not differential, discrimination. The fluorescent nature of such dies allowed inert state until interrogated by a laser, thus limiting accumulation of background signal (Selvin, 2000).

High precision meant that each locus could be called individually (not necessarily sequentially), and in *ad hoc* hierarchical patterns. A fine, focused beam was used for the excitation of the dyes, a feature poised to increase accuracy and resolution. The abovementioned process was termed "scanning." Alternatively, there was the interrogation by "imaging," where a whole (micro)array could be interrogated at once by a wider beam, to produce a unitary image, which was shot and saved in real time. The latter was deemed easier in interpretation and was definitely faster, especially in extremely dense and populous microarrays.

Fluorescent labels are widely used to detect the bound analyte in DNA microarrays. However, the low stability of most fluorescent dyes (which suffer photobleach when repetitively interrogated), their high cost, and the need for expensive scanning equipment call for the development of alternative labeling systems, which must be inexpensive and robust. The dying approach of biotin–peroxidase is halfway between the previous two methods: cheap, flexible, and accurate, although by definition single-channel and thus of lower resolution. It produces visual color that can be captured by simple optics, and, most importantly, without any deteriorating agent to prohibit repeated scans with different settings so as to optimize signal processing by locus or groups of loci (Wang et al., 2006; Sakai et al., 2014).

The use of image-producing solutions, by single-, bi-, or multicolor dyes ("color" used *sensu lato*) is not the only way to extract data from an array. Actually, it is quite questionable if it is even the best one, as image processing is a very complicated and expensive issue, which may mar the reliability of massive arrays, once the human factor cannot be called upon to assist in tackling issues. More elaborate approaches, such as composite dye systems with exciters, resonators, or quenchers coupled to the dyes, are one solution. Especially quenchers may be used to create "dark-positive" formats, where a reporter loaded with a captured analyte turns negative in signal production. This concept may be very helpful in high-clutter samples and is conceptually similar to reverse imaging, which is applicable in many cases, including some applications in microscopy—the India Ink negative staining being a prime example (Guess et al., 2018).

Similarly, three-party formats, with a reporter anchoring the analyte and a separate, free probe attaching to a different part of the captured analyte to label it, increase the reliability by introducing safety features. Approaches implementing contiguous stacking hybridization (CSH) (Parinov et al., 1996), by using probes complementary to the analyte to the previous or from the next nucleotide of the reporter oligonucleotide binding area, introduce dissimilarly high stability, due to stacking effect, and sensitivity, due to the shorter individual attachments. All such approaches, though, by definition increase cost and complexity. Multi-thousand- to million-feature assays become cost-ineffective.

A wholly different approach is to read the array by analyzing the mass of the loci. Captured analytes make the reporters heavier and bigger—a feature applicable to many microarray types other than nucleic acid ones. By using MALDI TOF MS (matrix-assisted laser desorption ionization-time-of-flight) mass spectrometry or any other applicable mass spectrometry solution, a digital effect of positive (with captured analyte) or negative (without analyte) loci may be compiled, preferably in comparison with a whole-negative array as the background/negative control. The mass difference, though, of the four nucleotides are at minimum 9 Da, comfortably within the discriminatory power of MALDI TOF MS. Thus single-base substitutions or deletions can be determined in short, stringent hybridization events (Stomakhin et al., 2000) allowing differential discrimination in addition to digital one.

More elaborate formats, as to the nature/composition of the captured analytes, are conceivable, depending on cost, level of technology, and computational resources. The paradigm in this case is mass labels attached onto very short oligonucleotide probes, which augment MALDI TOF MS discriminatory efficiency by hybridizing onto the captured analytes (Stomakhin et al., 2000) in CSH (Parinov et al., 1996).

Surface plasmon resonance imaging is another way to determine the loci with bound analytes. In this case, though, more elaborate substrates are needed and the application is mostly favored in non = genomic arrays (Hook et al., 2009).

AMPLIFICATION

A central issue to be determined in an array design, or an array application, is the amplification step. The sample usually needs amplification or enrichment, in order to achieve binding of enough analytes to the reporters of a locus so as to produce/emit identifiable signal over a sensitivity threshold (Sakai et al., 2014). In order to amplify the sample before or within the signal production step, microarray protocols for nucleic acid analysis utilize thermal cycling, usually in the form or a PCR spin-off, for sample enrichment in analytes, with or without simultaneous labeling them and/or reverse transcription if RNA targets are implicated. Specific, targeted PCR formats present problems in multiplexing their primers, especially in high-order multiplexing (Kambouris, 2016; Wang et al., 2005); on the other hand, generic, whole genome amplification produces extremely high levels of noise and is costly (Lee et al., 2008) and rather poised for NGS.

Isothermal strategies allow considerable savings as they do not require expensive precision instrumentation such as the cyclers and can be performed literally at bedside, or, more interestingly, in austere conditions in point-of-need. The rather short amplification products of isothermal formats (100–500 bp) require, of course, a higher level of discrimination built in the microarray proper, to compensate for their insufficient robustness or a more elaborate isothermal approach, such as the recombinase polymerase amplification (RPA)—which makes assays truly point-of-need or bedside-rated (Sakai et al., 2014). RPA isothermal technique has been used for rapid identification of fungi, viruses, and bacteria (Euler et al., 2012; Sakai et al., 2014; Glais and Jacquot, 2015) and is compatible with single channel, though affordable and directly readable, biotin labeling of sample DNA.

Another approach is the implementation of amplification on the microarray. In such a case, labeling and amplification may be combined in one step and integrated with the binding (hybridization) step. The obvious advantage is that the targets will have been resolved before being turned to analytes, a feature extremely valuable in complicated samples where significant differences in representation are to be expected among different targets. Contrary to a multiplex free-flow amplification, where more abundant species shadow the less abundant ones in terms of accessibility to enzyme and to other reagents, the array resolves spatially the different species and allows equal access in reagents, but it usually conserves the imbalance all the way to the products; this offers a notion of quantitative output on the original target population. This "notion" may or may not become fully quantitative, by the use of standard curves' logic.

This *in situ*, or, rather, "on-chip" amplification (Sosnowski et al., 2002; Strizhkov et al., 2000) found its logical conclusion as the "on-array sequencing". For historical accuracy, the concept of on-array sequencing has been instrumental for the advent of second-generation sequencing (the original NGS). The Genome Analyzer®, Solexa®, and Hiseq 2000® NGS platforms and their updates, presented by Illumina and their peer RS® of Pacific Biosciences, all use spin-offs of planar arrays (Quail et al., 2012). At the same time the SOLiD® platform of Applied Biosystems and the GS FLX 454® of Roche are based on suspension/self-assembled arrays (Mardis, 2013). The on-array sequencing process should not be confused with (re)sequencing on tiled microarrays (Diamandis, 2000), an enzyme-less approach where differential hybridization reveals the allelic status of a nucleotide and may do so for some hundred thousand bases (Wang et al., 2006; Lin et al., 2006; Malanoski et al., 2006).

On-array sequencing can be implemented by immobilizing unresolved sequences at homogenous loci and conducting sequencing by generic priming and reading the incorporation order either by fluorescent dyes or by pyrosequencing approaches. More orthodox, and less reminiscent of old hybridization and blotting experiments, is the use of known capture oligonucleotides/reporters, which bind targets and serve as primers to explore the target sequence upstream of the binding area. This approach requires the reporters having their 3′ end free, a rather tricky issue for both synthesized and spotted arrays, but eventually resolved (Lindroos et al., 2001); its application can be restricted in single-base extension/"minisequencing" format (Lindroos et al., 2001)—essentially SNE as described during the late 20th century (Shumaker et al., 1996; Goelet et al., 1999)—or may evolve to longer reads needing real-time imaging to register.

THE MICROBIOMIC ASPECT OF MICROARRAY CONCEPTS

THE GENOMIC ASPECT

For microbiomic applications, DNA microarrays have been determined as the optimal array–based solution. Microbial epitopes change regularly and intermittently, thus negating a stable target for microarrays using antibodies as reporters. It is an undeniable fact that immunoassays have been one of the most elaborate tools in medical microbiology due to specificity, sensitivity, and streamlined technology (Andreotti et al., 2003); still, cross-reactions with nonpathogenic near relatives and pathogenic but phylogenetically distant agents make the approach impractical for large omic-scale interrogations and/or applications where multiple targets/pathogens may be expected (Malanoski et al., 2006; Liu et al., 2009; Sakai et al., 2014); this will be so at least until the emergence of "immunomic" profiles (Kambouris et al., 2018a,d), similar to highly degenerate DNA inquiry, exemplified by highly degenerate arrays with 2–50 capture probes for each target species/genus (Sakai et al., 2014; Liu et al., 2009; Gardner et al., 2010; Chou et al., 2006).

At the same time, expression profiles vary greatly by cell phase and status. Thus, proteomic and metabolomic solutions are suboptimal for cellular microbiota and practically inapplicable for acellular ones. Cultures, on the other hand, are time-consuming, unsuitable when biocontainment issues arise, and their applicability is not universal. Uncultivable and fastidious microorganisms may only be recovered by elaborate culturomic processes (see Chapter 8) and even this does not guarantee results (Kambouris et al., 2018b,c; Lagier et al., 2012; Palacios et al., 2007). This leaves the nucleic acid microarrays as the most suitable option (Palacios et al., 2007), although the phenotypic arrays have some positive score and promising potential in identification settings.

In the context of microbiome, microarrays containing probes for microbial detection, discovery, functionality or a combination of them are at a premium. The approaches may be distinguished according to the range of the targeted pathogens, the probe design strategy, the objectives served, and the microarray platform used (Gardner et al., 2010). The usual objectives focus on detecting either presence or functionality (Table 6.2).

When tackling the former, target selection and probe design must balance between uniqueness and conservation (Gardner et al., 2010). Detection arrays can tell what organisms are present with rather high resolution (lower taxon, usually species, level), and discovery arrays determine rather if there are agents present, with a low resolution (higher taxon); while functional arrays can tell what

Table 6.2 Microarrays of Microbiomic Application.

Type	Target	Purpose
Cognitive	Viruses	Detection
• Target, purpose	Bacteria	Discovery
Technical	Fungi	Presence
• Platform	Specific taxa (genera/species)	Functionality
• Reporter principle		
• Reporter design		

Function, purpose, and (sub)microbiomic targets of microarray categories and kinds potentially applicable in microbiomic studies.

qualities, capabilities, and traits these organisms might possess, such as virulence factors, resistance genes (Chizhikov et al., 2001; Gardner et al., 2010), and other translational parameters, which define a microbiote or predict its attitude in different conditions.

(Micro)arrays can be designed with a combination of detection and discovery probes. The former target species-specific regions (uniqueness), for precise characterization of known pathogens; the latter target more conserved regions (conservation), to enable detection of novel organisms with some homology to previously sequenced organisms (Chou et al., 2006; Sakai et al., 2014; Gardner et al., 2010). Still, there is an issue with truly novel, naturally occurring or engineered microbiota: probes designed from known sequences are unlikely to detect truly novel organisms lacking homology to such sequences (Gardner et al., 2010). Highly novel targets with no similarity to genomes deposited in databases or to probes on the microarray will not be detected (Gardner et al., 2010), and this is a strong argument in favor of NGS and other agnostic-oriented methods (Kambouris et al., 2018a,d). As new sequence data become available from newly discovered or newly sequenced organisms, continuous updates to incorporate probes for newly sequenced microbes will be required (Gardner et al., 2010).

This situation arose before, at a different level: when microarrays were coming to vogue, genomic data were sparse, and many applications were seriously handicapped by this fact (Gardner et al., 2010). Specific oligonucleotide probes for pathogen identification depend on assumptions regarding target sequence composition (Lin et al., 2006). Once massive sequencing, possibly in some cases NGS, alleviated this issue, better panmicrobial arrays could be designed (Gardner et al., 2010); but, at the same time, the NGS started making them redundant in terms of methodology, although not in terms of operations; when massive samples in a population or in an area (ethno- and geo-epidemiology, respectively) must be analyzed affordably, microarrays are at their best.

The applications of detection arrays are mostly comparative; they tackle diverse microbiota with any genomic or translational similarity, found in a certain environment/habitat or causing a certain (group of) syndromes. Respiratory pathogens in general and the respective microarrays are an eloquent example (Wang et al., 2006; Malanoski et al., 2006; Lin et al., 2006; Palacios et al., 2007). This approach is suitable for direct diagnostics.

A different approach provides for taxonomically similar targets, usually at the level of kingdom (fungi, bacteria, and viruses) but possibly at lower or higher levels (Sakai et al., 2014;Warsen et al., 2004; Roth et al., 2004; Palacios et al., 2007; Chou et al., 2006; Wang et al., 2002, 2004). In such applications, microbes can be detected when melting temperatures are at a level that allows hybridization, despite a lack of precise complementarity between probe and target (Palacios et al., 2007). Thus a microarray is only able to resolve identity to the level of divergence represented by the diversity of probes present (Wang et al., 2006).

Viral probes were designed to target a minimum of three genomic regions for each family or genus, including at least one highly conserved region coding for polymerase, which performs viral detection, and two or more variable regions, which enact the identification (Gardner et al., 2010; Palacios et al., 2007).

Microarray protocols and methods have been developed for the identification of a variety of pathogens, including viruses, bacteria, parasites, and fungi (Palacios et al., 2007; Sato et al., 2010; Sakai et al., 2014; Wang et al., 2004; Hong et al., 2004; Chou et al., 2006; Liu et al., 2009). Such arrays are better suited for surveillance, for seeking unknown or possibly novel agents or whenever broad classification is redundant or already implemented. The advent of "panmicrobial microarrays"

(Palacios et al., 2007) implies reporter oligonucleotides for agents from all kingdoms, while other approaches used reporters from different, but not all, kingdoms and were still referred to as "panmicrobial" (Gardner et al., 2010) instead of "multimicrobial" as would have been better. The exact target mixture is always an issue, as it defines target range and specificity.

In both functional and taxonomic applications the used microarrays may be designed for either detection or discovery or both (Gardner et al., 2010). Detection functions by rather stringent hybridization conditions, which may discriminate subtle differences among similar taxa, be they species or other. Conclusive microarray architectures, with interrelated loci and redundant reporters to increase accuracy and discrimination are the preferred formats (Palacios et al., 2007; Sakai et al., 2014; Chou et al., 2006). Individual strains are represented by multiple probes targeted to different and distinct genes for increased precision (Liu et al., 2009).

In discovery mode, probes are designed to tolerate some sequence variation to enable biding of divergent species with homology to sequenced organisms (Gardner et al., 2010). Thus microbes can be detected when melting temperatures are high to the point that allows hybridization, despite a lack of precise complementarity between probe and target (Palacios et al., 2007). These generic reporters bind all similar targets, be they known or unknown. Groups of stringent reporters affiliated to different generic ones reveal more exact sequence characteristics of the targets in the former case or establish the novel but relative status in the latter case (Palacios et al., 2007; Sakai et al., 2014; Chou et al., 2006).

On the other hand, tiled arrays able to actually (re)sequence fragments of interest can be even more descriptive (Malanoski et al., 2006), but, at the same time, of lower overall scope or, alternatively, extremely massive in number of features and thus requiring more sample or elaborate incubation protocols. Tiled resequencing arrays contain sets of 10^5–10^6 probes of either 25 or 29 mers and are extremely degenerate: for reading just one base in a sequence, up to 10 loci must be engaged: One for each possible base per strand ($2 \times 4 = 8$); 2 of the 8 probes represent perfect matches, while the others correspond to all possible mismatches at the central position of the probe. In addition, one more probe may—or may not—be added for a deletion per strand (Diamandis, 2000; Lin et al., 2006; Hacia, 1999), thus forming in essence a small, exclusive subarray.

Tiled sequencing arrays (or subarrays) come handier in the detection of phylogenetically associated agents, which share a differential but common nucleus of homology. The ribosomal DNA sequences in prokaryotes and eukaryotes are such cases; viruses are devoid of such, and different core genomic areas must be determined at the basis of large groups and/or higher taxa, exemplified by, but not restricted to, Baltimore divisions (Gardner et al., 2010).

All in all, there are some issues of strategy concerning the utility of panmicrobial arrays. The first is the use of long capture probes, 50–70 nucleotides long (Palacios et al., 2007; Lin et al., 2006; Heller, 2002), which increase accuracy but also the false-positive capture and thus compromise assay specificity; one nucleotide mismatch bears less impact on the binding affinity. Short probes (14–25 nucleotides) were the original approach (Heller, 2002) but grew to <40 (Malanoski et al., 2006) and work the other way: they are much less prone to mishybridization, or rather non-perfect match capture. But they are less specific, making it necessary to target multiple markers and rely on hybridization patterns; thus highly degenerate arrays, with many more loci, each populated with short reporters, are needed to increase specificity (Lin et al., 2006). As an example, probes designed against the 16S rRNA variable bacterial regions for a given species frequently cross-hybridized across distant bacterial genera, so that some bacteria could only be identified at family or class resolution (Gardner et al., 2010); in a fungal application 3–12 capture probes for each species/genus were deemed necessary, plus 6 universal fungal probes that would produce signal even if the fungal DNA in the sample was not anticipated and thus no specific respective probe had been bound on the microarray slide (Sakai et al., 2014).

The second issue is the use of instrumentation for reading an array or the development of instrumentation-free formats, for point-of-need use, an advantageous field for microarrays compared to other technologies. Pregnancy tests paved the way, and high densities can be developed in similar contexts, with sliding magnification glasses/plastics embedded at the assay bed so as to read the high density, small size, visually inquired loci.

PHENOTYPIC MICROARRAYS

Phenotypic (micro)arrays (PMs) came of age in early 2000s and introduced the concept of phenomics, allowing a breathing space to conventional microbiology and microbiologists during the genomic tsunami, which threatened almost all other approaches. Both the concept and the object bear somewhat misleading names, which is expected as they were corporate spawn to allow legacy biochemical assays to ride the wave of omics and high-throughput formats. Misleading, because PMs are neither microarrays (at the beginning the loci, wells of 96-well plate, were comfortably observable and resolvable by naked eye) nor phenotypic. Their advent focuses on physiology processes (practically assimilation and inhibition profiles), homogenized through respiration tagging, thus leaving morphology, the leading phenotypical cluster, totally out (Bochner et al., 2001).

Phenomics present by definition some infringement with culturomics (see Chapter 8), although the latter introduce some features of cultures, such as sample recovery and sample segregation and purification, absent from phenomics. In addition, the culturomics allows observation, recording and data extraction for every kind of growth dynamics, morphology included, but with a premium on growth no growth (Kambouris et al., 2018b).

Phenotype microarrays, as was the original name, where arrays by the loosest use of the term; their features were individually compartmentalized, aliquots of the sample had to be siphoned in each different compartment, and the signal moiety was not immobilized, but in liquid form in each compartment. Still, they allowed massive biochemical interrogation of strains, which, in most cases of prokaryotes, due to compromised or even uniform morphology, is the only way to discriminate taxa. And this holds true despite the heavy interference from lateral gene transfer, most of the times implemented by conjugation, but also by transformation and transduction (Stokes and Gillings, 2011). On top of that, the PMs allowed elaborate schemes for the testing of engineered mutants and for discriminating similar strains (Bochner, 2009).

LIVE CELL MICROARRAYS

Cell microarrays are a spin-off of phenotype microarrays, in that living cells and their physiology consist the *processing unit*; this is a much more elaborate, functional mechanism than the mostly analytical biochemical reactions of other, molecular microarray types (antibody-, enzyme-, or oligonucleotide-based).

In cell microarrays the cells are immobilized, and this is a defining difference from PMs where cells *are* the analyte. Cell arrays can be spotted or synthesized with photolithography or similar technologies onto slides for planar formats (or even, *sensu latto*, in microplate wells) or can be molded, encapsulated, or attached on bead-like structures for suspension formats. But their potential develops more when they are combined with microfluidic technology. Microfluidics are used in other (micro)arrays as well, and, along with some active array concepts, they progress beyond parallel formats to sequential and/or divergent, convergent, and other spatiotemporal arrangements aiming to emulate more complicated processes, such as physiology interactions (Hong et al., 2017). This aim led first to 3D cell loci, which better emulate actual dispositions and microenvironments of cells (Hong et al., 2017), and then to organoids, as the ultimate step (Powell, 2018).

Tissue (meta)arrays have been invented but are referring to histological samples. Areas of interest are collected in the form of cylindrical cuts out of histological blocks, then arrayed into *metablocks*, cut into thin sections, and prepared in densities of thousand samples per slide treated with absolute uniformity and interrogated simultaneously and uniformly (Jawhar, 2009). This approach allows significant savings in preparation and processing costs and labor over conventional section preparation of blocks of embedded tissues. There is unrealized potential in microbiomics for dissecting various dynamics and kinetics aspects of endocellular pathogens (as are prions, viruses, and mycoplasmata among others), from the location and form to the extent of damage. Application in both environmental/agriculture and medicinal/public health contexts will be a most valuable tool, especially as high-throughput microscopy and dye chemistries improve as well.

Basic cell arrays are used for screening, which means to determine toxic and septic effects, and thus allow studies on drug efficiency on pathogens and host cells; toxic effects of pollutants, drugs, biotoxins, radiomaterials, and from any other source. In this function the cell array usually is a complementary single-feature microarray to be attached to proper, true microarrays, where different stimuli are embedded, usually in non-degenerate formats. Other approaches include spotting adhesive material mixed with different transfective DNA sequences or viral vectors onto a slide, dipping it into cell dispersion and letting it to rest. Cells adhere and propagate onto the spotted material and become transfected by the embedded DNA or infected by the viruses. Thus different cells/reporters emerge to different loci in order to test a single analyte (Hong et al., 2017). In microbiomic applications, such formats can be used for drug and toxicity tests in the presence, absence, and combination of transfection elements and septic/toxic effect of transfected microbiota onto different host cell populations.

Cell arrays are easier to handle than liquid, multiwell plate tests (Hong et al., 2017), but their uniformity is far inferior to the respective of molecular microarrays. Cell populations are less manageable and disciplined than molecular populations (Yarmush and King, 2009). Different kinds of treatments have been developed in order to bind, signal, and stimulate cells for single-cell features, or to homogenize and synchronize cells for developing similar multicell loci: chemical, mechanical, and electromagnetic patternings are the essentials of microfabricated cell arrays, a methodology one step beyond spotting and printing.

And lastly come the microfluidic microarrays, characterized by the closed, controlled environment. All other approaches constitute open-volume formats with free liquid–air interfaces, resulting in actual inability to experimentally control the soluble environment in space and time over different, open-air loci. The microfluidic spin-off of cell microarrays, on the other hand, is characterized by the inherent provisions for controllable, dynamic flow of the liquid phase, allowing phased interrogation of the reporter cells, that is, in the presence of chemotactic factors to study thresholds and the molecular mechanisms in general (Yarmush and King 2009). Most microfluidic devices are controlled through the use of external pumps and tubing; however, recent advances have enabled design of integrated valve and pumping strategies (Unger et al., 2000). In addition, efforts have been made to adapt microfluidic devices to achieve compatibility with conventional microtiter plate laboratory automation equipment (Unger et al., 2000).

The other function of living cell arrays is to dissect cell-to-cell or virus-to-cell interaction. In this way, they may assist in microbiomic studies much more than culturomics and, even more, phenomics. Both of them focus on growth in more or less defined environments, while cell arrays may come handy in elucidation of virulence in terms of direct interactions (pathogen- to- host and immune response agent -to- pathogen); in analytically dissecting microecology, by tracking changes in dynamically changing microenvironments; in establishing functionality patterns, including exogenome–microbiome relations; and in determination (detection–identification) of uncultivable strains or dormant forms while offering a higher biocontainment level (Yarmush and King, 2009; Hong et al., 2017).

CONCLUSION

The microarray, once developed, allows precisely targeted, massive inquiries and very high content of information; much of its cost is upfront, at the R&D stage and at the acquisition of the needed hardware (scanner/reader, and perhaps incubator) and software (image processing and analysis, scoring, allele calling and deconvolution). The more the number of loci increases, the design and development become vastly more challenging, especially in the "conclusive" kind of microarrays where the combined differential signal of all loci produces the result (Cremonesi et al., 2007). To resolve interactions and resulting low specificity, one can always revert to redundancy, adding multiple features per target and detailing software to make decisions (Liu et al., 2009; Gardner et al., 2010). However, this approach increases cost and complexity to levels where cost-effectiveness might be cast in doubt.

Still, microarrays have considerable potential in microbiome-related studies compared to all and any other methodologies, a fact that suggests keeping a research and even an industrial base alive for improving and evolving them. They are polyvalent; which means there are many different kinds of arrays, with different reporting principles, targets, and applications, but with relatively similar support and infrastructure requirements in design, production, reading, and analysis. They are discrete, and this discretion also saves cost and time per assay.

They also, and most importantly, massively perform similar assays at a fraction of the time and cost of accuracy diagnostics, especially in molecular genetics and in direct diagnosis and fungal identification of onychomycosis agents (Han et al., 2014). DNA microarrays achieve this by simultaneously interrogating hundreds to thousands of immobilized DNA oligonucleotides, where each reporter oligonucleotide provides a single query for a known sequence that is unique for an organism or trait (nondegenerate arrays). The alternative was to perform repeatedly the tests for one biomarker at a time (Lin et al., 2006), which had been the concept of first-generation, serial interrogation, low-throughput molecular genetics. Even more interesting is the fact that this parallel, massive format is attainable without resolving the whole genome as is the concept of NGS, a procedure that is costly, time-consuming, and fraught with algorithmic guesswork; educated, and even deep-learning compliant, but still guesswork. And the microarrays can do so reproducibly with little concern for depot conditions and logistics for at least as long as there is stock of the same production batch (Kambouris, 2016). Such attributes are precious for whole new fields of research, as are the electrogenomics, which need to pinpoint the genes implicated in responses to electric and magnetic amenities (Gao et al., 2005; Zhao et al., 2006).

In cases where slight differences in DNA entail significant biological divergence, as in pathovars and pathotypes and in some particular varieties, subtypes, serotypes, and strains in both plant protection and clinical microbiology (Kim and Mudgett, 2013; Palacios et al., 2007; Martinez et al., 2001; Clermont et al., 2000, respectively), an array can produce fast, targeted resolution after a simplex PCR round instead of PCR sequencing per sample in a massive sample influx. Contrary to NGS that produces the ultimate volume and context of information in a given, simple or complicated sample, the microarray can resolve cost-effectively large numbers of samples, each of them for quite a number of biomarkers, without any dependence on their respective positioning and possibly do so in the field, as mentioned in the following paragraph.

On the same page, some kinds of arrays (including DNA microarrays) can be used without elaborate equipment, becoming point-of-need assays. To screen nucleic acids from an uncharacterized sample, sequencing remains a favored choice as it provides the most in-depth and unbiased information. However, the expense and time required for sequencing using high-throughput methods such as 454® (Roche), Solexa® (Illumina), or SOLiD® (Life Technologies) can make these methods prohibitive for routine or even for expedited use. Until the processing time and cost of high-throughput sequencing (including data analysis) decreases enough to be feasible for large numbers of samples at sufficient depth, microarrays will continue to be a valuable tool (Gardner et al., 2010). They are likely to carry on being so whenever time-related and infrastructure-independent logistics burden the choice of approach (Kambouris, 2016; Kambouris et al., 2018a; Palacios et al., 2007; Gardner et al., 2010; Malanoski et al., 2006; Lin et al., 2006).

Furthermore, outbreaks caused by different pathogens may also overlap in time and geography (Palacios et al., 2007). In such conditions, multiple targets, possibly co-existing, must be resolved without starting titers becoming an issue, due mainly to amplification bias in the random PCR sample preparation protocols (Gardner et al., 2010). It is well established that with unique/pure samples involving nucleic acids from an uncharacterized sample, sequencing provides the most in-depth and unbiased information (Gardner et al., 2010). NGS outputs definite results, with less problems and fuss. Though mixed samples are another issue altogether as NGS relies mostly on software than wet procedure to make sense and calls, and there lie many uncertainties, and possibly some extrapolations as well. Arrays, on the other hand, are more objective, need less interpretation, even if signal processing can cause some (systematic, not sample-specific) issues. And this was firmly established in 2007, which marked the year of their proof-of-concept: in two documented case reports, multiplexed genomic arrays, namely the panmicrobial GreeneChip (Palacios et al., 2007) and the panviral ViroChip (Chiu et al., 2007) detected and classified an infectious agent when traditional methods of investigation had failed (Liu et al., 2009).

Inexpensive array chips can be used in many scenarios, in adverse, austere, primitive, or even negated conditions; but, the most important is that they can evolve into biosensors, not simple assays. In particular, instrument-free microarrays, which also require minimal analyte pretreatment, as are the cellular arrays and the antibody arrays, can produce massively actionable information in analytic (the latter) and in both analytic and functional (the former) queries in either pure or complex environmental or clinical samples. Thus, especially some more elaborate designs, using microfluidics and possibly implementing not exclusively parallel processing, can be the front end of complex instruments and procedures that would explore or survey inaccessible, denied or contested environments (i.e., in biosecurity and biosafety applications, as mentioned in Chapter 15) without explicit need for active human interference. Robotic, educated mission solutions or remotely controlled arms/probes, vehicles and vessels will be able to persistently and/or promptly survey space of interest, in micro- or macro-environments.

The microarrays in such a context can be reusable, permanent, or expendable, depending on technologies and cost. But modular solutions, expanding as and if required and replaced by different, alternative combinations of features or even kinds of microarrays, probably designed and produced on-demand (Fig. 6.3), seem to be the most promising approach.

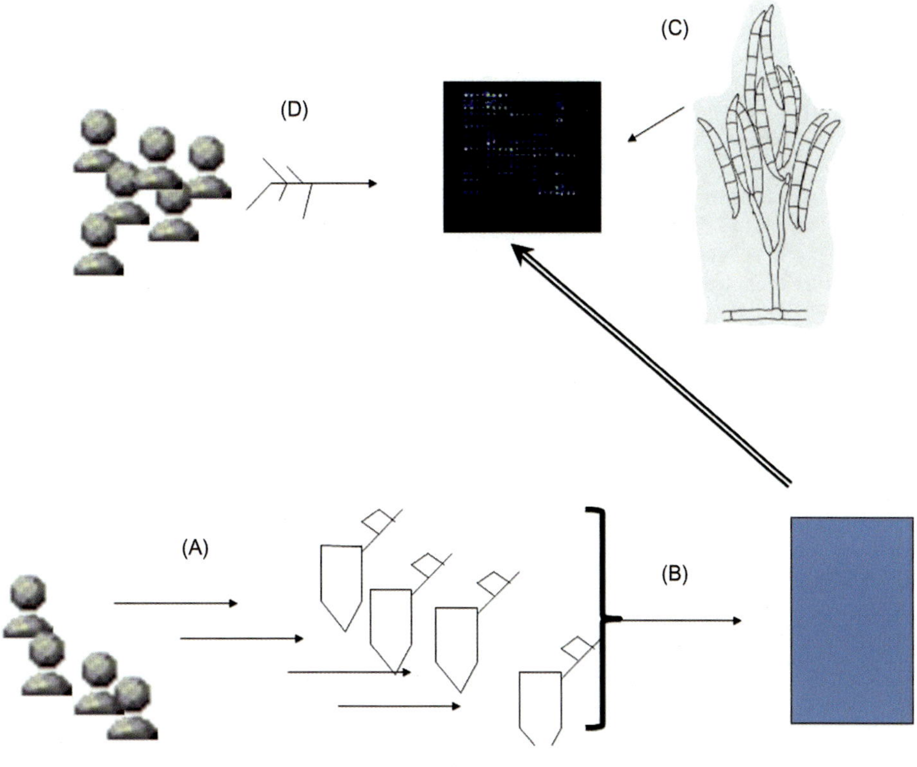

FIGURE 6.3

The concept of on-demand, ad hoc screening microarrays: the population of interest is sampled, possibly recorded, or randomized or/and deidentified, according to concept. The samples are treated individually (A), possibly amplified, and spotted (B) to a surface (glass/plastic slide or other) to produce a microarray. Each sample can be spotted once or in multiple, to create different degrees of redundancy, as required. The microarray is exposed to similar or dissimilar interrogation, which might be chemical, physical, or biological in nature (i.e., binding, hybridization, and fluorescence). After calling the loci, conclusions can be drawn on the population sampled (D) which may, or may not, become individualized, depending on the exact proceedings during and provisions of step (A).

These biosensor platforms will be focused on supervision and intervention and deployed only as and when needed. Novel pathogens emerge in new contexts and treatment strategies are beginning to be tailored to specific infectious agents (Palacios et al., 2007). Especially after the rushed fielding into the real world in 2007 as mentioned earlier, the primary service of arrays in microbial context has shifted: from characterization of agents propagated to high titer *in vitro* (Palacios et al., 2007), the new focus in the area of microbiomics seems to be on the resolution of complicated samples *in situ*, in both environmental and clinical contexts.

REFERENCES

Andreotti, P.E., Ludwig, G.V., Peruski, A.H., Tuite, J.J., Morse, S.S., Peruski, L.F., 2003. Immunoassay of infectious agents. Biotechniques 35, 850–859.

Bochner, B.R., 2009. Global phenotypic characterization of bacteria. FEMS Microbiol. Rev. 33, 191–205.

Bochner, B.R., Gadzinski, P., Panomitros, E., 2001. Phenotype microarrays for high-throughput phenotypic testing and assay of gene function. Genome Res. 11, 1246–1255.

Chang, T.W., 1983. Binding of cells to matrixes of distinct antibodies coated on solid surface. J. Immunol. Methods 65, 217–223.

Chiu, C.Y., Alizadeh, A.A., Rouskin, S., Merker, J.D., Yeh, E., Yagi, S., et al., 2007. Diagnosis of a critical respiratory illness caused by human metapneumovirus by use of a pan-virus microarray. J. Clin. Microbiol. 45, 2340–2343.

Chizhikov, V., Rasooly, A., Chumakov, K., Levy, D.D., 2001. Microarray analysis of microbial virulence factors. Appl. Environ. Microbiol. 67, 3258–3263.

Chou, C.-C., Lee, T.-T., Chen, C.-H., Hsiao, H.-Y., Lin, Y.-L., Ho, M.-S., et al., 2006. Design of microarray probes for virus identification and detection of emerging viruses at the genus level. BMC Bioinformatics 7, 232.

Clermont, O., Bonacorsi, S., Bingen, E., 2000. Rapid and simple determination of the *Escherichia coli* phylogenetic group. Appl. Environ. Microbiol. 66, 4555–4558.

Cremonesi, L., Ferrari, M., Giordano, P.C., Harteveld, C.L., Kleanthous, M., Papasavva, T., et al., 2007. An overview of current microarray-based human globin gene mutation detection methods. Hemoglobin 31, 289–311.

Diamandis, E.P., 2000. Sequencing with microarray technology--a powerful new tool for molecular diagnostics. Clin. Chem. 46, 1523–1525.

Euler, M., Wang, Y., Otto, P., Tomaso, H., Escudero, R., Anda, P., et al., 2012. Recombinase polymerase amplification assay for rapid detection of *Francisella tularensis*. J. Clin. Microbiol. 50, 2234–2238.

Gao, W., Liu, Y., Zhou, J., Pan, H., 2005. Effects of a strong static magnetic field on bacterium *Shewanella oneidensis*: an assessment by using whole genome microarray. Bioelectromagnetics 26, 558–563.

Gardner, S.N., Jaing, C.J., McLoughlin, K.S., Slezak, T.R., 2010. A microbial detection array (MDA) for viral and bacterial detection. BMC Genomics 11, 668.

Gkantouna, V.A., Kambouris, M.E., Viennas, E.S., Ioannou, Z.-M., Paraskevas, M., Lagoumintzis, G., et al., 2015. Introducing dAUTObase: a first step towards the global scale geoepidemiology of autoimmune syndromes and diseases. Bioinformatics 31, 581–586.

Glais, L., Jacquot, E., 2015. Detection and characterization of viral species/subspecies using isothermal recombinase polymerase amplification (RPA) assays. Methods Mol. Biol. 1302, 207–225.

Goelet, P., Knapp, M.R., Anderson, S., 1999. U.S. Patent No 5,888,819. U.S. Patent and Trademark Office, Washington, DC.

Guess, T., Lai, H., Smith, S.E., Sircy, L., Cunningham, K., Nelson, D.E., et al., 2018. Size matters: measurement of capsule diameter in *Cryptococcus neoformans*. J. Vis. Exp. 132. Available from: https://doi.org/10.3791/57171.

Hacia, J.C., 1999. Resequencing and mutational analysis using oligonucleotide microarrays. Nat. Genet. 21, 42–47.

Han, H.W., Hsu, M.M., Choi, J.S., Hsu, C.K., Hsieh, H.Y., Li, H.C., Chang, H.C., Chang, T.C., 2014. Rapid detection of dermatophytes and *Candida albicans* in onychomycosis specimens by an oligonucleotide array, BMC Infect Dis 14, 581.

Heller, M.J., 2002. DNA microarray technology: devices, systems, and applications. Annu. Rev. Biomed. Eng. 4, 129–153.

Hong, B.-X., Jiang, L.-F., Hu, Y.-S., Fang, D.-Y., Guo, H.-Y., 2004. Application of oligonucleotide array technology for the rapid detection of pathogenic bacteria of foodborne infections. J. Microbiol. Methods 58, 403–411.

Hong, H., Koom, W., Koh, W.-G., 2017. Cell microarray technologies for high-throughput cell-based biosensors. Sensors 17, 1293.

Hook, A.L., Thissen, H., Voelcker, N.H., 2009. Surface plasmon resonance imaging of polymer microarrays to study protein – polymer interactions in high throughput. Langmuir 25, 9173–9181.

Huguenin, A., Moutte, L., Renois, F., Leveque, N., Talmud, D., Abely, M., et al., 2012. Broad respiratory virus detection in infants hospitalized for bronchiolitis by use of a multiplex RT-PCR DNA microarray system. J. Med. Virol. 84, 979–985.

Jawhar, N.M., 2009. Tissue microarray: a rapidly evolving diagnostic and research tool. Ann. Saudi Med. 29 (2), 123–127.

Jiang, L., Tsubakihara, M., Heinke, M.Y., Yao, M., Dunn, M.J., Phillips, W., et al., 2001. Heart failure and apoptosis: electrophoretic methods support data from micro- and macro-arrays. A critical review of genomics and proteomics. Proteomics 1, 1481–1488.

Kambouris, M.E., 2016. Population screening for hemoglobinopathy profiling: is the development of a microarray worthwhile? Hemoglobin 40, 240–246.

Kambouris, M.E., Kantzanou, M., Arabatzis, M., Velegraki, A., Patrinos, G.P., 2018a. Rebooting bioresilience: a multi-OMICS approach to tackle Global Catastrophic Biological Risks (GCBRs) and next generation biothreats. OMICS-JIB 22, 35–51.

Kambouris, M.E., Pavlidis, C., Skoufas, E., Arabatzis, M., Kantzanou, M., Velegraki, A., et al., 2018b. Culturomics: a new kid on the block of phenomics and pharmacomicrobiomics for personalized medicine. OMICS-JIB 22, 108–118.

Kambouris, M.E., Manousopoulos, Y., Kritikou, S., Milioni, A., Mantzoukas, S., Velegraki, A., 2018c. Towards decentralized agrigenomics surveillance? A PCR-RFLP approach for adaptable and rapid detection of user-defined fungal pathogens in potato crops. OMICS-JIB 22, 264–273.

Kambouris, M.E., Gaitanis, G., Manoussopoulos, Y., Arabatzis, M., Kantzanou, M., Kostis, G.D., et al., 2018d. Humanome versus microbiome: games of dominance and pan-biosurveillance in the Omics universe. OMICS 22, 528–538.

Kim, J.-G., Mudgett, M.B., 2013. XopD peptidase. In: Rawlings, N.D., Salvesen, G. (Eds.), Handbook of Proteolytic Enzymes. Elsevier, pp. 2382–2385.

Lagier, J.-C., Armougom, F., Million, M., Hugon, P., Pagnier, I., Robert, C., et al., 2012. Microbial culturomics: paradigm shift in the human gut microbiome study. Clin. Microbiol. Infect. 18, 1185–1193.

Lee, Y.-S., Tsai, C.-N., Tsai, C.-L., Chang, S.-D., Hsueh, D.-W., Liu, C.-T., et al., 2008. Comparison of whole genome amplification methods for further quantitative analysis with microarray-based comparative genomic hybridization. Taiwan. J. Obstet. Gynecol. 47, 32–41.

Lin, B., Wang, Z., Vora, G.J., Thornton, J.A., Schnur, J.M., Thach, D.C., et al., 2006. Broad-spectrum respiratory tract pathogen identification using resequencing DNA microarrays. Genome Res. 16, 527–535.

Lindroos, K., Liljedahl, U., Raitio, M., Syvänen, A.C., 2001. Minisequencing on oligonucleotide microarrays: comparison of immobilisation chemistries. Nucleic Acids Res. 29, E69–E99.

Liu, Y., Sam, L., Li, J., Lussier, Y.A., 2009. Robust methods for accurate diagnosis using pan-microbiological oligonucleotide microarrays. BMC Bioinformatics 10 (Suppl. 2), S11.

Malanoski, A.P., Lin, B., Wang, Z., Schnur, J.M., Stenger, D.A., 2006. Automated identification of multiple micro-organisms from resequencing DNA microarrays. Nucleic Acids Res. 34, 5300–5311.

Mardis, E.R., 2013. Next-generation sequencing platforms. Annu. Rev. Anal. Chem. 6, 287–303.

Martinez, L.R., Garcia-Rivera, J., Casadevall, A., 2001. *Cryptococcus neoformans* var. *neoformans* (serotype D) strains are more susceptible to heat than *C. neoformans* var. *grubii* (serotype A) strains. J. Clin. Microbiol. 39, 3365–3367.

Palacios, G., Quan, P., Jabado, O.J., Conlan, S., Hirschberg, D.L., Liu, Y., et al., 2007. Panmicrobial oligonucleotide array for diagnosis of infectious diseases. Emerg. Infect. Dis. 13, 73–81.

Parinov, S., Barsky, V., Yershov, G., Kirillov, E., Timofeev, E., Belgovskiy, A., et al., 1996. DNA sequencing by hybridization to microchip octa-and decanucleotides extended by stacked pentanucleotides. Nucleic Acids Res 24, 2998–3004.

Patirupanusara, P., Jackrit Suthakorn, J., 2012. Introduction of an active DNA microarray fabrication for medical applications. In: International Conference on Advances in Electrical and Electronics Engineering (ICAEEE'2012). April 13–15, 2012, Pattaya, pp. 75–79.

Powell, M., 2018. Organoids: A New Model to Study Infectious Diseases?. Infectious Diseases Hub. Available from: <https://www.id-hub.com/2018/06/14/organoids-new-model-study-infectious-diseases/> (accessed 17.06.18.).

Peytavi, R., Raymond, F.R., Gagné, D., Picard, F.J., Jia, G., Zoval, J., et al., 2005. Microfluidic device for rapid (15 min) automated microarray hybridization. Clin. Chem. 51, 1836–1844.

Quail, M.A., Smith, M., Coupland, P., Otto, T.D., Harris, S.R., Connor, T.R., et al., 2012. A tale of three next generation sequencing platforms: comparison of Ion Torrent, Pacific Biosciences and Illumina MiSeq sequencers. BMC Genomics 13, 341.

Roth, S.B., Jalava, J., Ruuskanen, O., Ruohola, A., Nikkari, S., 2004. Use of an oligonucleotide array for laboratory diagnosis of bacteria responsible for acute upper respiratory infections. J. Clin. Microbiol. 42, 4268–4274.

Sakai, K., Trabasso, P., Moretti, M.L., Mikami, Y., Kamei, K., Gonoi, T., 2014. Identification of fungal pathogens by visible microarray system in combination with isothermal gene amplification. Mycopathologia 178, 11–26.

Sato, T., Takayanagi, A., Nagao, K., Tomatsu, N., Fukui, T., Kawaguchi, M., et al., 2010. Simple PCR-Based DNA microarray system to identify human pathogenic fungi in skin. J. Clin. Microbiol. 48, 2357–2364.

Selvin, P.R., 2000. The renaissance of fluorescence resonance energy transfer. Nat. Struct. Biol. 7, 730–734.

Shin, J., 2008. Electrically Active Microarray of 3D Carbon Mems Electrodes for Pathogen Detection Systems (M.Sc. thesis). San Diego State University. Available from: <http://www.digitaladdis.com/sk/Phoebe_Shin_Thesis_Kassegne_Lab.pdf> (accessed 22.12.18.).

Shumaker, J.M., Metspalu, A., Caskey, C.T., 1996. Mutation detection by solid phase primer extension. Hum. Mutat. 7, 346–354.

Sosnowski, R., Heller, M.J., Tu, E., Forster, A.H., Radtkey, R., 2002. Active microelectronic array system for DNA hybridization, genotyping and pharmacogenomic applications. Psychiatr. Genet. 12, 181–192.

Stokes, H.W., Gillings, M.R., 2011. Gene flow, mobile genetic elements and the recruitment of antibiotic resistance genes into Gram-negative pathogens. FEMS Microbiol. Rev. 35, 790–819.

Stomakhin, A.A., Vasiliskov, V.A., Timofeev, E., Schulga, D., Cotter, R.J., Mirzabekov, A.D., 2000. DNA sequence analysis by hybridization with oligonucleotide microchips: MALDI mass spectrometry identification of 5mers contiguously stacked to microchip oligonucleotides. Nucl Acid Res. 28, 1193–1198.

Strizhkov, B.N., Drobyshev, A.L., Mikhailovich, V.M., Mirzabekov, A.D., 2000. PCR amplification on a microarray of gel-immobilized oligonucleotides: detection of bacterial toxin- and drug-resistant genes and their mutations. Biotechniques 29, 844–848. 850–852, 854 passim.

Unger, M.A., Chou, H.P., Thorsen, T., Scherer, A., Quake, S.R., 2000. Monolithic microfabricated valves and pumps by multilayer soft lithography. Science 288, 113–116.

Unknown, 1994. ICH Harmonised Tripartite Guideline. Validation of Analytical Procedures, Text And Methodology Q2(R1) Step 4, Version 13. Available from: <https://www.ich.org/fileadmin/Public_Web_Site/ICH_Products/Guidelines/Quality/Q2_R1/Step4/Q2_R1__Guideline.pdf> (accessed 21.12.18.).

Walt, D.R., 2000. Techview: molecular biology. Bead-based fiber-optic arrays. Science 287, 451–452.

Wang, D., Coscoy, L., Zylberberg, M., Avila, P.C., Boushey, H.A., Ganem, D., et al., 2002. Nonlinear partial differential equations and applications: microarray-based detection and genotyping of viral pathogens. Proc. Natl. Acad. Sci. U.S.A. 99, 15687–15692.

Wang, Z., Vora, G.J., Stenger, D.A., 2004. Detection and genotyping of *Entamoeba histolytica*, *Entamoeba dispar*, *Giardia lamblia*, and *Cryptosporidium parvum* by oligonucleotide microarray. J. Clin. Microbiol. 42, 3262–3271.

Wang, H.-Y., Luo, M., Tereshchenko, I.V., Frikker, D.M., Cui, X., Li, J.Y., et al., 2005. A genotyping system capable of simultaneously analyzing >1000 single nucleotide polymorphisms in a haploid genome. Genome Res. 15, 276–283.

Wang, Z., Daum, L.T., Vora, G.J., Metzgar, D., Walter, E.A., Canas, L.C., et al., 2006. Identifying influenza viruses with resequencing microarrays. Emerg. Infect. Dis. 12, 638–646.

Warsen, A.E., Krug, M.J., LaFrentz, S., Stanek, D.R., Loge, F.J., Call, D.R., 2004. Simultaneous discrimination between 15 fish pathogens by using 16S ribosomal DNA PCR and DNA microarrays. Appl. Environ. Microbiol. 70, 4216–4221.

Yarmush, M.L., King, K.R., 2009. Living-cell microarrays. Annu. Rev. Biomed. Eng. 11, 235–257.

Zhao, M., Song, B., Pu, J., Wada, T., Reid, B., Tai, G., et al., 2006. Electrical signals control wound healing through phosphatidylinositol-3-OH kinase-gamma and PTEN. Nature 442, 457–460.

Zlatanova, J., Mirzabekov, A., 2001. Gel-immobilized microarrays of nucleic acids and proteins: production and application for macromolecular research. DNA Arrays. Humana Press, New Jersey, pp. 17–38.

CHAPTER

METAGENOMICS IN MICROBIOMIC STUDIES

7

Martin Laurence

Shipshaw Labs, Montreal, QC, Canada

INTRODUCTION

In 2005, the introduction of high-throughput DNA sequencing enabled new techniques for the detection of nonabundant microbes in clinical specimens, including unculturable, fastidious, and novel species. Prior to high-throughput DNA sequencing, the detection of novel species from extracted DNA was done using consensus PCR and single-read Sanger sequencing (Sanger et al., 1977), which only revealed the most abundant microbe whose genome matched chosen PCR primers. By substituting single-read Sanger sequencing with 454 pyrosequencing, thousands of amplicons could be sequenced at once, enabling the detection of less abundant microbes (Dowd et al., 2008). This method is colloquially called "16S," which refers to the sedimentation rate of the amplified *rrs* gene's RNA transcript in bacteria. The technically correct term for this method is a mouthful: "high-throughput sequencing of consensus PCR-amplified small ribosomal subunit RNA gene (*rrs*)." This method was a major step forward in microbe detection technology which enabled the microbiome research era. However, it came with two important limitations.

First, consensus PCR amplification is *not universal*. A pair of consensus PCR primers needs to be selected to amplify a small region of a microbe's genome—usually the small ribosomal subunit RNA gene (*rrs*). Though *rrs* contains highly conserved sequences suitable for consensus PCR, primers matching *all known* species do not exist—let alone undiscovered species! Typically used V34 primers (`CCTACGGGNGGCWGCAG` and `GGACTACHVGGGTATCTAATCC`) amplify most known bacterial species' *rrs* gene, but miss much of the tree of life, including fungi, protists, and archaea. Unlike cellular microbes, viruses do not have *rrs* genes and cannot be detected using this method.

Second, consensus PCR amplification is *biased*. Amplicons containing either a very low or a very high fraction of guanine and cytosine (G/C) bases are underrepresented in the final product. Amplicons with high G/C content form strong hairpins (the DNA strand being copied folds back onto itself) which DNA polymerase enzymes have difficulty copying. Amplicons with low G/C content tend to denature (the two DNA strands separate) before DNA polymerase enzymes have time to complete the copy. Mismatches between consensus primers and target genes introduce a second important source of bias. Though such mismatches can be partially addressed using degenerate bases (underlined in V34 primers earlier), priming bias due to mismatches cannot be eliminated completely.

Since the consensus PCR step is causing all this trouble, why not skip it? This new method was initially coined "unbiased high-throughput sequencing" to emphasize its low bias and universality

Microbiomics. DOI: https://doi.org/10.1016/B978-0-12-816664-2.00007-4

(Lipkin, 2010). Today, it is usually referred to as metagenomics, which captures both differences (meta) and similarities (genomics) with whole genome shotgun sequencing on which it was initially based. The simplest metagenomics approach to study the human microbiome uses the same laboratory protocol that is typically used to sequence the human genome; only the bioinformatics analysis differs.

The first step in standard genomic bioinformatics workflows is to find host sequences and discard unaligned sequences. Unaligned sequences are those that do not seem to belong to the host and likely originate from laboratory contaminants or the host's microbiome. In microbiome metagenomics workflows, the first bioinformatics step is the exact opposite: discard host sequences and keep unaligned sequences for microbiome analysis. These unaligned sequences are then aligned to some or all GenBank sequences, resulting in taxonomical classification for each non-host sequence. A similar workflow can be used to detect microbes by sequencing extracted RNA rather than extracted DNA (Cottier et al., 2018). RNA-based methods typically outperform DNA-based methods on all metrics except bias, and can detect RNA-only viruses which are completely missed by DNA-based methods.

Metagenomics scored two early wins in 2008, when 454 pyrosequencing was used to identify novel viruses in human clinical specimens using deep RNA sequencing (Feng et al., 2008; Palacios et al., 2008). The first study detected two viral sequences out of 395,734 total sequences (0.0005%) (Feng et al., 2008), while the second detected 14 viral sequences out of 103,632 (0.0135%) (Palacios et al., 2008). In both cases the vast majority of sequences found were human transcripts, mainly ribosomal and messenger RNA, which were discarded during data analysis. This highlights the two key differences between consensus PCR and metagenomic assays: the main advantage of metagenomics is the detection of all microbes and viruses, including novel species. The main drawback is cost, because a large number of human sequences are discarded during data analysis—sequencing them is both uninformative and expensive! Consensus PCR assays are still commonly used because they efficiently eliminate most human DNA during the PCR amplification step, meaning that less sequences are wasted. As DNA sequencing technology becomes cheaper, the case for using metagenomics instead of consensus PCR improves. Today, metagenomics studies are mainly based on Illumina technology, which is largely credited for the more than 1000-fold decrease in sequencing cost over the last decade. This decrease in cost is expected to continue.

COMMENSALS AND INFECTIOUS AGENTS

One of the main goals of metagenomics is universality: the ability to detect all microbes and viruses, both known and unknown (Fig. 7.1). It is worthwhile to review the main characteristics of known microbial and viral classes, as these affect library preparation flows and assay sensitivity. Only one class of known self-replicating agents is undetectable using metagenomics: prions. These are misfolded proteins, and metagenomic assays cannot detect protein folding.

All known cellular microbes store their genes in very large double-stranded DNA molecules called chromosomes. Most microbes have a single DNA copy of their genome in each cell, whereas most animal cells have two. When calculating the minimum amount of DNA from a microbe present in a clinical specimen, the genome size is used as a lower bound. It is not possible to have

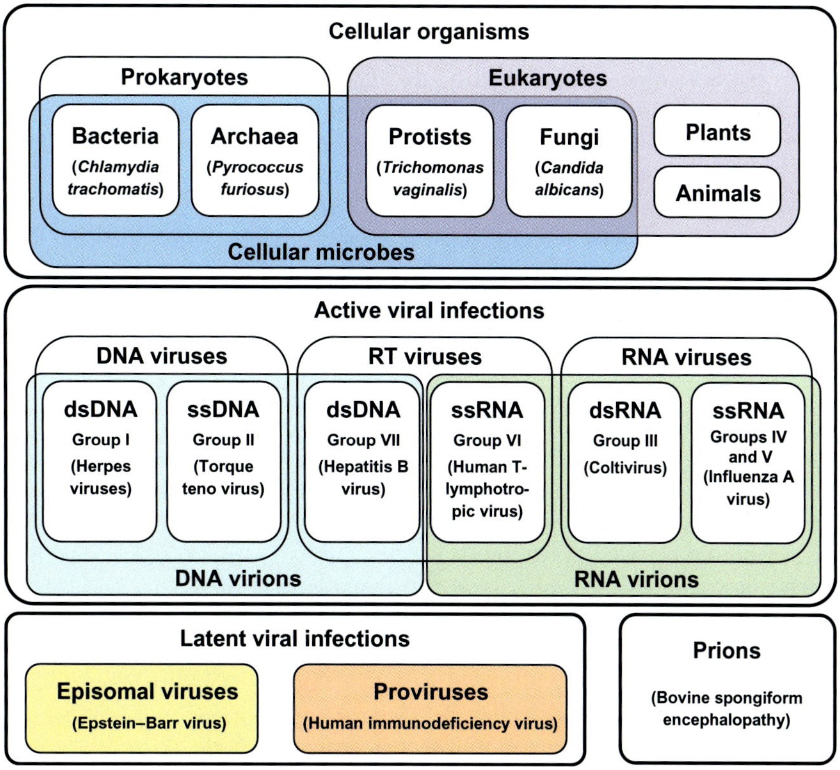

FIGURE 7.1

Venn diagram of self-replicating commensals and infectious agent types, with examples for each type. Prions are only composed of protein, thus cannot be directly detected using DNA or RNA sequencing.

fractional genomes in a clinical specimen because entire microbial cells are being collected! Only after DNA extraction—which naturally shears chromosomes into ~30,000 base pair DNA fragments—and aliquoting do fractional genomes occur. The record for the smallest medically important cellular microbe genome currently belongs to *Mycoplasma genitalium* (~580,000 base pairs). Obligate symbionts such as mitochondria can have much smaller genomes, because many of their genes and functions have been subcontracted to the host cell: using their genome size as a lower bound for metagenomic assays does not makes sense. Thus 580,000 base pairs is the most conservative (smallest) genome size and is used to estimate assay sensitivity.

All known cellular microbes synthesize proteins using ribosomes made mostly of two RNA strands transcribed from genes *rrs* and *rrl*. The number of ribosomes per cell varies between species and growth environments: ribosome counts mainly track cell size and metabolic activity. *Spiroplasma melliferum*, one of the smallest known bacterial species, can have as few as ~1000 ribosomes per cell (Ortiz et al., 2006). In contrast, a *Saccharomyces cerevisiae* yeast cell growing in optimal culture conditions contains ~178,000 ribosomes, which drops by two-thirds in slow growth conditions (von der Haar, 2008). An *Escherichia coli* bacterium cell contains ~55,000 ribosomes in optimal growth conditions (Bakshi et al., 2012). In contrast with minimum genome sizes, it is difficult to place an exact lower bound on the minimum number of ribosomes per cell, which makes sensitivity and cell count analysis less accurate when using RNA as the sequencer's input.

Viruses are acellular microbes and lack many conserved characteristics of cellular microbes. Their genomes are much smaller than those of cellular microbes, and they lack *rrs* and *rrl* genes. To replicate, viruses use ribosomes in the cells they infect, feeding them messenger RNA molecules that encode viral proteins. Within infected cells, viruses can insert DNA within the host's chromosomes (proviruses) or be found as DNA plasmids separate from the host's chromosomes (episomal viruses). Some virus types do not make use of DNA at all and operate with only RNA. RNA-only viruses cannot be detected using DNA extracted from clinical specimens. DNA-only viruses do not exist because it would be impossible for them to synthesize their proteins, as this requires messenger RNA.

However, some DNA viruses—such as the human immunodeficiency virus and the Epstein–Barr virus (EBV)—can enter latency. This means that for a time, they do not produce much RNA in infected cells and lay dormant. Detecting dormant viruses is challenging because unlike active infections, only a few cells in a specimen might be infected. For example, EBV infects hosts for life by persisting in memory B cells. Detecting EBV DNA in blood is very challenging because only ~1 infected memory B cell is present per 5 mL of blood (Thorley-Lawson, 2015).

When transiting between cells or hosts, viruses pack their DNA or RNA genomes into virions, the viral equivalent of a cell. As one would expect, detecting DNA-containing virions requires DNA extraction/sequencing, and detecting RNA-containing virions requires RNA extraction/sequencing. The sequencing depth needed to detect as little as a *single* virion in a clinical specimen is much greater than detecting a *single* microbial cell, because each virion carries much less DNA/RNA than a cellular microbe. The largest known medically important virions are herpes viruses, containing 125,000 (HHV-3) to 236,000 (HHV-5) base pairs of DNA. One of the smallest known medically important viruses is hepatitis D, which contains only 1700 bases of RNA. Many medically important viruses have genome sizes of about 10,000 bases.

FIVE KEY METRICS

Consensus PCR and metagenomic assay performance can be measured using five key metrics: sensitivity, efficiency, bias/universality, ability to taxonomically classify novel microbes, and susceptibility to contamination.

SENSITIVITY

The sensitivity of an assay is defined as the minimum number of microbial cells of a given species which can be detected reliably. Once all microbial cells present in a clinical specimen are detected reliably, sensitivity can no longer be improved. Increasing sequencing depth beyond this point is wasteful, except to refine microbial abundance estimates.

EFFICIENCY

The efficiency of an assay is defined as the cost to achieve a given sensitivity level. The main determinant of efficiency is the host-to-microbe sequence ratio. Both host and microbial cells are present in clinical specimens, and their DNA/RNA is extracted simultaneously, ending-up thoroughly mixed together in the eluate (suspended in water or TE buffer). Host sequences are extremely abundant in some sample types such as biopsies. Host sequences are not informative in microbiome analysis, so assays which partially eliminate host DNA/RNA strands before loading them into the sequencer reduce the cost of achieving a given sensitivity level. For example, V34 primers used in consensus PCR-based microbiome assays were chosen to *not* amplify human DNA, making this assay highly efficient. In contrast, metagenomic studies such as those described in the introduction (Feng et al., 2008; Palacios et al., 2008) are more than a 1000-fold less efficient, because more than 99.9% of the DNA/RNA loaded into the sequencer is human. Many different techniques can be used to increase the efficiency of metagenomic assays. As the cost of sequencing drops, these techniques gradually become less useful because maximum sensitivity can be reached at a reasonable price.

BIAS/UNIVERSALITY

Bias occurs when a microbe present in a clinical specimen is underrepresented in the sequencing results. Though there are many sources of bias, the most important ones are: (1) consensus PCR priming bias due to mismatches with target genes (metagenomic approaches avoid this entirely); (2) DNA/RNA extraction bias, due to hard-to-lyse microbial cells (most microbes are easy to crack open like peanuts, but some are tough to crack like Brazil nuts); and (3) PCR amplification bias due to high or low G/C content within amplicons (this bias increases exponentially with the number of PCR cycles run, so assays requiring many PCR cycles are very susceptible, and minimizing the number of PCR cycles is always advisable). Other forms of bias are specific to RNA-based assays and occur during the reverse transcription step used to convert RNA into DNA: (1) random hexamer-based reverse transcription favors G/C-rich transcripts; (2) modified bases can block reverse transcription entirely and are very abundant in ribosomal RNA gene transcripts; and (3) the amount of RNA present in microbial cells varies and cannot be known precisely. This means microbial cell counts based on RNA are less accurate than those based on DNA: each microbial cell's DNA consists of exactly one copy of its genome, while most of its RNA consists of ribosomes whose quantities vary based on metabolic requirements. In some cases, bias caused by consensus PCR primer mismatches, modified bases which block reverse transcription or inadequate cell lysis (Vesty et al., 2017; Dupuy et al., 2014) can be so great that microbes are missed altogether, resulting in a loss of universality.

TAXONOMIC CLASSIFICATION OF NOVEL MICROBES

During data analysis, taxonomic classification is trivial when base sequences are found to match exactly with taxonomically labeled reference sequences in GenBank. However, not all species or strains have been discovered and placed in GenBank. This means some sequences do not align well with *any* GenBank entry and are labeled as taxonomically unknown. If metagenomics had existed in the early 1980s, Barry Marshall, who discovered *Helicobacter pylori* in the stomach using microscopy, could have used this method instead; however *H. pylori*'s complete genome was not known at the time, so its sequences would have been classified as taxonomically unknown and would have been indistinguishable from sequences originating from food. To get around this limitation, modern taxonomical classification uses two ribosomal RNA genes (*rrs* and *rrl*) that evolve slowly and are never horizontally transferred between distant species. With only a few hundred reference sequences, it is possible to coarsely taxonomically classify *rrs* or *rrl* genes from all species, both known and unknown. Conveniently, ~75% of RNA strands present in microbes are *rrs* and *rrl* transcripts, whereas less than 1% of DNA strands are *rrs* and *rrl* genes; this means when RNA is used as starting material, most of the sequences can be taxonomically classified, even if they originate from novel species. Conversely, metagenomic assays cannot efficiently classify novel species when DNA is used as starting material. Viruses do not have *rrs* and *rrl* genes, and taxonomical identification of novel viruses generally requires manual review of alignments. When searching for novel viruses, base sequences are typically translated into amino acids and aligned to known virus proteins, because viral protein amino acid sequences are better conserved than viral DNA/RNA base sequences.

CONTAMINATION

Contamination occurs when DNA/RNA from microbes that were *not* present in the clinical specimen appear in sequencing results (Laurence et al., 2014; Grens, 2014; Strong et al., 2014; Salter et al., 2014). Contamination usually occurs due to instruments or reagents containing microbial DNA/RNA from the laboratory or manufacturing environment. Unlike cell culture, where sterilization techniques such as autoclaving can completely eliminate contamination by killing microbes, DNA is extremely resistant to destruction and survives most sterilization methods. Sequencing workflows are typically quite long, and troubleshooting contamination at each step is labor intensive. Some metagenomic assays label sequences early in the workflow, which protects the protocol from unlabeled contaminant insertion in downstream steps. Unfortunately, steps upstream of labeling such as DNA/RNA extraction cannot be protected from contamination using this method. Exposure to UVC light modifies some bases (especially consecutive uracils or thymines) in a way that blocks polymerase enzymes, rendering whole DNA/RNA strands unsequenceable (Borst et al., 2004; Besaratinia et al., 2011). It is thus possible to "genetically sterilize" reagents and laboratory equipment used in metagenomic assays by disabling DNA/RNA contaminant strands using UVC light. Note that PCR primers are made of DNA and cannot be sterilized in this manner because they would also be disabled! Contaminants can also plague reference sequence databases, when GenBank entries are taxonomically mislabeled (Gruber, 2015; Laurence et al., 2014). This *in silico* contamination greatly complicates data analysis, often requiring manual curation of sequence databases. Smaller databases such as SILVA (Quast et al., 2012) are better curated and contain few mislabeled entries.

TOTAL DNA OR RNA SEQUENCING USING ILLUMINA

This section is a walkthrough of steps needed to perform simple metagenomic assays using Illumina's next-generation sequencing platform (Fig. 7.2).

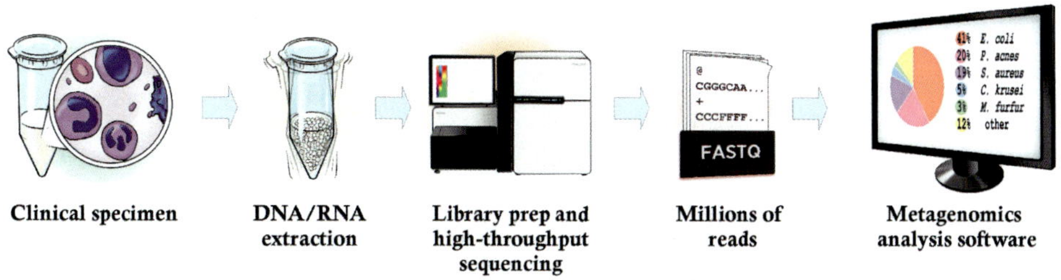

FIGURE 7.2

Metagenomic workflow overview.

SPECIMEN COLLECTION AND STORAGE

The first step is to collect a specimen for analysis, placing either host tissue or fluids (or an environmental specimen) in a collection tube. Collection tubes are provided with most DNA/RNA extraction kits (e.g., the Zymo Research Quick-DNA Fungal/Bacterial and Quick-RNA Fungal/Bacterial Kits). Instruments and containers used to manipulate the specimen (syringes, swabs, tweezers, or scalpels) need to be DNA- and RNA-free, not simply sterilized by autoclaving. If DNA/RNA extraction is not performed immediately, the specimen needs to either be stored at $-80°C$ or preserved with a special reagent (such as the Zymo Research DNA/RNA Shield). These reagents lyse host cells and disable proteins that can degrade DNA/RNA such as DNases and RNases. They also prevent growth of microbes present in the specimens. DNA typically survives inadequate storage conditions, but RNA does not. Poorly stored specimens are the main reason to choose DNA rather than RNA input for metagenomic assays. Integrity can be checked by running extracted DNA/RNA on an agarose gel or a bioanalyzer (Fig. 7.3). High-integrity DNA will produce a single high molecular weight band with little smearing. High-integrity RNA extracted from host specimens will produce two bright bands representing the two main host transcripts (*rrs* and *rrl*).

DNA/RNA EXTRACTION

DNA/RNA extraction is a straightforward process which is usually performed following the kit manufacturer's instructions. It is important to use kits that efficiently lyse microbes which have thick and disruption-resistant cell walls (such as the fungus *Malassezia*). Generally, long mechanical disruption using high-density beads and a homogenizer (such as the Benchmark Scientific BeadBug) is the preferred approach. It is important to note that microbes which are dividing or those in hyphal form are much easier to lyse mechanically than non-dividing microbes or small spherical microbes; this means control assays that use *in vitro* cultures as an input may misleadingly show high lysis efficiency, while actual efficiency might be much lower for the same microbe in clinical specimens or in the environment. It is also important to note that kits optimized for fungi, bacteria, or viruses very efficiently extract host DNA/RNA as well, so they can reduce bias but do not improve assay efficiency much. The DNA/RNA produced by these kits is called total DNA or total RNA, to indicate that no chemical selection methods were used to isolate a subset of strand types.

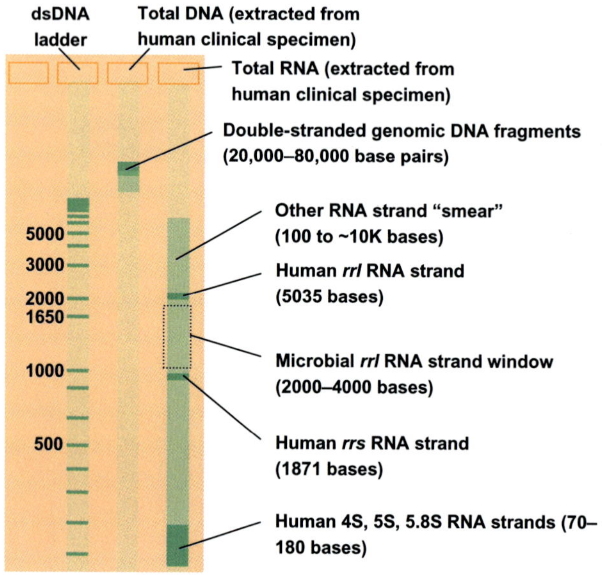

FIGURE 7.3

Total DNA and total RNA extracted from human clinical specimen run on a 1.2% agarose gel. *Total DNA*: Genomic DNA from chromosomes is always sheared during DNA extraction, producing double-stranded DNA fragments ~30,000 base pairs long. Human mitochondrial DNA strands are sufficiently short to not fragment during extraction (16,569 base pairs). However, the human mitochondrial DNA band is typically not visible because it represents less a 1% of total DNA. Rough mechanical extraction methods such as bead beating can produce shorter DNA fragments. The length of DNA fragments has little impact on metagenomic workflows. *Total RNA*: Human RNA transcripts are of different lengths, producing a complex pattern on the gel. The three bright bands are the ribosomal RNA strands (*rrs*, *rrl*, 5S, 5.8S) and transfer RNA strands (4S). All microbes, plants, and animals produce these strands, as they are essential for protein synthesis. Human mitochondria also produce these strands, though they are not visible on the gel, because they represent only ~1% of total RNA. The *rrl* RNA strand of mammals is much longer than those of microbes, so a simple method to improve assay efficiency is to run a gel and extract RNA between the human *rrl* and *rrs* bands; this eliminates about 90% of human RNA while retaining about 50% of microbial RNA (mainly microbial *rrl* strands), increasing assay efficiency approximately five-fold. *Gel migration speed*: It is important to note that RNA strands do not migrate at exactly the same speed as DNA strands of the same mass because they form sequence-dependent secondary structures. This is because secondary structures form in single DNA or RNA strands when bases on the same strand bind to each other, but these secondary structures cannot form when DNA or RNA is double stranded (each base is already paired up). In practice, DNA is almost always double stranded and RNA is almost always single stranded, which is why their migration speeds differ. The only relevant molecular difference between DNA and RNA is an oxygen molecule in the backbone, which does not affect gel migration speed much for double-stranded DNA or RNA—however, it does affect base binding strength and secondary structures formation for single strands, which means single DNA and RNA strands of identical sequences will not migrate at the same speed, as secondary structures will form differently in each.

ISOLATION OF RELEVANT DNA/RNA

Many techniques can be used to remove unwanted host DNA/RNA in order to improve assay efficiency (Gaudin and Desnues, 2018). The simplest metagenomic methods skip this step entirely, using total DNA or total RNA as in input for sequencing. Skipping this step ensures minimal bias and reduces the risk of contamination. When using DNA as an input, human DNA contains many CpG methylation sites and Alu element sequences, whereas bacterial and fungal DNA do not; paramagnetic bead capture (e.g., the NEBNext Microbiome DNA Enrichment Kit) can be used to remove such strands by binding to CpG methylation or Alu sites, improving efficiency of the assay. When using an RNA input, most sequencing centers assume the goal is to measure messenger RNA, so they remove ribosomal RNA by default (e.g., using the NEBNext rRNA Depletion Kit or NEBNext Poly(A) mRNA Magnetic Isolation Module). Yet ribosomal RNA is the most interesting transcript in metagenomic assays, so it must not be removed! It is important to clearly indicate that total RNA must be sequenced (including all ribosomal RNA) and that RNA isolation or purification steps must be skipped. If the RNA is of high quality (not fragmented), most human RNA strands can be removed using indirect target capture with biotinylated probes hybridizing specifically with human ribosomal RNA (by targeting poorly conserved sequences of *rrs* and *rrl* which are specific to humans). It is important to remember that these techniques greatly increase the risk of contamination and bias, and should only be used if absolutely necessary.

LIBRARY PREPARATION

DNA/RNA cannot simply be fed into an Illumina sequencer as is. RNA needs to first be converted into DNA. DNA strands need to be fragmented into ~300 bp segments. Adaptor sequences must then be added to both ends of these short DNA fragments, allowing them to bind to nanowells within the sequencer (see Table 7.1 row 3 for a completed Illumina library DNA strand example). Sequencing centers typically offer adequate library preparation services. However, standard library preparation is optimized for genomics or transcriptomics projects, not metagenomics. This means custom library preparation can improve efficiency, lower bias, and minimize contamination. It is not unusual for contamination to occur during library preparation. Typical library preparation protocols are shown in Figs. 7.4 and 7.5.

SEQUENCING

Sequencing centers usually sell their services in units of one "lane." Several lanes, each containing a different sample, are typically run at the same time in one DNA sequencing machine. Lanes cost between 500 and 10,000 USD, depending on how many nanowells are present. Each nanowell reads the sequence of a single DNA strand (Table 7.1). As of December 2018, the smallest flow cell lane (Illumina iSeq 100) typically produces ~4 million sequence reads; the largest flow cell lane (Illumina NovaSeq 6000 S4) typically produces ~5 billion sequence reads. Current metagenomic projects usually use an Illumina HiSeq 4000 flow cell lane, which produces ~300 million sequence reads. Note that it is also possible to pool specimens into the same lane using adaptors which contain a short label sequence (barcode) to identify the original specimen. The sequencing step

Table 7.1 Example of a DNA Fragment at Each Step in the Workflow

```
5' GCGTTGATTACGTCCCTGCCCTTTGTACACACCGCCCGTCGCTACTACCGATTGAATGGCTTAGTGAGCCTTTGGGATTGGCAGCTTGTACCGGCAACGGTAGCAGGCAGCCGAAAACTTAGGCAAACTTGGTCATTTAGAGGAAGTAAAAGTCGTAACAAGGTTTCTGTAGGTGAACCTGCAGAAGG 3'
   :::::::::::::::::::::::::::::::::::::::::::::::::::::::::::::::::::::::::::::::::::::::::::::::::::::::::::::::::::::::::::::::::::::::::::::::::::::::::::::::::::::::::::::::::::::::::::
3' CGCAACTAATGCAGGGACGGGAAACATGTGTGGCGGGCAGCGATGATGGCTAACTTACCGAATCACTCGGAAACCCTAACCGTCGAACATGGCCGTTGCCATCGTCCGTCGGCTTTTGAATCCGTTTGAACCAGTAAATCTCCTTCATTTTCAGCATTGTTCCAAAGACATCCACTTGGACGTCTTCC 5'
```

```
                                                         5' GCTTAGTGAGCCTTTGGGATTGGCAGCTTGTACCGGCAACGGTAGCAGGCAGCCGAAAACTTAGG 3'
                                                            :::::::::::::::::::::::::::::::::::::::::::::::::::::::::::::::::
                                                         3' CGAATCACTCGGAAACCCTAACCGTCGAACATGGCCGTTGCCATCGTCCGTCGGCTTTTGAATCC 5'
```

```
5' AATGATACGGCGACCACCGAGATCTACACTCTTTCCCTACACGACGCTCTTCCGATCTGCTTAGTGAGCCTTTGGGATTGGCAGCTTGTACCGGCAACGGTAGCAGGCAGCCGAAAACTTAGGAGATCGGAAGAGCACACGTCTGAACTCCAGTCACATCACGATCTCGTATGCCGTCTTCTGCTTG 3'
   :::::::::::::::::::::::::::::::::::::::::::::::::::::::::::::::::::::::::::::::::::::::::::::::::::::::::::::::::::::::::::::::::::::::::::::::::::::::::::::::::::::::::::::::::::::::::::
3' TTACTATGCCGCTGGTGGCTCTAGATGTGAGAAAGGGATGTGCTGCGAGAAGGCTAGACGAATCACTCGGAAACCCTAACCGTCGAACATGGCCGTTGCCATCGTCCGTCGGCTTTTGAATCCTCTAGCCTTCTCGTGTGCAGACTTGAGGTCAGTGTAGTGCTAGAGCATACGGCAGAAGACGAAC 5'
```

```
@K00162:137:HH2NGBBXX:4:1220:4564:23892 1:N:0:CGATGT
GCTTAGTGAGCCTTTGGGATTGGCAGCTTGTACCGGCAAC
+
AAFFFJJJJJJJJJJJJJJJFJJJJFJJFJJJJJJFJJJ
```

```
@K00162:137:HH2NGBBXX:4:1220:4564:23892 1:N:0:CGATGT
CCTAAGTTTTCGGCTGCCTGCTACCGTTGCCGGTACAAGC
+
AAFFFJJJJJJJJJJJJJJJJJJFJJJJJFJJJJAJJJJ
```

```
@000000080_1_1@DNA3S2_step07.fx@K00162:137:HH2NGBBXX:4:1220:4564:23892 1:N:0:CGATGT
GCTTAGTGAGCCTTTGGGATTGGCAGCTTGTACCGGCAAC
CCTAAGTTTTCGGCTGCCTGCTACCGTTGCCGGTACAAGC
+
AAFFFJJJJJJJJJJJJJJJFJJJJFJJFJJJJJJFJJJ
AAFFFJJJJJJJJJJJJJJJJJJFJJJJJFJJJJAJJJJ
```

```
    **** Group 80 (read pair 1 of 1) **** @000000080_1_1@DNA3S2_step07.fx@K00162:137:HH2NGBBXX:4:1220:4564:23892 1:N:0:CGATGT
     1639 -> 1703 "silva_ssu.fa>ti|76773|LFDC01000031.1 Malassezia globosa" Consensus[ 100%"Malassezia globosa(76773)"LFDC01000031  97%"Malassezia(55193)"LFGB01000057]
     AAFFFJJJJJJJJJJJJJJJFJJJJFJJFJJJJJJFJJJ              JJJJAJJJJFJJJJJFJJJJJJJJJJJJJJJJJJJFFFAA
     GCTTAGTGAGCCTTTGGGATTGGCAGCTTGTACCGGCAAC ----------> GCTTGTACCGGCAACGGTAGCAGGCAGCCGAAAACTTAGG
100% ||||||||||||||||||||||||||||||||||||||||            ||||||||||||||||||||||||||||||||||||||||
     GCTTAGTGAGCCTTTGGGATTGGCAGCTTGTACCGGCAAC --  -15 --> GCTTGTACCGGCAACGGTAGCAGGCAGCCGAAAACTTAGG
```

```
290858583 274516459 total spots (spot = paired-end read)
137276280 130694398 low quality spots
      507      1279 duplicate spots
     4343      6253 low entropy spots
   DNA3S2    DNA5S2
        7         9 (root)
       19        29 cellular organisms;
        8        12 cellular organisms;Bacteria;
       19       368 cellular organisms;Bacteria;Bacteroidetes;Bacteroidia;Bacteroidales;Prevotella;
       14         7 cellular organisms;Bacteria;Bacteroidetes;Bacteroidia;Bacteroidales;Prevotella;Prevotella buccalis;
       18        47 cellular organisms;Bacteria;Gammaproteobacteria;Enterobacterales;
        6        11 cellular organisms;Bacteria;Gammaproteobacteria;Enterobacterales;Escherichia;Escherichia coli;
        7      4064 cellular organisms;Bacteria;Actinobacteria;Bifidobacteriales;Gardnerella;Gardnerella vaginalis;
      117       223 cellular organisms;Bacteria;Actinobacteria;Corynebacteriales;Corynebacterium;
        3        69 cellular organisms;Bacteria;Actinobacteria;Corynebacteriales;Corynebacterium;Corynebacterium pseudogenitalium;
        2        49 cellular organisms;Bacteria;Actinobacteria;Corynebacteriales;Corynebacterium;Corynebacterium pyruviciproducens;
       42       310 cellular organisms;Bacteria;Actinobacteria;Propionibacteriales;
        6       222 cellular organisms;Bacteria;Firmicutes;Bacilli;Bacillales;Staphylococcus;
        2        28 cellular organisms;Bacteria;Firmicutes;Bacilli;Bacillales;Staphylococcus;Staphylococcus capitis;
        1        77 cellular organisms;Bacteria;Firmicutes;Bacilli;Bacillales;Staphylococcus;Staphylococcus epidermidis;
        0        18 cellular organisms;Bacteria;Firmicutes;Bacilli;Bacillales;Staphylococcus;Staphylococcus haemolyticus;
       14        17 cellular organisms;Bacteria;Firmicutes;Bacilli;Lactobacillales;Streptococcus;
       25         0 cellular organisms;Bacteria;Firmicutes;Tissierellia;Tissierellales;Finegoldia;Finegoldia magna;
        3        32 cellular organisms;Eukaryota;Opisthokonta;Fungi;Basidiomycota;Ustilaginomycotina;Malasseziales;Malassezia;Malassezia globosa;
        1         9 cellular organisms;Eukaryota;Opisthokonta;Fungi;Basidiomycota;Ustilaginomycotina;Malasseziales;Malassezia;Malassezia restricta;
     1617      2227 cellular organisms;Eukaryota;Opisthokonta;Metazoa;Bilateria;Deuterostomia;Vertebrata;Mammalia;Primates;
       57        50 Low homology (0%-9%)
      117       104 Low homology (50%-59%)
       77        75 Low homology (60%-69%)
      128       113 Low homology (70%-79%)
      187       242 Low homology (80%-89%)
```

An unusually small insert length of 65 bases, and two 40 base reads are used in this example due to a lack of space (we would need a much bigger page for more typical 300 base inserts and 150 base reads!). Illumina universal primer (IU) and index adaptor 1 (II1) were used in this example (see Table 7.4). Row 1: unfragmented Malassezia globosa *genomic DNA. Row 2: fragmented DNA (insert). Row 3: fragmented DNA with Illumina sequences appended to both ends, which is fed into the sequencer. Row 4: FASTQ file (read 1). Row 5: FASTQ file (read 2). Row 6: Leif Microbiome Analyzer merged paired-end read (.fx file format); Row 7: Leif Microbiome Analyzer aligned paired-end read (.qb file format); Row 8: tabulated consensus file (.csv file format).*

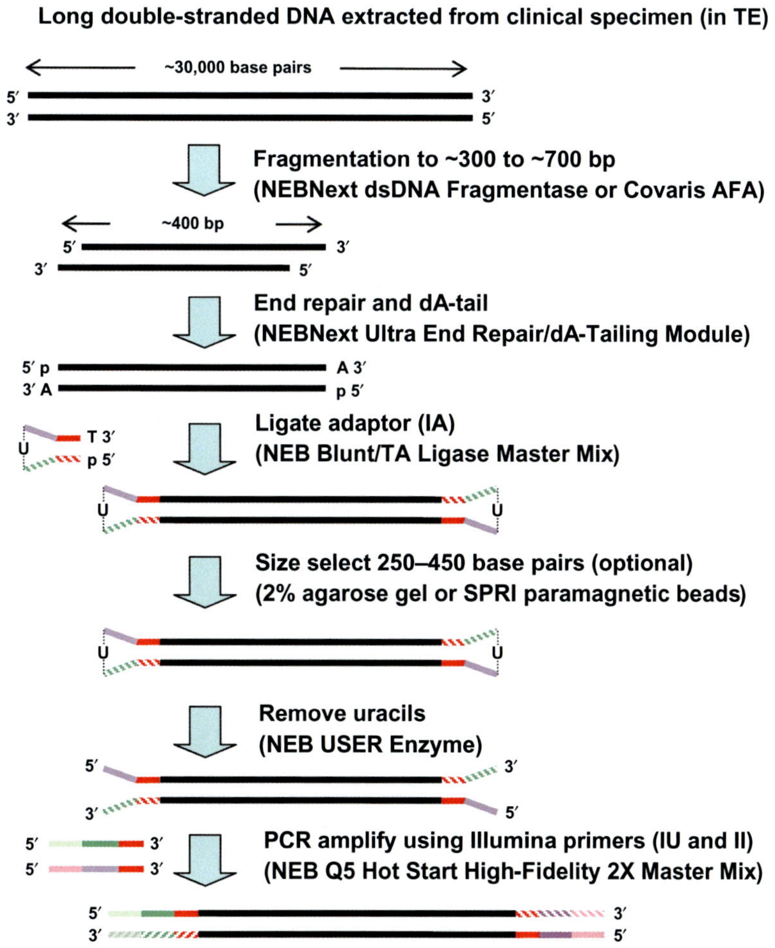

FIGURE 7.4

Standard Illumina library preparation protocol starting from DNA. This protocol can be simplified by combining the fragmentation, end repair, and dA-tailing steps using the NEBNext Ultra II FS DNA Module. The size selection step is recommended, because it increases the yield of long high-quality reads. Reagents to perform this library preparation flow can be purchased as a kit (e.g., NEBNext Ultra II DNA Library Prep Kit for Illumina).

produces a pair of very large FASTQ files that contain the base string and quality score of all the read sequences (Table 7.1 rows 4 and 5). Each strand is sequenced from the left side (typically producing 150 bases) and then from the right side (also typically producing 150 bases), which is why two FASTQ files are output. When the sequence strand is less than 300 bases long, some bases located in the middle of the strand will appear in both FASTQ files.

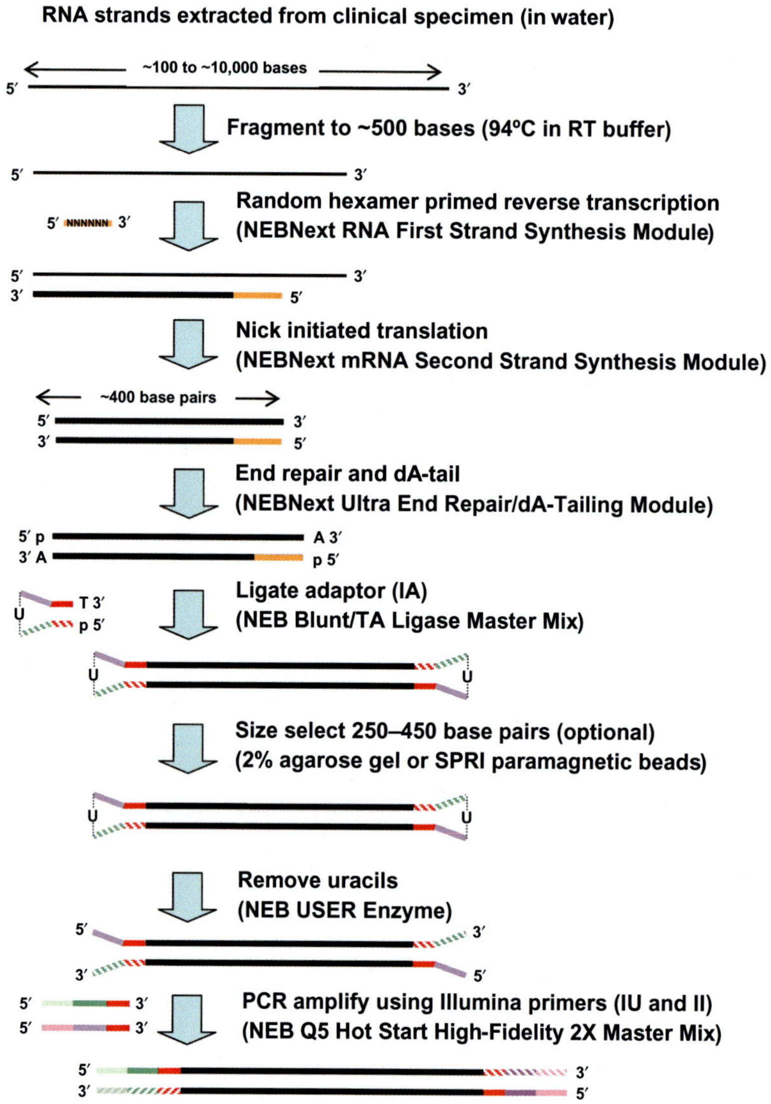

FIGURE 7.5

Standard Illumina library preparation protocol starting from RNA. The size selection step is recommended, because it increases the yield of long high-quality reads. Reagents to perform this library preparation flow can be purchased as a kit (e.g., NEBNext Ultra II RNA Library Prep Kit for Illumina).

ALIGNMENT

Alignment is the general term used to describe bioinformatics workflows that use sequencing results as an input (FASTQ files) and compare them to a reference database such as GenBank (FASTA files). Each read in the FASTQ file is broken into short "words" of say ~11 bases, and anywhere such a word appears in the reference database, the read and database are compared using an algorithm such as NCBI BLASTN. The best matches are retained and reported in a text file. This process is described in greater detail in the Bioinformatics section, and an example alignment can be seen in Table 7.1. Though a single alignment pass comparing to all reference sequences would work, alignment is usually done in multiple passes to accelerate the process. The initial pass aligns only to the host genome, discarding matches which approach 100% identity with known host sequences—these sequences originated from host cells and are not informative for microbiome workflows. The second pass aligns to microbes expected to be present in the specimen; sequences approaching 100% identity do not need to be aligned again. In subsequent passes, remaining sequences are aligned to most or all of GenBank—these final alignment passes are very slow due to the much larger reference database and to alignment parameters which allow detection of low identity matches.

TABULATION

Once all sequences in the FASTQ file have been aligned to a reference database, results are typically presented in a spreadsheet, where the number of reads attributed to a given taxon is shown on the same row. Multiple specimens can be combined in the same spreadsheet for ease of comparison (Table 7.1 row 8). Taxonomic precision (genus, species, or strain) and ambiguity tolerance must be chosen to produce this spreadsheet. Ambiguous sequences which match well with multiple genera, species, or strains end up reported in higher taxa. For example, a highly conserved sequence which matches well with humans and fungi will be reported under the taxon "Opisthokonta," because it is not possible to know which species or genus it is from within Opisthokonta. Sufficiently long sequence reads are essential to achieve accurate taxonomic classification—the Illumina default of 150 base sequences works well.

RIBOSOMAL RNA GENES *RRS* AND *RRL*

Ribosomal RNA genes are present in all cellular microbes and are never horizontally transferred between distant species, making them ideal for taxonomical classification. The main ribosomal RNA genes are *rrs* (Fig. 7.6) and *rrl*. They are sometimes called 16S/18S/18S and 23S/26S/28S, respectively. These S values represent their sedimentation rate following RNA extraction from bacteria/fungi/humans, respectively. These two RNA strands form the core of the ribosome, which performs the main metabolic activity of cells: converting messenger RNA into proteins. In all cellular microbes and animals, about 25% of RNA by mass is *rrs* transcripts and about 50% is *rrl* transcripts. *rrs* and *rrl* form such a large part of total RNA that messenger RNA assays remove *rrs* and *rrl* transcripts chemically prior to sequencing to avoid wasting reads. In microbiome studies, *rrs* and *rrl* are the most informative targets and are already dominant in total RNA: removing

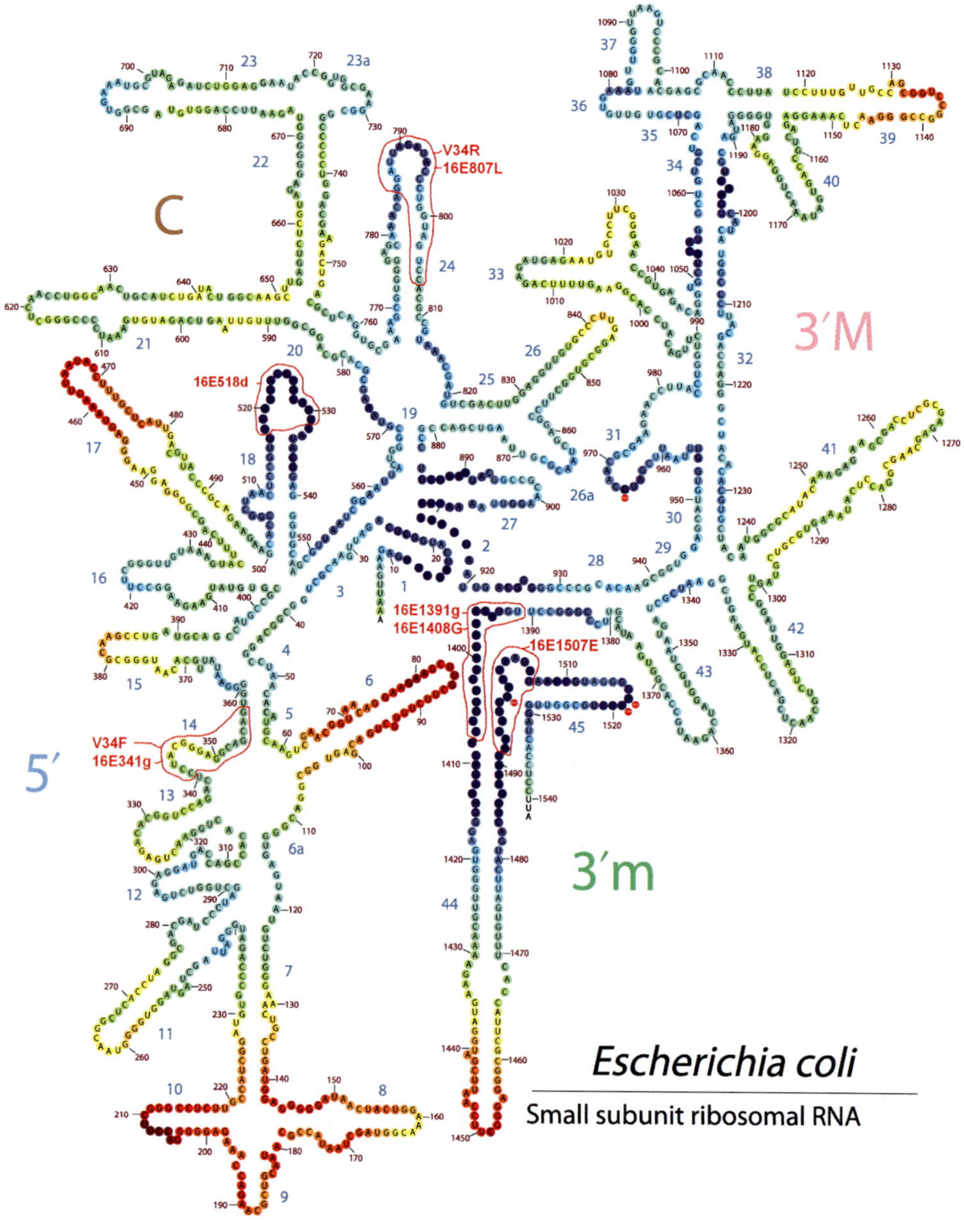

FIGURE 7.6

Escherichia coli rrs gene (RNA transcript secondary structure). Well-conserved sequences used in consensus PCR are circled in red. Stop signs indicate modified bases which block reverse transcription.

Illustration downloaded from http://apollo.chemistry.gatech.edu/RibosomeGallery/index.html.

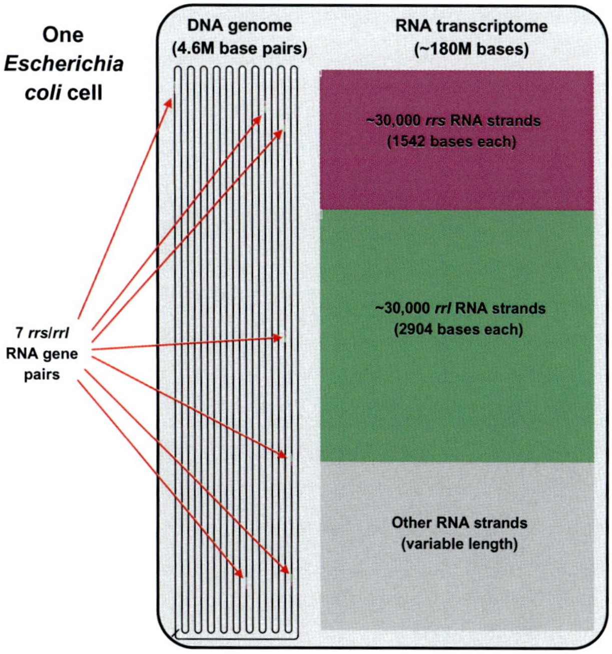

FIGURE 7.7

The seven *rrs/rrl* ribosomal RNA genes in *Escherichia coli*'s circular double-stranded DNA genome (left, per GenBank U00096.3) and ~30,000 *rrs/rrl* ribosomal single-stranded RNA strands present in one *E. coli* cell (right). Both are to scale (area = mass). *rrs* is pink (1542 DNA base pairs or RNA bases); *rrl* is green (2904 DNA base pairs or RNA bases). Amount of RNA varies somewhat based on metabolic activity. When sequencing a randomly chosen fragment of DNA or RNA (shotgun sequencing), finding *rrs/rrl* RNA strands is *much* easier than finding *rrs/rrl* genes (about 100-fold easier in this case as $(7 \times (1542 + 2904))/4.6\ \text{M} = 0.7\%$). Many microbes have only a single *rrs/rrl* gene pair in their entire genome, such as *Mycobacterium tuberculosis* (per GenBank AP018033.1).

messenger RNA chemically prior to sequencing does not improve efficiency enough to be worth the trouble. Though *rrs* and *rrl* genes also show up in total DNA sequencing results, they are not very abundant, representing less than 1% of sequences read (Fig. 7.7). This means sequencing total RNA is a much more efficient way of taxonomically identifying novel microbes in metagenomic assays. Nature has conveniently preamplified the genes best suited for taxonomic classification: *rrs* and *rrl* are highly expressed as compared to all other genes!

CONSERVED SEQUENCES

Genes *rrs* and *rrl* are transcribed into ribozymes whose active sites contain highly conserved base sequences required to catalyze chemical reactions. This allows consensus PCR primers to match genes present in the entire tree of life—though there are some holdouts which have somewhat diverged from consensus sequences (Table 7.2). In addition, not all mismatches are born equal: some mismatches are not very destabilizing (meaning that priming still occurs efficiently despite the mismatch), while other mismatches can prevent priming altogether. These are very important and complex issues that must be addressed when designing consensus PCR assays; fortunately, metagenomic assays do not use consensus PCR primers, greatly simplifying protocol design and enhancing robustness.

DIVERGENT SEQUENCES

Consensus PCR assays amplifying parts of the *rrs* or *rrl* gene can easily detect and taxonomically classify novel microbes by aligning amplicon sequence reads to a database of known microbes such as SILVA (Quast et al., 2012). This works because bases positioned *between* conserved

Table 7.2 Alignments of Select Cellular Microbes to the Three Best Conserved Consensus Sequences in the *rrs* Gene

***Escherichia coli* position** ***E. coli* sequence**	**Type**	***rrs* 518–529** CCAGCAGCCGCGGT	***rrs* 1390–1407** TGTACACACCGCCCGTC	***rrs* 1492–1506** AAGTCGTAACAAGGT
Human	Mammal	....	....	....
Babesia microti	Protist	....	....	....
Blastocystis hominis	Protist	....	....	....
Cyclospora cayetanensis	Protist	....	....	...........C...
Cryptosporidium hominis	Protist	....	....	....
Giardia lamblia	Protist	....	....	....
Plasmodium falciparum	Protist	....	....	....
Theileria parva	Protist	....	....	....
Toxoplasma gondii	Protist	....	....	....
Trichomonas vaginalis	Protist	........T.....	....	....
Trypanosoma brucei	Protist	......C.......	....	....
Aspergillus fumigatus	Fungi	....	....	....
Blastomyces dermatitidis	Fungi	....	....	....
Candida albicans	Fungi	....	....	....
Coccidioides immitis	Fungi	....	....	....
Cryptococcus neoformans	Fungi	.......T......	....	....
Encephalitozoon cuniculi	Fungi	....	....	...........T..C
Histoplasma capsulatum	Fungi	....	....	....
Hortaea werneckii	Fungi	....	....	....
Malassezia globosa	Fungi	....	....	....
Paracoc. brasiliensis	Fungi	....	....	....
Pneumocystis jirovecii	Fungi	....	....	....
Sporothrix schenckii	Fungi	....	....	....
Trichophyton tonsurans	Fungi	....	....	....
Trichosporon asahii	Fungi	....	....	....
Acinetobacter baumannii	Bacteria	....	....	....

Table 7.2 Alignments of Select Cellular Microbes to the Three Best Conserved Consensus Sequences in the *rrs* Gene *Continued*

***Escherichia coli* position *E. coli* sequence**	**Type**	***rrs* 518–529** CCAGCAGCCGCGGT	***rrs* 1390–1407** TGTACACACCGCCCGTC	***rrs* 1492–1506** AAGTCGTAACAAGGT
Bartonella henselae	Bacteria	....	....	....
Bordetella pertussis	Bacteria	....	....	....
Borrelia burgdorferi	Bacteria	....	....	....
Burkholderia pseudomallei	Bacteria	....	....	....
Campylobacter jejuni	Bacteria	....	.....T...........	....
Chlamydia trachomatis	Bacteria	........T.....	....	....
Chlamydophila pneumoniae	Bacteria	........T.....	....	....
Clostridium perfringens	Bacteria	....	....	....
Coxiella burnetii	Bacteria	....	....	....
Ehrlichia ruminantium	Bacteria	....	.........T.......	....
Enterococcus faecalis	Bacteria	....	....	....
Francisella tularensis	Bacteria	....	....	....
Haemophilus ducreyi	Bacteria	....	....	....
Haemophilus influenzae	Bacteria	....	....	....
Helicobacter pylori	Bacteria	....	.....T...........	....
Lactobacillus acidophilus	Bacteria	....	....	....
Legionella pneumophila	Bacteria	....	....	....
Listeria monocytogenes	Bacteria	....	....	....
Mycobacterium leprae	Bacteria	....	....	....
Mycoplasma genitalium	Bacteria	.......T......	....	....
Mycoplasma pneumoniae	Bacteria	.......T......	....	....
Neisseria gonorrhoeae	Bacteria	....	....	....
Prevotella buccalis	Bacteria	....	....	....
Propionibacterium acnes	Bacteria	....	....	....
Rickettsia rickettsii	Bacteria	....	.........T.......	....
Salmonella enterica	Bacteria	....	....	....
Staphylococcus epidermis	Bacteria	....	....	....
Streptococcus pyogenes	Bacteria	....	....	....
Treponema pallidum	Bacteria	....	....	....
Vibrio cholerae	Bacteria	....	....	....
Yersinia pestis	Bacteria	....	....	....

rrs or *rrl* sequences mutate quickly enough to differentiate most species. The detection of novel species in metagenomic assays works in a similar way but is much more accommodating to consensus sequence mismatches than consensus PCR. The reason for this is simple: target sequences hybridize with consensus PCR primers *chemically* during the PCR reaction, so this process cannot be precisely adjusted to handle mismatches. PCR reactions require *two* primers to match, one at each end of the amplicon, rather than just one. In metagenomic workflows, comparison with consensus sequences is done *computationally* by scanning each read sequence and comparing with only *one* consensus sequence at a time; for example, a program can independently search for each sequence circled in Fig. 7.6. This means microbes which have diverged from consensus sequences (such as *M. genitalium*, *Trichomonas vaginalis*, and *Encephalitozoon cuniculi*, see Table 7.2) can be detected with certainty and with extremely low bias using metagenomics, but not using consensus PCR. This is one of the main advantages of metagenomic assays over consensus PCR assays.

MODIFIED BASES

In DNA, there are only a small number of modified base types, such as CpG methylation where the cytosine is changed to 5-methylcytosine. These modified bases typically do not affect PCR amplification or DNA sequencing, thus have no impact on metagenomic assays. In RNA, post-transcription base modifications are very common and varied in nature, especially in *rrs* and *rrl* transcripts (Limbach et al., 1994; Agris, 2018). While most modified bases do not affect reverse transcription or sequencing, some can block reverse transcription altogether (Motorin et al., 2007). For example, *E. coli rrs* 1518 and 1519 are a pair of adenines, which are post-transcriptionally modified by a dimethyladenosine transferase enzyme to N^6,N^6-dimethyladenosines; this modified base blocks reverse transcription. Modified bases have not been studied in all species, and their locations and types can vary between species. The effect of modified bases on sequencing results can be seen in the unequal coverage of *Malassezia globosa rrs* transcripts from a total RNA sequencing run (Fig. 7.8). If consensus RT-PCR was used to detect microbes from RNA, modified

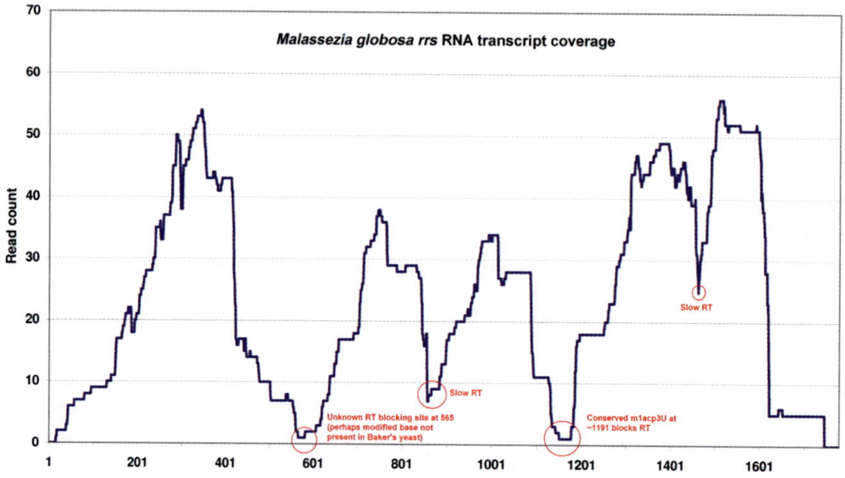

FIGURE 7.8

Coverage of *Malassezia globosa rrs* RNA transcripts from human total RNA, sequenced using a single Illumina HiSeq 4000 lane. Reverse transcription was primed with random hexamers. Insert lengths ranged from 100 to 250 base pairs. Uneven coverage is caused by modified bases in the RNA strand which block reverse transcription (RT).

bases which block reverse transcription located between priming sites would introduce unacceptable levels of bias. In contrast, total RNA metagenomic assays use a shotgun approach to cover entire *rrs* and *rrl* genes, so they are very resilient to modified bases which block reverse transcription.

INTRONS

In some species, *rrs* and *rrl* genes contain introns. Such introns can greatly increase amplicon lengths (Hedberg and Johansen, 2013) and even straddle consensus sequences (e.g., *E. coli*'s *rrs* 2576–2602 conserved sequence is split by an intron in bacterium *Thermosynechococcus elongatus*). In both cases, a high level of bias against such species' amplicons can occur when using consensus PCR assays. In contrast, metagenomic assays are unaffected by introns.

SENSITIVITY (METAGENOMICS)

When running a metagenomics study, it is important to choose the right sequencing depth. Choosing too many sequence reads per specimen can get expensive; choosing too few sequence reads means less abundant microbes will be missed—and these microbes matter!

For example, EBV is a medically important virus that greatly increases multiple sclerosis risk (Laurence and Benito-León, 2017) and can cause mononucleosis (Thorley-Lawson, 2015). Detecting EBV in blood using metagenomics requires a very high sequencing depth: only ~1 human cell per 5 mL of blood contains EBV DNA (Thorley-Lawson, 2015)! The probability of catching an EBV-infected cell is thus 20% per 1 mL of blood. This means 14 mL of blood must be drawn to have a 96% chance of catching an infected cell [(1.0%–20%)^14 = 0.04]. This amount of blood contains ~500,000 ng of human DNA. An infected cell contains ~6 EBV episomes of 172,000 base pairs each, totaling 0.000001 ng. This means each sequence read has a 500,000,000,000:1 chance of detecting EBV (500,000/0.000001 ng = 2e − 12 = 500,000,000,000:1). To have a 95% chance of detecting EBV—this means finding one EBV read within sequencing results—about 1.5 trillion sequences must be read [(1.0–2e − 12)^1.5e12 = 0.05]. Using currently available sequencing technology, this would cost about 3,000,000 USD.

EBV detection in blood is an extreme case, because this virus persists in the human body despite infecting few human cells. Fortunately, EBV has already been discovered, isolated, and cultured without the help of metagenomics. Today, testing for antibodies against EBV is a much more cost-effective method to determine infection status than metagenomics. How many other viruses like EBV are there in humans that have escaped detection so far? Metagenomics is the most reliable way to find out.

Specimens analyzed by metagenomic assays come in all shapes and sizes. For most clinical applications, extracted DNA originates mainly from host cells and totals between 1000 and 100,000 ng. Human cells contain about 0.007 ng of DNA each, implying that these specimens contain between 142,000 and 14,200,000 human cells each. The smallest known medically important bacterial genome is ~580,000 base pairs (*M. genitalium*), which means each microbial cell contains at least 0.0000006 ng of DNA. When starting with a typical clinical specimen producing 10,000 ng of DNA

(1,420,000 human cells), each sequence read has a 16,000,000,000:1 chance of detecting a single *M. genitalium* cell within this specimen (10,000/0.0000006 ng = 6e − 11 = 16,000,000,000:1). To have a 95% chance of detecting a lone *M. genitalium* cell in the specimen—this means finding one *M. genitalium* read within sequencing results—about 50 billion sequences must be read [(1.0−6e − 11)^5e10 = 0.05]. Using the best currently available sequencing technology, this would cost about 100,000 USD.

The preceding calculations can be adjusted based on desired sensitivity (usually 95% probability of detecting one microbe cell is used), minimum genome size (usually 10,000 base pairs for viruses, 580,000 base pairs for bacteria, 3,000,000 base pairs for eukaryotic microbes such as fungi), and the number of human cells present (this can be estimated by measuring the total amount of DNA extracted). Even if hundreds of cells of each microbial species are expected to be present in a clinical specimen, it is best to aim for the maximum possible sensitivity, which is to detect a lone microbial cell. An important reason to aim for such high sensitivity is to partially compensate for inefficient cell lysis, which is expected to occur for some hard-to-crack microbes. For example, if 500 *Malassezia* cells are present in a clinical specimen, but lysis efficiency is a meager 1% due to their thick cell wall, then only a few *Malassezia* cells' DNA will be extracted. Inefficient DNA extraction decreases the relative amount of *Malassezia* as compared with other microbes—it introduces bias—but it is still valuable to know that *Malassezia* are present at all in the specimen, even if the exact amount is difficult to estimate due to poor cell lysis. If DNA extraction completely fails on some microbial species (no cells are lysed), increasing sequencing depth cannot help detect these species at all.

The same calculations are applicable when extracting and sequencing RNA instead of DNA. However, the amount of RNA per human cell varies substantially per cell type (Table 7.3), meaning that the sensitivity and efficiency of the assay will also change considerably. For example, when extracting DNA or RNA from blood, the same specimen will yield ~20 times less RNA than DNA. This is because the bulk of human DNA/RNA in blood originates from white blood cells

Table 7.3 DNA and RNA Content of Some Cell Types.

Description	Genome Size (Mbp)	DNA Per Cell (fg)	RNA Per Cell (fg)	Reference
Average human cell	3300.0	6600.0	~25,000	
Human T cell or B cell		6600.0	~1100	Glen (1967)
Human leukocyte in blood		6600.0	~2000	Chomczynski et al. (2016)
Human spermatozoon		3300.0	~37	Cappallo-Obermann et al. (2011)
Saccharomyces cerevisiae cell	12.0	12.0	~500	von der Haar (2008)
Escherichia coli cell	5.1	5.1	~100	Bakshi et al. (2012)
Spiroplasma melliferum cell	1.3	1.3	~5	Ortiz et al. (2006)

The RNA/DNA ratio typically means microbe detection is more efficient when sequencing RNA rather than DNA. Efficiency depends on the types of host cells and microbes present.

that contain little RNA (Table 7.3). In addition, the RNA-to-DNA ratio in microbial cells is typically greater than that of human cells, further increasing sensitivity (Table 7.3). This means RNA-based metagenomic assays are >20-fold more efficient than DNA-based assays when specimens consist of blood (Table 7.3).

For most metagenomic applications, attaining the maximum possible sensitivity by sequencing total DNA or total RNA remains expensive. This means techniques to chemically remove human DNA or RNA *before* sequencing remain useful, and consensus PCR assays remain an attractive alternative to metagenomics, due to lower cost. Techniques to improve sensitivity of metagenomic assays, and also to improve the universality of consensus PCR assays, are described in the following sections. These techniques will remain relevant until sequencing costs are so low that the simplest metagenomic assays outperform all other options on every metric, including cost.

SENSITIVITY (ALIQUOTING AND CONSENSUS PCR)

Given that consensus PCR is a commonly used alternative to metagenomics; it is useful to compare the sensitivity of these two techniques under similar conditions. Consensus PCR assays typically amplify the *rrs* gene. Unlike most genes, there are often multiple copies of *rrs* in each genome. The *E. coli* genome has seven copies of this gene (Fig. 7.7), the human genome has about 300 copies and the *M. genitalium* genome has only one.

Library preparation kits for Illumina typically handle a maximum of 1000 ng of input DNA. Consensus PCR reactions are usually loaded with at most 100 ng of input DNA. DNA molecules sequester polymerase enzymes (such as *Taq*), which significantly slows PCR reactions when present in large amounts, so input DNA is typically capped to ~100 ng. When ~10,000 ng of extracted DNA are *aliquoted* to perform library preparation or consensus PCR, targeted microbial strands can be missed altogether due to sampling—they were present in the extracted DNA, but the pipetting step to load the PCR reaction did not capture *any* targeted microbial strand! A single *M. genitalium* cell chromosome is broken up during DNA extraction into ~20 fragments of ~30,000 base pairs each. Any of these fragments can be detected by metagenomic assays, but only the fragment on which the *rrs* gene is located can be detected by consensus PCR. Furthermore, PCR reactions typically use 10 times less input DNA than Illumina library prep kits. At high sequencing depth the PCR assay is thus 200-fold less sensitive than the metagenomics assay for the detection of a single *M. genitalium* cell, due to aliquoting and targeting of a single gene (*rrs*). Sensitivity of the metagenomics assay is constrained by the probability of selecting one or more of the ~20 *M. genitalium* fragments when pipetting 1000 ng of DNA out of the 10,000 ng total for Illumina library preparation [0.2% chance per ng (1.0–0.2%)^1000 = 0.14 = 86% sensitivity]. The sensitivity of the consensus PCR assay is constrained by the probability of selecting the *M. genitalium* fragment that contains the *rrs* gene when pipetting 100 ng of DNA out of the 10,000 ng total [0.01% chance per ng (1.0–0.01%)^100 = 0.99 = 1% sensitivity]. Sensitivity of the metagenomics assay can be further improved by changing the first step of the library preparation protocol: performing fragmentation on the entire 10,000 ng and then pipetting 1000 ng for library preparation. This increases the number of *M. genitalium* fragments from ~20 to ~2000, making it extremely unlikely that it would be missed due to aliquoting [20% chance per ng (1.0%–10%)^1000 = 1e − 97 = 100% sensitivity].

Fragmenting DNA does not help the consensus PCR assay at all, because it does not increase the number of fragments containing the *rrs* gene.

When using RNA as input material, each lysed cell releases a thousand or more *rrs* transcript strands, which means the aliquoting issues described earlier for DNA are not relevant. The sensitivity of metagenomics and consensus PCR assays starting from RNA is thus not materially affected by routine aliquoting. However, modified bases which block reverse transcription mean that consensus RT-PCR from RNA risks missing microbes altogether and is not thus recommended. This is especially true when producing long amplicons that cover most of the *rrs* gene: longer amplicons imply a higher probability of spanning a modified base that blocks reverse transcription. Metagenomic assays starting from RNA use a shotgun approach that is minimally affected by modified bases, so this approach is the most sensitive and robust microbe detection method available.

NEARLY UNIVERSAL CONSENSUS PCR

Metagenomics is a more reliable, more sensitive, more universal, and less biased approach for the detection of microbes and viruses than consensus PCR. Widely used consensus PCR techniques require one reaction for bacteria and another for fungi—protists and archaea are usually ignored, and viruses cannot be detected because they lack *rrs* genes. Because of its lower cost, consensus PCR remains an attractive and commonly used alternative to metagenomics. How universal can a consensus PCR assay be?

Widely used consensus PCR primers such as V34 can detect most bacteria but not the rest of the tree of life. The conserved sequences chosen for V34 priming are such that they mismatch with host sequences to avoid amplifying the human *rrs* gene (Table 7.4). This greatly reduces the number of sequence reads wasted on host amplicons, but also limits microbiome coverage to bacteria.

Table 7.4 Primer Sequences (Highlighted) Used to Perform Consensus PCR (Fig. 7.9)

<table>
<tr><th></th><th>Primer/Sequence name</th><th>Alignment (5′ → 3′)</th><th>T_m</th></tr>
<tr><td rowspan="4">**Consensus primers (bacteria)**</td><td>V34F (aka 16E341g)</td><td>CCTACGGGNGGCWGCAG
| | || |||||||||</td><td>~66.5°C</td></tr>
<tr><td>18H461g</td><td>TCCAAGGAAGGCAGCAG</td><td>~47.7°C</td></tr>
<tr><td>V34R (aka 16E807L)</td><td>GGACTACHVGGGTATCTAATCC
||||| | ||||||| |||</td><td>~62.2°C</td></tr>
<tr><td>18H1075L</td><td>GAACTACGACGGTATCTGATCG</td><td>~31.7°C</td></tr>
<tr><td rowspan="4">**Consensus primers (all life)**</td><td>16E518d</td><td>CCAGCAGCCGCGGT
||||||||||||||</td><td>67.9°C</td></tr>
<tr><td>18S565d (aka 18H614d)</td><td>CCAGCAGCCGCGGT</td><td>67.9°C</td></tr>
<tr><td>16E1408G</td><td>GACGGGCGGTGTGTACA
|||||||||||||||||</td><td>66.2°C</td></tr>
<tr><td>18S1645G (aka 18H1709G)</td><td>GACGGGCGGTGTGTACA</td><td>66.2°C</td></tr>
<tr><td rowspan="4">**Blocking primers (human)**</td><td>18H760m</td><td>TCGATGCTCTTAGCUGAGTGTCC
| || | | | |</td><td>67.1°C</td></tr>
<tr><td>18S708m</td><td>CCTTGAGTCCTTGTGGCTCTTGG</td><td><0°C</td></tr>
<tr><td>18H912L</td><td>GCCTCAGTUCCGAAAACCAACA
| || | |||||||||</td><td>66.8°C</td></tr>
<tr><td>18S855L</td><td>ACGATGGTCCTAGAAACCAACA</td><td>26.4°C</td></tr>
</table>

Better conserved *rrs* sequences exist; however, they are very similar or identical to human sequences, so it is much more difficult to find primers that amplify all cellular microbes' DNA but not human DNA. As shown in Table 7.2, primers targeting the highly conserved *E. coli* sequences *rrs* 518–529, *rrs* 1390–1407, and *rrs* 1492–1506 match exactly or nearly exactly with all known medically important microbial species, including bacteria, fungi, and protists. They also match exactly with human sequences. If we were to use primers matching these sequences to amplify the *rrs* gene, the PCR reaction would produce mainly human amplicons. Let us compare this consensus PCR approach to the metagenomics approach.

This consensus PCR method is nearly universal, covering all parts of the tree of life (but not viruses). It is more biased than the metagenomics approach because more PCR cycles need to be run for consensus PCR than during Illumina library preparation (to amplify the consensus PCR gene of interest *rrs*), and each PCR cycle increases bias (as explained in the "Bias/ Universality" section). In addition, some species' conserved sequences mismatch the consensus primers by one base, leading to underrepresentation in the PCR product. This consensus PCR method is 20 times more efficient at detecting *M. genitalium* than metagenomics, despite the fact that it also amplifies the human *rrs* gene [there are 600 *rrs* genes per diploid human cell vs 1 *rrs* gene in a *M. genitalium* cell, and there are 0.007 ng total DNA in a human cell vs 0.0000006 ng in a *M. genitalium* cell, so $(0.007/0.0000006)/(600/1) = 20$]. This is a decent improvement in efficiency. However, it is far short of the efficiency achieved with the V34 primers that do not amplify the human *rrs* gene at all! Though a 20-fold improvement in efficiency might seem like a good reason to choose this consensus PCR method over metagenomics, it is important to remember that metagenomics is more universal, less biased, and more sensitive at high sequencing depths due to the aliquoting and gene targeting issues described in the previous section. As with metagenomics, absolute quantification of microbial cells per human cell is possible using this consensus PCR method, because human DNA is amplified and sequenced along with microbial DNA.

NEARLY UNIVERSAL CONSENSUS PCR WITH BLOCKING PRIMERS

At high sequencing depth (more than 1 billion sequence reads per specimen), metagenomics is superior to V34 consensus PCR, because if enough reads are analyzed, these two techniques become equally sensitive—and metagenomics remains more universal and less biased. At low sequencing depth (less than 1 million sequence reads per specimen), V34 bacterial consensus PCR is preferable to metagenomics, because it is much more efficient and sensitive, due to the elimination of human DNA prior to sequencing. Would it not be great if we could avoid amplifying the human *rrs* gene with primers targeting the nearly universal sequences *rrs* 518–529 and *rrs* 1390–1407, keeping V34 consensus PCR's high efficiency while expanding it to cover all microbe types? There is actually an easy way of doing this by using blocking primers. This method fixes V34 consensus PCR's lack of coverage of fungi and protists, while retaining its low cost, high efficiency, and simplicity.

Vestheim and Jarman (2008) ran into a very similar problem when characterizing the diet of krill using consensus PCR. Krill eat eukaryotic species such as algae, so consensus PCR primers targeting conserved algae *rrs* sequences also amplify the krill *rrs* gene. To avoid amplifying the krill gene, they looked for a sequence in the krill amplicon which was not present in its diet. Such sequences are not difficult to find because some regions of the *rrs* gene are not well conserved and vary significantly between species. They added a *third* primer to the PCR reaction, which matches with this krill-specific sequence. They called this an "elongation arrest" primer, which prevented the *Taq* DNA polymerase enzyme from copying krill amplicons during the PCR reaction—while amplicons from the krill's diet were efficiently copied. The framework for this approach was first described by Lewis et al. (1994). Getting this method to work was tricky, because the *Taq* DNA polymerase can remove primers that are in its way using strand displacement and its 5′ exonuclease domain.

When preparing DNA for metagenomics or consensus PCR assays, it is best to avoid sequence errors introduced while DNA strands are copied. Newer high-fidelity DNA polymerase enzymes are used for this purpose, such as ThermoFisher Scientific Phusion and NEB Q5 DNA polymerases. These are several hundred times more accurate than *Taq* DNA polymerase, greatly reducing the error rate during PCR. Conveniently, these enzymes are much better suited for the "elongation arrest" technique described earlier, because they simply stop copying DNA when they encounter a primer that is in their way! They also stop copying DNA when they encounter a uracil—this is an RNA base typically not present in DNA, but which can easily be added to PCR primers to suppress amplification (as described next). These two properties enable simple, efficient, and nearly universal consensus PCR assays (Fig. 7.9).

Briefly, a PCR reaction is setup using an equal concentration of consensus primers 16E518d/16E1408G and uracilated human-specific blocking primers 18H760m/18H912L (Fig. 7.10 and Table 7.4). These blocking primers largely prevent the amplification of human *rrs* amplicons, while amplification of microbial *rrs* amplicons remains unhindered. Uracils are added to these primers to prevent them from producing amplicons themselves. If blocking primers are omitted, the main amplicon will originate from the human *rrs* gene (1095 base pairs long); if blocking occurs, the main amplicons will originate from bacterial *rrs* genes that are around 900 base pairs long. The PCR product is then purified, and standard Illumina library preparation is used to prepare it for sequencing (Fig. 7.4).

Absolute quantification of microbes per human cell does not occur naturally using this method, because blocking efficiency is very sensitive to PCR cycling conditions, and thus the number of human amplicons which slip through the blocking process can vary considerably between runs. To get around this limitation, adding a known quantity of a control microbe as a spike-in to the PCR reaction is recommended. For example, in addition to the 100 ng of DNA pipetted from the clinical specimen and fed into the PCR reaction, 0.1 ng of DNA from a control microbe such as *E. coli* can be added—with the caveat that levels of the control organism in the clinical specimen can no longer be detected by the assay. It is usually best to choose a spike-in species that is not present in the clinical specimen. Accurate DNA quantification can be performed with a fluorometer such as the ThermoFisher Scientific Qubit or Promega QuantiFluor.

NEARLY UNIVERSAL CONSENSUS RT-PCR WITH BLOCKING PRIMERS

When consensus PCR is applied to extracted RNA instead of DNA, this is called consensus RT-PCR. RT means "reverse transcription," the extra step required to convert RNA into DNA

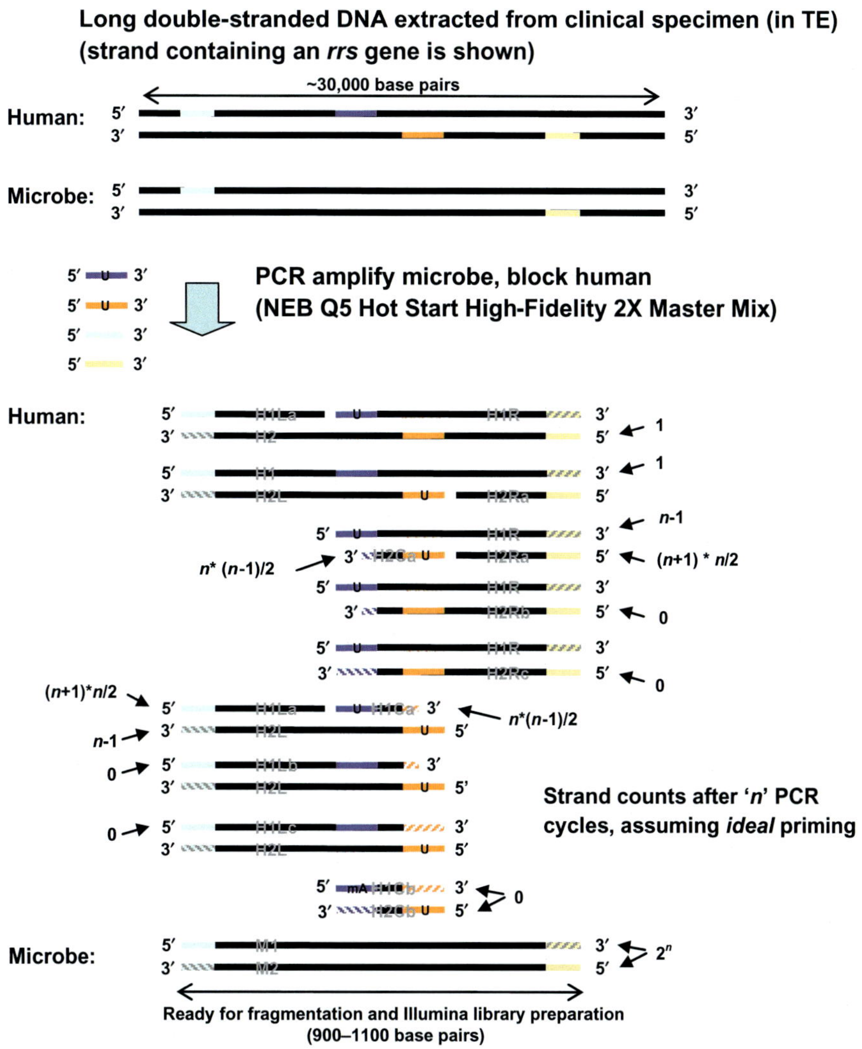

FIGURE 7.9

Consensus PCR protocol with blocking primers, primed using the highly conserved *Escherichia coli* sequences *rrs* 518–529 and *rrs* 1390–1407.

before the PCR reaction can begin. This technique is more sensitive than consensus PCR because ribosomal RNA genes are transcribed into RNA much more often than other genes by all cellular microbe species. This means the *rrs* target is naturally "preamplified" within cells and the equivalent of ~10 PCR cycles have already been run *in vivo*!

However, this technique has one very important drawback beyond PCR bias and lack of true universality: RNA strands of interest (*rrs* and *rrl* transcripts) contain modified bases that block

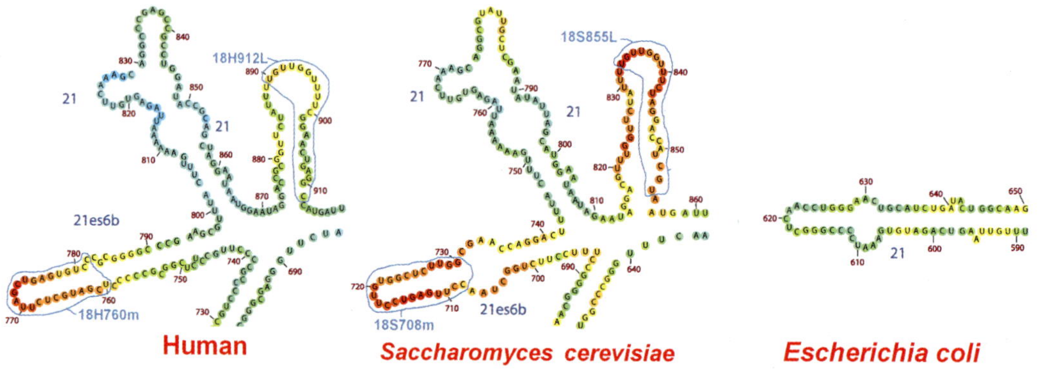

FIGURE 7.10

Two highly specific human sequences located in the C domain of the *rrs* gene: 18H_760w, 18H_890v.

reverse transcription. A consensus RT-PCR assay attempting to amplify an *rrs* segment containing such a modified base will be highly biased. The exact locations of blocking bases are not known for all species, so it is difficult to estimate how universal consensus RT-PCR assays really are. That said, consensus RT-PCR with blocking primers is the simplest, most sensitive, and most efficient microbiomic analysis method available (Fig. 7.11 and Table 7.5).

Consensus RT-PCR remains a viable alternative to metagenomics because a short poorly conserved sequence is flanked by the two best conserved *rrs* sequences (*rrs* 1390–1407 and *rrs* 1492–1506) as shown in Fig. 7.6. This *rrs* segment does not contain modified bases that block reverse transcription in model organisms *E. coli* and *S. cerevisiae* (note that this might not hold for all species). The primer pair used in the previous consensus PCR section cannot be used for RT-PCR, because they amplify a longer region of the *rrs* transcript that contains modified bases that block reverse transcription (Fig. 7.8). The amplified region for the consensus RT-PCR assay described here is very short, which means taxonomical classification is not as precise (closely related species cannot be distinguished) and Illumina library preparation is much simpler (fragmentation is not required because amplicons are short).

Briefly, an RT-PCR is setup using up to 1000 ng of RNA from a clinical specimen and a consensus RT primer targeting *rrs* 1492–1506. This primer has a 5′ extension matching with the Illumina Il1 primer, which eliminates the adaptor ligation step usually required during library preparation. It can optionally contain a long random base sequence which can be used to detect PCR duplicates during data analysis. This reaction produces a complementary DNA (cDNA) strand hybridized to each human and microbial *rrs* transcript. Then, a PCR reaction is setup to amplify this cDNA using a primer matching the second conserved region (*rrs* 1390–1407) and a truncated Illumina Il1 primer. An elongation arrest primer is added to the reaction to prevent amplification of human *rrs* amplicons. Rather than using the uracilated blocking primer method from the consensus PCR section, this blocking primer is extended at the 3′ end to *mismatch* with the human sequence and thus prevent it from priming. Because this blocking primer is longer and contains enough G/C bases, it can block the DNA polymerase despite not be being extended. Only a single blocking primer is required here, because cDNA is not double stranded (unlike genomic DNA). If the

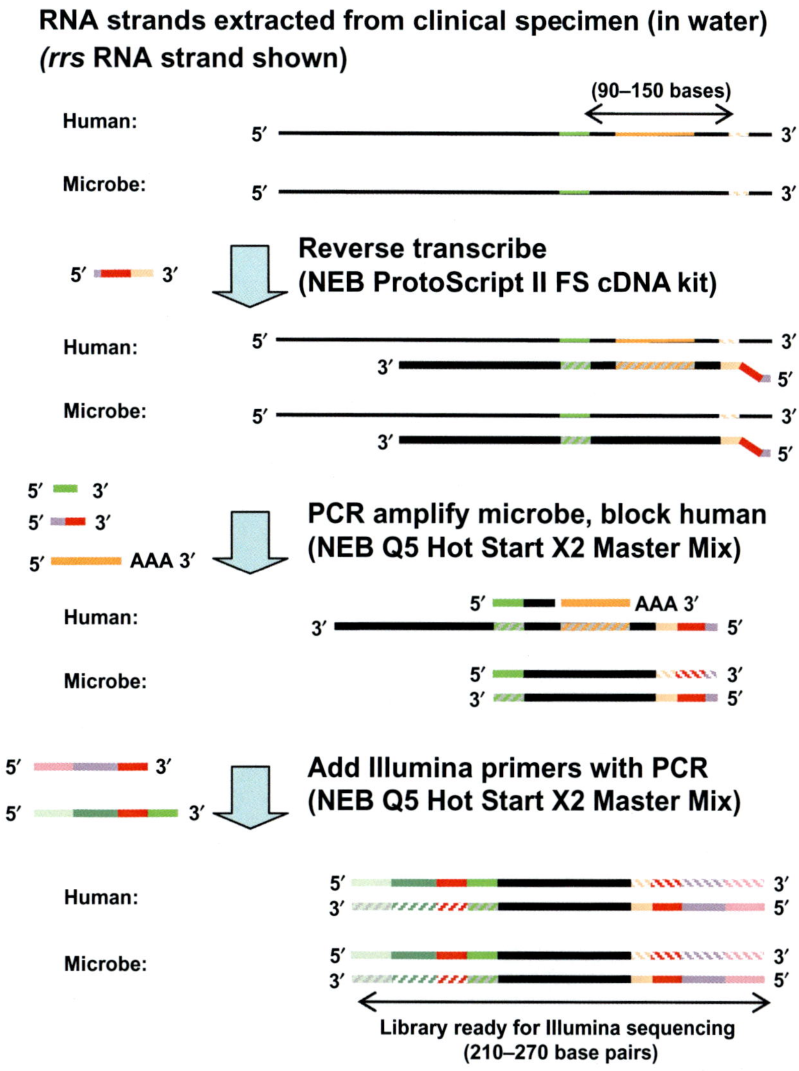

FIGURE 7.11

Consensus RT-PCR protocol with blocking primer, primed using the highly conserved *Escherichia coli* sequences *rrs* 1390–1407 and *rrs* 1492–1506.

blocking primer is omitted, the main amplicon will be the human *rrs* transcript (193 base pairs long); if blocking occurs, the main amplicons will be bacterial *rrs* transcripts (~162 base pairs long) and the human mitochondrial *rrs* transcript (143 base pairs long). Finally, a second short PCR reaction is run to complete the library using full length Illumina primers (IU_16E1391g and II1).

Table 7.5 Primer Sequences (Highlighted) Used to Perform Consensus RT-PCR (Fig. 7.11)

<table>
<tr><th></th><th>Primer/Sequence name</th><th>Alignment (5′ → 3′)</th><th>T_m</th></tr>
<tr><td rowspan="2">Consensus primers (all life)</td><td>16E1390e</td><td>TTGTACACACCGCCC
|||||||||||||||</td><td>61.0°C</td></tr>
<tr><td>16E1390f (aka 18S1627f)</td><td>TTGTACACACCGCCCG</td><td>61.0°C</td></tr>
<tr><td rowspan="6">Illumina compatible primers and adaptors</td><td>IU_16E1391g</td><td>AATGATACGGCGACCACCGAGATCTACACTCTTTCCCTACACGACGCTCTTCCGATCTTGTACACACCGCCCGT*C
|||||||||||||||||| |</td><td>~79.0°C</td></tr>
<tr><td></td><td>TTGTACACACCGCCCGT-C</td><td>66.5°C</td></tr>
<tr><td>II1</td><td>CAAGCAGAAGACGGCATACGAGATCGTGATGTGACTGGAGTTCAGACGTGTGCTCTTCCGATC*T
||||||||||||||||||||| |</td><td>~78.0°C</td></tr>
<tr><td>II33v</td><td>AGACGTGTGCTCTTCCGATC-T</td><td>66.8°C</td></tr>
<tr><td>II44h</td><td>AGACGTGTGCTCTTCCGA
|||||||||||||||</td><td>65.3°C</td></tr>
<tr><td>II46f</td><td>ACGTGTGCTCTTCCGAT</td><td>62.8°C</td></tr>
<tr><td rowspan="2">RT primer (all life)</td><td>II46i_25N_T_16E1507E</td><td>ACGTGTGCTCTTCCGATCTNNNNNNNNNNNNNNNNNNNNNNNNNTACCTTGTTACGACTT
|||||||||||||||</td><td>n/a</td></tr>
<tr><td>16E1507E</td><td>ACCTTGTTACGACTT</td><td>50.8°C</td></tr>
<tr><td rowspan="2">Blocking primers (human)</td><td>18H1721!s_16A</td><td>TGGATGGTTTAGTGAGGCCCTCGGATCGGCCCCGCCGGGGTCGGCCCACGGCCCTAAAAAAAAAAAAAAAA
|| |||| ||||||||||| ||||| || | || || | || || |</td><td>~85.0°C</td></tr>
<tr><td>18S1657!p</td><td>TGAATGGCTTAGTGAGGCCTCAGGATCTGCTTAGAGAAGG--GGGCAAC-TCCAT</td><td>54.5°C</td></tr>
</table>

A microbial control spike-in is recommended for the same reasons explained in the consensus PCR section. For example, 1000 ng of RNA can be taken from the clinical specimen, and 1 ng of RNA can be taken from a control microbe such as *E. coli*, again with the caveat that the control species can no longer be detected by the assay.

CUSTOM ILLUMINA LIBRARY PREPARATION

Standard Illumina library preparation methods used for metagenomic assays are shown in Figs. 7.4 and 7.5. These protocols are usually executed as a service by sequencing centers. Unfortunately, control of microbial contaminants is not required for routine genomics or transcriptomics, so such contaminants are often inadvertently inserted at the sequencing center during library preparation. These contaminants are not important for projects which only care about host sequences, and sequencing centers have generally not optimized their workflow for metagenomic assays. This contamination can make the interpretation of metagenomics results difficult.

Library preparation workflows can be slightly modified to add a "signature sequence" early on; sequence reads lacking this signature sequence can be detected during data analysis and eliminated as contamination which occurred downstream of the step where the signature sequence was added. Contamination occurring upstream cannot be identified and eliminated using his method. For example, random hexamers used to prime reverse transcription in Fig. 7.5 can be extended to include a short signature sequence (Table 7.6). Sequence reads lacking this known sequence can be identified during data analysis and discarded.

As another example, a restriction enzyme can be used to fragment input DNA at specific sequences, instead of other methods which cut DNA randomly (Fig. 7.4). This means sequence reads which do not begin with a restriction enzyme sequence very likely originated elsewhere and

Table 7.6 Primer and Adaptor Sequences (Highlighted) Used in Illimuna Library Preparation for Metagenomic Assays (Figs. 7.4 and 7.5)

<table>
<tr><th></th><th>Primer/Sequence name</th><th>Alignment (5′ → 3′)</th><th>T_m</th></tr>
<tr><td rowspan="10">Illumina compatible primers and adaptors</td><td>IA (aka II64U_IU25x)</td><td>/5Phos/GATCGGAAGAGCACACGTCTGAACTCCAGTCUACACTCTTTCCCTACACGACGCTCTTCCGATC*T</td><td>n/a</td></tr>
<tr><td>IU</td><td>AATGATACGGCGACCACCGAGATCTACACTCTTTCCCTACACGACGCTCTTCCGATC*T</td><td>~78.0°C</td></tr>
<tr><td></td><td>|||||||||||||||||||||||||||||||| |</td><td></td></tr>
<tr><td>IU26w</td><td>ACACTCTTTCCCTACACGACGCTCTTCCGATC-T</td><td>74.6°C</td></tr>
<tr><td>II1</td><td>CAAGCAGAAGACGGCATACGAGATCGTGATGTGACTGGAGTTCAGACGTGTGCTCTTCCGATC*T</td><td>~78.0°C</td></tr>
<tr><td></td><td>||||||||||||||||||||||||||||||| |</td><td></td></tr>
<tr><td>II33v</td><td>GACTGGAGTTCAGACGTGTGCTCTTCCGATC-T</td><td>74.0°C</td></tr>
<tr><td>II2</td><td>CAAGCAGAAGACGGCATACGAGATACATCGGTGACTGGAGTTCAGACGTGTGCTCTTCCGATC*T</td><td>~78.0°C</td></tr>
<tr><td>II3</td><td>CAAGCAGAAGACGGCATACGAGATGCCTAAGTGACTGGAGTTCAGACGTGTGCTCTTCCGATC*T</td><td>~78.0°C</td></tr>
<tr><td>II4</td><td>CAAGCAGAAGACGGCATACGAGATTGGTCAGTGACTGGAGTTCAGACGTGTGCTCTTCCGATC*T</td><td>~78.0°C</td></tr>
<tr><td rowspan="5">RT primers</td><td>6N</td><td>NNNNNN</td><td>n/a</td></tr>
<tr><td>6N1</td><td>CGTGATNNNNNN</td><td>n/a</td></tr>
<tr><td>6N2</td><td>ACATCGNNNNNN</td><td>n/a</td></tr>
<tr><td>6N3</td><td>GCCTAANNNNNN</td><td>n/a</td></tr>
<tr><td>6N4</td><td>TGGTCANNNNNN</td><td>n/a</td></tr>
</table>

Primer names assigned by the command "leif primer0 <sequence > ". Melting temperatures are calculated with the command "leif tm0 <seq1> <seq2> 0.5 0 2 0.8."

should be discarded. A restriction enzyme which works well for this purpose is HpyCH4IV (A/CGT); it cuts microbial genomes into relatively short fragments suitable for Illumina sequencing, while most human fragments remain too long to be sequenced. This method is thus biased against human DNA, which increases assay efficiency by partially eliminating human DNA before sequencing. Note that it also inserts a *known* bias for each microbial species, which means an adjustment is needed to correct the relative amount of each microbe.

BIOINFORMATICS

MULTIPLE ALIGNMENT PASSES

Three variables affect the speed of sequence alignments: (1) the number of sequence reads to align (how big are the FASTQ files?); (2) the number of reference sequences (how big are the FASTA files?); (3) the alignment algorithm and parameters (how close must these sequences match for alignment to occur?). In the first alignment pass, the goal is to quickly eliminate uninteresting sequence reads, especially those originating from the host. Due to the large number of sequence reads, it is best to minimize the reference database size (human sequences only) and use a very simple alignment algorithm. This step cuts the number of sequences reads which need further processing by orders of magnitude by discarding human reads. In the second alignment pass, the goal is to taxonomically classify remaining sequence reads which match very well with common microbes. This is typically done with a larger FASTA file containing the genomes of well-curated microbes (e.g., using the "nt" database, supplemented by manually downloading additional microbe genomes from GenBank if necessary). Sequence reads which do not align well in the second pass require additional alignment passes, which are substantially slower. These reads are usually in one of the following categories: (1) they contain a sequencing error such as an insertion, a deletion, or a long stretch of wrongly called bases; (2) they are a chimera of two DNA fragments inadvertently ligated together during library preparation; (3) they originate from a microbe present in GenBank which was not included in the database used for the second alignment pass; and (4) they originate from a novel microbe which is not in GenBank yet. The third alignment pass is usually setup to align to all GenBank sequences, in order to detect known microbes which were not present in the smaller database used in the second pass (they should be added!). The fourth and final alignment pass is typically run with a more sophisticated alignment algorithm which can handle indels (insertions or deletions) and chimeras. It is much slower, and the results usually need to be manually curated in an attempt to figure out if there are novel species present in the specimen.

GAPPED ALIGNMENT

Illumina sequencing produces few errors and very few indels. Such low error rates are very important for microbiome assays: (1) they are required to distinguish closely related species and (2) they also greatly accelerate data analysis by allowing the use of less computationally intensive alignment algorithms. In particular, gapped alignment algorithms (which can handle indels) are substantially slower than non-gapped alignment algorithms. Sequences containing indels align poorly with non-gapped algorithms, ending up taxonomically labeled as "unknown." The vast majority of reads will have been classified during non-gapped alignment and will not need gapped alignment.

WORD LENGTH

Alignment algorithms typically use a two-step approach to accelerate the process. The first step is to find alignment "seeds," which means scanning both FASTQ and FASTA files looking for short base strings which match exactly. The length of these strings is called the "word length" and is typically chosen to be between 7 and 28 bases long. The longer the word length, the more difficult it is to find a seed location, so fewer (mostly spurious) alignment seeds are produced; this means alignment will occur faster, but low homology alignments might be missed altogether. The second alignment step calculates how homologous the FASTQ and FASTA sequences are; this second step can be quite slow and must be invoked for each seed that is found.

HOST VERSUS NON-HOST

The first alignment step used to eliminate host sequences can be greatly accelerated by omitting the second part of alignment algorithms, simply discarding all sequence reads which contains a "seed" matching with the human genome. A long word length must be used to avoid accidentally discarding microbial reads coincidentally matching with the human genome. Word lengths of 32–64 work well for this purpose. However, the most interesting genes for taxonomic classification (*rrs*/*rrl*) contain long conserved sequences which can match with the human *rrs*/*rrl* sequence even at these longer word lengths. For this part of the genome, this simple "discard if seeded" method does not work. This means *rrs*/*rrl* sequences must be removed from the human genome FASTA files before performing the first alignment pass, so that both microbial and human *rrs*/*rrl* reads will be carried forward. Most reads will then be identified as human in a downstream alignment step which

Table 7.7 Important Files Used in the Bioinformatics Workflow

	Hyperlink	Size (G)	Comment
Taxonomy	ftp.ncbi.nih.gov/pub/taxonomy/taxdump.tar.gz	0.04	NCBI Taxonomy database (numerical taxid → taxonomic name, e.g., 9606 → "Homo sapiens")
	ftp.ncbi.nih.gov/pub/taxonomy/accession2taxid/nucl_est.accession2taxid.gz	0.53	File containing numerical taxids for each GenBank sequence (accession id → numerical taxid, e.g., "NR_145820.1" → 9606)
	ftp.ncbi.nih.gov/pub/taxonomy/accession2taxid/nucl_gb.accession2taxid.gz	1.00	
	ftp.ncbi.nih.gov/pub/taxonomy/accession2taxid/nucl_gss.accession2taxid.gz	0.20	
	ftp.ncbi.nih.gov/pub/taxonomy/accession2taxid/nucl_wgs.accession2taxid.gz	3.00	
	ftp.ncbi.nih.gov/pub/taxonomy/accession2taxid/prot.accession2taxid.gz	4.10	
	ftp.ncbi.nih.gov/pub/taxonomy/accession2taxid/dead_nucl.accession2taxid.gz	0.13	
	ftp.ncbi.nih.gov/pub/taxonomy/accession2taxid/dead_prot.accession2taxid.gz	0.55	
FASTA nucleotide	ftp.ncbi.nlm.nih.gov/blast/db/FASTA/nt.gz	44	Well-curated nucleotide sequences and genomes
	ftp.ncbi.nlm.nih.gov/blast/db/FASTA/other_genomic.gz	226	
	ftp.ncbi.nlm.nih.gov/sra/wgs	>500	Directory containing less well curated, usually incompletely assembled genomes
	www.arb-silva.de/no_cache/download/archive/current/Exports (rrs)	0.572	Directory containing well-curated nucleotide sequences of rrs and *rrl* genes
	www.arb-silva.de/no_cache/download/archive/current/Exports (rrl)	0.115	
FASTA protein	ftp.ncbi.nlm.nih.gov/blast/db/FASTA/nr.gz	40	Well-curated protein sequences
Websites	ncbi.nlm.nih.gov/taxonomy	n/a	NCBI Taxonomy website
	blast.ncbi.nlm.nih.gov	n/a	NCBI online alignment tool (BLAST +)

The size of these files was measured on November 5, 2018.

performs a complete alignment algorithm, thus can distinguish between closely matching human and microbial *rrs*/*rrl* sequences.

DATABASES

Databases used for metagenomic assays are mostly maintained by the NCBI (Table 7.7). A non-NCBI database known as SILVA (Quast et al., 2012) is useful to align sequence reads originating from ribosomal RNA genes or transcripts (e.g., *rrs*/*rrl* sequences). The main database used in metagenomic assays is called "nt." It contains a human genome and sequences from common microbes. It does *not* contain the genomes of all microbes in GenBank. A much larger database of genomes is called "other_genomic"; while more comprehensive, it does *not* contain the genomes of all microbes in GenBank either. The databases mentioned so far are well curated, and few of their entries contain taxomically mislabeled sequences. The rest of GenBank is usually referred to as "wgs," a very large database which contains partly assembled genomes of all species present in GenBank. It is not well curated, which means many sequences in "wgs" are taxonomically inaccurate. This is usually the result of microbial contaminants inadvertently inserted in the laboratory which sequenced a given microbe, animal, or plant genome. Contaminant sequences in "wgs" tend to be short (less than 2000 bases long), because the number of reads originating from contaminants are usually insufficient to reassemble the microbe's genome; this is an easy way of eliminating most contaminants while retaining most relevant genome sequences.

ALIGNING RIBOSOMAL RNA AGAINST SILVA

The simplest and most universal alignment method compares ribosomal RNA (*rrs*/*rrl*) sequence reads to the well-curated SILVA database (Quast et al., 2012). When sequencing total RNA, *rrs*/*rrl* reads comprise about 75% of total reads. The remaining 25% of reads will align poorly to SILVA sequences, because they are not ribosomal RNA strands (most are messenger RNA strands).

How can one distinguish a novel species whose *rrs*/*rrl* sequence matches poorly with SILVA, from messenger RNA reads which also match poorly with SILVA? This is not a trivial problem. *rrs*/*rrl* genes contain a few dozen well-conserved regions that can be automatically detected in software (such as those shown in Table 7.2), allowing *rrs*/*rrl* sequence reads to be separated from the rest—even if they originate from a novel species. This is the most efficient method to detect novel cellular microbe species using metagenomics.

CONCLUSION

Many details and considerations must be taken into account when designing metagenomic assays. It may seem overwhelming at first, but it is not that difficult. The easiest metagenomic assay is sequencing total RNA as shown in Fig. 7.5 and aligning the results to the well-curated SILVA database only. If this approach proves too expensive, consensus PCR methods (Figs. 7.9 and 7.11) are nearly universal and significantly cheaper due to the elimination of human DNA/RNA prior to sequencing. The same bioinformatics flow can be used for both approaches.

DISCLOSURES

The Leif Microbiome Analyzer, used in some of the examples, is a commercial product of Shipshaw Labs.

REFERENCES

Agris, P., 2018. The RNA Modification Database. The RNA Institute, Albany, NY. Available From: <http://mods.rna.albany.edu> (accessed 11.08.18.).

Bakshi, S., Siryaporn, A., Goulian, M., Weisshaar, J.C., 2012. Superresolution imaging of ribosomes and RNA polymerase in live *Escherichia coli* cells. Mol. Microbiol. 85, 21–38.

Besaratinia, A., Yoon, J.-I., Schroeder, C., Bradforth, S.E., Cockburn, M., Pfeifer, G.P., 2011. Wavelength dependence of ultraviolet radiation-induced DNA damage as determined by laser irradiation suggests that cyclobutane pyrimidine dimers are the principal DNA lesions produced by terrestrial sunlight. FASEB J. 25, 3079–3091.

Borst, A., Box, A., Fluit, A., 2004. False-positive results and contamination in nucleic acid amplification assays: suggestions for a prevent and destroy strategy. Eur. J. Clin. Microbiol. Infect. Dis. 23, 289–299.

Cappallo-Obermann, H., Schulze, W., Jastrow, H., Baukloh, V., Spiess, A.-N., 2011. Highly purified spermatozoal RNA obtained by a novel method indicates an unusual 28S/18S rRNA ratio and suggests impaired ribosome assembly. Mol. Hum. Reprod. 17, 669–678.

Chomczynski, P., Wilfinger, W.W., Eghbalnia, H.R., Kennedy, A., Rymaszewski, M., Mackey, K., 2016. Inter-individual differences in RNA levels in human peripheral blood. PLoS One 11, e0148260.

Cottier, F., Srinivasan, K.G., Yurieva, M., Liao, W., Poidinger, M., Zolezzi, F., et al., 2018. Advantages of meta-total RNA sequencing (MeTRS) over shotgun metagenomics and amplicon-based sequencing in the profiling of complex microbial communities. NPJ Biofilms Microbiomes 4, 2.

Dowd, S.E., Wolcott, R.D., Sun, Y., Mckeehan, T., Smith, E., Rhoads, D., 2008. Polymicrobial nature of chronic diabetic foot ulcer biofilm infections determined using bacterial tag encoded FLX amplicon pyrosequencing (bTEFAP). PLoS One 3, e3326.

Dupuy, A.K., David, M.S., Li, L., Heider, T.N., Peterson, J.D., Montano, E.A., et al., 2014. Redefining the human oral mycobiome with improved practices in amplicon-based taxonomy: discovery of Malassezia as a prominent commensal. PLoS One 9, e90899.

Feng, H., Shuda, M., Chang, Y., Moore, P.S., 2008. Clonal integration of a polyomavirus in human Merkel cell carcinoma. Science 319, 1096–1100.

Gaudin, M., Desnues, C., 2018. Hybrid capture-based next generation sequencing and its application to human infectious diseases. Front. Microbiol. 9, 2924.

Glen, A., 1967. Measurement of DNA and RNA in human peripheral blood lymphocytes. Clin. Chem. 13, 299–313.

Grens, K., 2014. DNA extraction kits contaminated. The Scientist .

Gruber, K., 2015. Here, there, and everywhere. EMBO Rep. 16, 898–901.

Hedberg, A., Johansen, S.D., 2013. Nuclear group I introns in self-splicing and beyond. Mob. DNA 4, 17.

Laurence, M., Benito-León, J., 2017. Epstein–Barr virus and multiple sclerosis: updating Pender's hypothesis. Mult. Scler. Relat. Disord. 16, 8–14.

Laurence, M., Hatzis, C., Brash, D.E., 2014. Common contaminants in next-generation sequencing that hinder discovery of low-abundance microbes. PLoS One 9, e97876.

Lewis, A.P., Sims, M.J., Gewert, D.R., Peakman, T.C., Spence, H., Crowe, J.S., 1994. Taq DNA polymerase extension of internal primers blocks polymerase chain reactions allowing differential amplification of molecules with identical 5′ and 3′ ends. Nucleic Acids Res. 22, 2859.

Limbach, P.A., Crain, P.F., Mccloskey, J.A., 1994. Summary: the modified nucleosides of RNA. Nucleic Acids Res. 22, 2183–2196.

Lipkin, W.I., 2010. Microbe hunting. Microbiol. Mol. Biol. Rev. 74, 363–377.

Motorin, Y., Muller, S., Behm-Ansmant, I., Branlant, C., 2007. Identification of modified residues in RNAs by reverse transcription-based methods. Methods Enzymol. 425, 21–53.

Ortiz, J.O., Förster, F., Kürner, J., Linaroudis, A.A., Baumeister, W., 2006. Mapping 70S ribosomes in intact cells by cryoelectron tomography and pattern recognition. J. Struct. Biol. 156, 334–341.

Palacios, G., Druce, J., Du, L., Tran, T., Birch, C., Briese, T., et al., 2008. A new arenavirus in a cluster of fatal transplant-associated diseases. N. Engl. J. Med. 358, 991–998.

Quast, C., Pruesse, E., Yilmaz, P., Gerken, J., Schweer, T., Yarza, P., et al., 2012. The SILVA ribosomal RNA gene database project: improved data processing and web-based tools. Nucleic Acids Res. 41, D590–D596.

Salter, S.J., Cox, M.J., Turek, E.M., Calus, S.T., Cookson, W.O., Moffatt, M.F., et al., 2014. Reagent and laboratory contamination can critically impact sequence-based microbiome analyses. BMC. Biol. 12, 87.

Sanger, F., Nicklen, S., Coulson, A.R., 1977. DNA sequencing with chain-terminating inhibitors. Proc. Natl. Acad. Sci. U.S.A. 74, 5463–5467.

Strong, M.J., Xu, G., Morici, L., Bon-Durant, S.S., Baddoo, M., Lin, Z., et al., 2014. Microbial contamination in next generation sequencing: implications for sequence-based analysis of clinical samples. PLoS Pathog. 10, e1004437.

Thorley-Lawson, D.A., 2015. EBV persistence—introducing the virus, Epstein Barr Virus, Vol. 1. Springer.

Vestheim, H., Jarman, S.N., 2008. Blocking primers to enhance PCR amplification of rare sequences in mixed samples–a case study on prey DNA in Antarctic krill stomachs. Front. Zool. 5, 12.

Vesty, A., Biswas, K., Taylor, M.W., Gear, K., Douglas, R.G., 2017. Evaluating the impact of DNA extraction method on the representation of human oral bacterial and fungal communities. PLoS One 12, e0169877.

von der Haar, T., 2008. A quantitative estimation of the global translational activity in logarithmically growing yeast cells. BMC Syst. Biol. 2, 87.

CHAPTER

CULTUROMICS: THE ALTERNATIVE FROM THE PAST 8

Manousos E. Kambouris
The Golden Helix Foundation, London, United Kingdom

INTRODUCTION

The appearance of Culturomics (Lagier et al., 2012) reversed the widespread practice in modern microbiote research to focus on genomic and structural studies, the latter exemplified by MALDI-TOF (Lay, 2001; Ragoussis et al., 2006; Kolecka et al., 2014; Becker et al., 2014) and disregard humble and messy microbiology proper (Fig. 8.1). Thus, in practice, Koch's postulates were dismissed as well. Genomics had the advantage of fast turnover, compared to cultures (Velegraki et al., 1999). Structural studies had none and were widely using as a focal argument the - undeniably factual but probably exaggerated - concept of uncultivable microbiota (Orjala, 2008; Zengler et al., 2002; Greub, 2012; Dubourg et al., 2013; Puspita et al., 2012; Cavalier-Smith, 2004).

Among a number of flaws the involved methods bring along (Long et al., 2013; Peng et al., 2012; Zhou et al., 2014), is the - also undeniable – fact that they do not further downstream interrogation as they fail to furnish living samples; the same holds true for projections of adaptability and growth dynamics in diverse sets of conditions. Simultaneously, the concept of "uncultivable microbiota" was more and more proven to be a generalization, and possibly erratic: given the proper conditions for a precise simulation of their habitat, uncultivable microbiota could be cultured (Kaeberlein et al., 2002; Singh et al., 2013; Puspita et al., 2012).

Spectroscopy needs pure cultures with a reasonable yield (Becker et al., 2014). These require microbiologic criteria, and single-cell colonies bypass the issue only up to a point; *one* pure culture per sample is hardly the context of microbiome study and qualitative analysis: any kind of microbiomic sample most probably contains more than one species, and these must be segregated to pure cultures for identification. Spectroscopy works with pure samples and compares them to metrics of pure cultures in standardized conditions by compiling plots (Verwer et al., 2014; Becker et al., 2014). Needless to say, if a cell cannot be cultured, being "uncultivable," it cannot be interrogated by structural assays.

Furthermore, metagenomic approaches fall short in complicated samples, as microbiota subrepresented in the population become overshadowed by more abundant ones and thus underscored or unscored due to insufficient threshold depth (Greub, 2012). The depth indicates roughly the number of target copies available and affects most of all exponential amplification techniques such as the PCR—the selectivity of the used primers substitutes the enrichment step of conventional microbiology. Once this selectivity is aborted and generic procedures are used to enhance all present

Microbiomics. DOI: https://doi.org/10.1016/B978-0-12-816664-2.00008-6

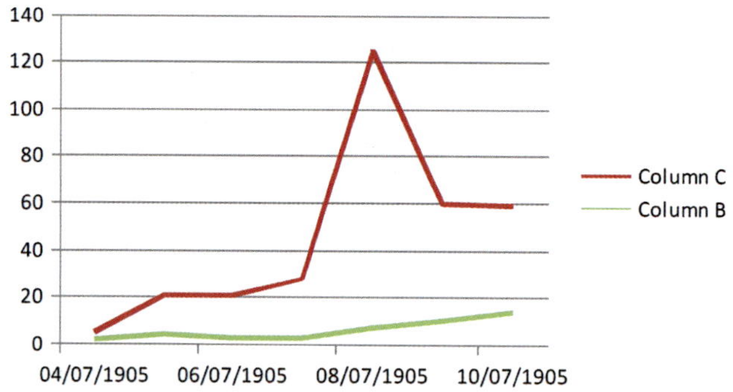

FIGURE 8.1

Dissemination of culturomics. PubMed returns to the word "culturomics" and returns with the term appearing in the title. The latter shows a steady but slow increase, the former peaked in 2016 and dropped in 2017, without recuperating in 2018 (as to 31/10/18).

sequences to detectable levels, more abundant ones have a distinct advantage in exponential applications, leading scarcer sequence populations to apparent extinction through threshold failure. Linear applications are affected less: scarce targets may be shadowed in a probabilistic competition for amplification resources. Even if such a template produces signal at any stage, it might fail the detection threshold of the method either in absolute terms or due to rejection during signal processing as background noise (Zhou et al., 2014).

A (meta)genomic analysis is a 2D projection of life—the latter being a dynamic and complex procedure unfolding in 3D. It is self-evident that if the signature of the life phenomenon is harnessed by objective criteria and standards, the descriptive power attainable skyrockets and usable data, such as toxin and other secondary metabolite production, may be readily available at an identification process and not hidden craving to be mined through DNA sequence deciphering, as proposed the concept of phenomics (Schilling et al., 1999).

CULTUROMICS: INVENTING OR RECASTING?

The culturing of a mixed sample to a series of media so as to enhance different populations has been routine in medical microbiology and usually termed "culture" or "growth profiles" of mixed samples. For their sake, enrichment and differentiating/selective media were discovered and synthesized for at least two centuries now (Beijerinck, 1901), and conventional microbiology considers routine the multiple cultures per sample (Weinstein, 1996). The extremely high number of substrates and conditions used by culturomics, many of which innovative and some exotic, do not earn "-omics" suffix by themselves but introduce the concept into the club of highly parallel and high-throughput methods. The key issue that grants the "-omics" status is the concept of a multiaspect growth analysis providing a more dynamic, detailed, and extensive description of the microbiote than other "-omics." Many among the latter, that is, genomics, transcriptomics, proteomics, and metabolomics are massively detailed but one-dimensional; they produce analytic descriptions in individual or collective terms of their analytes within the spatiotemporal limitations of an organism. With life unfolding mainly into time (ontogeny) such a context is a frame of a movie. It can be *in* or *out of* context, but it is not *the* context.

On the contrary, culturomics may morph to suit different needs and inquiries, but by definition they result in a continuous, "motion" picture of the subject. This sequence of events may develop over time and this differentiation may be captured, studied, analyzed, described, and used for different purposes.

Culturomics infringes with phenomics (Houle et al., 2010); they do overlap, but partially: phenomics includes *in situ*, nonculture-dependent phenotype characteristics and are mostly restricted to frame capture. Contrarily, the most basic use of culturomics (with or without the temporal dimension) remains the digital growth/no growth without venturing into description of form and function (Fig. 8.2).

Finally, culturomics in many cases intersects with the array methodology, with or without the prefix "micro." There is, though, a clear-cut difference: (micro)arrays are a methodology, a way to organize and implement tasks. Culturomics is a concept, the standardization and introduction of highly parallel format in the culture principle. Implementation is entirely possible through arrays; "phenotypic arrays" are just one example (Bochner et al., 2001); "living-cell microarrays" are even more prominent a case (Yarmush and King, 2009). But alternative solutions, such as stacks of conventional dishes, or—even better, compartmentalized petri dishes (Fig. 8.3)—are just as eligible for Culturomics. Thus the two entities are neither inclusive with each other, nor exclusive, they simply intersect: there are culturomic approaches without reverting to the use of arrays, as already mentioned, and of course, there are arrays, which apply to sectors completely irrelevant to culturomics.

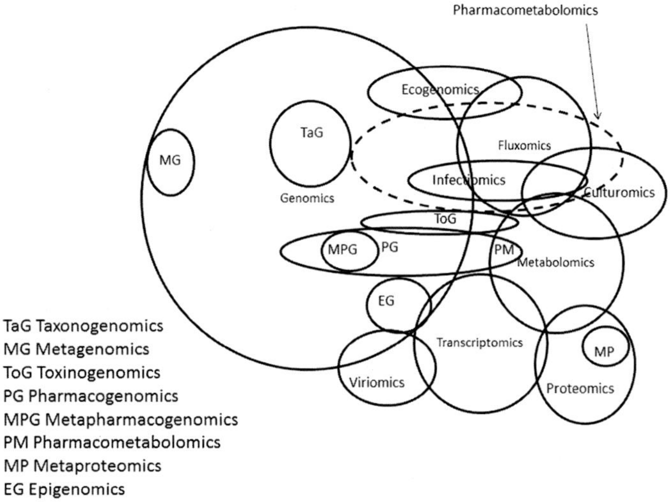

FIGURE 8.2

Interaction and interspace of culturomics with other descriptive and functional -Omics.

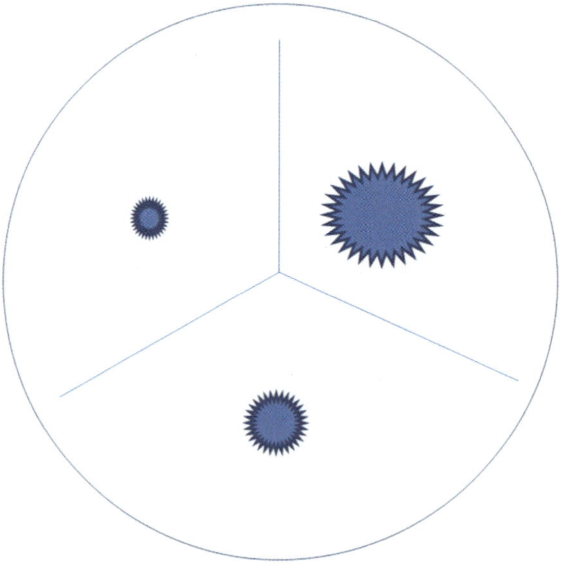

FIGURE 8.3

Low-throughput culturomics format: compartmentalized petri dish loaded with three different substrates and inoculated with the same sample, in order to produce simultaneous and microenviromentally identical growth curves and phenotypes.

PHYLOGENESIS OF CULTUROMICS

Culturomics far exceeds conceptually the format and application that introduced the sector (Lagier et al., 2012) and which may be termed *analytical culturomics* (Table 8.1). The operative idea there and then had been the growth of the maximum possible number of subpopulations in a multitude of media and conditions in order to enrich the content in subrepresented but cultivable microbiota and to grow on culture alleged uncultivable microbiota.

Diverse, highly specific media, compositions, and sets of conditions are needed, in order to impede growth of ubiquitous microbiota to many different cultures, so as to enhance differentiation and discrimination. Colonies from different culture units can be studied and analyzed by a host of other modalities and methods, which is the defining advantage of culture methodology. In this way, different colonies are fused into a species (or other taxa) while developing its analytical culturomic profile (growth/no growth in a series of culture media and conditions). The opposite result is to segregate colonies to different species (or taxa) by *comparative* profiling.

Analytical culturomics ventures deep into extending culture conditions and formats but is restrictive in the collection and use of data. The inherent information context of living microbiota is left untapped, and that occurs at a time when tracking technologies as are multispectral imaging, spectroscopy and successive generations of software tools to fuse, mine, collect, interpret and explore respective data evolve by the day (Maia et al., 2016; Smolina et al., 2010; Buchan and Ledeboer, 2014). Conceptually, its advance is marginal at best over blood cultures or multiple urine

Table 8.1 Breeds and Types of Culturomics.

Culturomics Category	Sample Type	Substrates/Conditions	Output	Aim/Use	Nature (Breed/Type)	Specific Requirements of Hardware
Analytical	Mixed	Multiple	Digital growth	Segregate populations	Cognitive (S)	None
Comparative	Pure	Multiple	Digital growth	Differentiate taxa	Cognitive (S)	None
Descriptive	Pure	Multiple	Growth curves	Define taxa	Cognitive (S)	None
Electro-	Any	Multiple/differential electromagnetic treatment	Growth/metabolism	Discrimination/optimization for applications	Functional (DS)	Conductivity/transparency
Radio-	Any	Multiple/differential irradiation exposure (IR, visual/colors, beta/gamma/X-ray)	Growth/metabolism	Discrimination/optimization for applications	Functional (DS)	Transparency
Thermo-	Any	Multiple/differential thermal treatment	Growth/metabolism	Discrimination/optimization for applications	Functional (DS)	Tolerance in extreme temperatures
Bari-	Any	Multiple/differential pressure application (explosive, sonic, hydrostatic, osmotic, atmospheric). hyper/hypobaric oxygen	Growth/metabolism	Discrimination/optimization for applications	Functional (DS)	Tolerance and conductivity of pressure effects
Immuno-	Pure	Multiple	Immuophenotypes/immunoprofile	Differentiate taxa	Cognitive (S)	None
Hetero-	Mixed	Multiple	Symbiotic dynamics		Functional (S)	None

Breed refers to the use of produced data; type to the different parameters, mostly physical rather than chemical, factored in a culturomic format. S (for similar) refers to types achievable in standard plastic-/glassware; DS (for dissimilar) stands for formats requiring modifications in culture vessels.

or fecal cultures; still it must be recognized that the range of detectable microbiota expanded significantly and microbiological methods are reinvigorated if not totally resurrected.

Cultures produce a multitude of differential characters among which the digital growth (growth/no-growth) is only one. Creating and comparing growth profiles in hundreds of different substrates defines the second—cognitive—iteration of culturomics, *comparative culturomics* (Table 8.1). It originates from the decades-old practice of growth profiling assays of pure samples, such as Analytical Profile Index (API) (Ieven et al., 1995), Biolog (Romagnoli et al., 1995), and phenotypic arrays (Orro et al., 2015, Bochner et al., 2001). They all have been used initially to compile a database by testing known microbiotes and describing and defining them in order to proceed to detection/identification of unknown/suspect samples by comparison to database entries.

A culturomic profile may include, without being restricted to, assimilation tests as is the 2,3,5-TriphenylTetrazolium Chloride (TTC) for *Candida dubliniensis* (Velegraki and Logotheti, 1998); secondary metabolite detection, as is the CO_2 production (Sperber and Swan, 1976); macroscopic morphological characters such as color, size, and texture of a colony (McDonald, 2003). All the above can be assayed in different media and incubation conditions, such as humidity, temperature, constitution of atmosphere, and pressure, to mention just a few. The culturomic profile would characterize type strains very similarly to an ID card. But its prospect is to exclude from, or include in a given taxon unknown samples, as its descriptive power far exceeds the respective of current culture methodology. Similarly, its potential in predicting physiology peculiarities of an isolate outdoes molecular approaches.

A third possible iteration of culturomics, still cognitive in nature, may be the *descriptive culturomics* (Table 8.1), which fully introduces the temporal parameter. A pure strain or isolate is not simply cultured in a wide range of media and conditions so as to develop phenotypic characteristics; the progress, development, and differentiation of the growth are constantly monitored and plotted to curves relative to incubation time. Growth curves are well-known for such indices but of low information content, irrespective of the countable parameter (circumference, surface, or radius) of the colony. Quantitative and qualitative profiling of morphology, metabolomic content and output, and formation of dormant or hypobiotic bodies (as in sporulation and endocystosis) against time provides robust discrimination and individualization approaches. Some microbiota are characterized by elaborate morphology. The mycelia of molds, for example, offer a range of morphologic indices in addition to growth rates (Kambouris and Velegraki, 2016), which can be plotted as a function of time in different culture conditions, or as a function of certain conditions. These plots, especially their growth component, cannot be predicted by current genomics, and enhance the discrimination potential inherent in culturomics for a wide scope of applications in multisectorial health, industry, and environment (Kambouris et al., 2011; Houle et al., 2010).

Descriptive culturomics enables the most valuable and rather unique ability of the new sector, which is the testing of responses and susceptibilities to combinations of similar or dissimilar stimuli. Simulating responses and determining susceptibilities are routine tasks for conventional culture formats, let alone for improved, high-density ones—possibly guised as "living-cell microarrays" (Yarmush and King, 2009). But descriptive culturomics provides a coherent conceptual basis, amenable to expansion and standardization in microbiote–substrate–conditions triangular interactions, thus paving the way for interactome analysis (Barabási et al., 2011) in microbiote research.

THE TECHNICAL DIMENSION: INSTRUMENTATION AND DEVICES

As with many other cases of new scientific disciplines, the study of culturomics, as mentioned earlier, is a concept, and thus, by definition, not linked to specific technical solutions, both in terms of hardware and software. Any technology bound to satisfy specifications and criteria laid down for the method is likely to be tested for compliance and suitability and possibly adopted for standard—and, in due time, routine and massive use.

A historical paradigm on the issue has been the "household PCR" (Levene and Zechiedrich, 2009), with a range of water baths, a timer, and a flying lab coat moving stacks of tubes. Taking the analogy to culturomics, different techniques, conditions, and hardware are available to handle, treat, and inoculate a sample in different media. Similarly, data output synchronous or asynchronous in nature can be fused and processed; or processed and the results fused to compile a wider picture.

Hardware and software evolution allows handling of hundreds of different substrates either in petri dishes, or in multiwell plates. The former result in high volumes of adaptable and flexible culture formats. The latter tend to achieve low-volume, cost-effective, massive, and standardized testing. In these conditions the advent of The Total Lab Automation approach seems perfectly timed. It provides for casting/pouring, inoculating, marking/printing, moving, incubating, and scanning with digital imagers different culture containers, that is, flasks, dishes, or plates (Iversen et al., 2016), while also delineating, collecting, and handling grown colonies (Brown, 2011; Turunen et al., 2009).

Slightly less advanced and more affordable, the newer liquid handling systems may also oblige; the main problems in the use of such automation in microbiology had been the viscosity and leftovers of liquefied solid media and the sterilization of tubing. Such concerns are currently effectively addressed by translating solutions from simpler media dispensers to complex and automated liquid handlers (Tecan, 2012).

Once a methodology gains acceptance and becomes widespread, the quest for standardization ensues in two levels: the initial level is among different users of the same modality or rather methodology, to attain comparable results in an interlaboratory setting, which allows for comparisons, fusion, or division of work and a better, flexible collective workflow in general. The next level comprises standardization among different assays of the method, because simple details in hardware and procedures, exemplified by, but not restricted to, exact dimensions and specifications of plasticware and different heating protocols respectively, may lead to important differentiations of the experimental output. And a possible third level will refer to temporal consistency: asynchronous tests, which are the norm today, will be discouraged once the comparative fusion and the resulting metadata of growth curves' analysis become commonplace.

Due to such concerns, a repetition of the events of the technological evolution of genomics is likely to take place. This means baseline hardware, by a limited number of suppliers, will be widely adopted and perhaps identified with the method; genomics and nucleic acid microarrays had been - erroneously - linked to identity even in references and texts intended for nonexpert audiences (Wikipedia, 2016) before new generation sequencing (NGS) proved the link erratic.

In realistic terms a high-throughput culturomic format would develop around a robotic device implementing media preparation and pouring distribution and sample inoculation. It would be complemented by a robotic stacker and able to access imaging and incubation devices. In the near future, options would abound: chemical sensors, exemplified by, but not restricted to, pH meters, pressure, and atmospheric content sensors are valid candidates.

Processing high numbers of individual queries simultaneously in a single assay requires some degree of aggregation: the physical condition of the media (solid and liquid) is less restrictive than might be expected; the wells of a plate can be inoculated and output differential optical signatures of a digital nature preferably if filled with liquid media, but with solid ones as well. The real issue is the diverse sets of incubation conditions, including, without being restricted to, temperature, humidity, atmosphere context, pressure, and illumination/irradiation. A culturomics setting must provide for separate plates for different combinations of conditions, with enough growth margin to incorporate new media.

An important challenge will appear with the diversity required less for culturomics media and more for incubation conditions. Most probably, a next generation of consumables will be needed, satisfying a very diverse set of specifications, or perhaps, as happens today, parts of this extended range of specifications. If the first approach is selected by manufacturers, improved materials will be needed to produce disposable and recyclable culture containers combining affordability with manufacturing precision, rigidity in stress and tear for mechanical handling; and ruggedness and tolerance for corrosive and destructive stimuli.

The concept of preformed platforms, taking minimal space in storage and brought to shape by an impulse (pneumatic, pressure, or current discharge) delivered on demand, might point at a solution employing, among other elements, nanomaterials morphing into shape. A most intriguing dimension of these novel expendables is the need to both accommodate and endure a variety of stimuli, which shall define a host of functionally different types of culturomics (Table 8.1), probably using different prefixes. The ruggedness and tolerance mentioned earlier will be indispensable for experiments with electromagnetic exposures as dictates the concept of electroceuticals (Kambouris et al., 2014) but also with diverse treatments such as high doses of light (including IR and UV), sonic (including infra- and ultrasounds), pressure (static or in waves), or other forms of irradiation (Bettner et al., 1998; Hoover et al., 2002).

Some of the abovementioned stimuli, the electroceuticals being the paradigm, can be successfully applied through the use of improvisations (Fig. 8.4). They are not accurate, nor standardized enough for massive use, but for low throughputs they may be an affordable solution. Still, electroceuticals can be applied massively only if complying wells and covers are used. The former require integrating the function of an electrode at their bottom, to collect electric charges delivered either by a cover- implanted or imprinted electrode or by spraying (as in noncontact current transfer-NCCT). An elaboration may be the use of plastic-like but conditionally conductive materials, exemplified by the intrinsically conductive polymers (Inzelt, 2008*)* for the manufacture of expendable culture containers.

Containers, especially plates, transparent to different electromagnetic fields and conceptually and technologically similar to the dielectric radomes of airborne radars (Meyer, 2015) may result in exposing different media and cultures to such fields to pinpoint differentiation in metabolomic and phenomic aspects under different modalities and settings. Thus description of new and perhaps cryptic taxa is a possibility which may alter our picture for the microbiota at large (Kambouris et al., 2014; Kambouris and Velegraki, 2016). Starting from the *electroculturomics* (Table 8.1) different experimentation lines may include other challenges, such as, but not restricted to, pressure

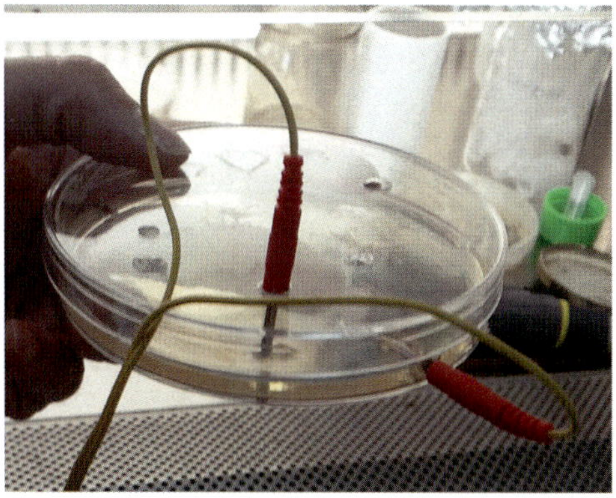

FIGURE 8.4

Improvised culture set for electroculturomics. Pin electrodes can be inserted through drilled holes either perpendicularly or horizontally into the agar base of the dish.

application in wave or in steady form—*bariculturomics*, see Table 8.1—to hyperbaric/hypobaric atmosphere or to heat spikes, delivered individually, in combination or along with differentiated media chemistries, as with the inclusion of antibiotics and biocides.

Software tools will be needed and these in iterations as well: the simple task is to associate assay positions to samples and condition sets, similarly to microarrays assignment software. Tracking the assay for the duration of the experiment is within the specifications, and such functions have been used in drug susceptibility assays and are included as fully integrated application to current releases (www.Watson LIMS for Bioanalytical Laboratories.htm). Much more challenging is the compilation of image processing software for calling growth and phenotype indices per assay position (Durai et al., 2013; Szybut et al., 2013), a task requiring an additional module for automated comparisons with large databases of high-definition phenomes of reference and type strains. The comparison module will use decision-making algorithms, possibly through machine learning, to assign a sample phenome to a known taxon or project a possible next-of-kin taxon.

The ultimate step will be a version of tracking software and able to stack frames taken from each individual assay locus (i.e., well or plate) by the image capturing device (camera) in temporal sequence, producing a quasicontinuity that would allow plotting of different indices, such as color, texture, and other morphic shifts against time (Reverberi et al., 2013). Of course, the drop of prices may make video cameras practical in continuously monitoring the growth, movie-like, with the ability to extract frames and respective time values.

The above refer to a more or less expected line of evolution for the routine application of culturomics. There are more exotic ones: as technology evolves, micro- (and nano-) materials are becoming more affordable and available, thus bearing hope for highly differentiated and dedicated, though cost-effective culture formats. High-viscosity liquid cultures are one, and rather primitive at that, such development. The advantage is that cells will be fixed in spatial terms, so as to develop solid culture-like morphology and also facilitate sampling. Concomitantly, nutrients, metabolites, and other, needed or toxic (by)products will be diffused freely, thus enhancing the sustainability of the culture (Gepp et al., 2009; Zengler et al., 2002). Moreover, encapsulation in high-viscosity environment improves biosafety and biosecurity; local spilling and airborne contamination of the wider environment become much more difficult, which is a must in order to allow working with, in general. and handling of, in particular, highly hazardous microbiota.

The most promising and dedicated development is the explosive growth of microfluidics. Occasionally used for advanced microarray formats, the microfluidics allow innovative culture concepts, including but not restricted to sequential culture schemes; selective intermittent interventions to the culture medium; and hydrodynamic simulation of non-stale liquid environments (Zengler et al., 2002). Such prospects create optimism for cultivating even more currently uncultivable microbiota (Greub, 2012) and elucidating dose–response curves of microbial growth, which show time- and/or phase-dependent variability (Ranalli et al., 2002). Within the state-of-the-art level is the continuous monitoring by microcameras that would allow the capture of anomalous growth phases. The latter either cause loss of important timeframes and data collection, or otherwise require very frequent reading sessions that may be impracticable and also may infringe with the proper incubation conditions.

SIMULATING INFECTIVITY: LEGACY AND INNOVATIVE APPLICATIONS

Widespread and standardized adoption of culturomics in optimized settings will result in decrease of the direct and indirect costs. As a result, the new approach should become affordable for multiple applications where multilevel interrogation of microbiota is required to determine their dynamic phenome rather than extrapolate it by deliberating on genomic entries (Kaeberlein et al., 2002). The industrial potential of such an analytic and predictive tool is obvious for a wide spectrum of applications—from environmental intervention to specific metabolite production.

With a focus to health, though, culture data tend to be inadequate for deducing infection dynamics and treatment opportunities. Thus if culturomics is to become an approach directly involved to the new era of medicine, namely, the precision medicine, it must increase the width, depth and relevance of its informative context. One way to such an end is to incorporate different concepts of enhanced realism; from high-fidelity simulation (Zengler et al., 2002) to biomimetic cultures (Guyette et al., 2016). The cost of the latter may remain prohibitive for quite sometime, thus making organoids, a 3D cell culture which emulates live tissue much better than conventional, 2D cell cultures, a valid alternative (Powell, 2018a). The obvious end is to incur a realistic environment as of the infection sequence of events and the determination of possible tropisms, which either through big data or through network theory (or any other predictive means) may reveal optimized treatment opportunities, as is the prerogative of precision medicine (Mitropoulos et al., 2017; Ginsburg et al., 2018).

Using specific organ/tissue extracts for natural or synthetic culture media, it is possible to attain axenic cultures of intracellular/fastidious/obligate parasitic microbiotes (Singh et al., 2013; Renesto et al., 2003; Ogata and Claverie, 2005). This leads to another end of precision medicine, directly involving culturomics. The culturing and subsequent study of elusive microbiota implicated in disease, not necessarily as causative agents. The idea is not new; natural media predate synthetic ones by far. But its combination with cell cultures, actual dissected tissue parts and a robust, culturomic high-throughput format may address new needs, beyond current aims and ends. Realistic infectivity profiles may be compiled for known, for spontaneously emerging, for genetically modified and for fully artificial microbiotes (Gibson et al., 2010; Hutchison et al., 2016), to provide insight to infection risks, development and growth patterns, substrate/host preferences, affinities and tropisms, and perhaps highly robust pharmacodynamics.

A development that increases the detection potential of culture-based approaches enhances their usefulness by improving the awareness on infectious events, and by affording more time for early intervention and accurately targeted, suitable countermeasures of both preventive and therapeutic nature. Its importance will increase as industry intensifies exploitation of natural and engineered microbiota, and as custom-made microbioat, with composite or compiled genomes (Gibson et al., 2010; Hutchison et al., 2016) become available to any interested party. The biorisk involved in their accidental or intended release is obvious, and the same applies to the culturomic dimension of containment efforts as mentioned earlier. But in terms of pure regulatory function, such strains would have to be quality-tested and safety/security-cleared by qualified agents and agencies before released for industrial use. And the more dependable and accurate way to do so is through culturomics.

The current field of application of culturomics is, as mentioned earlier, the massive setup of conventional and even exotic, culture conditions, to better describe microbiota. The obvious field is the advanced culture setups, based on microfluidics and other novel concepts, which may assist in describing more microbiota *and* introduce important levels of complexity into the study. But the innovative field of application comes with the rise of the microbiome as a central concept in microbiology. Accurate infectivity, growth, and expression profiles may assist in a wholly different way the broadening of precision medicine.

The use of living organisms for therapeutic purposes gains momentum after a time of distrust, especially in the West. The probiotics are assisting in good health by regulating appendage microbiomes to low aggressiveness (Borchert et al., 2008; Cogen et al., 2008) and by assisting in the metabolomic network. The pharma(co)biotics (Vitetta et al., 2012) are to preferentially exert the former function, and in a wider setup. Controlled, intentional infections with specially prepared microbiota, accurately applied and closely monitored show considerable therapeutic potential (Borchert et al., 2008; Powell, 2018b; Shaffer, 2015). The regeneration/replenishment of appendage microbiomes depleted by disease or therapy has long been used in medicine (Verna and Lucak, 2010); active management of microbiomic content of different anatomical sites, bioaugmentation (Nzila et al., 2016) and biorestoration (Satijn and de Boks, 1988) in environmental applications, become routine practice.

The innovation refers to much more aggressive uses of the concept, which may mature quickly and safely through culturomics and lead to faster releases (Fig. 8.5). The most well-known such concept, which may benefit from culturomics testing, is phage therapy (Abedon et al., 2011), which uses the inherent specificity of phages to selectively neutralize pathogenic bacteria while sparing host cells and benign microbiota. The lytic cycle of the phage produces follow-up batches of the drastic element, in stark contrast with micro- or macromolecular compositions, which wash away and decay; these renewed batches still target the same molecular receptors. Once the pathogenic bacterium is extinct, the phage does not multiply any further and is cleared by the immune system as a low-priority antigen. As phage functions are sensitive to environmental factors, their delivery, and the means to implement it, to different anatomic sites are focal; culturomics may assist in developing the optimal balance of strains, formulations and conditions.

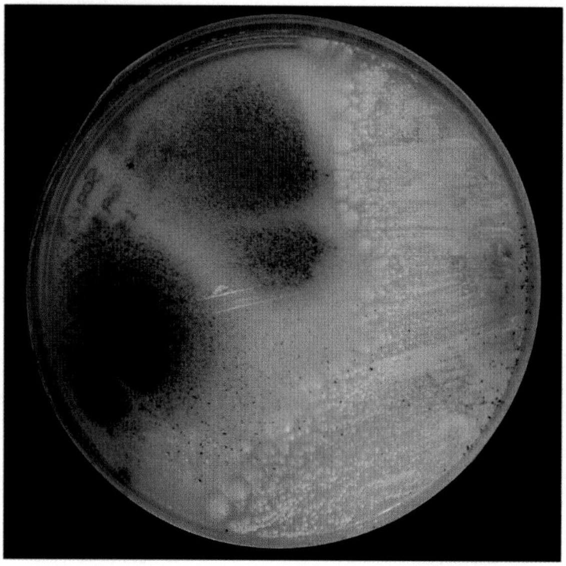

FIGURE 8.5

Culturomics in similar (pictured here) or dissimilar heterogenous formats are the best if not the only approach for determining the dynamics and even assessing approximate kinetics of antagonistic/hostile coexistence of different microbiotes (here *Candida lusitaniae* vs *Penicillium* sp.) so as to expedite the development of biologicals for any antimicrobial application in a sustainable context.

Another concept is therapeutic infectivity (Borchert et al., 2008; Shaffer, 2015), which might exhibit a degree of spontaneous pathogenicity (similar to the side effects of current drug formulations). Its use against more immediately threatening conditions may be considered a last resort solution, but it is a solution where there had been none. Modulation of the immune system in some autoimmune syndromes by microbiota has been observed (Zaccone et al., 2006) and is likely to be better studied and exploited, with the culturomics posing as the perfect enabler, which can create simulations for selection of the most promising strain; implement quality control (QC) clearing of the development and production and perform safety checks to determine the innate aggressiveness and health risks from the use of the strain.

Targeted oncolytic viruses (Howells et al., 2017) and selective bacteria (Price, 2017; Ben-Jacob, 2013; Massa et al., 2013; Kramer et al., 2018), along with the stimulation of respective immune elements (Borchert et al., 2008; Shaffer, 2015), may apply to containment and clearance of tumors, too. Existing and engineered strains may be tested for preference and tropism for different tumor types, and then for safety against healthy tissues and for effectiveness against the targeted tumors/cells.

The prospect of modulating the immune system with a xenobiote to somehow more effectively detect, classify, and dispose of malignant cells, at least at the first stages of the development of a tumor, may lead to earlier and even prophylactic interventions, thus saving on cost and toxicity, which usually shadow belated, aggressive regimens. As before, culturomics may offer the simulation basis to establish accurate prediction of the dynamics of the stimulatory infection.

AFFILIATIONS, OPPORTUNITIES, AND IMPACT

Microbiomes are unstable, dynamic entities and the appendage ones can be as unique and descriptive as a signature, but valid for a certain spatiotemporal window, which is not consistent even for a given (human or other) host. Thus in medical applications the analyses, may that be in the coarsest of terms, consist sensitive data and occasionally essential in precision medicine.

Although somewhat less stringent, similar restrictions and legal limitations will appear for the whole microbiomic universe, especially for taxa usable in, or essential for, lucrative applications. Discretion of an analytical method will be a prime concern, and descriptive culturomics allows an exhaustive description of a microbiote without involving any decipherment of genomic entries, which are prone to be the most well-kept proprietary information in an upcoming era, where bioengineered microbiotes will reign supreme from foodstuff production and processing to energy and medicament production.

The legal concerns notwithstanding, the importance of genomics, both descriptive and functional, along with other downstream -omics disciplines, are elemental aspects in microbiome studies and applications and supportive to culturomics. If metagenomic approaches result in high-quality and satisfactory volumes of data fast—and this is a big "if" (Bardet et al., 2018)—the uncertainties to be tackled by a culturomic analysis diminish, resulting in a more cohesive, focused, and limited set of parameters with lower cost and most probably faster results generation in both wet and *in silico* phases.

This combined approach, provisionally termed "assisted culturomics," is already proposed in principle, with mixed, genomic and culturomic profiles being submitted for novel organisms (Bardet et al., 2018). Unfortunately, the required splicing of the different data sources might not be straightforward, nor even possible in some occasions. As a consequence, culturomic analyses might be performed individually, either wholly independently or to confirm/round up previous (meta) genomic assays but without taking any cues from the latter. In such unaided conditions (unassisted culturomics), the full extent of the potential of culturomics might be sought for, as it taps on an unprecedented wealth of information.

The routine use of comparative and descriptive culturomics by itself is likely to produce data in the league of Big Data, without mobilizing any additional resources. Further quest for usable data generation may multiply such massive output. Indirect or unorthodox ways, at least in microbiological terms and combined actual and *in silico* resources, may be exploited. An example of indirect data sets may be the relaxing of the usual causality and optimization clauses of culture options, which will result in microbiota inoculated on suboptimal media, or media incubated at suboptimal conditions. The resulting growth curves may be highly informative, especially if taken synchronous and combined (rather than as independent characters, which is the usual practice) projecting thus, precise dynamics and kinetics of the isolate. The basic idea is once more known, with minimal media used for highly informative characterization, as happens with the Czapek Dox Agar in mycology—an excellent but not singular example of this concept.

The fact that culturomics is expected to come of age as the only practical testing and developing ground for combined treatments further presses home the argument. The use of chemicals and biochemicals along with other, physical amenities to control (not necessarily to suppress) a microbial population, and, even more important, to reshape a microbiome, requires massive tests, which are practical only through culturomic approaches. Some candidates mentioned by Giladi et al. (2008) are various light sources (tried as photodynamic therapy in dentistry and dermatology); ultrasounds (used for dental plaque removal and, in combination with antibiotics, for the eradication of bacterial biofilms *in vitro* and *in vivo*); and thermotherapy (originally developed against cancerous tumors, has been found since to be effective against cutaneous leishmaniasis). Even more so, in soft materials of food industry the use of high pressure stresses *Escherichia coli*, *Staphylococcus aureus*, and *Penicillium roqueforti*, possibly by denying exchanges through the membrane and compromising the functional aspects of the cell morphology (O'Reilly et al., 2000). But the most promising area is the combination of antimicrobial substances with electromagnetic amenities.

Portions of the abovementioned characteristic data will complement all strain circulation and exchange either for research or for profit purposes, forming a network of safety features tuned against intellectual property infringements in biotechnological context and malevolent practices in biosecurity context.

A genome-agnostic, translation-focused identification setup, as the unassisted culturomics format mentioned above, may revolutionize the terms of the centuries-old war of humanity against pathogens; it constitutes the currently missing biosecurity amenity for unbiased, massive and polyvalent, diverse risk assessment of existing, occurring and artificial microbiotes. Its application in multisectoral spontaneous biosafety issues simply increases its versatility, which addresses the various biodefense concerns currently emerging due to the diffusion of organism engineering and synthetic biology (Gibson et al., 2010; Hutchison et al., 2016; Jackson et al., 2001; Smith et al., 2003). Depots of biological agents, deployed bioweapons, and natural reservoirs of pathogens will be assayed promptly and reliably to determine (not promptly but very reliably) their actual tropisms, toxicity, growth mode and rate, and possible vulnerabilities so as to develop and deploy preventive, suppressive, and therapeutic countermeasures—some of which may be tested and optimized thanks to the massive testing formats provided by the same culturomic infrastructure and setup.

REFERENCES

Abedon, S.T., Kuhl, S.J., Blasde, B.G., Kutte, E.M., 2011. Phage treatment of human infections. Bacteriophage 1 (2), 66–85.

Barabási, A.-L., Gulbahce, N., Loscalzo, J., 2011. Network medicine: a network-based approach to human disease. Nat. Rev. 12, 56–68.

Bardet, L., Cimmino, T., Buffet, C., Michelle, C., Rathored, J., Tandina, F., et al., 2018. Microbial culturomics application for global health: noncontiguous finished genome sequence and description of *Pseudomonas massiliensis* strain CB-1T sp. nov. in Brazil. OMICS 22 (2), 164–175.

Becker, P.T., de Bel, A., Martiny, D., Ranque, S., Piarroux, R., Cassagne, C., et al., 2014. Identification of filamentous fungi isolates by MALDI-TOF mass spectrometry: clinical evaluation of an extended reference spectra library. Med. Mycol. 52, 826–834.

Beijerinck, M.W., 1901. Anhaufungsversuche mit Ureumbakterien. Centralblatt f. Bakteriologie II 7, 33–61.

Ben-Jacob, E., 2013. Engineering Trojan-horse bacteria to fight cancer. Blood 122 (5), 619–621.

Bettner, M.D., Beiswanger, M.A., Miller, C.H., Palenik, C.J., 1998. Effect of ultrasonic cleaning on microorganisms. Am. J. Dent. 11 (4), 185–188.

Bochner, B.R., Gadzinski, P., Panomitros, E., 2001. Phenotypic microarrays for high-throughput phenotypic testing and assay of gene function. Genome Res. 11, 1246–1255.

Borchert, D., Sheridan, L., Papatsoris, A., Faruquz, Z., Barua, J.M., Junaid, I., et al., 2008. Prevention and treatment of urinary tract infection with probiotics: review and research perspective. Indian J. Urol. 24 (2), 139–144.

Brown, T., 2011. Gene cloning and colony picking. Gen. Eng. Biotechnol. News 31 (19). Available from: <http://www.genengnews.com/gen-articles/gene-cloning-and-colony-picking/3904>. (accessed 03.11.18.).

Buchan, B.W., Ledeboer, N.A., 2014. Emerging technologies for the clinical microbiology laboratory. Clin. Microbiol. Rev. 27 (4), 783–822.

Cavalier-Smith, T., 2004. Only 6 kingdoms of life. Proc. R. Soc. Lond. 271, 1251–1262.

Cogen, A.L., Nizet, V., Gallo, R.L., 2008. Skin microbiota: a source of disease or defence? Br. J. Dermatol. 158 (3), 442–455.

Dubourg, G., Lagier, J.C., Armougom, F., Robert, C., Hamad, I., Brouqui, P., et al., 2013. The proof of concept that culturomics can be superior to metagenomics to study atypical stool samples. Eur. J. Clin. Microbiol. Infect. Dis. 32 (8), 1099.

Durai, A.P., Sankaran, K., Muttan, S., 2013. An image based microtiter plate reader system for 96-well format fluorescence assays. Eur. J. Biomed. Informat. 9 (2), 58–68.

Gepp, M.M., Ehrhart, F., Shirley, S.G., Howitz, S., Zimmermann, H., 2009. Dispensing of very low volumes of ultra high viscosity alginate gels: a new tool for encapsulation of adherent cells and rapid prototyping of scaffolds and implants. Biotechniques 46 (1), 31–43.

Gibson, D.G., Glass, J.I., Lartigue, C., Noskov, V.N., Chuang, R.Y., Algire, M.A., et al., 2010. Creation of a bacterial cell controlled by a chemically synthesized genome. Science 329 (5987), 52–56.

Giladi, M., Porat, Y., Blatt, A., Wasserman, Y., Kirson, E.D., Dekel, E., Palti, Y., et al., 2008. Microbial growth inhibition by alternating electric fields. Antimicrob. Agents Chemother. 52 (10), 3517–3522.

Ginsburg, G.S., McCarthy, J.J., Patrinos, G.P. (Eds.), 2018. A note from the editors. Per. Med. 15 (4), 237–238.

Greub, G., 2012. Culturomics: a new approach to study the human microbiome. Clin. Microbiol. Infect. 18 (2), 1157–1159.

Guyette, J.P., Charest, J.M., Mills, R.W., Jank, B.J., Moser, P.T., Gilpin, S.E., et al., 2016. Bioengineering human myocardium on native extracellular matrix. Circul. Res. 118, 56–72.

Hoover, K., Bhardwaj, M., Ostiguy, N., Thompson, O., 2002. Destruction of bacterial spores by phenomenally high efficiency non-contact ultrasonic transducers. Mat. Res. Innovat. 6 (5), 291–295.

Houle, D., Govindaraju, D.R., Omholt, S., 2010. Phenomics: the next challenge. Nat. Rev. Gen. 11 (12), 855–866.

Howells, A., Marelli, G., Lemoine, N.R., Wang, Y., 2017. Oncolytic viruses—interaction of virus and tumor cells in the battle to eliminate cancer. Front. Oncol. 7, 195.

Hutchison, C.A., Chuang, R.-Y., Noskov, V.N., Assad-Garcia, N., Deerinck, T.J., Ellisman, M.H., et al., 2016. Design and synthesis of a minimal bacterial genome. Science 351, aad6253/1–aad6253/11.

Ieven, M., Verhoeven, J., Pattyn, S.R., Goossens, H., 1995. Rapid and economical method for species identification of clinically significant coagulase-negative staphylococci. J. Clin. Microbiol. 33, 1060–1063.

Inzelt, G., 2008. Introduction. In: Scholz, F. (Ed.), Conducting Polymers: A New Era in Electrochemistry. Monographs in Electrochemistry. Springer-Verlag Berlin, Heidelberg, pp. 1–6.

Iversen, J., Stendal, G., Gerdes, C.M., Meyer, C.H., Andersen, C.Ø., Frimodt-Møller, N., 2016. Comparative evaluation of inoculation of urine samples with the copan WASP® and BD Kiestra™ InoqulA™ instruments. J. Clin. Mcrobiol. 54 (2), 328–332.

Jackson, R.J., Ramsay, A.J., Christensen, C.D., Beaton, S., Hall, D.F., Ramshaw, I.A., 2001. Expression of mouse interleukin-4 by a recombinant ectromelia virus suppresses cytolytic lymphocyte responses and overcomes genetic resistance to mouse pox". J. Virol. 75 (3), 1205–1210.

Kaeberlein, T., Lewis, K., Epstein, S.S., 2002. Isolating "uncultivable" microorganisms in pure culture in a simulated natural environment. Science 296 (5570), 1127–1129.

Kambouris, M.E., Velegraki, A., 2016. Essentials in Mycology. Parisianos, Athens, pp. 91–94 (in Greek).

Kambouris, M.E., Matsianikas, G., Velegraki, A., 2011. Methodology and determining criteria for growth phase of simultaneous fungal solid cultures. e-J. Sci. Technol. 4 (6), 1–8.

Kambouris, M.E., Zagoriti, Z., Lagoumintzis, G., Poulas, K., 2014. From therapeutic electrotherapy to electroceuticals: formats, applications and prospects of electrostimulation. Annu. Res. Rev. Biol. 4 (20), 3054–3070.

Kolecka, A., Khayhan, K., Arabatzis, M., Velegraki, A., Kostrzewa, M., Andersson, A., et al., 2014. Efficient identification of Malassezia yeasts by matrix-assisted laser desorption ionization-time of flight mass spectrometry (MALDI-TOF MS). Br. J. Dermatol. 170, 332–341.

Kramer, M.G., Masner, M., Ferreira, F.A., Hoffman, R.M., 2018. Bacterial therapy of cancer: promises, limitations, and insights for future directions. Front. Microbiol. 9, 16.

Lagier, J.-C., Armougom, F., Million, M., Hugon, P., Pagnier, I., Robert, C., et al., 2012. Microbial culturomics: paradigm shift in the human gut microbiome study. Clin. Microbiol. Infect. 18, 1185–1193.

Lay, J.J., 2001. MALDI-TOF mass spectrometry of bacteria. Mass. Spectrom. Rev. 20, 172–194.

Levene, S.D., Zechiedrich, L., Nick Cozzarelli, N., 2009. Zechiedrich L. Nick Cozzarelli: A personal remembrance. In: Benham, C.J., Harvey, S., Olson, W.K., Sumners, D.W., Swigon, D. (Eds.), Mathematics of DNA Structure, Function and Interactions. Springer-Verlag, New York, p. 2.

Long, S.W., Williams, D., Valson, C., Cantu, C.C., Cernoch, P., Musser, J.M., et al., 2013. A genomic day in the life of a clinical microbiology laboratory. J. Clin. Microbiol. 51 (4), 1272–1277.

Maia, M.R., Marques, S., Cabrita, A.R., Wallace, R.J., Thompson, G., Fonseca, A.J., et al., 2016. Simple and versatile turbidimetric monitoring of bacterial growth in liquid cultures using a customized 3D printed culture tube holder and a miniaturized spectrophotometer: application to facultative and strictly anaerobic bacteria. Front. Microbiol. 7, 1381.

Massa, P.E., Paniccia, A., Monegal, A., de Marco, A., Rescigno, M., 2013. Salmonella engineered to express CD20-targeting antibodies and a drug-converting enzyme can eradicate human lymphomas. Blood 122 (5), 705–714.

McDonald, W., 2003. Morphology of Medically Important Fungi, second ed. Available from: <http://labmed.ucsf.edu/education/residency/fung_morph/launchpage.html> (accessed 03.11.18.).

Meyer, G.J., 2015. Polyurethane Foam: Dielectric Materials for Use in Radomes and Other Applications. General Plastics Manufacturing Company. Available from: <https://www.generalplastics.com/technical-papers/dielectric-materials-use-radomes> (accessed 03.11.18).

Mitropoulos, K., Cooper, D.N., Mitropoulou, C., Agathos, S., Reichardt, J.K.V., Al-Maskari, F., et al., 2017. Genomic medicine without borders: which strategies should developing countries employ to invest in precision medicine? A new "fast-second winner" strategy. OMICS 11, 647–657.

Nzila, A., Razzak, S.A., Zhu, J., 2016. Bioaugmentation: an emerging strategy of industrial wastewater treatment for reuse and discharge. Int. J. Environ. Res. Public. Health 13 (9), 846.

Ogata, H., Claverie, J.M., 2005. Metagrowth: a new resource for the building of metabolic hypotheses in microbiology. Nucleic Acids Res. 33 (Database issue), D321–D324.

O'Reilly, C.E., O'Connor, P.M., Kelly, A.L., Beresford, T.P., Murphy, P.M., 2000. Use of hydrostatic pressure for inactivation of microbial contaminants in cheese. Appl. Environ. Microbiol. 66 (11), 4890–4896.

Orjala, J., 2008. Exploring the Diversity and Biopharmaceutical Potential of Uncultivable Bacteria from Lake Michigan Sediments. Final Report. Univ Illinois, pp. 1–11. Available from: <http://www.iisgcp.org/research/reports/Orjala_final.pdf> (accessed 08.09.16.).

Orro, A., Cappelletti, M., D'Ursi, P., Milanesi, L., Di Canito, A., Zampolli, J., et al., 2015. Genome and phenotype microarray analyses of *Rhodococcus* sp. BCP1 and *Rhodococcus opacus* R7: genetic determinants and metabolic abilities with environmental relevance. PLoS One 10, e0139467.

Peng, Y., Leung, H.C.M., Yiu, S.M., Chin, F.Y.L., 2012. IDBA-UD: a de novo assembler for single-cell and metagenomic sequencing data with highly uneven depth. Bioinformatics 28, 1420–1428.

Powell, M., 2018a. Organoids: A New Model to Study Infectious Diseases? Infectious Diseases Hub. Available from: <https://www.id-hub.com/2018/06/14/organoids-new-model-study-infectious-diseases/> (accessed 03.11.18.).

Powell, M., 2018b. Microbiome Therapeutics – The Pipeline for *C. difficile* Infection. Infectious Diseases Hub. Available from: <https://www.id-hub.com/2018/02/20/microbiome-therapeutics-the-pipeline-for-c-difficile-infection/> (accessed 03.11.18.).

Price, M., 2017. Scientists Turn Food Poisoning Microbe Into Powerful Cancer Fighter. Research. Available from: <http://www.sciencemag.org/news/2017/02/scientists-turn-food-poisoning-microbe-powerful-cancer-fighter> (accessed 03.11.18.).

Puspita, I.D., Kamagata, Y., Tanaka, M., Asano, K., Nakatsu, C.H., 2012. Are uncultivated bacteria really uncultivable? Microbes Environ. 27 (4), 356–366.

Ragoussis, J., Elvidge, G.P., Kaur, K., Colella, S., 2006. Matrix-assisted laser desorption/ionisation, time-of-flight mass spectrometry in genomics research. PLoS Genet. 2, e100.

Ranalli, G., Iorizzo, M., Lustrato, G., Zanardini, E., Grazia, L., 2002. Effects of low electric treatment on yeast microflora. J. Appl. Microbiol. 93, 877–883.

Renesto, P., Crapoulet, N., Ogata, H., Vestris, G., et al., 2003. Genome-based design of a cell-free culture medium for *Tropheryma whipplei*. Lancet 362 (9382), 447–449.

Reverberi, M., Punelli, M., Scala, V., Scarpari, M., Uva, P., Mentzen, W.I., et al., 2013. Genotypic and phenotypic versatility of *Aspergillus flavus* during maize exploitation. PLoS One 8 (7), e68735.

Romagnoli, M., Osterhout, J., Merz, W.G., 1995. Evaluation of the biolog system for the identification of medically important yeast. In: 95th General Meeting of the American Society for Microbiology Abstracts, p. 107.

Satijn, H.M.C., de Boks, P.A., 1988. Biorestoration, a technique for remedial action on industrial sites. In: Contaminated Soil'88, pp. 745–753.

Schilling, C.H., Edwards, J.S., Palsson, B.O., 1999. Toward metabolic phenomics: analysis of genomic data using flux balances. Biotechnol. Prog. 15, 288–295.

Shaffer, L., 2015. DIY Parasite Infection Treats Autoimmune Disorders. Discover Available from: <http://discovermagazine.com/2015/july-aug/4-take-worms-call-me> (accessed 03.11.18.).

Singh, S., Eldin, C., Kowalczewska, M., Raoult, D., 2013. Axenic culture of fastidious and intracellular bacteria. Trends Microbiol. 21 (2), 92–99.

Smith, H.O., Hutchison III, C.A., Pfannkoch, C., Venter, J.C., 2003. Generating a synthetic genome by whole genome assembly: X174 bacteriophage from synthetic oligonucleotides. Proc. Natl. Acad. Sci. U.S.A. 100, 15440–15445.

Smolina, I., Miller, N.S., Frank-Kamenetskii, M., 2010. PNA-based microbial pathogen identification and resistance marker detection: an accurate, isothermal rapid assay based on genome-specific features. Artif. DNA PNA XNA 1 (2), 1–7.

Sperber, W.H., Swan, J., 1976. Hot-loop test for the determination of carbon dioxide production from glucose by lactic acid bacteria. Appl. Environ. Microbiol. 31 (6), 990–991.

Szybut, C., Alcantara, S.L., O'Callaghan, T., Artymovich, K., O'Clair, B., Appledorn, D., et al., 2013. Miniaturised Live Cell Kinetic Imaging Assays in 384-Well Format. Available from: <https://www.researchgate.net/publication/323612871_Miniaturised_live_cell_kinetic_imaging_assays_in_384-well_format> (accessed 03.11.18.).

Tecan, 2012. A new dimension in plate preparation for 3D cell assays. Tecan J. 1 (2012), 22–23. Available from: <http://www.tritechresearch.com/pourboy.html>. (accessed 08.09.16.).

Turunen, L., Takkinen, K., Söderlund, H., Pulli, T., 2009. Automated panning and screening procedure on microplates for antibody generation from phage display libraries. J. Biomol. Scr. 14 (3), 282–293.

Velegraki, A., Logotheti, M., 1998. Presumptive identification of an emerging yeast pathogen: *Candida dubliniensis* (sp. nov.) reduces 2,3,5-triphenyltetrazolium chloride. FEMS Immunol. Med. Microbiol. 20, 239–241.

Velegraki, A., Kambouris, M., Kostourou, A., Chalevelakis, G., Legakis, N.J., 1999. Rapid extraction of fungal DNA from clinical samples for PCR amplification. Med. Mycol. 37, 69–73.

Verna, E.C., Lucak, S., 2010. Use of probiotics in gastrointestinal disorders: what to recommend? Therap. Adv. Gastroenterol. 3 (5), 307–319.

Verwer, P.E., van Leeuwen, W.B., Girard, V., Monnin, V., van Belkum, A., Staab, J.F., 2014. Discrimination of *Aspergillus lentulus* from *Aspergillus fumigatus* by Raman spectroscopy and MALDI-TOF MS. Eur. J. Clin. Microbiol. Infect. Dis. 33, 245–251.

Vitetta, L., Briskey, D., Hayes, E., Shing, C., Peake, J., 2012. A review of the pharmacobiotic regulation of gastrointestinal inflammation by probiotics, commensal bacteria and prebiotics. Inflammopharmacology 20 (5), 251–266.

Weinstein, M.P., 1996. Current blood culture methods and systems: clinical concepts, technology, and interpretation of results. Clin. Infect. Dis. 23, 40–46.

Wikipedia, 2016. Genomics/functional genomics. In: The Omics Revolution. Available from: <https://en.wikipedia.org/wiki/Genomics> (accessed 08.09.16.).

Yarmush, M.L., King, K.R., 2009. Living-cell microarrays. Annu. Rev. Biomed. Eng. 11, 235–257.

Zaccone, P., Fehervari, Z., Phillips, J.M., Dunne, D.W., Cooke, A., 2006. Parasitic worms and inflammatory diseases. Parasite Immunol. 28, 515–523.

Zengler, K., Toledo, G., Rappe, M., Elkins, J., Mathur, E.J., Short, J.M., et al., 2002. Cultivating the uncultured. Proc. Natl. Acad. Sci. U.S.A. 99 (24), 15681–15686.

Zhou, Q., Su, X., Ning, K., 2014. Assessment of quality control approaches for metagenomic data analysis. Sci. Rep. 4, 6957/1–6957/11.

FURTHER READING

Bomar, L., Maltz, M., Colston, S., Graf, J., 2011. Directed culturing of microorganisms using metatranscriptomics. mBio 2 (2), e00012–11.

Cunefare, K.A., Carter, S.D., Ahern, D., 2002. Enhancement of the biocidal efficacy of a mild disinfectant through enhanced transient cavitation. J. Acoust. Soc. Am. 112 (5), 2371.

Karpova, V., Kudryavtseva, A.D., Lednev, V.N., Mironova, T.V., Oshurko, V.B., Pershin, S.M., et al., 2016. Stimulated low-frequency Raman scattering in a suspension of tobacco mosaic virus. Laser Phys. Lett. 13 (8), 085701/1–085701/4.

Kromer, J.O., Sorgenfrei, O., Klopprogge, K., Heinzle, E., Wittmann, C., 2004. In-depth profiling of lysine-producing *Corynebacterium glutamicum* by combined analysis of the transcriptome, metabolome, and fluxome. J. Bacteriol. 186, 1769–1784.

Kurland, I.J., Accili, D., Burant, C., Fischer, S.M., Kahn, B.B., Newgard, C.B., et al., 2013. Application of combined omics platforms to accelerate biomedical discovery in diabesity. Ann. N. Y. Acad. Sci. 1287, 1–16.

Lorca, T.A., Claus, J.R., Eifert, J.D., Marcy, J.E., Sumner, S.S., 2016. Effects of Explosively-Generated Hydrodynamic Shock Wave Treatments on the Microbial Flora of Beef Steaks and Ground Beef, and *Listeria innocua*. Available from: <https://vtechworks.lib.vt.edu/handle/10919/11046> (accessed 03.11.16.).

Martirosyan, V., Markosyan, L., Hovhanesyan, H., Hovnanyan, K., Ayrapetyan, S., 2012. The frequency-dependent effect of extremely low-frequency electromagnetic field and mechanical vibration at infrasound frequency on the growth, division and motility of *Escherichia coli* K-12. Environmentalist 32, 157–165.

Novak, K.F., Govindaswami, M., Ebersole, J.L., Schaden, W., House, N., Novak, M.J., 2008. Effects of low-energy shock waves on oral bacteria. J. Dent. Res. 87 (10), 928–931.

Palla, G., Derenyi, I., Farkas, I., Vicsek, T., 2005. Uncovering the overlapping community structure of complex networks in nature and society. Nature 435, 814–818.

Petrini, L.E., Petrini, O., 2013. Identifying moulds: a practical guide. J. Cramer in der Gebruder Borntraeger Verlagsbuchhandlung, Stuttgart, pp. 40–88.

Sarvaiya, N., Kothari, V., 2015. Effect of audible sound in form of music on microbial growth and production of certain important metabolites. Microbiology 84 (2), 227–235.

Tavanti, A., Davidson, A.D., Gow, N.A.R., Maiden, M.C.J., Odds, F.C., 2005. Candida orthopsilosis and *Candida metapsilosis* spp. nov. to replace *Candida parapsilosis* Groups II and III. J. Clin. Microbiol. 43 (1), 284–292.

Tsen, S.-W.D., Wu, T.C., Kiang, J.G., Tsen, K.-T., 2012. Prospects for a novel ultrashort pulsed laser technology for pathogen inactivation. J. Biomed. Sci. 19 (1), 62–72.

Zhang, W., Li, F., Nie, L., 2010. Integrating multiple 'omics' analysis for microbial biology: application and methodologies. Microbiology 156, 287–301.

CHAPTER

NEXT-GENERATION SEQUENCING: THE ENABLER AND THE WAY AHEAD

9

Sonja Pavlovic, Kristel Klaassen, Biljana Stankovic, Maja Stojiljkovic and Branka Zukic

Institute of Molecular Genetics and Genetic Engineering, University of Belgrade, Belgrade, Serbia

INTRODUCTION

Microbiome research has long been hindered by the fact that the vast majority of microbiota are uncultivable by standard laboratory techniques (Riesenfeld et al., 2004; Schloss and Handelsman, 2005). When possible, the identification of species was achieved through culturing, followed by biochemical characterization, but for the far greater number of microbiota a whole range of culture-independent methods was developed to assess the biodiversity from various types of samples, such as environmental (soil, seawater, freshwater, etc.) (Fierer, 2017; Moran, 2015; Besemer, 2015), plant (Berg et al., 2016), as well as animal and human (including skin, gastrointestinal, urogenital, and from other biocompartments) (Kong et al., 2017; Barko et al., 2018; Aragon et al., 2018).

In the new age of -omes and -omics, it became possible to investigate microbiota in their natural environments. The fascinating development of high-throughput technology has introduced the science of metagenomics in the research of genetic basis of complex communities in which microbiota normally live. Metagenomics, defined as the analysis of genomes extracted directly from an environmental sample, aims at understanding biology of microbiota at the aggregate level. Its focus is on the genes in the community and the influence of the genes on each other's activities in serving collective functions.

There are two main principles in genomic research when it comes to the characterization of microbiota without prior cultivation: amplicon sequencing of marker genes, such as 16S rRNA for bacteria and archaea, 18S rRNA for eukaryotic organisms, and internal transcribed spacer (ITS) for fungi [also known as "consensus polymerase chain reaction (PCR)"]; and shotgun metagenomic studies, where the whole DNA of a given microbial community is subjected to shotgun sequencing (Rondon et al., 2000). In 16S rRNA gene studies, which are by far the most frequently used amplicon studies, sequences from rRNA are used as a "molecular clock"; that is, as stable phylogenetic markers of more or less linear association to the phylogenetic distance, suitable for the identification of the organisms in a sample and for assessing the taxonomic and phylogenetic identity of new entries. The 16S rRNA gene of prokaryotes is composed of scattered conserved and variable regions, which enable PCR amplification and subsequent analysis. Conserved regions give evidence on early evolutionary events, while less conserved regions testify of recent changes. The degree of divergence of present day rRNA sequences enables an assessment of their phylogenetic distance

Microbiomics. DOI: https://doi.org/10.1016/B978-0-12-816664-2.00009-8

(De Mandal et al., 2015). The 16S rRNA gene studies are considered to be "metagenomic" in terms of the analysis of a heterogeneous microbial community. One of the advantages of 16S rRNA gene studies is cost effectiveness. Microbial profiling using a single gene out of each genome in a community is more affordable than shotgun sequencing.

On the other hand, shotgun metagenomics implies random sequencing of the genomic DNA of the entire microbial community in a sample. Microbial profiles obtained using shotgun metagenomic studies are indispensable for the true understanding of the structure and function of those genomes, which will lead to gaining an insight into functional potential of microbial communities with respect to their translational implication.

Both the 16S rRNA gene studies and shotgun metagenomic studies have been reborn during the last two decades by the emergence of a potent novel technology that completely revolutionized the field of genomic research—Next-generation sequencing (NGS).

NEXT-GENERATION SEQUENCING: A GENERAL OVERVIEW

The pioneer work that enabled the first entire genomes to be sequenced took place in 1976. The first genome to be sequenced belonged to a small single-stranded RNA virus, bacteriophage MS2, with a genome size of just over 3.5 kbp (kilobase pair—1000 bp) (Fiers et al., 1976), followed by the first DNA-based genome, that of a PhiX174 virus to be sequenced in 1977 (Sanger et al., 1977). In 1995 the first organism to have its entire genome of 1.8 million bp sequenced was a small bacterium, *Haemophilus influenzae* (Fleischmann et al., 1995). Afterward, the genomes of other bacteria and archaea were successfully sequenced, mainly owing to their small genome size. As a result of decades of laborious scientific benchwork, the number of genomes sequenced continued to rise throughout the years, which enabled the worldwide expansion of genomics research on human and environmental microbiomes. This expensive international endeavor relied heavily on software and hardware automation of the chain termination sequencing method, first introduced by Frederick Sanger in 1977, by using fluorescently labeled terminators, fragment separation by means of capillary electrophoresis, and automated laser-based fluorescence detection. Although the successful use of this technology made it possible to generate various genomes to which later data would be aligned to, several big disadvantages, such as low throughput, cost limitations, and need for large quantity of starting material, created a pressing requirement for more potent sequencing technologies which would be able to produce larger quantities of data more quickly, with lower average cost per base pair and using smaller sample sizes (Rizzo and Buck, 2012). Another immense shortcoming of Sanger sequencing for microbial research applications is the heterogeneity of mixed microbial communities, since the extracted DNA has to be cloned into a bacterial vector, which considerably limits the metagenomic studies to low-diversity microbiomes. With the obligatory cloning step, only the organisms that can be cultured in a laboratory and subsequently successfully cloned into a vector were suitable for Sanger sequencing, thus significantly lowering the number of organisms that can be isolated and identified in a given microbiome. Therefore, a highly diverse microbiome may produce an oversimplified picture and not be adequately characterized. Also, one should have in mind the laborious benchwork necessary for the culturing, cloning, and sequencing of a multitude of samples in a high diversity microbiome, making Sanger sequencing challenging, to say the least, for metagenomic studies.

Having in mind all the limitations of Sanger method, the last decades have witnessed a fundamental shift away from the chain termination method of DNA sequencing, now referred to as the "first generation sequencing," for the analysis of diverse genomes. Despite its excellent accuracy and widespread availability, Sanger sequencing has restricted applications arising from technical bottlenecks inherent to its sequencing chemistry. The amount of DNA sequence information that can be acquired from a single sequencing reaction, that is, the throughput, is considered to be its biggest limitation. The need for electrophoretic separation of dideoxy-terminated elongated DNA fragments restricts the throughput of Sanger sequencing to a magnitude of approximately 1 kb per sequencing reaction. This fact, together with cost and time limitations, restricts the use of Sanger sequencing in modern genomics, thus limiting its translational applicability and underlining the necessity for high-throughput sequencing approaches (Rizzo and Buck, 2012).

NGS, sometimes referred to as the "massively parallel DNA sequencing" or "second-generation sequencing," is a blanket term used to collectively address diverse methods of high-throughput DNA sequencing which arose in the previous 15 years (Rizzo and Buck, 2012). With the burst of NGS technologies the prospect of sequencing data obtained fast and affordably has furnished unforetold accessibility, fidelity, and flexibility, thus expanding the experimental capacity beyond the plain determination of the sequence of base pairs (Metzker, 2010; Kahvejian et al., 2008).

NEXT-GENERATION SEQUENCING: GENERAL TECHNICAL ASPECTS

The newly developed NGS technologies consist of various strategies and methods broadly grouped as steps of: (1) template generation, (2) sequencing and imaging, and (3) data analysis (Metzker, 2010). Almost all NGS technologies cyclically monitor the addition of differently labeled nucleotides to immobilized and isotropic arrayed DNA templates. However, each of them is individually quite unique in how every step is performed (Rizzo and Buck, 2012). A specific combination of template generation steps, proprietary sequencing chemistry, and imaging process distinguishes each NGS manufacturer and sequencing platform from the others and determines the type and quality of data produced by each platform (Rizzo and Buck, 2012; Metzker, 2010).

Starting material for template generation in nearly all NGS experiments is double-stranded DNA; however, it can be acquired from various sources, such as DNA isolated from pure samples, DNA isolated from natural microbial communities of any ecosystem, or reversely transcribed RNA (Head et al., 2014; Edlund et al., 2018). Following template generation, in most NGS approaches, the sequencing process itself consists of consecutive cycles of enzyme-driven sequencing reactions collectively referred to as "sequencing by synthesis," which is a term used to describe either DNA polymerase- or DNA ligase-dependent methods (Rizzo and Buck, 2012; Metzker, 2010; Shendure and Ji, 2008). For most of the technologies, data are acquired by imaging bioluminescence or fluorescence produced at each cycle on a monomolecular nucleotide-by-nucleotide basis by all arrayed DNA templates simultaneously ("in parallel"), (Rizzo and Buck, 2012; Metzker, 2010; Shendure and Ji, 2008). Obtained short-sequence reads can further be processed by aligning them to a known reference genome, if available; alternatively a de novo bioinformatic assembly can be conducted, after which the data can be analyzed in research-oriented fashion (Rizzo and Buck, 2012).

In technical terms an overall advantage of NGS methods relative to Sanger sequencing is the higher throughput, as hundreds of millions of sequencing reads can be obtained by parallel processing, and thus almost simultaneously. This array-based technology circumvents the aforementioned bottleneck of discrete separation and detection requirements of Sanger sequencing, thus improving the throughput from the kilobase to the terabase order of magnitude per sequencing reaction (Rizzo and Buck, 2012; Metzker, 2010; Shendure and Ji, 2008). Great number of reads (in comparison with a single read in Sanger sequencing) enables the successful sequencing not only of pure samples but now also of complex mixtures—the main feature of diverse microbial communities—therefore setting the base for proper metagenomic studies. Another advantage is the progressing diminution of the sample size needed for sequencing, thus once again circumventing the cultivation step and enabling metagenomic studies of not only exceptionally diverse microbiomes but also of single cells (Xu and Zhao, 2018).

NEXT-GENERATION SEQUENCING PLATFORMS USED FOR METAGENOMICS

ROCHE 454 PYROSEQUENCING

The first massively parallel next-generation sequencer was based on pyrosequencing technology and launched in 2005 by "454," which was acquired by Roche in 2007 (Margulies et al., 2005). The pyrosequencing technology is based on the sequencing-by-synthesis principle, where the detection of signal is achieved in real time by detecting the nucleotide incorporated by a DNA polymerase. During the sequencing process, with the incorporation of every nucleotide, a pyrophosphate molecule is released, which in turn activates the enzyme luciferase, leading to the production of light. This bioluminescence signal is proportional to the amount of pyrophosphate produced, which is directly proportional to the number of added nucleotides. DNA fragments are first attached to microbeads and clonally amplified in an emulsion PCR. The beads are consequently spread into individual picolitre-sized wells and subjected to pyrosequencing. Pyrosequencing is fundamentally different from Sanger sequencing since bioluminescence results from strand elongation in real time, whereas in Sanger sequencing, fluorescence is detected as a separate step after chain termination, which makes pyrosequencing more quantitative (Harrington et al., 2013). Throughout the years the technique has continued to evolve and overcome the pitfalls encountered in the beginning, so the length of sequence reads as well as the output (yield per run) were constantly on the rise (from kilo to megabasepairs). The latest instruments actually overcame the initial limitation of NGS short reads in comparison to Sanger sequencing, achieving up to 1 kb read lengths (Siqueira et al., 2012). These long-read lengths have enabled better alignment to reference genomes, thus making 454 pyrosequencing NGS the method of choice in metagenomic studies for years (Harrington et al., 2013), along with sequencing of conserved marker gene tags, now commonly referred to as "pyrotag" sequencing (Bragg and Tyson, 2014). However, one of the limitations of pyrosequencing laid with its inaccuracy in the sequencing of strings of repeated nucleotides (homopolymers); more than five identical nucleotides cannot be detected efficiently (Fakruddin et al., 2013). Therefore, recent times have seen a rise in other NGS platforms, such as Illumina sequencing, which have successfully coped with some other weaknesses of the pyrosequencing as well, such as the high error rate, low yield, and high cost per bp (De Mandal et al., 2015).

ILLUMINA SEQUENCING

One of the most widely used NGS technologies at the moment is, without a doubt, Illumina's, which was launched in 2006. Initially developed as Solexa method and subsequently acquired by Illumina, it is based on the sequencing-by-synthesis principle as well. In this technology the template DNA fragments are joined with special adapters in library preparation, thus enabling their attachment to glass flow cells, practically being floating glass beads which bear anchored oligonucleotides complementary to the adapters. Each fragment, upon hybridization to an anchored oligonucleotide, serves as template for extending the primer and thus is clonally amplified. Elongated primers hybridize with other-end anchored oligos in a "bridge amplification" manner, recreating the original strand so as to form a cluster of like double-stranded DNA fragments anchored to the beads. Sequencing is then performed using fluorescently labeled reversible-terminator deoxyribonucleotide triphosphate-dNTPs, where the fluorophore is cleaved away from 3′ hydroxyl position of one base before the next one can bind, thus enabling incorporation of a single base in each cycle. The modified dNTPs and DNA polymerase are washed in cycles over the flow cell that has become coated with single-strand clusters. The adding of each new base at each cycle is followed by imaging, where the fluorophores are excited with lasers, which allows the sequencing to be recorded in a synchronous manner within a cluster (Bentley et al., 2008). The end result is exact base-by-base sequencing that enables accurate data and eliminates errors associated with homopolymers. Clonal bridge amplification brings another advantage to Illumina, a paired-end sequencing, which allows sequencing the desired DNA fragments from both ends, thus resulting in better coverage, greater numbers of reads and higher yield than in single-end sequencing. Paired-end sequencing provides finer alignment to a reference genome, especially across repetitive sequences. It is suitable for de novo sequencing because paired-end sequencing provides longer continuous sequences (contigs) by filling gaps in the consensus sequence and overcomes the fact that Illumina's read lengths are typically short (up to 2×300 b) (Ambardar et al., 2016). The high data yield has been one of the hallmarks of Illumina technology sequencing from the very beginning. The first Solexa sequencer, the Genome Analyzer, which was launched in 2006, yielded an astonishing 1 Gb of data in a single run. The technology has skyrocketed since then, with sequencers in 2014 outputting more than 1000 times more data, that is, 1.8 Tb in a single sequencing run (Park and Kim, 2016). Furthermore, low error rate makes this technology preferable in metagenomic studies, where the detection of single nucleotide variants in bacterial genomes is vital as these changes could lead to significant changes in phenotypic characters, such as antibiotic resistance and virulence (Schmieder and Edwards, 2011).

ION TORRENT SEQUENCING

Ion Torrent sequencing technology was developed by Ion Torrent Systems and swiftly acquired by Life Technologies (now ThermoFisher), when their first system, the Ion Personal Genome Machine sequencer, was launched in 2010 (Rothberg et al., 2011). As with the previously mentioned NGS technologies, 454 pyrosequencing and Illumina, Ion Torrent sequencing also uses sequencing-by-synthesis approach. In addition, it also shares the emulsion PCR amplification step with 454 pyrosequencing, after which the beads with amplified and attached DNA fragments (alias "Ion Sphere particles") are washed over microwells, so that each well contains a single Ion Sphere particle, and a single type of dNTP is added to the microwell at a time. But, contrary to other techniques, it does not use fluorescence or chemiluminescence for determining nucleotide incorporation by a DNA polymerase. Instead, it measures the H^+ ions released during base incorporation, hence referred to as the world's smallest solid-state pH meter. The detection of H^+ ions is accomplished using the sensing layer of microwell which translates the chemical signal into digital signal, a complementary metal-oxide-semiconductor technology, which is otherwise used in the microprocessor chips manufacturing (Rothberg et al., 2011). For that reason, this type of sequencing is also called semiconductor sequencing. The biggest advantage of this technology derives from its intrinsic chemistry. It allows for very rapid sequencing alongside the actual detection phase. In addition, with the omission of expensive incorporation of modified bases, Ion Torrent sequencing delivers the lowest run cost. With the run length of up to 400 bp and yield of up to 15 Gb, it represents a compromise between the low throughput of 454 pyrosequencing and ultrahigh throughput of Illumina (Bragg and Tyson, 2014; Besser et al., 2018). Nevertheless, as with 454 pyrosequencing, Ion Torrent is more error-prone in the interpretation of homopolymer sequences due to the loss of signal as multiple identical dNTPs are incorporated.

SEQUENCING BY OLIGONUCLEOTIDE LIGATION AND DETECTION

Sequencing by Oligonucleotide Ligation and Detection (SOLiD) was developed by Life Technologies and acquired by Applied Biosystems. The sequencer, launched in 2007, adopts the sequencing-by-ligation technology. First, template DNA fragments are merged with adapters, which allow their hybridization to special beads. Attached to beads, the fragments undergo the amplification step via emulsion PCR. Afterward, the beads carrying the amplified fragments are covalently attached to the surface of a glass slide. The sequencing process begins with the annealing of a universal sequencing primer onto the adapter sequence of the template DNA fragment (Mardis, 2008; Ambardar et al., 2016). Then, a specifically designed probe hybridizes to the fragment next to the primer and the ligase seals phosphate backbone between the probe and the primer. The probes are composed of eight bases, with the first two being actual bases, followed by three degenerate bases, and three universal bases which carry the fluorescent dye. Each of possible two-base combination corresponds to a fluorescent dye color. Therefore, after the probe was ligated to the primer, the appropriate fluorescent signal is measured and the universal bases at the end of the probe with the fluorophore are cleaved in each cycle. This leaves the now 5-base long, hybridized, and ligated probe ready for the attachment of another probe, at the next round of elongation, so the cycle can be repeated several times, after which the primer is removed. Thus a partial, intermittent read of the template is achieved (two bases followed by three-base intermissions). After the primer removal, another primer, longer by one base toward the 3′ end is introduced and hybridized, which will cover some of the gaps of the previous read, as it shifts the reading pattern by one base. For the complete target DNA to be sequenced, four such single base shifts are needed. Nucleotide sequence is encoded in a unique way, where four fluorescent colors are used to represent 16 possible combinations of two bases. Tens of millions of up to 85 bp reads are generated by a single run using this sequencing technology. The distinctive two-base encoding principle and the 8-bp labeled probe enable high quality and accuracy of 99.9% (Liu et al., 2012). SOLiD technology is not the best choice when palindromic sequences need to be sequenced but is used for whole-genome resequencing, targeted resequencing, transcriptome research (including gene expression profiling, small RNA analysis, and whole transcriptome analysis), and epigenome research (such as ChIPSeq and methylation) (Valouev et al., 2008; Ambardar et al., 2016).

THIRD-GENERATION SEQUENCING

When it comes to sequencing, science and technology strive for simpler, more efficient and rapid solutions. Third-generation sequencing opened a new chapter in sequencing technologies. Two main characteristics discriminate it from NGS platforms, leading to improvements in sequencing technology. First, PCR is not necessary to be done before sequencing; thus DNA preparation time for actual sequencing is reduced. Second, the produced signal, whether fluorescence or an electric current, is captured in real time and is monitored during the enzymatic reaction of adding nucleotide in the complementary strand—a feature seen in some, but not all, second-generation approaches.

SINGLE-MOLECULE REAL-TIME SEQUENCING

Single-molecule real-time (SMRT) sequencing, developed by Pacific Biosystems, is a third-generation technology based on sequencing-by-synthesis principle. During library preparation, template DNA fragments are provided with hairpin loop adapters on both sides, which make the template circular and also carry universal primer binding site and initiation sequence. Sequencing is performed in special nanophotonic visualization chambers referred to as zero-mode waveguide (ZMW). ZMW represents the lowest available volume for light detection, capable of detecting only a single nucleotide being incorporated by DNA polymerase (Fig. 9.1A). Sequencing is based on real-time imaging of distinct fluorescently labeled dNTPs as the polymerase synthesizes DNA along single template molecules (Eid et al., 2009). SMRT technology uses fluorophore (four different fluorophores, one for each dNTP) linked to terminal phosphate rather than the base itself, so when it is cleaved in the process of replication, the signal can be measured. A polymerase is attached to the bottom of the chamber. The DNA loop moves through the polymerase and every time a dNTP is detained by polymerase, a phosphate bond is broken, and a light pulse is produced, which is then interpreted in a sequence. Circular DNA template allows the polymerase to continue around to the second adapter sequence and then onto antisense strand, enabling the long reads. DNA polymerase φ29 is selected for use in this system; it is a stable, single subunit enzyme with site-specific mutations introduced, so as to endow it with a life expectancy of over 70,000 bases, high speed and efficacy, and a demonstrated capability for strand-displacement DNA synthesis in whole-genome amplification (Dean et al., 2002).

One of the most prominent advantages of SMRT sequencing is undoubtedly its long, continuous reads, with an average of 15 kb, but reaching 60 kb with novel systems, which make it a good choice in metagenomics, particularly for de novo assemblies of novel genomes and sequencing of full-length bacterial 16S rRNA (Roberts et al., 2013; Hebert et al., 2018; Wagner et al., 2016).

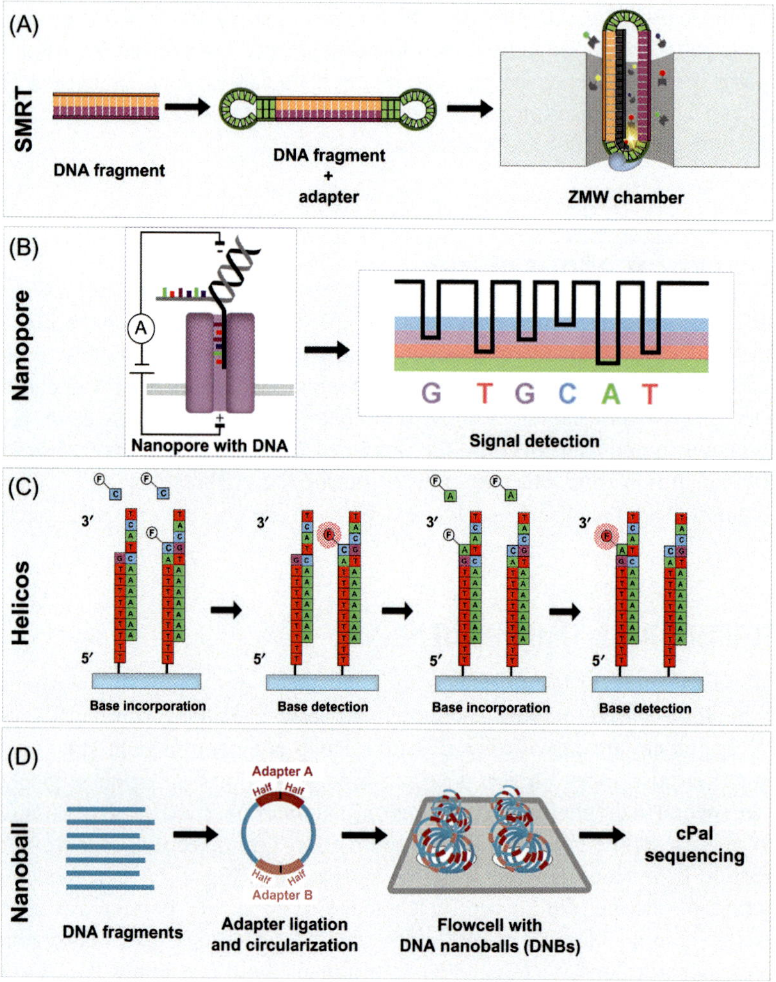

FIGURE 9.1

Schematic representation of third-generation NGS technologies. (A) SMRT sequencing: Template DNA fragments (one DNA strand denoted in *orange* and other DNA strand in *purple*) are provided with hairpin loop adapters (denoted in *green*) on both sides, creating circular DNA sequencing template. Sequencing is performed in ZMW chamber. DNA polymerase (in *light blue*) is attached at the bottom of the ZMW chamber. Universal primer binding site and initiation sequence are present in adapters. They allow the start of sequencing. The sequencing continues around to the second adapter sequence and then onto antisense strand. Distinct fluorescently labeled dNTPs (presented in *red*, *blue*, *green*, and *yellow*) are introduced and every time a dNTP is incorporated by polymerase, a phosphate bond is broken, and a light pulse is produced (denoted in *yellow*). (B) Nanosequencing: A nanopore (denoted in *purple*) in an electrically resistant membrane (denoted in *gray*) is a key element of nanosequencing technology. As the template DNA strand passes through a nanopore, the electrical

(*Continued*)

Also, since the base-calling step of the sequencing is implemented in real time, this sequencing technology is much faster compared to others. Furthermore, its ability to detect base modifications, such as methylation, opens wider possibilities in cancer research. Nevertheless, SMRT sequencing has several disadvantages, namely, the high error rate and low throughput, as well as high cost in comparison to other technologies (Rhoads and Au, 2015).

NANOSEQUENCING

Nanosequencing is a single-molecule sensing technology. Its principle is that a voltage-biased nanoscale pore in an electrically resistant membrane can measure differentially the passage of the different bases of a linear ssDNA or RNA molecule through that pore (Jain et al., 2016). Two main approaches have been proposed. In the first approach the excision of monomers from the DNA strand and their directing one-by-one through a nanopore is performed, as in NanoTag sequencing by Genia and Bayley Sequencing by Oxford Nanopore. The second approach implements strand sequencing wherein intact DNA is ratcheted through the nanopore base by base (Oxford Nanopore MinION). The latter method has been successfully implemented and commercialized. As the DNA sequence passes through a nanopore of an internal diameter of 1 nm, the electrical conductance of the pore is altered, and signal is detected (Fig. 9.1B) (Wanunu, 2012). The huge advantage of this technology is the making away with synthesis enzymes or chemical fluorophores/chromophores. Moreover, there is also no need for sample preparation, as cell lysate can be directly sequenced; and, finally, there is the benefit of relatively low cost.

The pocket-sized sequencing machine MinION may be used not only in the laboratory but also outside the lab. It has 512–2000 nanopores with each nanopore yielding 120–1000 base calls per minute (Branton et al., 2008). However, it can be used only one time and the error rate of 38.2% is pretty much inadequate. The nanosequencing technology is currently developing toward optimizing the type of nanopore used, considering biological nanopores, solid-state nanopores, and multichannel nanopore sensing approaches (Maitra et al., 2012).

◀ conductance of the pore is altered, ionic current is measured (represented with ampere meter labeled as A). Each nucleotide has its unique ionic current level (denoted in *purple*, *red*, *blue*, and *green*) and the signal is detected as DNA sequence. (C) Helicos sequencing: Individual DNA fragments with poly-A sequence (denoted as *green blocks*) are hybridized to oligo-dT sequences (denoted as *red blocks*) which are immobilized on flow cell (denoted in *light blue*). In each sequencing cycle the first step is base incorporation, followed by the wash and the base detection, imaging of the fluorophore lighting up (denoted as a *pink circle*). In the first cycle, cytosine is incorporated in one of the fragments (denoted as a *blue block*). In the next cycle, a different fluorescently labeled dNTP is incorporated, namely, adenosine (denoted as a *green block*), and subsequently detected by imaging of the fluorophore lighting up (denoted as a *pink circle*). (D) Nanoball sequencing: Genomic DNA is being sheared into fragments (denoted as *blue lines*). Half-adaptors (denoted in *red* and *pink*) are inserted to each end of the sheared fragments. Fragments are circularized through intramolecular ligation and complete adaptors are formed (denoted as adaptors A and B). Circular DNA library was clonally amplified and modified to produce DNBs via rolling circle amplification. Distinct DNBs are introduced to the surface of a flow cell (denoted in *gray*). Sequencing is performed using cPAL technology. *cPAL*, Combinatorial Probe-Anchor Ligation; *DNB*, DNA nanoballs; *NGS*, next-generation sequencing; *SMRT*, single molecule real-time; *ZMW*, zero-mode waveguide.

HELICOS SEQUENCING

The first single-molecule sequencing platform was developed by Helicos BioSciences Corporation in 2003. Helicos sequencing shares the sequencing-by-synthesis approach with previous technologies, but it does not make the use of PCR amplification or ligation (Voelkerding et al., 2009; Ambardar et al., 2016). First, DNA template is fragmented, followed by the addition of a poly-A to the 3′ end of each fragment. The final adenosine in poly-A sequence is fluorescently labeled. The fragments are then hybridized to the flow cell that contains immobilized oligo-dT sequences. Prior to sequencing, the exact positional coordinates of the template fragments on the flow cell are recorded with a CCD camera when a laser illuminates their fluorescent labels. The label is cleaved and washed away before the sequencing process begins. Each individual template hybridized to the flow cell surface will produce its own sequencing reaction. Sequencing begins with adding a polymerase and labeled dNTPs (referred to as Virtual terminator nucleotides) (Bowers et al., 2009) to the flow cell. Only one of four dNTPs is added with the polymerase at each cycle, and it is incorporated in a template-directed manner. The wash step removes polymerase and any unincorporated nucleotides. In an imaging step that follows all the individual templates which incorporated that specific nucleotide are recorded. After the imaging, the fluorescent label is cleaved, and the process can be repeated with the addition of another labeled nucleotide, until the desired read length is accomplished (Fig. 9.1C). Read lengths in Helicos sequencing are rather short and go up to 45–50 bases. One of the advantages of Helicos sequencing is that it allows sequencing and quantitation of RNA molecules directly, without converting them into cDNA, by hybridizing polyadenylated RNAs to the oligo-dTs of the flow cell (Hart et al., 2010).

GNUBIO SEQUENCING

Bio-Rad Laboratories has purchased NGS technology firm GnuBIO in 2014. GnuBIO has developed a droplet-based DNA platform for targeted sequencing that incorporates all the functions of DNA sequencing into a single integrated workflow. The complex sequencing reactions, including target selection and enrichment, DNA amplification, and DNA sequencing, are performed in single-use cartridge-based consumables using microfluidic and emulsion technology, which provides a scalable desktop sequencing system that allows interrogation of single genes, gene panels, and whole genomes. Sequencing reactions take place in a cascading manner inside tiny picoliter-sized aqueous droplets which flow through microfluidic channels, each injected with a single DNA fragment/PCR amplicon to be sequenced. Each of these nanodroplets has a single type of codified hexamer probe and using the extension assay it is determined if the particular hexamer sequence hybridizes or not in a given amplicon. When a hexamer probe binds to the DNA fragment/amplicon within its droplet, a quencher bound to the flour at the DNA amplicon terminal end is released and fluorescence is detected. Each droplet results in single DNA sequence read up to 1000 nucleotides long.

The advantage of using GnuBIO platform is that total sequencing run time is less than 4 h. Also, average accuracy is 99.991% and sequence is covered 100%. GnuBIO sequencing technology allows cost per sample to be a function of genomic region. There is no need for batching up the samples. Whether a user runs 1 sample or 1000 samples across a specific genomic region, the cost per sample will remain the same (Ambardar et al., 2016). As many nanodrops as needed can be created and analyzed in serial fashion for each sample.

DNA NANOBALL SEQUENCING

Complete Genomics company (founded in 2006, acquired by BGI-Shenzhen in 2013, and from 2018 a part of MGI, which is the instruments manufacturing business of BGI) focuses mainly on human whole-genome resequencing. Main characteristics of this sequencing technology are the creation of DNA nanoballs (DNBs) and Combinatorial Probe-Anchor Ligation (cPAL) sequencing technology. Sequencing process begins with library preparation and genomic DNA recursively being cut into fragments using type IIS restriction enzymes (*Acu*I and *Eco*P15). Half-adapters are inserted to each end of the sheared fragments and the fragments are circularized through intramolecular ligation resulting in formation of a complete adapter. DNA circles are then linearized and shortened. A second set of half-adapters is added, and the fragments are recircularized once more. The resulting DNA library was clonally amplified with Phi29 polymerase and modified to produce DNBs via rolling circle amplification. DNBs are small particles, hundreds of tandem copies of the sequencing substrate in palindrome-promoted coils of single-stranded DNA, distinct in size, density, and binding affinity properties. DNBs are introduced to the patterned surface of a flow cell and they "self-assemble" into high-density DNB arrays, with >95% occupancy of flow cell spots occupied by a single DNB (Fig. 9.1D). High-accuracy cPAL technology is then used for sequencing. A sequencing cycle begins with the hybridization of an anchor sequence to one of the introduced adapters. A pool of degenerate 9-mer oligonucleotide probes, labeled with specific fluorophores that correspond to a specific nucleotide (A, C, G, or T), is introduced to the reaction and the right matching probe is hybridized to a template and ligated to the anchor. The unligated probes are washed away and the flow cell is imaged. The ligated anchor-probe molecules are denatured. Cycles of anchor hybridization, probe ligation, imaging, and removal of the anchor-probe complex are repeated five times using different combination of fluorescently labeled 9-mer probes with known bases at the $n+1$, $n+2$, $n+3$, and $n+4$ positions. Another five cycles of sequencing continue by using a new set of probes and anchors to interrogate a different base position. The independent cyclic sequencing of 10 bases can be repeated up to eight times, starting at each of the unique anchors. Each DNB results in 62–70 base long reads. Advantage of using cPAL sequencing technology is that a base-read cycle is independent on the completeness of any of the previous cycles. Moreover, since the same original DNA template is used again and again in a linear amplification, rather than in exponential amplification as in PCR, errors introduced in the PCR process do not propagate, making the whole DNB sequencing more accurate. Also, exclusive assembly algorithms produce high-quality data and accurate variant calls (Drmanac et al., 2010).

The summary of the NGS platforms is presented in Table 9.1.

Table 9.1 Summary of Next-Generation Sequencing Platforms.

Platform	Sequencing Chemistry	Detection Method	Yield per Run	Read Length	Characteristics	Year Launched
454 (Roche) pyrosequencing	Sequencing by Synthesis	Bioluminescence	700 Mb	up to 1000 bp (average 800 bp)	Long reads High error rate in homopolymer sequencing	2005
Illumina	Sequencing by Synthesis	Fluorescently labeled	1.8 Tb	2×300 bp	High yield Paired-end sequencing Low error rate	2006
Ion Torrent	Sequencing by Synthesis	H^+ ions	15 Gb	400 bp	Low cost per bp Fast run High error rate in homopolymer sequencing	2010
SOLiD	Sequencing by ligation and two-base coding	Fluorescently labeled	155 Gb	50 + 35 bp	Short reads High accuracy	2006
SMRT	Sequencing by Synthesis	Fluorescently labeled	0.1 Gb	60 kbp	Extralong reads	2010
Nanosequencing	No chemistry	Electrical impulse	130 Gb	100 kbp	Pocket-size machine High throughput Ultralong reads	2012
Helicos	Sequencing by Synthesis	Fluorescently labeled	35 Gb	45–50 bp	DNA and RNA sequencing	2003
GnuBIO	Sequencing by hybridization	Specific dye	1.2 Gb	1000 bp	Use of droplets Decreased sequencing price	2014
DNA nanoball sequencing	Combination of sequencing by hybridization and sequencing by ligation	Four-color detection	120 Gb	up to 124 bp	Base-read cycle does not depend on the completeness of any of the previous cycles	2010

SMRT, *single molecule real-time;* SOLiD, *Sequencing by Oligonucleotide Ligation and Detection.*

BIG DATA IN GENOMICS

Generation of large datasets, specifically by high loads of genomics information, is a consequence of the newly developed high-throughput sequencing technologies. The explosion of available information pushed life sciences into the era of Big Data. The rate of biomedical data generation in particular is increasing exponentially.

Thus sequencing technologies with specific performance in terms of accuracy and depth of reading are used in a wide range of applications during the human lifespan (Morganti et al., 2019), from prenatal diagnostics and newborn screening to rare diseases diagnosis (Skakic et al., 2018; Komazec et al., 2018; Andjelkovic et al., 2018; Stojiljkovic et al., 2016). Big Data became extremely important in oncology (Marjanovic et al., 2016; Todorovic Balint et al., 2016), as well as for the development of pharmacogenomics tests (Mizzi et al., 2016) referring to predisposition for a number of diseases (Balasopoulou et al., 2016). As for bacterial or viral DNA found in human microbiome, novel genomes can easily be assembled de novo, particularly when using the third-generation sequencing technology, providing once again new data. Plainly, the Big Data revolution brought us to the doorstep of data-driven, evidence-based, truly precision medicine.

Similarly, an enormous amount of sequencing data in microbiomics has led to achieving spectacular progress in this field.

For example, it is possible to perform large-scale alignment and evolutionary studies. Furthermore, affordable sequencing of vast numbers of different microbial genomes, considered as one of the prime applications of NGS in microbiomic research, generates Big Data. Large datasets lead to the development of insightful approaches for unveiling the diversity patterns for myriads of microorganisms, thus shedding new light on human microbiome and its relationship with the host, and in discovering novel pathways in metabolism, as well as metabolites with potential as new therapeutic agents. Also, data on microbial gene expression, horizontal gene transfer, and other bio-functional information may affect antibiotic resistance in clinical settings. In environmental microbiomic research, the use of NGS will provide data of microbial communities in most diverse settings, from ice floes at the North Pole to desolated deserts (Hauptmann et al., 2014; Zhang et al., 2016).

However, the transition from the data generated using NGS to data usable in distinct fields of life sciences requires data analysis. The interdisciplinary field of bioinformatics combines biology and computer science and is thus endowed with a crucial role in analyzing, interpreting, and finally translating into applications biological data; according to most researchers, its actual status and performance represents the current bottleneck in the field of life sciences. Still, bioinformatics has already contributed greatly in handling and exploiting Big Data related to omics, microbiomics included.

BIOINFORMATIC METHODS FOR ANALYZING METAGENOMIC DATA

The two approaches for processing data produced by microbiome metagenomics using high-throughput sequencing (either the marker gene amplicon sequencing or the whole-genome shotgun metagenomics) provide different quantity and quality of data. Marker gene approach focuses on the sequencing of one particular gene (16S rRNA in bacteria/archea, ITS in fungi, and 18S rRNA in eukaryotes) which identifies the respective genome without analyzing it in its entirety. This method is suitable for identification of the makeup of a microbiome in terms of taxonomy but produces no data on other genes which can describe the underlying niches and interactions or, in simpler terms, the micro- and macro-ecological functionalities. This function-centered knowledge is revealed by the whole-genome shotgun analysis where sequencing data uncover the full set of microbial genes present, thus providing the information not only about *who/what* is present but also *what they are doing* and *how they interconnect and interact.*

The bioinformatic analyses used in these two approaches have different objectives and challenges, therefore applying different workflows for analysis of NGS sequencing data. The standard procedure for marker gene analysis consists of the following steps: preprocessing of raw sequencing reads, binning (grouping similar sequences into units that represent taxonomic clusters), taxonomic classification, and statistical analysis (Junemann et al., 2017; Oulas et al., 2015). In the case of shotgun approach, the standard workflow involves: preprocessing, the assembly of short sequences into longer continuous or semicontinuous genome fragments (contigs and scaffolds), binning (grouping reads/contigs/scaffolds into clusters of related fragments), the prediction of genes and putative proteins, and prediction of proteins' domains, functions, and pathways (Oulas et al., 2015; Roumpeka et al., 2017).

PREPROCESSING OF SEQUENCE DATA

Preprocessing of sequence reads is crucial in the assessment of metagenome (complete genomes of all microbes in the sample) as it can improve analysis accuracy and prevent overestimation of the species diversity (Kunin et al., 2010; Bokulich et al., 2013). Common preprocessing steps include removal of the artificial duplicates introduced during amplification process (deduplication), trimming of adapters/identifiers/primers, as well as trimming and filtering of low-quality reads. Reads that become too short after trimming should be removed, since they have low sequence complexity and might map incorrectly (Martin et al., 2018). Decontamination process, in which reads that do not belong to the studied genome(s) are filtered out from the analyzed data, is often included in the preprocessing workflow (Schmieder and Edwards, 2011). Denoising procedure is specific for 454 technology and it corrects intrinsic errors generated during pyrosequencing that may result in clustering erroneous reads and incorrect taxonomic classification (Huse et al., 2010; Quince et al., 2011). Also, the choice of a particular metagenomic approach may require additional steps while preprocessing NGS data. For instance, marker gene analysis involves identification and removal of chimera sequences. Chimeras are generated of two or more parent sequences as a result of cross hybridization during the PCR step. Their presence could increase diversity estimation and they should be removed from the dataset (Quince et al., 2011; Wright et al., 2012). Detection of chimeras can be accomplished using chimera-free reference databases or, when using de novo approach, by assuming that the parent sequences will be more common than any chimeric sequence within the products of a single PCR. Since no reference database is completely free from chimera sequences, the latter approach is advisable (Singh et al., 2017). After each preprocessing step, quality-control metrics should be observed in order to examine effectiveness in generating sequences of the highest possible quality. There is a plethora of tools developed for preprocessing of NGS data, such as FastQC (Andrews, 2010), FASTX toolkit (Gordon and Hannon, 2010),

PRINSEQ (Schmieder and Edwards, 2011), CD-HIT (Li and Godzik, 2006), BBTools (Bushnell, 2015), AmpliconNoise (Quince et al., 2011), UCHIME (Edgar et al., 2011), and many others.

16S RRNA ANALYSIS

In microbiome analysis, 16S rRNA is used as a gene marker, since it is present in a wide range of bacterial organisms and it contains both highly conserved as well as variable regions which evolve at a constant rate. 16S rRNA sequencing is a relatively simple method that describes the bacteriome as a set of 16S sequences and their respective frequencies (Morgan and Huttenhower, 2012).

After preprocessing steps, 16S rRNA sequences are clustered into "bins" called operational taxonomic units (OTUs) based on sequence similarity. The resulting OTUs can be used to estimate species diversity, composition, and richness of the analyzed bacteriome(s). There are three OTU clustering approaches: the de novo approach, which is purely based on pair-wise similarity among all sequences; the closed-reference approach, which is based on similarity to a reference database and it discards the reads that do not match any entry within the reference database; and open-reference approach, which is a hybrid of the previous two. According to open-reference approach, sequences are first clustered using a reference database, then those sequences that do not match with any reference entries are clustered de novo (Navas-Molina et al., 2013). Similarity cutoff for classifying sequences into the same OTU is arbitrary, but for bacteria it is widely accepted that similarity greater than 97% corresponds to the same species (Konstantinidis and Tiedje, 2005). A single sequence is selected from each OTU as a representative (consensus) and is subsequently annotated, so that all sequences within the same OTU get the same annotation (Nguyen et al., 2016). Taxonomic assignment of the representative consensus sequence (annotation) can be performed by searching against the 16S rRNA reference databases, such as Greengene database (Desantis et al., 2006), Ribosomal Database Project (RDP) (Cole et al., 2014), or SILVA (Pruesse et al., 2007), using available taxonomic classification algorithms (BLAST, RDP classifier, RTAX, and UCLUST). Approaches based on direct similarity to reference sequences are affected by the completeness, integrity, and accuracy of the database. Instead, methods which rely on sequence composition features extracted from the reference dataset and used in machine-learning processes to infer the taxonomy of the analyzed sequences may be applied (Wang et al., 2007). Statistical analysis is preformed to evaluate population diversity, using metrics such as alpha diversity (diversity within the sample) and beta diversity (diversity across the samples); to analyze samples by means of principal coordinate analysis or nonmetric multidimensional scaling; and to determine phylogenetic relationships (Jovel et al., 2016). Commonly used software suites for 16S rRNA data analysis such as QIIME (Caporaso et al., 2010) and Mothur (Schloss et al., 2009) combine algorithms and tools for all described steps into one platform.

Although marker gene 16S rRNA approach is simple and cheap, several factors, able to affect the resolution of the microbiome composition, should be taken into account. Some of these are the horizontal gene transfer, the existence of multiple and perhaps variable copies of 16S rRNA within a cell or a homogenous cell population, the choice of hypervariable region, nonadequate sequence similarity cutoffs for OTU clustering, and PCR amplification errors. Finally, 16S rRNA marker analysis is dedicated for prokaryotes (mainly bacteria) and does not allow analysis of other organisms present in the community, such as viruses and eukaryotes.

Therefore, the whole-genome shotgun approach that allows more comprehensive analysis has become the approach of choice.

WHOLE-GENOME SHOTGUN ANALYSIS

In shotgun analysis, sequenced DNA fragments are used for in silico reconstruction of the bacterial/archean/eukaryotic/viral genomes found in the studied community. This process is called "genome assembly" and it involves "stitching" of the short reads together into longer continuous sequences (contigs) in order to acquire a meaningful length for alignments to known sequences so as to perform taxonomic and functional annotations. Shotgun analysis gives more detailed and concise overview of the studied microbiome and enables recovery and characterization of both known and novel genes and genomes. However, this is more challenging than marker gene approach, mostly because of the low and incomplete coverage of the sequenced genomes, and of the high quantity of data, which requires intensive computational processing.

The process of assembling reads into contigs may be implemented by two different methods, either by the reference-guided (comparative) approach or by the de novo approach. The choice depends on the studied microbiome, availability of reference databases, and available computational power (Oulas et al., 2015; Wajid and Serpedin, 2012). In comparative assembly a reference genome is used as a "map" that guides reconstruction of contigs. This method is less computationally demanding than de novo assembling and applicable in metagenomic analysis of microbiomes containing known microorganisms (Cepeda et al., 2017; Wajid and Serpedin, 2012). Also, guided assembly is suggested when sequence coverage is low due to either low cellular abundance or high intraspecies variability (Bragg and Tyson, 2014). Available tools used for reference-guided assembly are ABBA (Salzberg et al., 2008), AMOScmp (Pop et al., 2004b), and MetaCompass (Cepeda et al., 2017). They process data with optimal speed and memory requirements, which makes them suitable for regular desktop or laptop hosting.

The de novo assembly method uses reference-free strategies for the assembly of sequence reads into contigs, commonly through the use of overlap–layout–consensus (OLC) or de Bruijn graph methods (Nagarajan and Pop, 2013). Both algorithms are based on the graphs construction consisting of nodes (which represent short sequences), connected with edges (which represent overlaps). However, these two algorithms differ in their logic; in OLC method, all reads are compared pair-wise to find regions with significant overlaps, while de Bruijn graph is constructed by reading consecutive *k*-mers (nucleotide sequence of a *k* length) within each read without the need for pair-wise comparison (Nagarajan and Pop, 2013). Assembly is the result of traversing through these overlaps so that each node is linked in the correct order into contiguous sequence. The OLC method is more accurate and highly suited for the assembly of the long reads, while de Bruijn graph enables computationally efficient assembly of the data consisting of short reads (Vollmers et al., 2017). In de Bruijn graph approach the choice of *k*-mer size is essential, since longer *k*-mers tend to facilitate the reconstruction of highly abundant genomes, whereas shorter *k*-mers are better suited for recovering rare ones (Sczyrba et al., 2017). Most current metagenome assemblers, such as MetaVelvet (Namiki et al., 2012), IDBA-UD (Peng et al., 2012), MetaSPAdes (Nurk et al., 2017), SOAPdenovo2 (Luo et al., 2012), and Ray Meta (Boisvert et al., 2012), implement the de Bruijn graph method.

In subsequent steps the contigs are organized into scaffolds based on the position of the paired-end and mate-paired sequences, their orientation, and expected insert size. Even though many assembly tools have a scaffolding module, there are also stand-alone software called "scaffolders" such as Bambus (Pop et al., 2004a), SOPRA (Dayarian et al., 2010), and Opera (Gao et al., 2011) which offer greater flexibility, particularly when combining data from different sequencing platforms.

The quality of assembly can be assessed by different metrics such as the number of generated contigs/scaffolds, their length, coverage, or N50 size. N50 is a statistical metric used for assessing the contiguity of a genome assembly. If contigs are sorted by size, N50 is defined as the length of the shortest contig at the 50% of the total genome length or greater (El-Metwally et al., 2013). Contig and scaffold lengths are important metrics as these should be longer than gene-length to enable full recovery of the gene sequence (Roumpeka et al., 2017).

The number of assembly errors should be examined in order to make downstream analysis more accurate. Remapping of all reads on the assembly can identify incorrect assembling if mislocated paired-end/mate-paired reads occur. Also, erroneous assembly can be detected by inspecting the regions with unusual depth of coverage. If coverage is too high, this may indicate collapse of the repeat (two adjacent repeated sequences detected as one) and if it is too low, it may indicate incorrect stitching of unrelated genomic regions (Nagarajan and Pop, 2013). If a reference genome is available, the accuracy of the assembled sequences can be assessed by aligning the draft genome and reference genome (El-Metwally et al., 2013). MetaQuest is a tool designed for quality assessment of metagenomics assemblies (Mikheenko et al., 2016).

There are several concerns regarding de novo assembly of metagenomic data such as difficult reconstruction of rare taxa, the presence of artificial contigs (chimeras) as a result of incorrect assembly, as well as problematic assembly of the repetitive regions. Combining long-read (Pacific Bioscience) and short-read (Illumina) sequences in the same assembly might help researchers to overcome some of these limitations. Also, one more drawback of the de novo assembly is its high computational requirements. However, there are ways to alleviate this, such as applying preassembly binning to reduce data complexity and then assembling bins independently (Sharpton, 2014).

Binning steps can be implemented before as well as after the assembly. There are two binning methods. One is based on sequence similarity to the reference database. The other is based solely on sequence composition features such as GC content, read coverage, codon usage, oligonucleotide frequencies, and periodic sequence signatures (Mrazek, 2009; Albertsen et al., 2013). The latter, the composition-based method, does not require reference dataset and therefore provides insight into the presence of novel genomes, which are difficult to identify otherwise. Also, it is much faster and less computationally demanding compared to the former, the similarity-based method. However, the similarity binning may deliver higher annotation accuracy and taxonomic resolution (Sharpton, 2014). Examples of the available tools that use composition-based binning are PhyloPythiaS (Patil et al., 2012), PhymmBL (Brady and Salzberg, 2009), and NBC (Rosen et al., 2011), while similarity-based binning tools are MEGAN (Huson et al., 2011), CARMA (Gerlach and Stoye, 2011), MetaPhyler (Liu et al., 2011), etc.

Further steps involve identification of genes within the unassembled and/or assembled data, a process known as "gene calling." Using assembled data provides more accurate prediction of genes but this approach is biased toward sequences that are more abundant in the sample and get

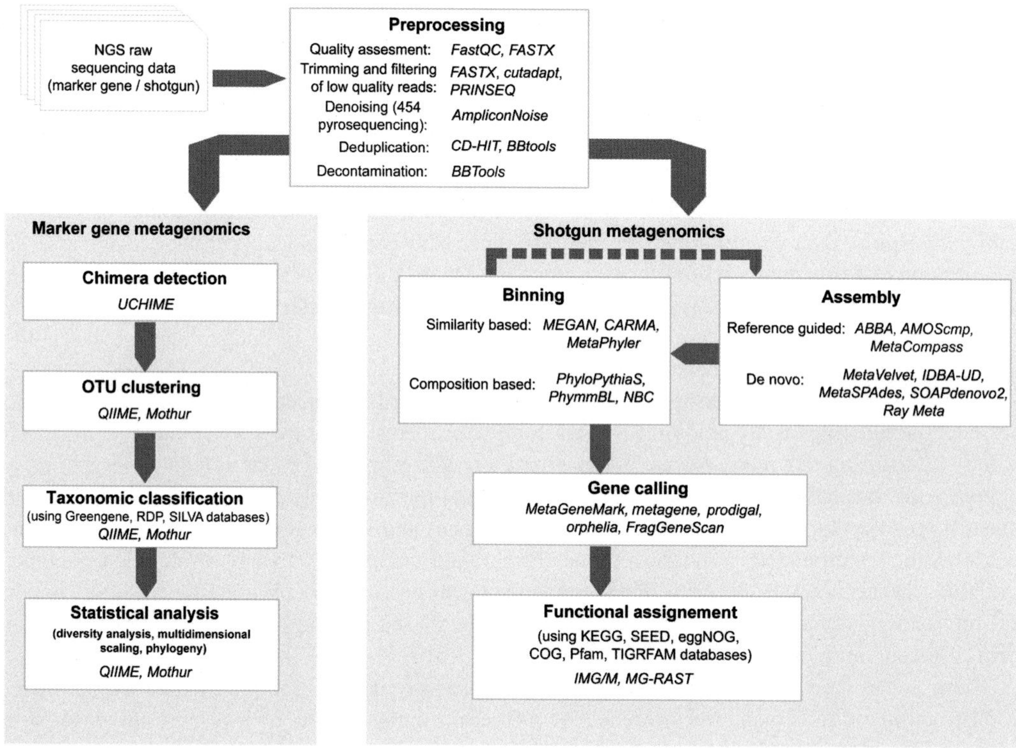

FIGURE 9.2

Flowchart of steps and tools used in metagenomic data analysis. Flowchart shows basic steps included in the metagenomic data analysis. Analysis starts with the common preprocessing steps but afterward, choice of the metagenomic approach (marker gene or whole-genome shotgun analysis) directs the bioinformatic pipeline. Flowchart displays some popular tools (indicated in italic) used in processing of the presented data-analysis steps.

assembled very easily. On the other hand, using unassembled data approach is more sensitive to rare sequences, but less reliable (Singh et al., 2017). Gene-prediction tools use codon information to identify potential open reading frames and hence label sequences as coding or noncoding (Oulas et al., 2015). Available tools for identifying protein-coding sequences are MetaGeneMark (Zhu et al., 2010), Metagene (Noguchi et al., 2006), Prodigal (Hyatt et al., 2012), Orphelia (Hoff et al., 2009), and FragGeneScan (Kim et al., 2015). Subsequent analysis consists of functional assignment of the predicted protein-coding genes either by using homology-based searches against databases containing known functional and/or taxonomic information such as KEGG (Kanehisa et al., 2012), SEED (Overbeek et al., 2005), eggNOG (Huerta-Cepas et al., 2016), and COG (Tatusov et al., 2000) or by applying Hidden Markov Models to assign protein sequences to protein families such as in Pfam (Bateman et al., 2000) and TIGRFAM (Haft et al., 2003).

IMG/M and MG-RAST are widely used online platforms for metagenomics analysis that offer fully automated pipelines to conduct gene prediction, the database search, protein family classification, and annotation (Glass et al., 2010; Markowitz et al., 2014). There are also stand-alone downloadable pipelines, such as MetAMOS (Treangen et al., 2013), which can be installed on the local system and used with more analytical flexibility.

Most of the described steps of metagenomic analysis apply equally well to viruses. However, there are technical challenges that limit the virus-analysis process. Abundance of viral genomes in clinical and environmental samples is markedly lower compared to host (in the clinical context) and bacterial genomes. This fact limits the amount of signal obtained for analysis after sequencing. In silico decontamination step should be therefore applied to discard unneeded sequences through alignment with one or more nonviral reference genomes leaving only viral sequences in the dataset (Daly et al., 2015). Moreover, viruses show extremely high genetic diversity and divergence because of the high mutation rate, which makes their detection more difficult even when it is known that a certain virus is present in the sample. Thus virus metagenomics requires more tolerant to mismatches alignment algorithms so as to enable virus identification. Also, similarity-based searches use translated nucleotide sequences alignments in order to bypass synonymous mutations and facilitate virus classification (Fancello et al., 2012). These and other issues mean that methods for virus analysis require bioinformatic approaches and tools specifically adjusted for viral metagenomics. Various software tools have been developed to satisfy such specifications and requirements, including but not restricted to VirusSeeker (Zhao et al., 2017), MetaVir (Roux et al., 2011), VirSorter (Roux et al., 2015), ViromeScan (Rampelli et al., 2016), and VIP (Li et al., 2016). In addition, the fact that databases are incomplete regarding viral reference sequences significantly reduces our ability to detect known and discover novel viruses. Establishment of database(s) containing assembled and annotated viral data as well as unclassified sequences from different projects is suggested as a solution for this limitation. The unclassified viral sequences, often called "dark matter," are most likely the signature of hitherto unknown viral pathogens. Availability of and accessibility to such data (unclassified reads originating from diverse, multiple studies) may determine elusive trends and enable more sensitive and robust identification of new viruses (Rose et al., 2016).

Flowcharts of steps and tools used in metagenomic data analysis are presented in Fig. 9.2.

CONCLUSION

Metagenomic analysis sheds light on the complex biodiversity patterns of microbiota throughout the most versatile ecosystems, resulting in discovery of novel organisms with advantageous features. Microbiomics research has already produced promising data to be exploited in medical and veterinary science. In order to realize the full potential of translational research in microbiomics, data, information, and knowledge about various functional genomic elements, individual species genome variations, gene expression patterns, and complex gene-environment association must be acquired. Such amounts of data can only be obtained using affordable and highly efficient NGS technologies. In the years to come a big crescendo in NGS usage is prone to happen, as the new sequencing technologies improve on many features of current sequencing methods. Novel high-throughput methods will result in unveiling the diversity, structure, and function of microbiomes worldwide. Generation of Big Data must be accompanied by powerful software tools for the subsequent analysis. Consequently, the progress of bioinformatics is destined to play a crucial role in the microbiomics, as with all other -omics.

ACKNOWLEDGMENT

This work has been funded by grant from the Ministry of Education, Science and Technological Development, Republic of Serbia (III 41004) given to SP.

REFERENCES

Albertsen, M., Hugenholtz, P., Skarshewski, A., Nielsen, K.L., Tyson, G.W., Nielsen, P.H., 2013. Genome sequences of rare, uncultured bacteria obtained by differential coverage binning of multiple metagenomes. Nat. Biotechnol. 31, 533–538.

Ambardar, S., Gupta, R., Trakroo, D., Lal, R., Vakhlu, J., 2016. High throughput sequencing: an overview of sequencing chemistry. Indian J. Microbiol. 56, 394–404.

Andjelkovic, M., Minic, P., Vreca, M., Stojiljkovic, M., Skakic, A., Sovtic, A., et al., 2018. Genomic profiling supports the diagnosis of primary ciliary dyskinesia and reveals novel candidate genes and genetic variants. PLoS One 13, e0205422.

Andrews, S., 2010. FastQC: a quality control tool for high throughput sequence data [Online]. Available from: <http://www.bioinformatics.babraham.ac.uk/projects/fastqc/>.

Aragon, I.M., Herrera-Imbroda, B., Queipo-Ortuno, M.I., Castillo, E., Del Moral, J.S., Gomez-Millan, J., et al., 2018. The urinary tract microbiome in health and disease. Eur. Urol. Focus 4, 128–138.

Balasopoulou, A., Stankovic, B., Panagiotara, A., Nikcevic, G., Peters, B.A., John, A., et al., 2016. Novel genetic risk variants for pediatric celiac disease. Hum. Genomics 10, 34.

Barko, P.C., Mcmichael, M.A., Swanson, K.S., Williams, D.A., 2018. The gastrointestinal microbiome: a review. J. Vet. Intern. Med. 32, 9–25.

Bateman, A., Birney, E., Durbin, R., Eddy, S.R., Howe, K.L., Sonnhammer, E.L., 2000. The Pfam protein families database. Nucleic Acids Res. 28, 263–266.

Bentley, D.R., Balasubramanian, S., Swerdlow, H.P., Smith, G.P., Milton, J., Brown, C.G., et al., 2008. Accurate whole human genome sequencing using reversible terminator chemistry. Nature 456, 53–59.

Berg, G., Rybakova, D., Grube, M., Koberl, M., 2016. The plant microbiome explored: implications for experimental botany. J. Exp. Bot. 67, 995–1002.

Besemer, K., 2015. Biodiversity, community structure and function of biofilms in stream ecosystems. Res. Microbiol. 166, 774–781.

Besser, J., Carleton, H.A., Gerner-Smidt, P., Lindsey, R.L., Trees, E., 2018. Next-generation sequencing technologies and their application to the study and control of bacterial infections. Clin. Microbiol. Infect. 24, 335–341.

Boisvert, S., Raymond, F., Godzaridis, E., Laviolette, F., Corbeil, J., 2012. Ray Meta: scalable de novo metagenome assembly and profiling. Genome Biol. 13, R122.

Bokulich, N.A., Subramanian, S., Faith, J.J., Gevers, D., Gordon, J.I., Knight, R., et al., 2013. Quality-filtering vastly improves diversity estimates from Illumina amplicon sequencing. Nat. Methods 10, 57–59.

Bowers, J., Mitchell, J., Beer, E., Buzby, P.R., Causey, M., Efcavitch, J.W., et al., 2009. Virtual terminator nucleotides for next-generation DNA sequencing. Nat. Methods 6, 593–595.

Brady, A., Salzberg, S.L., 2009. Phymm and PhymmBL: metagenomic phylogenetic classification with interpolated Markov models. Nat. Methods 6, 673–676.

Bragg, L., Tyson, G.W., 2014. Metagenomics using next-generation sequencing. Methods Mol. Biol. 1096, 183–201.

Branton, D., Deamer, D.W., Marziali, A., Bayley, H., Benner, S.A., Butler, T., et al., 2008. The potential and challenges of nanopore sequencing. Nat. Biotechnol. 26, 1146–1153.

Bushnell, B., 2015. BBMap short-read aligner, and other bioinformatics tools [Online]. Available from: <https://sourceforge.net/projects/bbmap/>.

Caporaso, J.G., Kuczynski, J., Stombaugh, J., Bittinger, K., Bushman, F.D., Costello, E.K., et al., 2010. QIIME allows analysis of high-throughput community sequencing data. Nat. Methods 7, 335–336.

Cepeda, V., Liu, B., Almeida, M., Hill, C.M., Koren, S., Treangen, T.J., et al., 2017. MetaCompass: reference-guided assembly of metagenomes. bioRxiv 212506.

Cole, J.R., Wang, Q., Fish, J.A., Chai, B., Mcgarrell, D.M., Sun, Y., et al., 2014. Ribosomal Database Project: data and tools for high throughput rRNA analysis. Nucleic Acids Res. 42, D633–D642.

Daly, G.M., Leggett, R.M., Rowe, W., Stubbs, S., Wilkinson, M., Ramirez-Gonzalez, R.H., et al., 2015. Host subtraction, filtering and assembly validations for novel viral discovery using next generation sequencing data. PLoS One 10, e0129059.

Dayarian, A., Michael, T.P., Sengupta, A.M., 2010. SOPRA: scaffolding algorithm for paired reads via statistical optimization. BMC Bioinformatics 11, 345.

Dean, F.B., Hosono, S., Fang, L., Wu, X., Faruqi, A.F., Bray-Ward, P., et al., 2002. Comprehensive human genome amplification using multiple displacement amplification. Proc. Natl Acad. Sci. U.S.A. 99, 5261–5266.

De Mandal, S., Panda, A., Bisht, S., Kumar, N., 2015. Microbial ecology in the era of next generation sequencing. Gener. Seq. Appl. S1, 2.

Desantis, T.Z., Hugenholtz, P., Larsen, N., Rojas, M., Brodie, E.L., Keller, K., et al., 2006. Greengenes, a chimera-checked 16S rRNA gene database and workbench compatible with ARB. Appl. Environ. Microbiol. 72, 5069–5072.

Drmanac, R., Sparks, A.B., Callow, M.J., Halpern, A.L., Burns, N.L., Kermani, B.G., et al., 2010. Human genome sequencing using unchained base reads on self-assembling DNA nanoarrays. Science 327, 78–81.

Edgar, R.C., Haas, B.J., Clemente, J.C., Quince, C., Knight, R., 2011. UCHIME improves sensitivity and speed of chimera detection. Bioinformatics 27, 2194–2200.

Edlund, A., Yang, Y., Yooseph, S., He, X., Shi, W., Mclean, J.S., 2018. Uncovering complex microbiome activities via metatranscriptomics during 24 hours of oral biofilm assembly and maturation. Microbiome 6, 217.

Eid, J., Fehr, A., Gray, J., Luong, K., Lyle, J., Otto, G., et al., 2009. Real-time DNA sequencing from single polymerase molecules. Science 323, 133–138.

El-Metwally, S., Hamza, T., Zakaria, M., Helmy, M., 2013. Next-generation sequence assembly: four stages of data processing and computational challenges. PLoS Comput. Biol. 9, e1003345.

Fakruddin, M., Chowdhury, A., Hossain, N., Mahajan, S., Islam, S., 2013. Pyrosequencing: a next generation sequencing technology. World Appl. Sci. J. 24, 1558–1571.

Fancello, L., Raoult, D., Desnues, C., 2012. Computational tools for viral metagenomics and their application in clinical research. Virology 434, 162–174.

Fierer, N., 2017. Embracing the unknown: disentangling the complexities of the soil microbiome. Nat. Rev. Microbiol. 15, 579–590.

Fiers, W., Contreras, R., Duerinck, F., Haegeman, G., Iserentant, D., Merregaert, J., et al., 1976. Complete nucleotide sequence of bacteriophage MS2 RNA: primary and secondary structure of the replicase gene. Nature 260, 500–507.

Fleischmann, R.D., Adams, M.D., White, O., Clayton, R.A., Kirkness, E.F., Kerlavage, A.R., et al., 1995. Whole-genome random sequencing and assembly of *Haemophilus influenzae* Rd. Science 269, 496–512.

Gao, S., Sung, W.K., Nagarajan, N., 2011. Opera: reconstructing optimal genomic scaffolds with high-throughput paired-end sequences. J. Comput. Biol. 18, 1681–1691.

Gerlach, W., Stoye, J., 2011. Taxonomic classification of metagenomic shotgun sequences with CARMA3. Nucleic Acids Res. 39, e91.

Glass, E.M., Wilkening, J., Wilke, A., Antonopoulos, D., Meyer, F., 2010. Using the metagenomics RAST server (MG-RAST) for analyzing shotgun metagenomes. Cold Spring Harb. Protoc. 2010, pdb prot5368.

Gordon, A., Hannon, G., 2010. Fastx-toolkit [Online]. Available from: <http://hannonlab.cshl.edu/fastx_toolkit>.

Haft, D.H., Selengut, J.D., White, O., 2003. The TIGRFAMs database of protein families. Nucleic Acids Res. 31, 371–373.

Harrington, C.T., Lin, E.I., Olson, M.T., Eshleman, J.R., 2013. Fundamentals of pyrosequencing. Arch. Pathol. Lab. Med. 137, 1296–1303.

Hart, C., Lipson, D., Ozsolak, F., Raz, T., Steinmann, K., Thompson, J., et al., 2010. Single-molecule sequencing: sequence methods to enable accurate quantitation. Methods Enzymol. 472, 407–430.

Hauptmann, A.L., Stibal, M., Baelum, J., Sicheritz-Ponten, T., Brunak, S., Bowman, J.S., et al., 2014. Bacterial diversity in snow on North Pole ice floes. Extremophiles 18, 945–951.

Head, S.R., Komori, H.K., Lamere, S.A., Whisenant, T., Van Nieuwerburgh, F., Salomon, D.R., et al., 2014. Library construction for next-generation sequencing: overviews and challenges. Biotechniques 56, 61–64. 66, 68, passim.

Hebert, P.D.N., Braukmann, T.W.A., Prosser, S.W.J., Ratnasingham, S., Dewaard, J.R., Ivanova, N.V., et al., 2018. A Sequel to Sanger: amplicon sequencing that scales. BMC Genomics 19, 219.

Hoff, K.J., Lingner, T., Meinicke, P., Tech, M., 2009. Orphelia: predicting genes in metagenomic sequencing reads. Nucleic Acids Res. 37, W101–W105.

Huerta-Cepas, J., Szklarczyk, D., Forslund, K., Cook, H., Heller, D., Walter, M.C., et al., 2016. eggNOG 4.5: a hierarchical orthology framework with improved functional annotations for eukaryotic, prokaryotic and viral sequences. Nucleic Acids Res. 44, D286–D293.

Huse, S.M., Welch, D.M., Morrison, H.G., Sogin, M.L., 2010. Ironing out the wrinkles in the rare biosphere through improved OTU clustering. Environ. Microbiol. 12, 1889–1898.

Huson, D.H., Mitra, S., Ruscheweyh, H.J., Weber, N., Schuster, S.C., 2011. Integrative analysis of environmental sequences using MEGAN4. Genome Res. 21, 1552–1560.

Hyatt, D., Locascio, P.F., Hauser, L.J., Uberbacher, E.C., 2012. Gene and translation initiation site prediction in metagenomic sequences. Bioinformatics 28, 2223–2230.

Jain, M., Olsen, H.E., Paten, B., Akeson, M., 2016. The Oxford Nanopore MinION: delivery of nanopore sequencing to the genomics community. Genome Biol. 17, 239.

Jovel, J., Patterson, J., Wang, W., Hotte, N., O'keefe, S., Mitchel, T., et al., 2016. Characterization of the gut microbiome using 16S or shotgun metagenomics. Front. Microbiol. 7, 459.

Junemann, S., Kleinbolting, N., Jaenicke, S., Henke, C., Hassa, J., Nelkner, J., et al., 2017. Bioinformatics for NGS-based metagenomics and the application to biogas research. J. Biotechnol. 261, 10–23.

Kahvejian, A., Quackenbush, J., Thompson, J.F., 2008. What would you do if you could sequence everything? Nat. Biotechnol. 26, 1125–1133.

Kanehisa, M., Goto, S., Sato, Y., Furumichi, M., Tanabe, M., 2012. KEGG for integration and interpretation of large-scale molecular data sets. Nucleic Acids Res. 40, D109–D114.

Kim, D., Hahn, A.S., Wu, S.-J., Hanson, N.W., Konwar, K.M., Hallam, S.J., 2015. FragGeneScan-Plus for scalable high-throughput short-read open reading frame prediction. In: Computational Intelligence in Bioinformatics and Computational Biology (CIBCB), 2015 IEEE Conference on. IEEE, pp. 1–8.

Komazec, J., Zdravkovic, V., Sajic, S., Jesic, M., Andjelkovic, M., Pavlovic, S., et al., 2018. The importance of combined NGS and MLPA genetic tests for differential diagnosis of maturity onset diabetes of the young. Endokrynol. Pol. 70, 28–36.

Kong, H.H., Andersson, B., Clavel, T., Common, J.E., Jackson, S.A., Olson, N.D., et al., 2017. Performing skin microbiome research: a method to the madness. J. Invest. Dermatol. 137, 561–568.

Konstantinidis, K.T., Tiedje, J.M., 2005. Genomic insights that advance the species definition for prokaryotes. Proc. Natl. Acad. Sci. U.S.A. 102, 2567–2572.

Kunin, V., Engelbrektson, A., Ochman, H., Hugenholtz, P., 2010. Wrinkles in the rare biosphere: pyrosequencing errors can lead to artificial inflation of diversity estimates. Environ. Microbiol. 12, 118–123.

Li, W., Godzik, A., 2006. Cd-hit: a fast program for clustering and comparing large sets of protein or nucleotide sequences. Bioinformatics 22, 1658–1659.

Li, Y., Wang, H., Nie, K., Zhang, C., Zhang, Y., Wang, J., et al., 2016. VIP: an integrated pipeline for metagenomics of virus identification and discovery. Sci. Rep. 6, 23774.

Liu, B., Gibbons, T., Ghodsi, M., Treangen, T., Pop, M., 2011. Accurate and fast estimation of taxonomic profiles from metagenomic shotgun sequences. BMC Genomics 12 (Suppl. 2), S4.

Liu, L., Li, Y., Li, S., Hu, N., He, Y., Pong, R., et al., 2012. Comparison of next-generation sequencing systems. J. Biomed. Biotechnol. 2012, 251364.

Luo, R., Liu, B., Xie, Y., Li, Z., Huang, W., Yuan, J., et al., 2012. SOAPdenovo2: an empirically improved memory-efficient short-read de novo assembler. Gigascience 1, 18.

Maitra, R.D., Kim, J., Dunbar, W.B., 2012. Recent advances in nanopore sequencing. Electrophoresis 33, 3418–3428.

Mardis, E.R., 2008. The impact of next-generation sequencing technology on genetics. Trends Genet. 24, 133–141.

Margulies, M., Egholm, M., Altman, W.E., Attiya, S., Bader, J.S., Bemben, L.A., et al., 2005. Genome sequencing in microfabricated high-density picolitre reactors. Nature 437, 376–380.

Marjanovic, I., Kostic, J., Stanic, B., Pejanovic, N., Lucic, B., Karan-Djurasevic, T., et al., 2016. Parallel targeted next generation sequencing of childhood and adult acute myeloid leukemia patients reveals uniform genomic profile of the disease. Tumour Biol. 37, 13391–13401.

Markowitz, V.M., Chen, I.M., Chu, K., Szeto, E., Palaniappan, K., Pillay, M., et al., 2014. IMG/M 4 version of the integrated metagenome comparative analysis system. Nucleic Acids Res. 42, D568–D573.

Martin, T.C., Visconti, A., Spector, T.D., Falchi, M., 2018. Conducting metagenomic studies in microbiology and clinical research. Appl. Microbiol. Biotechnol. 102, 8629–8646.

Metzker, M.L., 2010. Sequencing technologies – the next generation. Nat. Rev. Genet. 11, 31–46.

Mikheenko, A., Saveliev, V., Gurevich, A., 2016. MetaQUAST: evaluation of metagenome assemblies. Bioinformatics 32, 1088–1090.

Mizzi, C., Dalabira, E., Kumuthini, J., Dzimiri, N., Balogh, I., Basak, N., et al., 2016. A European spectrum of pharmacogenomic biomarkers: implications for clinical pharmacogenomics. PLoS One 11, e0162866.

Moran, M.A., 2015. The global ocean microbiome. Science 350, aac8455.

Morgan, X.C., Huttenhower, C., 2012. Chapter 12: Human microbiome analysis. PLoS Comput. Biol. 8, e1002808.

Morganti, S., Tarantino, P., Ferraro, E., D'amico, P., Viale, G., Trapani, D., et al., 2019. Complexity of genome sequencing and reporting: Next generation sequencing (NGS) technologies and implementation of precision medicine in real life. Crit. Rev. Oncol. Hematol. 133, 171–182.

Mrazek, J., 2009. Phylogenetic signals in DNA composition: limitations and prospects. Mol. Biol. Evol. 26, 1163–1169.

Nagarajan, N., Pop, M., 2013. Sequence assembly demystified. Nat. Rev. Genet. 14, 157–167.

Namiki, T., Hachiya, T., Tanaka, H., Sakakibara, Y., 2012. MetaVelvet: an extension of Velvet assembler to de novo metagenome assembly from short sequence reads. Nucleic Acids Res. 40, e155.

Navas-Molina, J.A., Peralta-Sanchez, J.M., Gonzalez, A., Mcmurdie, P.J., Vazquez-Baeza, Y., Xu, Z., et al., 2013. Advancing our understanding of the human microbiome using QIIME. Methods Enzymol. 531, 371–444.

Nguyen, N.P., Warnow, T., Pop, M., White, B., 2016. A perspective on 16S rRNA operational taxonomic unit clustering using sequence similarity. NPJ Biofilms Microbiomes 2, 16004.

Noguchi, H., Park, J., Takagi, T., 2006. MetaGene: prokaryotic gene finding from environmental genome shotgun sequences. Nucleic Acids Res. 34, 5623–5630.

Nurk, S., Meleshko, D., Korobeynikov, A., Pevzner, P.A., 2017. metaSPAdes: a new versatile metagenomic assembler. Genome Res. 27, 824–834.

Oulas, A., Pavloudi, C., Polymenakou, P., Pavlopoulos, G.A., Papanikolaou, N., Kotoulas, G., et al., 2015. Metagenomics: tools and insights for analyzing next-generation sequencing data derived from biodiversity studies. Bioinform Biol. Insights 9, 75–88.

Overbeek, R., Begley, T., Butler, R.M., Choudhuri, J.V., Chuang, H.Y., Cohoon, M., et al., 2005. The subsystems approach to genome annotation and its use in the project to annotate 1000 genomes. Nucleic Acids Res. 33, 5691–5702.

Park, S.T., Kim, J., 2016. Trends in next-generation sequencing and a new era for whole genome sequencing. Int. Neurourol. J. 20, S76–S83.

Patil, K.R., Roune, L., Mchardy, A.C., 2012. The PhyloPythiaS web server for taxonomic assignment of metagenome sequences. PLoS One 7, e38581.

Peng, Y., Leung, H.C., Yiu, S.M., Chin, F.Y., 2012. IDBA-UD: a de novo assembler for single-cell and metagenomic sequencing data with highly uneven depth. Bioinformatics 28, 1420–1428.

Pop, M., Kosack, D.S., Salzberg, S.L., 2004a. Hierarchical scaffolding with Bambus. Genome Res. 14, 149–159.

Pop, M., Phillippy, A., Delcher, A.L., Salzberg, S.L., 2004b. Comparative genome assembly. Brief. Bioinform 5, 237–248.

Pruesse, E., Quast, C., Knittel, K., Fuchs, B.M., Ludwig, W., Peplies, J., et al., 2007. SILVA: a comprehensive online resource for quality checked and aligned ribosomal RNA sequence data compatible with ARB. Nucleic Acids Res. 35, 7188–7196.

Quince, C., Lanzen, A., Davenport, R.J., Turnbaugh, P.J., 2011. Removing noise from pyrosequenced amplicons. BMC Bioinformatics 12, 38.

Rampelli, S., Soverini, M., Turroni, S., Quercia, S., Biagi, E., Brigidi, P., et al., 2016. ViromeScan: a new tool for metagenomic viral community profiling. BMC Genomics 17, 165.

Rhoads, A., Au, K.F., 2015. PacBio sequencing and its applications. Genomics Proteomics Bioinformatics 13, 278–289.

Riesenfeld, C.S., Schloss, P.D., Handelsman, J., 2004. Metagenomics: genomic analysis of microbial communities. Annu. Rev. Genet. 38, 525–552.

Rizzo, J.M., Buck, M.J., 2012. Key principles and clinical applications of "next-generation" DNA sequencing. Cancer Prev. Res. (Phila.) 5, 887–900.

Roberts, R.J., Carneiro, M.O., Schatz, M.C., 2013. The advantages of SMRT sequencing. Genome Biol. 14, 405.

Rondon, M.R., August, P.R., Bettermann, A.D., Brady, S.F., Grossman, T.H., Liles, M.R., et al., 2000. Cloning the soil metagenome: a strategy for accessing the genetic and functional diversity of uncultured microorganisms. Appl. Environ. Microbiol. 66, 2541–2547.

Rose, R., Constantinides, B., Tapinos, A., Robertson, D.L., Prosperi, M., 2016. Challenges in the analysis of viral metagenomes. Virus Evol. 2, vew022.

Rosen, G.L., Reichenberger, E.R., Rosenfeld, A.M., 2011. NBC: the Naive Bayes classification tool webserver for taxonomic classification of metagenomic reads. Bioinformatics 27, 127–129.

Rothberg, J.M., Hinz, W., Rearick, T.M., Schultz, J., Mileski, W., Davey, M., et al., 2011. An integrated semiconductor device enabling non-optical genome sequencing. Nature 475, 348–352.

Roumpeka, D.D., Wallace, R.J., Escalettes, F., Fotheringham, I., Watson, M., 2017. A review of bioinformatics tools for bio-prospecting from metagenomic sequence data. Front. Genet. 8, 23.

Roux, S., Faubladier, M., Mahul, A., Paulhe, N., Bernard, A., Debroas, D., et al., 2011. Metavir: a web server dedicated to virome analysis. Bioinformatics 27, 3074–3075.

Roux, S., Enault, F., Hurwitz, B.L., Sullivan, M.B., 2015. VirSorter: mining viral signal from microbial genomic data. PeerJ 3, e985.

Salzberg, S.L., Sommer, D.D., Puiu, D., Lee, V.T., 2008. Gene-boosted assembly of a novel bacterial genome from very short reads. PLoS Comput. Biol. 4, e1000186.

Sanger, F., Nicklen, S., Coulson, A.R., 1977. DNA sequencing with chain-terminating inhibitors. Proc. Natl Acad. Sci. U.S.A. 74, 5463–5467.

Schloss, P.D., Handelsman, J., 2005. Metagenomics for studying unculturable microorganisms: cutting the Gordian knot. Genome Biol. 6, 229.

Schloss, P.D., Westcott, S.L., Ryabin, T., Hall, J.R., Hartmann, M., Hollister, E.B., et al., 2009. Introducing mothur: open-source, platform-independent, community-supported software for describing and comparing microbial communities. Appl. Environ. Microbiol. 75, 7537–7541.

Schmieder, R., Edwards, R., 2011. Quality control and preprocessing of metagenomic datasets. Bioinformatics 27, 863–864.

Sczyrba, A., Hofmann, P., Belmann, P., Koslicki, D., Janssen, S., Droge, J., et al., 2017. Critical Assessment of Metagenome Interpretation—a benchmark of metagenomics software. Nat. Methods 14, 1063–1071.

Sharpton, T.J., 2014. An introduction to the analysis of shotgun metagenomic data. Front. Plant. Sci. 5, 209.

Shendure, J., Ji, H., 2008. Next-generation DNA sequencing. Nat. Biotechnol. 26, 1135–1145.

Singh, B., Crippen, T.L., Tomberlin, J.K., 2017. An introduction to metagenomic data generation, analysis, visualization, and interpretation. In: Carter, D.O., Tomberlin, J.K., Benbow, M.E., Metcalf, J.L. (Eds.), Forensic Microbiology, Wiley.

Siqueira Jr., J.F., Fouad, A.F., Rocas, I.N., 2012. Pyrosequencing as a tool for better understanding of human microbiomes. J. Oral. Microbiol. 4.

Skakic, A., Djordjevic, M., Sarajlija, A., Klaassen, K., Tosic, N., Kecman, B., et al., 2018. Genetic characterization of GSD I in Serbian population revealed unexpectedly high incidence of GSD Ib and 3 novel SLC37A4 variants. Clin. Genet. 93, 350–355.

Stojiljkovic, M., Klaassen, K., Djordjevic, M., Sarajlija, A., Brasil, S., Kecman, B., et al., 2016. Molecular and phenotypic characteristics of seven novel mutations causing branched-chain organic acidurias. Clin. Genet. 90, 252–257.

Tatusov, R.L., Galperin, M.Y., Natale, D.A., Koonin, E.V., 2000. The COG database: a tool for genome-scale analysis of protein functions and evolution. Nucleic Acids Res. 28, 33–36.

Todorovic Balint, M., Jelicic, J., Mihaljevic, B., Kostic, J., Stanic, B., Balint, B., et al., 2016. Gene mutation profiles in primary diffuse large B cell lymphoma of central nervous system: next generation sequencing analyses. Int. J. Mol. Sci. 17.

Treangen, T.J., Koren, S., Sommer, D.D., Liu, B., Astrovskaya, I., Ondov, B., et al., 2013. MetAMOS: a modular and open source metagenomic assembly and analysis pipeline. Genome Biol. 14, R2.

Valouev, A., Ichikawa, J., Tonthat, T., Stuart, J., Ranade, S., Peckham, H., et al., 2008. A high-resolution, nucleosome position map of *C. elegans* reveals a lack of universal sequence-dictated positioning. Genome Res. 18, 1051–1063.

Voelkerding, K.V., Dames, S.A., Durtschi, J.D., 2009. Next-generation sequencing: from basic research to diagnostics. Clin. Chem. 55, 641–658.

Vollmers, J., Wiegand, S., Kaster, A.K., 2017. Comparing and evaluating metagenome assembly tools from a microbiologist's perspective – not only size matters!. PLoS One 12, e0169662.

Wagner, J., Coupland, P., Browne, H.P., Lawley, T.D., Francis, S.C., Parkhill, J., 2016. Evaluation of PacBio sequencing for full-length bacterial 16S rRNA gene classification. BMC Microbiol. 16, 274.

Wajid, B., Serpedin, E., 2012. Review of general algorithmic features for genome assemblers for next generation sequencers. Genomics Proteomics. Bioinformatics 10, 58–73.

Wang, Q., Garrity, G.M., Tiedje, J.M., Cole, J.R., 2007. Naive Bayesian classifier for rapid assignment of rRNA sequences into the new bacterial taxonomy. Appl. Environ. Microbiol. 73, 5261–5267.

Wanunu, M., 2012. Nanopores: a journey towards DNA sequencing. Phys. Life Rev. 9, 125–158.

Wright, E.S., Yilmaz, L.S., Noguera, D.R., 2012. DECIPHER, a search-based approach to chimera identification for 16S rRNA sequences. Appl. Environ. Microbiol. 78, 717–725.

Xu, Y., Zhao, F., 2018. Single-cell metagenomics: challenges and applications. Protein Cell 9, 501–510.

Zhang, B., Kong, W., Wu, N., Zhang, Y., 2016. Bacterial diversity and community along the succession of biological soil crusts in the Gurbantunggut Desert, Northern China. J. Basic Microbiol. 56, 670–679.

Zhao, G., Wu, G., Lim, E.S., Droit, L., Krishnamurthy, S., Barouch, D.H., et al., 2017. VirusSeeker, a computational pipeline for virus discovery and virome composition analysis. Virology 503, 21–30.

Zhu, W., Lomsadze, A., Borodovsky, M., 2010. Ab initio gene identification in metagenomic sequences. Nucleic Acids Res. 38, e132.

PART

III

NOVEL AND LEGACY FIELDS OF MICROBIAL APPLICATIONS

CHAPTER

CANCER MICROBIOMATICS?

10

Georgios Gaitanis[1] **and Martin Laurence**[2]

[1]*Department of Skin and Venereal Diseases, Faculty of Medical Sciences, School of Medicine, University of Ioannina, Ioannina, Greece* [2]*Shipshaw Labs, Montreal, QC, Canada*

INTRODUCTION

In the 1880s, Koch convincingly linked a bacterium to tuberculosis, opening the modern era of infectious disease medicine. The main techniques used at the time were light microscopy, cell culture, and animal models. In the following two decades a flurry of research linking novel microbes and viruses to diseases was performed—among others, tuberculosis (Koch, 1884), gonorrhea (Bockhart, 1883), and syphilis (Schaudinn and Hoffmann, 1905) finally had known causes. Infectious agents with distinctive histological signatures (inclusion bodies, corkscrew shapes, motility) were quickly discovered. Those which were always present in affected hosts but not in all healthy controls (Koch's first postulate), and which readily caused a similar disease when inoculated in healthy animals or humans (Koch's third postulate), were deemed causative (Evans, 1976).

Koch's postulates were applied with great success to infectious diseases, but they were much less successful in explaining the origins of cancer. The introduction of consensus PCR (16S) and metagenomics in the last decade has given researchers a powerful new tool to reassess the link between microbes and health, enabling a second prolific era of scientific research in the role of microbes in idiopathic diseases and cancers.

IMPORTANT HOLDOUTS

While Koch's work quickly resulted in attributing infectious etiologies to many diseases, there were important holdouts: some microbes' role in diseases eluded researchers for much of the 20th century and were eventually resolved by astute research approaches. Here are three examples.

In the first half of the 20th century, *Helicobacter pylori* colonized nearly everyone's stomach from infancy until death (Kusters et al., 2006). This bacterium has a distinctive shape and is motile, so it is easy to detect using light microscopy (Kusters et al., 2006). It does not grow well using routine cell culture and does not cause human-like symptoms in commonly used animal models (Kusters et al., 2006). Despite being ubiquitous in children, gastric ulcers are very rare in this population—in a natural setting, these ulcers typically do not arise immediately following the initial infection (Kusters et al., 2006). The pH of the stomach was considered too acidic for microbial

Microbiomics. DOI: https://doi.org/10.1016/B978-0-12-816664-2.00010-4

colonization to be possible. Because of these complicating factors, it took until the 1980s to confirm *H. pylori*'s presence in the stomach and to establish its primary role in gastric ulcers (Marshall et al., 1985a; Kusters et al., 2006). By that time, improvements in hygiene had substantially reduced *H. pylori*'s prevalence in the population, allowing small case–control studies to find an association with gastric ulcers (Kusters et al., 2006)—in earlier times, nearly all healthy controls would have had *H. pylori* too, so only high-powered studies could hope to find an association. Koch's third postulate was confirmed by inoculating a noninfected individual (Marshall et al., 1985a), and by showing that antibacterial drugs which cleared *H. pylori* cured gastric ulcers (Kusters et al., 2006). *H. pylori* is also a major risk factor for gastric cancer, increasing the risk of this cancer approximately 10-fold (Kusters et al., 2006).

A similar situation linked the Epstein–Barr virus (EBV) to multiple sclerosis (MS) in the early 21st century (Ascherio and Munger, 2010). In developing countries, EBV prevalence reaches nearly 100% in the first years of life: this virus is generally passed between family members through saliva (Hjalgrim et al., 2007). In contrast, EBV prevalence is only ~50% by early adolescence in Europe and in the United States (Hjalgrim et al., 2007). This difference is generally attributed to cultural behavior and hygiene (Hjalgrim et al., 2007). In adolescents and young adults, EBV is transmitted mainly through intimate contact (Hjalgrim et al., 2007), and its prevalence plateaus at about 95% (Ascherio and Munger, 2007). This means demonstrating an association between EBV and MS in adults required very large case–control studies, because age-matched controls who are EBV negative are rare (Almohmeed et al., 2013; Pakpoor et al., 2012; Ascherio and Munger, 2007). If EBV's prevalence had been ~100%, as it is in developing countries, finding a statistically significant association would have been impossible.

An even more daunting task awaited researchers attempting to attribute dandruff and seborrheic dermatitis to *Malassezia* (Shuster, 1984; Gupta and Bluhm, 2004). The main problem is that all humans are colonized—it is impossible to find a *Malassezia*-free individual! This means Koch's first and third postulates cannot be fulfilled at all. The strongest evidence that *Malassezia* are necessary factors in dandruff and seborrheic dermatitis consists mainly of two observations: (1) they are the only fungus present on affected skin (Findley et al., 2013) and (2) various topical compounds with antifungal properties reduce symptoms, as long as they are applied (Shuster, 1984; Gupta and Bluhm, 2004). Though psoriasis and atopic dermatitis have also been suspected of being caused by an adaptive immune response against *Malassezia* (Kanda et al., 2002), poor efficacy of topical antifungal drugs has prevented a similar conclusion from being reached. Because Koch's postulates cannot be directly fulfilled for any skin disease putatively caused by *Malassezia*, a consensus as to their role has yet to emerge.

MICROBIOMICS AND CANCER

In contrast to infectious diseases such as tuberculosis and syphilis, experiments attempting to replicate cancer in animals—generally following the transfer of putative cancer–causing microbes from humans to fulfill Koch's third postulate—failed (Ewing, 1908; Williams, 1908). A notable exception is Rous' sarcoma, where a retrovirus causes connective tissue cancer, which kills the host chicken within a few weeks (Rous, 1911). In the early 20th century, it was not yet understood that adult-onset cancers in humans only develop after an oncogenic factor has been present for many years—for example, cigarette smoke in lung cancer (Frank, 2007). Multistage models of oncogenesis were developed in the 1950s to account for the approximate cubic increase in incidence of these cancers according to age (Nordling, 1953; Armitage and Doll, 1954). Surprisingly, animal models for these cancers did not follow the dose responses typically seen in lethal-dose testing of chemicals: while a high dose of mutagenic factor (e.g., radiation) applied *once* had little effect on cancer risk, a much lower dose applied for a *sustained* period significantly increased cancer risk (Frank, 2007). To reconcile the cubic increase in risk and this nonlinear dose response, multistage theory proposes that cancer arises once several key mutations have occurred within a given cell lineage—each mutation being necessary to enable the next mutation (Frank, 2007).

As evident in the cases of *H. pylori* and human papillomaviruses (HPVs), the long lag between infection onset and cancer has made this link much more difficult to prove than for classical infectious diseases: from Koch's initial work in 1880s (Koch, 1884), it took a century of research to discover that these two important cancers are usually caused by infections (Nomura et al., 1991; Dürst et al., 1983).

KOCH'S BLIND SPOTS

Tying microbes to cancer is much harder than tying them to infectious diseases because cancer typically develops following *decades* of exposure to the mutagenic factor (Frank, 2007). This unfortunately means that (1) experiments replicating this process in animals are necessarily very long and (2) epidemiologic studies seeking signs of the initial infection must compare data collected many decades before cancer occurs; for example, hygiene in the first years of life and stomach cancer risk in late adulthood. Neither of these limitations applies to classical infectious diseases, where symptoms typically appear shortly following the initial infection, and where infection prevalence is low in the general population.

Microbes that persist in humans for decades are prime suspects in cancer development. However, they are unamenable to fulfilling Koch's postulates because: (1) they tend to have a prevalence of nearly 100% in adults, preventing case–control studies from reaching statistical significance, such as *H. pylori* and EBV in earlier times, and *Malassezia* now; and (2) they tend to be fastidious and stenoxenous, respectively, failing to grow using typical cell-culture methods and failing to cause disease in animal models—such as *H. pylori*.

H. pylori was established as a quasi-necessary factor in gastric ulcers by antibiotics that can clear this bacterium (Kusters et al., 2006); this method comes close to fulfilling Koch's third postulate. However, there are currently no drugs that can completely clear certain infections from the body, such as EBV and *Malassezia*. Even if such drugs existed, experimentally demonstrating that their targets are involved in cancer would require long prospective studies, similar to those run to demonstrate HPV vaccines' effectiveness in cervical cancer (Pollock et al., 2014).

BREAKTHROUGHS IN ESTABLISHING MICROBIOMIC CAUSALITY IN CANCER

The first major success in associating microbiota and cancer followed the discovery that the stomach can be colonized by *H. pylori*, resulting in chronic gastritis in a subset of the population (Robin Warren and Marshall, 1983). Conveniently, a strong association was found between gastric ulcers and this bacterium in small case–control studies (Marshall et al., 1985b), prompting much larger prospective studies necessary to link it with noncardia gastric cancer (HACC Group, 2001). *H. pylori* greatly increases the risk of noncardia gastric cancer (relative risk of ~5.9 in developed countries), though 10%–20% of cases are negative for this bacterium (Plummer et al., 2015).

The second major success in building such associations followed the discovery of oncogenic HPV strains and their strong association with squamous cell cervical carcinoma (Dürst et al., 1983). Oncogenic HPV strains are almost always present in cancerous lesions (relative risk of >100) (Muñoz et al., 2003). Unlike *H. pylori* and gastric ulcers/cancer, no known drug can clear HPV from the cervix, and developing vaccines to prevent oncogenic HPV infections took 25 years (Govan, 2008; Monie et al., 2008). This greatly delayed the deployment of effective interventions in cervical cancer, as compared to the *H. pylori*-related diseases.

A third notable success linking microbiota and cancer was the association of chronic hepatitis caused by hepatitis B and C viruses (HBV/HCV) with hepatocellular carcinoma (Arbuthnot and Kew, 2001). These account for approximately ~78% of hepatocellular carcinoma cases worldwide, though they account for a much lower fraction of cases in developed countries, where HBV is not as prevalent (Perz et al., 2006). Unlike the respective cases of *H. pylori* and HPV, a large majority of hepatocellular carcinoma cases in developed countries cannot be attributed to HBV or HCV (Altekruse et al., 2009).

Some less studied cancer types have also been convincingly linked to microbiota, including EBV and *Plasmodium falciparum* being linked to the endemic variant of Burkitt's lymphoma, Kaposi's sarcoma–associated herpes virus (KSHV) and human immunodeficiency virus (HIV) to Kaposi's sarcoma, *Candida albicans* to oral and esophageal squamous cell carcinomas (SCCs), and Merkel cell polyomavirus to Merkel cell carcinoma. The endemic variants of Burkitt's lymphoma and Kaposi's sarcoma are interesting because their risk is highest in individuals simultaneously harboring two infections: a herpes virus and either *P. falciparum* (Rochford et al., 2005) or HIV (Mesri et al., 2010). Individuals with rare genetic defects in the interleukin (IL)-17 pathway suffer from lifelong oral candidiasis, which likely explains their unusually high risk of SCC in this site (Rautemaa et al., 2007; Delsing et al., 2012). Merkel cell carcinoma's strong association with the then-unknown Merkel cells polyomavirus was found using high-throughput sequencing technology and metagenomics (Feng et al., 2008). This method is very promising for future studies of microbiomics and cancer because it can efficiently detect all microbes and viruses, both known and unknown. This is a marked improvement over previously used techniques such as microscopy and cell culture.

Some data suggest that eliminating *H. pylori* or oncogenic HPV, respectively, reduces the risk of gastric (Rokkas et al., 2017) and cervical cancers (Harper and Demars, 2017). This raises the possibility that cancers attributed to an infectious cause in the future will become preventable using simple interventions such as vaccination, improved hygiene, or antibiotic drugs.

BECOMING WISER

Four main lessons emerge from the above successes in establishing a causal relationship between microbiota and cancer.

All virus and microbe types need to be investigated. Historically, most researches have focused on linking viruses and cancer. As discussed in the previous section, at least one cancer type has been convincingly linked to a bacterium, fungus, protist, or virus. Modern microbiomics techniques based on metagenomics are truly universal and can detect all cellular microbe and virus types, so future studies are expected to sport a more systematic approach to linking microbial or viral contributors to cancer. This may well reveal that previous studies investigated a too narrow set of microbiota and missed important contributors to cancer.

A given cancer type is caused by few bacterial, fungal, protistan, or viral species. Until now, studies have linked either one or just a few species to a given cancer type. This might be due to technical limitations that prevented detection of all involved microbiota. Alternatively, it might mean that oncogenic processes are driven by very specific interactions of the host with specific microbes and that interactions involving many microbial species rarely cause cancer.

Causative microbiota are present at the cancer site for a long time. In all cancer types discussed above, causative infectious agents were directly present at the site of carcinogenesis and had persisted there for months or years before cancer developed. This concords with the multistage model's main prediction (Frank, 2007): cancer only arises after a mutagenic factor has been active for a prolonged period. It also means that finding candidate microbes should be as simple as applying metagenomic techniques to chronically inflamed organs.

Chronic inflammation is an important biomarker. Long-lasting chronic inflammation of a given organ often precedes cancer. The clearest examples of inflammation caused by microbiota and leading to cancer are *H. pylori*-induced gastritis and HBV/HCV-induced hepatitis. Chronic idiopathic gastritis and hepatitis were well known before the discovery of *H. pylori* and HBV/HCV. For example, in the 1970s, a chronic unexplained type of hepatitis called "non-HAV (hepatitis A virus), non-HBV hepatitis" was described (Feinstone et al., 1975); more than a decade later, it was realized that a third hepatitis virus (HCV) was their main cause (Weiner et al., 1990).

Chronic idiopathic inflammation occurs in many internal organs where cancer develops, such as the prostate (De Marzo et al., 2007). Are these organs sterile, or are they colonized by one or more elusive infectious agents like HCV? The skin, mouth, vagina, and intestines have long been known to harbor microbiomes. Historically, organs beyond these have been considered mostly sterile, with a few notable exceptions of long-term colonization by a handful of viruses, bacteria, protists, and fungi. These exceptions include HIV (AIDS), HPV (warts and cervical cancer), HBV/HCV (hepatitis and hepatocellular carcinoma), EBV (mononucleosis), herpes simplex viruses (HSV-1 and HSV-2) (cold sores and genital sores), KSHV (Kaposi's sarcoma), *Treponema pallidum* (syphilis), *M. tuberculosis* (tuberculosis), *Toxoplasma gondii* (toxoplasmosis), *Pneumocystis jirovecii* (pneumocystis pneumonia), *H. pylori* (gastric ulcers and noncardia gastric cancer), and finally *Plasmodium* (malaria and Burkitt's lymphoma). Is this the final list, or are prevalent microbes—possibly involved in cancers of internal organs—still missing?

The following section discusses the possibility that the genus *Malassezia*—historically considered commensal fungi restricted to the skin—might contain such microbes.

MALASSEZIA AS AN INDUCER

Malassezia are the most prevalent fungi in humans, colonizing all individuals' skin from birth (Nagata et al., 2012; Gross et al., 1992) and for life (Jo et al., 2016; Findley et al., 2013). They are lipophilic, relying on exogenous (host) lipids for growth (Theelen et al., 2018). They are difficult to detect because they grow poorly in culture—especially, *Malassezia restricta* and *Malassezia globosa*, the most prevalent species in humans (Guého-Kellermann et al., 2010); they have critical mismatches with commonly used PCR primers (Laurence, 2018); and they have a very thick cell wall that is resistant to lysis (Theelen et al., 2018). Their presence on the skin means that they are often dismissed as contamination when found in internal organs. They have been linked to various diseases of the skin, including dandruff, pityriasis versicolor, seborrheic dermatitis, folliculitis, psoriasis, and atopic dermatitis (Gupta et al., 2004). In rare cases, they can cause acute life-threatening inflammation following dissemination to internal organs (Shek et al., 1989; Redline et al., 1985).

SKIN MICROBIOME AND CARCINOGENESIS

As the outermost border of the body, the skin has been the area of extreme selection pressure. It primarily serves in the maintenance of key homeostatic functions such as water homeostasis and thermal regulation, and in the protection from infections and physical and chemical assaults (Brettmann and De Guzman Strong, 2018). This is eloquently depicted in cases of severe skin failure (thermal burns, toxic epidermal necrolysis, etc.), in which the first hours require supportive measures to keep water control and body temperature within limits that remain compatible to life (Haberal et al., 2010). Lethal systemic infections, often from innocuous skin commensals, such as *Staphylococcus epidermidis* (Ronat et al., 2014), may follow. In contrast to other large surfaces exposed to environmental agents (e.g., lungs to inhaled air or the gastrointestinal tract to ingested matter), the skin is subjected to completely unprocessed physical factors such as ultraviolet (UV) radiation, heat, air, and microbiota. The resident microbiota constitute the skin microbiome (virome, microbiome, mycobiome). Their translational (postgenomic in reality) adaptions to skin and environmental factors act as a transitional functional mantle, modulating the skin microenvironment in response to external stimuli. These responses include alterations in barrier function properties of the skin, modulation of epidermal proliferation, alterations in sebaceous gland function, and immune responses of the skin in total.

There is nowadays a large cohort of patients with iatrogenic immunosuppression of various intensities, such as rheumatology patients, inflammatory bowel disease (IBD) patients, as well as dermatological patients. Even marginally, the respective drug-induced immunosuppression favors the development of skin neoplasms (Chen et al., 2016; Bongartz et al., 2006). Immune modulating drugs such as azathioprine (Vos et al., 2018) and disease-modifying antirheumatic drugs or the newer antibodies that are used as pharmacological agents knock out particular aspects of the immune response—for example, antitumor necrosis factor and anti-IL-17 agents (Kaushik and Lebwohl, 2019). Certain representatives of the latter group favor the overgrowth of *Candida* yeasts (Saunte et al., 2017) and can exacerbate other immune-mediated diseases such as IBD (Orrell et al., 2018). Notably, their impact on skin microbial flora has been discussed, yet not extensively evaluated; the association of microbiome alterations and skin carcinogenesis in these cohorts of patients has not been evaluated yet.

Evidence supporting the participation of the skin virome in skin carcinogenesis is well attested. Kaposi's sarcoma, caused by the Kaposi sarcoma virus, flourished in the HIV epidemic and led to the discovery of the responsible virus (Katano, 2018). Likewise, the extensive skin carcinogenesis in solid organ transplant recipients is partly explained by HPV infections (Harwood et al., 2017), but this has not been confirmed in immunocompetent individuals. This confidence in pathogenesis attribution stems from the extension of Koch's postulates in virology research, in which when a virus is found to be associated with a neoplasm through genomic studies, a thorough screening of the infection process clarifies the key components of the carcinogenesis pathway and the factors that participate in tumorigenesis (Fig. 10.1A). However, when it comes to bacteria and eukaryotes, this is far more complex (Fig. 10.1B) due to the expected variance in the interactions of a multicellular organ, as is the skin with the population dynamics that determine the relative abundance and function of the microbiome that can subsequently impact skin carcinogenesis. Regarding skin, these processes take place under the constant presence of UV radiation, which affects both the skin and the microbiome.

Skin carcinogenesis evolves when relevant multipotent skin stem cells that reside in the interfollicular epidermis, the hair follicle, and the sebaceous gland acquire a significant number of tumor-initiating mutations, and a rare population of tumor-initiating cells with unlimited capacity for self-renewal and proliferation develops (Thieu et al., 2013). Subsequent modification of the tumor microenvironment is a prerequisite for this tumorigenic lineage to expand. For this the involvement of nondividing differentiated epidermal cells is required to modulate a protumorigenic epidermal microenvironment.

The most common keratinocytic tumors in Caucasians are basal cell carcinoma (BCC)—accounting for 80% of cases—and SCC—accounting for 20% of cases in the nonimmunosuppressed population (Alam and Ratner, 2001). Actinic keratoses (AK) are recognized precursor lesions of SCC. BCC and SCC differ in their prevalence, anatomical distribution, and association with UV exposure and immunosuppression. Thus BCC is mostly distributed on the face and less commonly on the trunk, usually sparing acral sites (Heckmann et al., 2002). There is an association with the quantity of UV exposure, but this association plateaus up to a point. However, SCC is mostly located on uncovered areas such as the face but does not spare the acral areas such as the back of the hands (Mackie and Quinn, 2004). In the general population of Caucasian ancestry the relative ratio BCC:SCC is 4:1 that is inversed in solid organ transplant recipients that are under constitutive immunosuppression (Zwald and Brown, 2011; Berg and Otley, 2002). AK follow the distribution of SCC and arise in cancerized areas of the skin, as determined in the context of skin cancerization (Szeimies et al., 2012). This suggests that there are large areas of mutated, not clinically evident, cellular clones in sun-exposed skin areas (Albibas et al., 2018). Pertinent to this fact is the concurrent alteration of the skin microbiome that could either be the result of the modified physiology of the sun-damaged, carcinogenic field, or act as a codriver of tumor promotion within this already primed skin area.

Along these suggestions a recent study defined the microbiome composition of AK, SCC, and proximal healthy sun-damaged skin in 13 SCC-prone human males and unveiled some interesting findings that corroborate the herein presented suggestions. It was recorded that AK and SCC are associated with more *Staphylococcus aureus* operational taxonomic units along with a specific

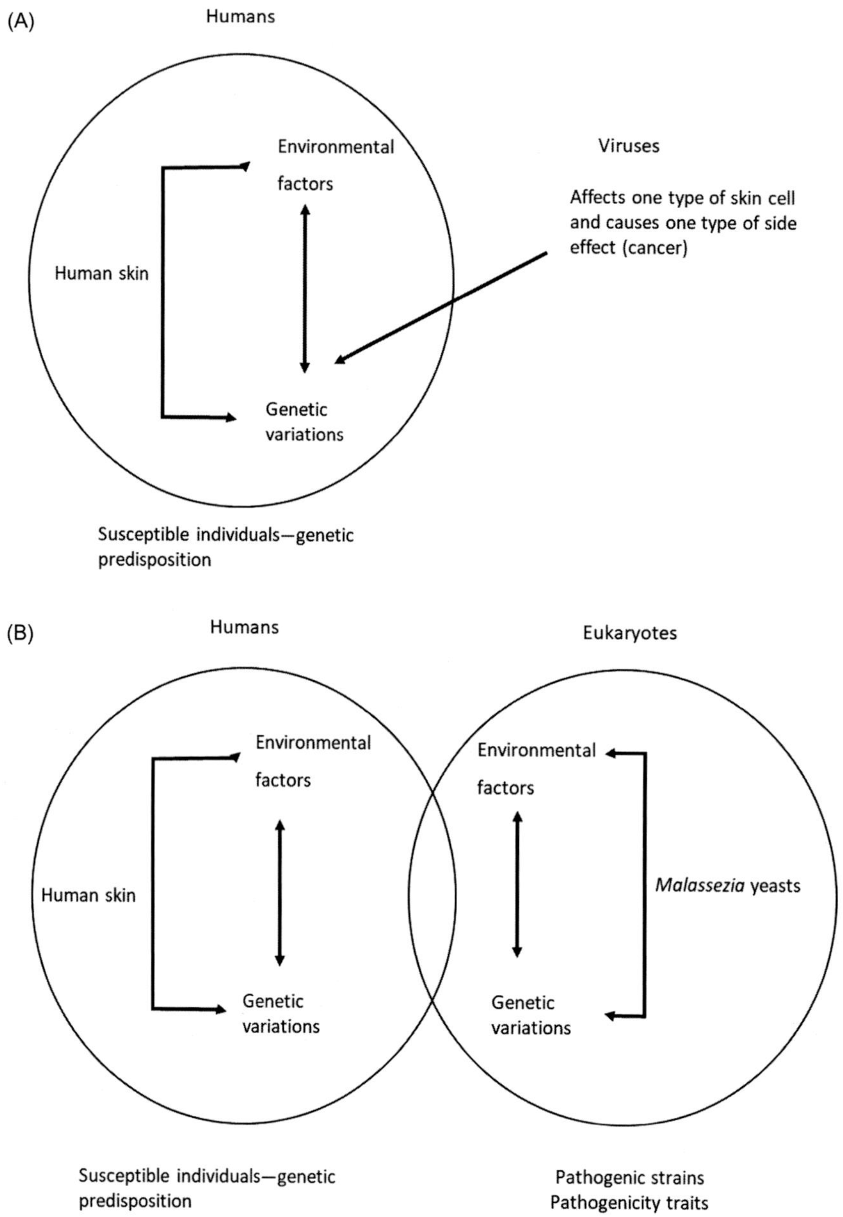

FIGURE 10.1

The impact of environmental factors, viruses, and eukaryotes (*Malassezia*) on the skin carcinogenesis and the relevant complexity in the recognition of the respective participation is depicted. (A) Viruses infect one specific cell type; available technology allows for in-depth evaluation of the alterations caused in susceptible individuals. (B) Evaluation of skin carcinogenesis is more complex when eukaryotes are involved, as multiple parameters participate and are all influenced by the fluctuating environmental factors, the principal one on the skin being ultraviolet radiation.

association of *S. aureus* with AK-to-SCC progression. Interestingly, nonlesional sun-damaged skin was associated with relatively larger populations of the lipophilic *Malassezia* and *Propionibacterium* genera (Wood et al., 2018). The uniqueness of each individual's skin microbiome (Grice and Segre, 2011) as well as the association of *S. aureus* with SCC is thus confirmed (Kullander et al., 2009), but it also raises the question of the participation of the respective microbiomes in the sequence sun-damaged skin—AK appearance—SCC progression as well as the factors that lead to AK regression. The absence of *S. aureus* from BCC (Kullander et al., 2009) and the suggestion of the participation of *Malassezia* yeasts in this tumor progression (Gaitanis et al., 2011) should also be taken into account. In order to further elaborate on these findings, we have to swiftly move from the big picture offered by metagenomic studies to the magnifying glass information provided by the study of individual cell's biology and the respective interactions at the cellular and tissue level.

In analogy to the gut epithelium, these phenomena take place on the polarized skin epithelium, where 1 million to billions of bacteria coexist with epidermal cells and innate and adaptive immune cells within a distance of <0.1 mm. This interaction takes place through the microbial transcriptome that includes both biologically active metabolites and structural components impacting the skin function and possibly modifying cancer initiation and progression. This happens for both the microbes and the skin cells under the omnipresent UV radiation. This is partly demonstrated by the significant alterations recorded in the *Malassezia* flora of astronauts or members of the Japanese Antarctic Research Expedition that have lived for a significant time in extreme weather conditions and restricted local hygiene conditions (Sugita et al., 2015, 2016).

An example of the potential of microbial metabolites to induce skin carcinogenesis is the production of potent indolic ligands of the aryl hydrocarbon receptor (AhR) by *Malassezia* yeasts. AhR is a ligand-activated transcription factor and acts as a sensor of endogenous and environmental cues (Bock, 2018). In healthy skin, it contributes to keratinocyte differentiation, skin barrier function, and skin pigmentation. However, there are differences in the downstream effects of AhR activation depending on the extent of the interaction (Bock, 2018). Well-studied AhR ligands such as dioxin lead to dysregulation of the downstream effects and cause inflammation and carcinogenicity. The genus *Malassezia* currently comprises 17 species (Theelen et al., 2018) and is the primary fungal eukaryote on human skin (Findley et al., 2013). Species that have been isolated from human skin include *M. globosa*, *M. restricta*, *M. furfur*, *M. sympodialis*, *M. slooffiae*, *M. obtusa*, *M. dermatis*, *M. japonica*, and *M. yamatoensis*. This genus is found in diverse environments and demonstrates fascinating host specificity. An example would be the recently described *Malassezia vespertilionis* that was isolated from buts and is a cold-tolerant species with optimal growth at 25°C (Lorch et al., 2018). As already mentioned, *Malassezia* cause in humans the relatively innocuous pityriasis versicolor and are also implicated in the cases of folliculitis (Gaitanis et al., 2012). It also participates in the exacerbations of the head and neck variants of atopic dermatitis through the production of an array of allergenic proteins and also aggravates seborrheic dermatitis (Gaitanis et al., 2013). Strains of *M. furfur* isolated from the latter condition demonstrate in vitro a stable characteristic ability to use L-tryptophan, when the latter is the single nitrogen source, to produce an array of bioactive indolic substances including malassezin, indirubin, indolo-[3,2b]-carbazole, pityriacitrin, and formylindolo[3,2b]-carbazole (Wille et al., 2001; Magiatis et al., 2013; Gaitanis et al., 2008; Mexia et al., 2015). This property is not restricted to *M. furfur* but also exists in other *Malassezia* species, such as *M. globosa* (Magiatis et al., 2013). From what is known from

M. furfur studies, this property is not species-specific but rather strain-specific, underscoring the significance of connecting *Malassezia* population dynamics and the relevant production of these bioactive indoles on the skin. Furthermore, there are exogenous physical factors such as UV radiation and pH that could further skew the preferential synthesis of AhR ligands after the initial deamination of L-tryptophan to indolepyruvate through the action of *M. furfur* aminotransferase 1 (Zuther et al., 2008; Preuss et al., 2013). The biological relevance of these substances on skin biology is underscored by the recent demonstration of the therapeutic potential of indirubin in psoriasis. Indirubin is the active ingredient of Danggui Longhui Wan, a traditional Chinese medicine originating from plants such as *Indigofera tinctoria* L. and *Isatis tinctoria* L. (Blažević et al., 2015). Indirubin has shown impressive efficacy in the treatment of skin (Lin et al., 2018) and nail psoriasis (Lin et al., 2014), the latter being notoriously difficult to treat due to its localization. During these studies the question of indirubin possibly restoring the normal *Malassezia* metabolome on psoriatic skin as well as on the modification of the skin microbiome arose (Gaitanis et al., 2018a,b; Lin and Lee, 2018). Interestingly, there is recent evidence that *Malassezia*-produced indoles have in vitro antifungal properties that can be extrapolated to in vivo concentrations against a variety of yeast and mold pathogens, including strains of the *Malassezia* genus (Gaitanis et al., 2018a,b). The latter suggests that on the skin, these indoles could have autoinhibitory effects to the producing *M. furfur* strains.

In an effort to define the possible associations of the skin mycobiome, which is principally composed by *Malassezia*, with cancer, one should try to investigate their participation in BCC initiation and progression. Significantly, more data attest in favor of the latter, occurring especially through the modification of the tumor microenvironment.

BCC is a tumor that carries a high number of mutations, and mutations activating the hedgehog pathway are present in ~85% of cases (Bonilla et al., 2016). The hedgehog pathway is one of the fundamental signaling pathways, which contributes to epidermal development, homeostasis, and repair, including hair follicle development and follicle bulge stem cell maintenance (Abe and Tanaka, 2017). Activating mutations of this pathway are key molecular events in the development of BCC, and modulation of the expression of this pathway by cancer cells is crucial for the maintenance of residual tumor during treatment with hedgehog inhibitors (Biehs et al., 2018). Mutations in the well-known tumor suppressor gene p53 are very often present as well as novel driver mutations in an array of established tumor-initiating genes (Bonilla et al., 2016; Pellegrini et al., 2017). Moreover, there is evidence that there is a noncanonical activation of the hedgehog pathway under UV radiation in BCC (Lesiak et al., 2018), highlighting the multifaceted action of UV radiation in this tumor. The tumor microenvironment is dominated by a Th2 skewed response with ample T-regulatory lymphocytes CD4 + CD25 + FOXP3 + cells around the tumor (Kaporis et al., 2007; Omland et al., 2016). Furthermore, there is a significant population of immature dendritic cells around the tumor that express IL-4, IL-10, CCL22, and IL-12/-23 (Kaporis et al., 2007). There is a similar immune state deviation in the *Malassezia*-associated seborrheic dermatitis lesions (Faergemann et al., 2001). The significance of these findings in the progression of BCC is underscored by the need to overcome this immunosuppressive environment to achieve tumor clearance in experimental and clinical models (De Giorgi et al., 2009; Urosevic et al., 2003; Bassukas and Gaitanis, 2009).

It should be stressed that available data do not support a role of AhR in BCC initiation within the framework of multistep carcinogenesis, despite the fact that continuous activation of this receptor by ligands such as dioxin definitely supports tumor promotion (IARC, 1997). The only indirect association is that indirubin can effectively downregulate glycogen synthase kinase 3 (Meijer et al., 2003), which could result in the modification of the Gli-binding protein SUFU through phosphorylation (Takenaka et al., 2007). The net effect of hedgehog signaling, that is, potentiation or attenuation, depends on the background activity of this pathway (Takenaka et al., 2007).

AhR activation produces hormetic-like dose–response curves with actual effects depending on the level of activation. Thus limited AhR activation induces proinflammatory immune responses, while stronger and sustained AhR activation results in immune suppression with the expansion of locally present Tregs (Ehrlich et al., 2018). The existence on the skin of a mixture of AhR ligands as part of the *Malassezia* metabolome is expected to result in variable AhR responses and in the modification of the inflammatory milieu in photodamaged skin with relevant functional regression (Vlachos et al., 2012).

In order to stress the effect a mixture of bioactive substances in skin carcinogenesis could have or not, one could use the paradigm of polycyclic aromatic hydrocarbons in dermatology. These have been used for many years in the form of coal tar treatments in psoriasis and atopic dermatitis patients. Yet, despite the fact that they possess significant AhR ligands and meaningful concentrations of established carcinogens, such as benzo-*a*-pyrene, they did not increase the incidence of skin cancers in the aforementioned subsets of patients (Roelofzen et al., 2010). Likewise, slowly metabolized xenobiotic AhR agonists such as dioxin may inhibit cell proliferation, whereas AhR stimulation by carbinol-derivatives promotes cell cycle progression and cell proliferation through interaction with the retinoblastoma protein Rb1 (Barhoover et al., 2010). Furthermore, AhR activation can interfere with cell cycle progression, modulate apoptosis (Elferink, 2003; Marlowe et al., 2008), and increase the expression of matrix metalloproteinase-1 in keratinocytes, the latter being a marker of enhanced tumor invasion (Murphy et al., 2004; Morita et al., 2009). In addition, AhR can impinge on the fate of initiated tumor cells (Schwarz et al., 2000) by inhibiting cell senescence (Esser et al., 2009).

Further data supporting the association of *Malassezia* yeasts with skin carcinogenesis originate from animals. Both *Malassezia* and BCCs coincide in dogs and cats (Rothwell et al., 1987) and are sparse to absent in lagomorphs and rodents (von Bomhard et al., 2007; Schrenk et al., 2004). Furthermore, in dogs and cats these tumors accrue in the head and neck regions overlapping with *Malassezia*'s niche (Rothwell et al., 1987). Among dogs, BCCs are more common in hairy animal breeds with long pendulous ears, as are the Wirehaired Pointing Griffons and the Kerry Blue and Wheaten Terriers (Villalobos, 2016), an anatomical trait that favors *Malassezia* overgrowth. Likewise in humans, BCCs concur in areas that are sun-protected but amply colonized by *Malassezia*, such as the eyelids, the inner canthuses, the retroauricular areas, and the seborrheic areas of the trunk (Heckmann et al., 2002).

MALASSEZIA IN INTERNAL ORGANS AND CANCER

Though *Malassezia* have historically been considered restricted to the skin, many recently published studies have reported their presence in internal organs. Detection of *Malassezia* in internal organs prior to the introduction of high-throughput DNA sequencing was extremely difficult, so the historical view that they are absent from internal organs was an unverified assumption. Recently published studies report that *Malassezia* are often present inside the body, including in the mouth (Abusleme et al., 2018; Dupuy et al., 2014), intestines (Sokol et al., 2016; Suhr et al., 2016; Liguori et al., 2015; Nash et al., 2017; Kellermayer et al., 2012; Richard, 2018; Hamad et al., 2018), pancreas (Aykut et al., 2019), cervix (Godoy-Vitorino et al., 2018), breast (Boix-Amorós et al., 2017), nose (Cleland et al., 2014; Gelber et al., 2016; Jung et al., 2015), central nervous system (Alonso et al., 2014a,b, 2018a,b), knee joint (Hammad and Tonge, 2018), and prostate (Laurence et al., 2018). The main reported species inside the body are the same as those found on the skin: *M. restricta* and *M. globosa.*

It is important to note that *Malassezia*'s presence on the skin raises the possibility that some of these reports might be the result of contamination. Some studies have reported very strong and specific associations between *Malassezia* inside the body and Crohn's disease (Kellermayer et al., 2012; Limon et al., 2019), ulcerative colitis (Richard, 2018), pancreatic cancer (Aykut et al., 2019), cervical cancer (Godoy-Vitorino et al., 2018), MS (Alonso et al., 2018a), and Alzheimer's disease (Alonso et al., 2018b). These studies show that *Malassezia* are likely present in internal organs, because laboratory contamination would have been expected to affect controls as well as cases. They also suggest that *Malassezia* might be directly implicated in these diseases (Laurence et al., 2018; Benito-León and Laurence, 2017, 2018; Limon et al., 2019; Aykut et al., 2019; Godoy-Vitorino et al., 2018).

Prostate cancer studies provide evidence of a fungal etiology in adenocarcinomas (Stott-Miller et al., 2013; Sutcliffe et al., 2014; Laurence et al., 2018). Single nucleotide polymorphism (SNP) rs10993994 is one of the most important genetic risk factors of prostate cancer (Eeles et al., 2008). It is located in the promoter region of gene *MSMB*, which encodes the protein PSP94. The risk allele reduces the amount of PSP94 in the prostate (Xu et al., 2010). In 2010 it was discovered that PSP94 is an extremely potent fungicide (Edström, 2010; Edstrom Hagerwall et al., 2012). This means prostate cancer risk is inversely related to a naturally produced fungicidal protein in the prostate, suggesting a fungal etiology (Sutcliffe et al., 2014).

An infectious cause for prostate cancer has long been suspected because this cancer has clear sexual risk factors (Dennis and Dawson, 2002; Dennis et al., 2009), and because chronic idiopathic prostate inflammation is common in men (Nickel et al., 2001; Korrovits et al., 2008). This inflammation is likely causing oncogenic mutations leading to prostate cancer (De Marzo et al., 2007). These observations suggest that prostate cancer has a sexually acquired infectious etiology, like cervical cancer before it (Strickler and Goedert, 2001). However, decades of research linking prostate cancer to known sexually transmitted infections have yielded only weak associations (Hrbacek et al., 2012). These studies have focused on bacteria and known sexually transmitted infections. *Malassezia* fit in none of these categories, so their presence in the prostate was not tested in these studies.

Chronic idiopathic prostate inflammation is strongly associated with spondyloarthritis symptoms (Laurence et al., 2018). Spondyloarthritis is strongly associated with an immune response against fungi and has clear sexual risk factors (Laurence et al., 2018). *Malassezia* are the prime suspects in spondyloarthritis, due to their involvement in Crohn's disease, ulcerative colitis, and psoriasis—three chronic inflammatory conditions strongly associated with spondyloarthritis (Laurence et al., 2018). These spondyloarthritis studies corroborate the link between a sexually acquired fungus and prostate cancer, and provide preliminary evidence that *Malassezia* might be the culprits in prostate cancer too.

Spondyloarthritis studies strongly suggest that the causative fungus is not restricted to the prostate but is also present in the gut, eyes, skin, cervix, joints, and spine (Laurence et al., 2018). PSP94 is not restricted to the prostate either, and is synthesized in other organs (Weiber et al., 1990; Ohkubo et al., 1995). Interestingly, PSP94 is present in the cell types susceptible to adenocarcinoma (Table 10.1). There is a significant energy cost in producing PSP94, so it is likely serving a similar antifungal purpose in organs other than the prostate—otherwise it would not be produced there. *Malassezia* have been reported in breast milk (Boix-Amorós et al., 2017), suggesting that they might be colonizing the ducts of the breast. Many groups now report that they are present in the gut, so they might be colonizing the epithelial lining of the intestines (Sokol et al., 2016; Suhr et al., 2016; Liguori et al., 2015; Nash et al., 2017; Kellermayer et al., 2012; Richard, 2018; Hamad et al., 2018). *Malassezia* have been singled out in Crohn's disease (Kellermayer et al., 2012; Limon et al., 2019) and ulcerative colitis (Richard, 2018) studies; these two chronic inflammatory diseases of the gut are known to greatly increase colon cancer risk (Canavan et al., 2006; Jess et al., 2012). *Malassezia* have been directly associated with cervical cancer risk (Godoy-Vitorino et al., 2018). Finally, a very recent study reported that *Malassezia* play a key role in pancreatic cancer, and that antifungal drugs are effective in preventing this cancer in mice (Aykut et al., 2019). Interestingly, antifungal drugs are known to be effective in advanced prostate cancer

Table 10.1 PSP94 Is Present in All Cell Types Involved in the Most Common Adenocarcinomas in Adults

Organ	Adenocarcinoma Cell Type	Reference for PSP94's Presence
Prostate	Prostate gland cell	Ohkubo et al. (1995)
Breast	Mammary nonstriated duct cell	Sheth et al. (1993)
	Mammary gland cell	Ohkubo et al. (1995)
Lung	Tracheobronchial mucous cell	Weiber et al. (1990)
	Tracheobronchial goblet cell	Weiber et al. (1990)
	Ciliated respiratory tract cell	Weiber et al. (1990)
Colon	Colorectal mucous cell	Weiber et al. (1990) and Ohkubo et al. (1995)
Uterus	Uterus endometrial cell	Baijal-Gupta et al. (2000) and Teni et al. (1992)
Stomach	Secretory stomach cell	Weiber et al. (1990)
Bladder	Bladder transitional cell	Xuan et al. (1995)
Kidney	Kidney proximal tubule cell	Ohkubo et al. (1995)
	Kidney distal tubule cell	Ohkubo et al. (1995)
	Kidney thin Henle's loop cell	Ohkubo et al. (1995)
	Kidney glomerular epithelial cell	Ohkubo et al. (1995)
Pancreas	Pancreas acinar cell	Weiber et al. (1990) and Ohkubo et al. (1995)
Liver	Liver hepatocyte cell	Ohkubo et al. (1995)
	Liver bile duct cell	Ohkubo et al. (1995)
Esophagus	Esophagus lower mucous cell	Weiber et al. (1990) and Ohkubo et al. (1995)
Cervix	Cervix ciliated cell	Weiber et al. (1990)

in humans (Mahler et al., 1993; Antonarakis et al., 2013), suggesting that they might help PSP94 fend-off *Malassezia* in all adenocarcinoma sites. Thus the possibility of *Malassezia* contributing to many adult-onset cancers by causing chronic inflammation in colonized organs should be considered and studied in depth.

CONCLUSION

Recent advances in high-throughput DNA sequencing have ushered a new era in microbiome research. Fastidious and low-abundance microbes—which were previously undetectable using cell culture and microscopy—can be now be reliably and inexpensively detected using either consensus PCR (16S) or metagenomics. Viruses can also be detected using metagenomics (DNA viruses) or metatranscriptomics (RNA viruses). Metabolic activity of microbial communities can now be measured inexpensively using metatranscriptomics. These new techniques have given researchers a torrent of new data that will hopefully allow strong new links to be established between idiopathic cancer types and microbes. The well-established links between infectious agents and cancers reviewed here suggest that research should focus on microbes and viruses that are highly prevalent and remain in the host for long periods of time—in particular *Malassezia* species.

DISCLOSURES

None.

REFERENCES

Abe, Y., Tanaka, N., 2017. Roles of the hedgehog signaling pathway in epidermal and hair follicle development, homeostasis, and cancer. J. Dev. Biol. 5, 12.

Abusleme, L., Diaz, P.I., Freeman, A.F., Greenwell-Wild, T., Brenchley, L., Desai, J.V., et al., 2018. Human defects in STAT3 promote oral mucosal fungal and bacterial dysbiosis. JCI Insight 3. Available from: https://doi.org/10.1172/jci.insight.122061.

Alam, M., Ratner, D., 2001. Cutaneous squamous-cell carcinoma. N. Engl. J. Med. 344, 975–983.

Albibas, A.A., Rose-Zerilli, M.J., Lai, C., Pengelly, R.J., Lockett, G.A., Theaker, J., et al., 2018. Subclonal evolution of cancer-related gene mutations in p53 immunopositive patches in human skin. J. Invest. Dermatol. 138, 189–198.

Almohmeed, Y.H., Avenell, A., Aucott, L., Vickers, M.A., 2013. Systematic review and meta-analysis of the sero-epidemiological association between Epstein Barr virus and multiple sclerosis. PLoS One 8, e61110.

Alonso, R., Pisa, D., Marina, A.I., Morato, E., Rábano, A., Carrasco, L., 2014a. Fungal infection in patients with Alzheimer's disease. J. Alzheimers Dis. 41, 301–311.

Alonso, R., Pisa, D., Rábano, A., Carrasco, L., 2014b. Alzheimer's disease and disseminated mycoses. Eur. J. Clin. Microbiol. Infect. Dis. 33, 1125–1132.

Alonso, R., Fernández-Fernández, A.M., Pisa, D., Carrasco, L., 2018a. Multiple sclerosis and mixed microbial infections. Direct identification of fungi and bacteria in nervous tissue. Neurobiol. Dis. 117, 42–61.

Alonso, R., Pisa, D., Fernández-Fernández, A.M., Carrasco, L., 2018b. Infection of fungi and bacteria in brain tissue from elderly persons and patients with Alzheimer's disease. Front. Aging Neurosci. 10, 159.

Altekruse, S.F., McGlynn, K.A., Reichman, M.E., 2009. Hepatocellular carcinoma incidence, mortality, and survival trends in the United States from 1975 to 2005. J. Clin. Oncol. 27, 1485.

Antonarakis, E.S., Heath, E.I., Smith, D.C., Rathkopf, D., Blackford, A.L., Danila, D.C., et al., 2013. Repurposing itraconazole as a treatment for advanced prostate cancer: a noncomparative randomized phase II trial in men with metastatic castration-resistant prostate cancer. The oncologist 18 (2), 163–173.

Arbuthnot, P., Kew, M., 2001. Hepatitis B virus and hepatocellular carcinoma. Int. J. Exp. Pathol. 82, 77–100.

Armitage, P., Doll, R., 1954. The age distribution of cancer and a multi-stage theory of carcinogenesis. Br. J. Cancer 8, 1–12.

Ascherio, A., Munger, K.L., 2007. Environmental risk factors for multiple sclerosis. Part I: the role of infection. Ann. Neurol. 61, 288–299.

Ascherio, A., Munger, K.L., 2010. Epstein-Barr virus infection and multiple sclerosis: a review. J. Neuroimmune Pharmacol. 5, 271–277.

Aykut, B., Pushalkar, S., Chen, R., Li, Q., Abengozar, R., Kim, J.I., et al., 2019. The fungal mycobiome promotes pancreatic oncogenesis via activation of MBL. Nature 1–4.

Baijal-Gupta, M., Clarke, M.W., Finkelman, M.A., Mclachlin, C.M., Han, V.K., 2000. Prostatic secretory protein (PSP94) expression in human female reproductive tissues, breast and in endometrial cancer cell lines. J. Endocrinol. 165, 425–433.

Barhoover, M.A., Hall, J.M., Greenlee, W.F., Thomas, R.S., 2010. Aryl hydrocarbon receptor regulates cell cycle progression in human breast cancer cells via a functional interaction with cyclin-dependent kinase 4. Mol. Pharmacol. 77, 195–201.

Bassukas, I.D., Gaitanis, G., 2009. Combination of cryosurgery and topical imiquimod: does timing matter for successful immunocryosurgery? Cryobiology 59, 116–117.

Benito-León, J., Laurence, M., 2017. The role of fungi in the etiology of multiple sclerosis. Front. Neurol. 8, 535.

Benito-León, J., Laurence, M., 2018. *Malassezia* in the central nervous system and multiple sclerosis. Infection 47, 135–136.

Berg, D., Otley, C.C., 2002. Skin cancer in organ transplant recipients: epidemiology, pathogenesis, and management. J. Am. Acad. Dermatol. 47, 1–20.

Biehs, B., Dijkgraaf, G.J.P., Piskol, R., Alicke, B., Boumahdi, S., Peale, F., et al., 2018. A cell identity switch allows residual BCC to survive Hedgehog pathway inhibition. Nature 562, 429–433.

Blažević, T., Heiss, E.H., Atanasov, A.G., Breuss, J.M., Dirsch, V.M., Uhrin, P., 2015. Indirubin and indirubin derivatives for counteracting proliferative diseases. Evid. Based Complement. Alternat. Med. 2015, 654098.

Bock, K.W., 2018. From TCDD-mediated toxicity to searches of physiologic AHR functions. Biochem. Pharmacol. 155, 419–424.

Bockhart, M., 1883. Beitrag zur Aetiologie und Pathologie des Harnröhrentrippers. Vierteljahresschr. Dermatol. Syph. 10, 3–18.

Boix-Amorós, A., Martinez-Costa, C., Querol, A., Collado, M.C., Mira, A., 2017. Multiple approaches detect the presence of fungi in human breast milk samples from healthy mothers. Sci. Rep. 7, 13016.

Bongartz, T., Sutton, A.J., Sweeting, M.J., Buchan, I., Matteson, E.L., Montori, V., 2006. Anti-TNF antibody therapy in rheumatoid arthritis and the risk of serious infections and malignancies: systematic review and meta-analysis of rare harmful effects in randomized controlled trials. JAMA 295, 2275–2285.

Bonilla, X., Parmentier, L., King, B., Bezrukov, F., Kaya, G., Zoete, V., et al., 2016. Genomic analysis identifies new drivers and progression pathways in skin basal cell carcinoma. Nat. Genet. 48, 398.

Brettmann, E.A., De Guzman Strong, C., 2018. Recent evolution of the human skin barrier. Exp. Dermatol. 27, 859–866.

Canavan, C., Abrams, K., Mayberry, J., 2006. Meta-analysis: colorectal and small bowel cancer risk in patients with Crohn's disease. Aliment. Pharmacol. Ther. 23, 1097–1104.

Chen, Y., Sun, J., Yang, Y., Huang, Y., Liu, G., 2016. Malignancy risk of anti-tumor necrosis factor alpha blockers: an overview of systematic reviews and meta-analyses. Clin. Rheumatol. 35, 1–18.

Cleland, E.J., Bassioni, A., Boase, S., Dowd, S., Vreugde, S., Wormald, P.J., 2014. The fungal microbiome in chronic rhinosinusitis: richness, diversity, postoperative changes and patient outcomes. Int. Forum Allergy Rhinol. 4, 259–265. Wiley Online Library.

De Giorgi, V., Salvini, C., Chiarugi, A., Paglierani, M., Maio, V., Nicoletti, P., et al., 2009. In vivo characterization of the inflammatory infiltrate and apoptotic status in imiquimod-treated basal cell carcinoma. Int. J. Dermatol. 48, 312–321.

Delsing, C., Bleeker-Rovers, C., Van De Veerdonk, F., Tol, J., Van Der Meer, J.W., et al., 2012. Association of esophageal candidiasis and squamous cell carcinoma. Med. Mycol. Case Rep. 1, 5–8.

De Marzo, A.M., Platz, E.A., Sutcliffe, S., Xu, J., Gronberg, H., Drake, C.G., et al., 2007. Inflammation in prostate carcinogenesis. Nat. Rev. Cancer 7, 256–269.

Dennis, L.K., Dawson, D.V., 2002. Meta-analysis of measures of sexual activity and prostate cancer. Epidemiology 13, 72–79.

Dennis, L.K., Coughlin, J.A., Mckinnon, B.C., Wells, T.S., Gaydos, C.A., Hamsikova, E., et al., 2009. Sexually transmitted infections and prostate cancer among men in the U.S. military. Cancer Epidemiol. Biomarkers. Prev. 18, 2665–2671.

Dupuy, A.K., David, M.S., Li, L., Heider, T.N., Peterson, J.D., Montano, E.A., et al., 2014. Redefining the human oral mycobiome with improved practices in amplicon-based taxonomy: discovery of *Malassezia* as a prominent commensal. PLoS One 9, e90899.

Dürst, M., Gissmann, L., Ikenberg, H., Zur hausen, H., 1983. A papillomavirus DNA from a cervical carcinoma and its prevalence in cancer biopsy samples from different geographic regions. Proc. Natl. Acad. Sci. U.S.A. 80, 3812–3815.

Edström, A., 2010. Antimicrobial Activity of Human Seminal Plasma and Seminal Plasma Proteins (Ph.D.). Lund University.

Edstrom Hagerwall, A.M., Rydengard, V., Fernlund, P., Morgelin, M., Baumgarten, M., Cole, A.M., et al., 2012. beta-Microseminoprotein endows post coital seminal plasma with potent candidacidal activity by a calcium- and pH-dependent mechanism. PLoS Pathog. 8, e1002625.

Eeles, R.A., Kote-Jarai, Z., Giles, G.G., Olama, A.A., Guy, M., Jugurnauth, S.K., et al., 2008. Multiple newly identified loci associated with prostate cancer susceptibility. Nat. Genet. 40, 316–321.

Ehrlich, A.K., Pennington, J.M., Bisson, W.H., Kolluri, S.K., Kerkvliet, N.I., 2018. TCDD, FICZ, and other high affinity AhR ligands dose-dependently determine the fate of CD4 + T cell differentiation. Toxicol. Sci. 161, 310–320.

Elferink, C.J., 2003. Aryl hydrocarbon receptor-mediated cell cycle control. Prog. Cell Cycle Res. 5, 261–267.

Esser, C., Rannug, A., Stockinger, B., 2009. The aryl hydrocarbon receptor in immunity. Trends Immunol. 30, 447–454.

Evans, A.S., 1976. Causation and disease: the Henle-Koch postulates revisited. Yale J. Biol. Med. 49, 175.

Ewing, J., 1908. Cancer problems. Arch. Intern. Med. I, 175–217.

Faergemann, J., Bergbrant, I.-M., Dohsé, M., Scott, A., Westgate, G., 2001. Seborrhoeic dermatitis and pityrosporum (*Malassezia*) folliculitis: characterization of inflammatory cells and mediators in the skin by immunohistochemistry. Br. J. Dermatol. 144, 549–556.

Feinstone, S.M., Kapikian, A.Z., Purcell, R.H., Alter, H.J., Holland, P.V., 1975. Transfusion-associated hepatitis not due to viral hepatitis type A or B. N. Engl. J. Med. 292, 767–770.

Feng, H., Shuda, M., Chang, Y., Moore, P.S., 2008. Clonal integration of a polyomavirus in human Merkel cell carcinoma. Science 319, 1096–1100.

Findley, K., Oh, J., Yang, J., Conlan, S., Deming, C., Meyer, J.A., et al., 2013. Topographic diversity of fungal and bacterial communities in human skin. Nature 498, 367–370.

Frank, S., 2007. Dynamics of Cancer. Princeton University Press, Princeton, NJ.

Gaitanis, G., Magiatis, P., Stathopoulou, K., Bassukas, I.D., Alexopoulos, E.C., Velegraki, A., et al., 2008. AhR ligands, malassezin, and indolo [3, 2-b] carbazole are selectively produced by *Malassezia furfur* strains isolated from seborrheic dermatitis. J. Invest. Dermatol. 128, 1620–1625.

Gaitanis, G., Velegraki, A., Magiatis, P., Pappas, P., Bassukas, I., 2011. Could *Malassezia* yeasts be implicated in skin carcinogenesis through the production of aryl-hydrocarbon receptor ligands? Med. Hypotheses 77, 47–51.

Gaitanis, G., Magiatis, P., Hantschke, M., Bassukas, I.D., Velegraki, A., 2012. The *Malassezia* genus in skin and systemic diseases. Clin. Microbiol. Rev. 25, 106–141.

Gaitanis, G., Velegraki, A., Mayser, P., Bassukas, I.D., 2013. Skin diseases associated with *Malassezia* yeasts: facts and controversies. Clin. Dermatol. 31, 455–463.

Gaitanis, G., Magiatis, P., Velegraki, A., Bassukas, I.D., 2018a. A traditional Chinese remedy points to a natural skin habitat: indirubin (indigo naturalis) for psoriasis and the *Malassezia* metabolome. Br. J. Dermatol. 179, 800.

Gaitanis, G.M., Magiatis, P., Mexia, N., Melliou, E., Efstratiou, M., Bassukas, I.D., et al., 2018b. Antifungal activity of selected *Malassezia* indolic compounds detected in culture. Mycoses 62, 597–603.

Gelber, J.T., Cope, E.K., Goldberg, A.N., Pletcher, S.D., 2016. Evaluation of *Malassezia* and common fungal pathogens in subtypes of chronic rhinosinusitis. Int. Forum Allergy Rhinol. 6, 950–955. Wiley Online Library.

Godoy-Vitorino, F., Romaguera, J., Zhao, C., Vargas-Robles, D., Ortiz-Morales, G., Vázquez-Sánchez, F., et al., 2018. Cervicovaginal fungi and bacteria associated with cervical intraepithelial neoplasia and high-risk Human Papillomavirus infections in a Hispanic population. Front. Microbiol. 9, 2533.

Govan, V.A., 2008. A novel vaccine for cervical cancer: quadrivalent human papillomavirus (types 6, 11, 16 and 18) recombinant vaccine (Gardasil®). Ther. Clin. Risk. Manag. 4, 65.

Grice, E.A., Segre, J.A., 2011. The skin microbiome. Nat. Rev. Microbiol. 9, 244.

Gross, G.J., Macdonald, N.E., Mackenzie, A.M., 1992. Neonatal rectal colonization with *Malassezia furfur*. Can. J. Infect. Dis. Med. Microbiol. 3, 9–13.

HACC Group, 2001. Gastric cancer and *Helicobacter pylori*: a combined analysis of 12 case control studies nested within prospective cohorts. Gut 49, 347–353.

Guého-Kellermann, E., Boekhout, T., Begerow, D., 2010. Biodiversity, phylogeny and ultrastructure. Malassezia and the Skin. Springer.

Gupta, A., Bluhm, R., 2004. Seborrheic dermatitis. J. Eur. Acad. Dermatol. Venereol. 18, 13–26.

Gupta, A.K., Batra, R., Bluhm, R., Boekhout, T., Dawson jr, T.L., 2004. Skin diseases associated with *Malassezia* species. J. Am. Acad. Dermatol. 51, 785–798.

Haberal, M., Abali, A.E.S., Karakayali, H., 2010. Fluid management in major burn injuries. Indian J. Plast. Surg. 43, S29.

Hamad, I., Abdallah, R.A., Ravaux, I., Mokhtari, S., Tissot-Dupont, H., Michelle, C., et al., 2018. Metabarcoding analysis of eukaryotic microbiota in the gut of HIV-infected patients. PLoS One 13, e0191913.

Hammad, D.B., Tonge, D.P., 2018. Molecular characterisation of the synovial fluid microbiome. bioRxiv .

Harper, D.M., Demars, L.R., 2017. HPV vaccines–a review of the first decade. Gynecol. Oncol. 146, 196–204.

Harwood, C., Toland, A., Proby, C., Euvrard, S., Hofbauer, G., Tommasino, M., et al., 2017. The pathogenesis of cutaneous squamous cell carcinoma in organ transplant recipients. Br. J. Dermatol. 177, 1217–1224.

Heckmann, M., Zogelmeier, F., Konz, B., 2002. Frequency of facial basal cell carcinoma does not correlate with site-specific UV exposure. Arch. Dermatol. 138, 1494–1497.

Hjalgrim, H., Friborg, J., Melbye, M., 2007. The epidemiology of EBV and its association with malignant disease. In: Arvin, A., Campadelli-Fiume, G., Mocarski, E. (Eds.), *Human Herpesviruses: Biology, Therapy, and Immunoprophylaxis*. Cambridge University Press, Cambridge.

Hrbacek, J., Urban, M., Hamsikova, E., Tachezy, R., Heracek, J., 2012. Thirty years of research on infection and prostate cancer: no conclusive evidence for a link. A systematic review. Urologic Oncology: Seminars and Original Investigations. Elsevier.

IARC, 1997. IARC Working Group on the evaluation of carcinogenic risks to humans: polychlorinated dibenzo-*para*-dioxins and polychlorinated dibenzofurans. Lyon, France, 4-11 February 1997. IARC Monogr. Eval. Carcinog. Risks Hum. 69, 1–631.

Jess, T., Rungoe, C., Peyrin–Biroulet, L., 2012. Risk of colorectal cancer in patients with ulcerative colitis: a meta-analysis of population-based cohort studies. Clin. Gastroenterol. Hepatol. 10, 639–645.

Jo, J.-H., Deming, C., Kennedy, E.A., Conlan, S., Polley, E.C., Ng, W.-L., et al., 2016. Diverse human skin fungal communities in children converge in adulthood. J. Invest. Dermatol. 136, 2356–2363.

Jung, W.H., Croll, D., Cho, J.H., Kim, Y.R., Lee, Y.W., 2015. Analysis of the nasal vestibule mycobiome in patients with allergic rhinitis. Mycoses 58, 167–172.

Kanda, N., Tani, K., Enomoto, U., Nakai, K., Watanabe, S., 2002. The skin fungus-induced Th1-and Th2-related cytokine, chemokine and prostaglandin E2 production in peripheral blood mononuclear cells from patients with atopic dermatitis and psoriasis vulgaris. Clin. Exp. Allergy 32, 1243–1250.

Kaporis, H.G., Guttman-Yassky, E., Lowes, M.A., Haider, A.S., Fuentes-Duculan, J., Darabi, K., et al., 2007. Human basal cell carcinoma is associated with Foxp3 + T cells in a Th2 dominant microenvironment. J. Invest. Dermatol. 127, 2391–2398.

Katano, H., 2018. Pathological features of Kaposi's sarcoma-associated herpesvirus infection. Human Herpesviruses. Springer.

Kaushik, S.B., Lebwohl, M.G., 2019. Psoriasis: which therapy for which patient: psoriasis comorbidities and preferred systemic agents. J. Am. Acad. Dermatol. 80, 27–40.

Kellermayer, R., Mir, S.A., Nagy-Szakal, D., Cox, S.B., Dowd, S.E., Kaplan, J.L., et al., 2012. Microbiota separation and C-reactive protein elevation in treatment naïve pediatric granulomatous Crohn disease. J. Pediatr. Gastroenterol. Nutr. 55, 243–250.

Koch, R., 1884. The etiology of tuberculosis. Mittheilungen aus dem Kaiserlichen Gesundheitsamte 2, 1–88.

Korrovits, P., Ausmees, K., Mandar, R., Punab, M., 2008. Prevalence of asymptomatic inflammatory (National Institutes of Health Category IV) prostatitis in young men according to semen analysis. Urology 71, 1010–1015.

Kullander, J., Forslund, O., Dillner, J., 2009. *Staphylococcus aureus* and squamous cell carcinoma of the skin. Cancer Epidemiol. Biomarkers Prev. 18, 472–478.

Kusters, J.G., Van Vliet, A.H., Kuipers, E.J., 2006. Pathogenesis of *Helicobacter pylori* infection. Clin. Microbiol. Rev. 19, 449–490.

Laurence, M., 2018. PSP94, What Is It Good For? Shipshaw Labs, Seattle, WA.

Laurence, M., Asquith, M., Rosenbaum, J.T., 2018. Spondyloarthritis, acute anterior uveitis and fungi: updating the Catterall-King hypothesis. Front. Med. 5, 80.

Lesiak, A., Sobolewska-Sztychny, D., Bednarski, I.A., Wódz, K., Sobjanek, M., Woźniacka, A., et al., 2018. Alternative activation of hedgehog pathway induced by ultraviolet B radiation: preliminary study. Clin. Exp. Dermatol. 43, 518–524.

Liguori, G., Lamas, B., Richard, M.L., Brandi, G., Da Costa, G., Hoffmann, T.W., et al., 2015. Fungal dysbiosis in mucosa-associated microbiota of Crohn's disease patients. J. Crohns Colitis 10, 296–305.

Limon, J.J., Tang, J., Li, D., Wolf, A.J., Michelsen, K.S., Funari, V., et al., 2019. Malassezia is associated with Crohn's disease and exacerbates colitis in mouse models. Cell Host Microbe 25 (3), 377–388.

Lin, Y.K., Lee, B.H., 2018. Does Lindioil (indirubin) treatment affect the composition of *Malassezia* species on psoriatic skin? Br. J. Dermatol. 179, 801.

Lin, Y.-K., See, L.-C., Huang, Y.-H., Chang, Y.-C., Tsou, T.-C., Lin, T.-Y., et al., 2014. Efficacy and safety of Indigo naturalis extract in oil (Lindioil) in treating nail psoriasis: a randomized, observer-blind, vehicle-controlled trial. Phytomedicine 21, 1015–1020.

Lin, Y.-K., See, L.-C., Huang, Y.-H., Chi, C.-C., Hui, R.-Y., 2018. Comparison of indirubin concentrations in indigo naturalis ointment for psoriasis treatment: a randomized, double-blind, dosage-controlled trial. Br. J. Dermatol. 178, 124–131.

Lorch, J.M., Palmer, J.M., Vanderwolf, K.J., Schmidt, K.Z., Verant, M.L., Weller, T.J., et al., 2018. *Malassezia vespertilionis* sp. nov.: a new cold-tolerant species of yeast isolated from bats. Persoonia 41, 56–70. 41, 56–70.

Mackie, R., Quinn, A., 2004. Nonmelanoma skin cancer and other epidermal skin tumors (basal cell carcinoma). In: Rook, A., Burns, T. (Eds.), Rook's Textbook of Dermatology, seventh ed. Blackwell Science, Malden, MA, p. 36.

Magiatis, P., Pappas, P., Gaitanis, G., Mexia, N., Melliou, E., Galanou, M., et al., 2013. *Malassezia* yeasts produce a collection of exceptionally potent activators of the Ah (dioxin) receptor detected in diseased human skin. J. Invest. Dermatol. 133, 2023–2030.

Mahler, C., Verhelst, J., Denis, L., 1993. Ketoconazole and liarozole in the treatment of advanced prostatic cancer. Cancer 71 (S3), 1068–1073.

Marlowe, J.L., Fan, Y., Chang, X., Peng, L., Knudsen, E.S., Xia, Y., et al., 2008. The aryl hydrocarbon receptor binds to E2F1 and Inhibits E2F1-induced apoptosis. Mol. Biol. Cell. 19, 3263–3271.

Marshall, B.J., Armstrong, J.A., Mcgechie, D.B., Glancy, R.J., 1985a. Attempt to fulfil Koch's postulates for pyloric *Campylobacter*. Med. J. Aust. 142, 436–439.

Marshall, B.J., Mcgechie, D., Rogers, P., Glancy, R., 1985b. Pyloric Campylobacter infection and gastroduodenal disease. Med. J. Aust. 142, 439–444.

Meijer, L., Skaltsounis, A.-L., Magiatis, P., Polychronopoulos, P., Knockaert, M., Leost, M., et al., 2003. GSK-3-selective inhibitors derived from Tyrian purple indirubins. Chem. Biol. 10, 1255–1266.

Mesri, E.A., Cesarman, E., Boshoff, C., 2010. Kaposi's sarcoma and its associated herpesvirus. Nat. Rev. Cancer 10, 707.

Mexia, N., Gaitanis, G., Velegraki, A., Soshilov, A., Denison, M.S., Magiatis, P., 2015. Pityriazepin and other potent AhR ligands isolated from *Malassezia furfur* yeast. Arch. Biochem. Biophys. 571, 16–20.

Monie, A., Hung, C.-F., Roden, R., Wu, T.C., 2008. Cervarix™: a vaccine for the prevention of HPV 16, 18-associated cervical cancer. Biologics 2, 107.

Morita, A., Torii, K., Maeda, A., Yamaguchi, Y., 2009. Molecular basis of tobacco smoke-induced premature skin aging. J. Investig. Dermatol. Symp. Proc. 14, 53–55.

Muñoz, N., Bosch, F.X., De Sanjosé, S., Herrero, R., Castellsagué, X., Shah, K.V., et al., 2003. Epidemiologic classification of human papillomavirus types associated with cervical cancer. N. Engl. J. Med. 348, 518–527.

Murphy, K.A., Villano, C.M., Dorn, R., White, L.A., 2004. Interaction between the aryl hydrocarbon receptor and retinoic acid pathways increases matrix metalloproteinase-1 expression in keratinocytes. J. Biol. Chem. 279, 25284–25293.

Nagata, R., Nagano, H., Ogishima, D., Nakamura, Y., Hiruma, M., Sugita, T., 2012. Transmission of the major skin microbiota, *Malassezia*, from mother to neonate. Pediatr. Int. 54, 350–355.

Nash, A.K., Auchtung, T.A., Wong, M.C., Smith, D.P., Gesell, J.R., Ross, M.C., et al., 2017. The gut mycobiome of the Human Microbiome Project healthy cohort. Microbiome 5, 153.

Nickel, J.C., Downey, J., Hunter, D., Clark, J., 2001. Prevalence of prostatitis-like symptoms in a population based study using the National Institutes of Health chronic prostatitis symptom index. J. Urol. 165, 842–845.

Nomura, A., Stemmermann, G.N., Chyou, P.-H., Kato, I., Perez-Perez, G.I., Blaser, M.J., 1991. *Helicobacter pylori* infection and gastric carcinoma among Japanese Americans in Hawaii. N. Engl. J. Med. 325, 1132–1136.

Nordling, C., 1953. A new theory on cancer-inducing mechanism. Br. J. Cancer 7, 68–72.

Ohkubo, I., Tada, T., Ochiai, Y., Ueyama, H., Eimoto, T., Sasaki, M., 1995. Human seminal plasma beta-microseminoprotein: its purification, characterization, and immunohistochemical localization. Int. J. Biochem. Cell. Biol. 27, 603–611.

Omland, S.H., Nielsen, P.S., Gjerdrum, L.M., Gniadecki, R., 2016. Immunosuppressive environment in basal cell carcinoma: the role of regulatory T cells. Acta Derm. Venereol. 96, 917–921.

Orrell, K.A., Murphrey, M., Kelm, R.C., Lee, H.H., Pease, D.R., Laumann, A.E., et al., 2018. Inflammatory bowel disease events after exposure to the IL-17 inhibitors, secukinumab and ixekizumab: a post-marketing analysis from the RADAR (Research on Adverse Drug events And Reports) Program. J. Am. Acad. Dermatol. 79, 777–778.

Pakpoor, J., Disanto, G., Gerber, J.E., Dobson, R., Meier, U.C., Giovannoni, G., et al., 2012. The risk of developing multiple sclerosis in individuals seronegative for Epstein-Barr virus: a meta-analysis. Mult. Scler. 19, 162–166. 1352458512449682.

Pellegrini, C., Maturo, M., Di nardo, L., Ciciarelli, V., Gutiérrez García-Rodrigo, C., Fargnoli, M., 2017. Understanding the molecular genetics of basal cell carcinoma. Int. J. Mol. Sci. 18, 2485.

Perz, J.F., Armstrong, G.L., Farrington, L.A., Hutin, Y.J., Bell, B.P., 2006. The contributions of hepatitis B virus and hepatitis C virus infections to cirrhosis and primary liver cancer worldwide. J. Hepatol. 45, 529–538.

Plummer, M., Franceschi, S., Vignat, J., Forman, D., De martel, C., 2015. Global burden of gastric cancer attributable to *Helicobacter pylori*. Int. J. Cancer 136, 487–490.

Pollock, K.G., Kavanagh, K., Potts, A., Love, J., Cuschieri, K., Cubie, H., et al., 2014. Reduction of low-and high-grade cervical abnormalities associated with high uptake of the HPV bivalent vaccine in Scotland. Br. J. Cancer 111, 1824.

Preuss, J., Hort, W., Lang, S., Netsch, A., Rahlfs, S., Lochnit, G., et al., 2013. Characterization of tryptophan aminotransferase 1 of *Malassezia furfur*, the key enzyme in the production of indolic compounds by *M. furfur*. Exp. Dermatol. 22, 736–741.

Rautemaa, R., Hietanen, J., Niissalo, S., Pirinen, S., Perheentupa, J., 2007. Oral and oesophageal squamous cell carcinoma–a complication or component of autoimmune polyendocrinopathy-candidiasis-ectodermal dystrophy (APECED, APS-I). Oral. Oncol. 43, 607–613.

Redline, R.W., Redline, S.S., Boxerbaum, B., Dahms, B.B., 1985. Systemic *Malassezia furfur* infections in patients receiving intralipid therapy. Hum. Pathol. 16, 815–822.

Richard, M., 2018. *Malassezia* in the human gut: friend or foe or artefact? ISHAM 2018 - Malassezia Workshop. Westerdijk Fungal Biodiversity Institute, Utrecht.

Robin Warren, J., Marshall, B., 1983. Unidentified curved bacilli on gastric epithelium in active chronic gastritis. Lancet 321, 1273–1275.

Rochford, R., Cannon, M.J., Moormann, A.M., 2005. Endemic Burkitt's lymphoma: a polymicrobial disease? Nat. Rev. Microbiol. 3, 182.

Roelofzen, J.H.J., Aben, K.K.H., Oldenhof, U.T.H., Coenraads, P.-J., Alkemade, H.A., Van De kerkhof, P.C. M., et al., 2010. No increased risk of cancer after coal tar treatment in patients with psoriasis or eczema. J. Invest. Dermatol. 130, 953–961.

Rokkas, T., Rokka, A., Portincasa, P., 2017. A systematic review and meta-analysis of the role of *Helicobacter pylori* eradication in preventing gastric cancer. Ann. Gastroenterol. 30, 414.

Ronat, J.-B., Kakol, J., Khoury, M.N., Berthelot, M., Yun, O., Brown, V., et al., 2014. Highly drug-resistant pathogens implicated in burn-associated bacteremia in an Iraqi burn care unit. PLoS One 9, e101017.

Rothwell, T.L.W., Howlett, C.R., Middleton, D.J., Griffiths, D.A., Duff, B.C., 1987. Skin neoplasms of dogs in Sydney. Aust. Vet. J. 64, 161–164.

Rous, P., 1911. A sarcoma of the fowl transmissible by an agent separable from the tumor cells. J. Exp. Med. 13, 397–411.

Saunte, D., Mrowietz, U., Puig, L., Zachariae, C., 2017. Candida infections in patients with psoriasis and psoriatic arthritis treated with interleukin-17 inhibitors and their practical management. Br. J. Dermatol. 177, 47–62.

Schaudinn, F., Hoffmann, E., 1905. Über Spirochaetenbefunde im Lymphdrüsensaft Syphilitischer. DMW Dtsch. Med. Wochenschr. 31, 711–714.

Schrenk, D., Schmitz, H.J., Bohnenberger, S., Wagner, B., Wörner, W., 2004. Tumor promoters as inhibitors of apoptosis in rat hepatocytes. Toxicol. Lett. 149, 43–50.

Schwarz, M., Buchmann, A., Stinchcombe, S., Kalkuhl, A., Bock, K.-W., 2000. Ah receptor ligands and tumor promotion: survival of neoplastic cells. Toxicol. Lett. 112-113, 69–77.

Shek, Y.H., Tucker, M.C., Viciana, A.L., Manz, H.J., Connor, D.H., 1989. *Malassezia furfur*—disseminated infection in premature infants. Am. J. Clin. Pathol. 92, 595–603.

Sheth, A.R., Chinoy, R.F., Garde, S.V., Panchal, C.J., Sheth, N.A., 1993. Immunoperoxidase localization and denovo biosynthesis of a 10.5-kDa inhibin in benign and malignant conditions of human breast. Cancer Lett. 72, 127–134.

Shuster, S., 1984. The aetiology of dandruff and the mode of action of therapeutic agents. Br. J. Dermatol. 111, 235–242.

Sokol, H., Leducq, V., Aschard, H., Pham, H.-P., Jegou, S., Landman, C., et al., 2016. Fungal microbiota dysbiosis in IBD. Gut 66, 1039–1048.

Stott-Miller, M., Wright, J.L., Stanford, J.L., 2013. MSMB gene variant alters the association between prostate cancer and number of sexual partners. Prostate 73, 1803–1809.

Strickler, H.D., Goedert, J.J., 2001. Sexual behavior and evidence for an infectious cause of prostate cancer. Epidemiol. Rev. 23, 144–151.

Sugita, T., Yamazaki, T., Yamada, S., Takeoka, H., Cho, O., Tanaka, T., et al., 2015. Temporal changes in the skin *Malassezia* microbiota of members of the Japanese Antarctic Research Expedition (JARE): a case study in Antarctica as a pseudo-space environment. Med. Mycol. 53, 717–724.

Sugita, T., Yamazaki, T., Makimura, K., Cho, O., Yamada, S., Ohshima, H., et al., 2016. Comprehensive analysis of the skin fungal microbiota of astronauts during a half-year stay at the International Space Station. Sabouraudia 54, 232–239.

Suhr, M.J., Banjara, N., Hallen-Adams, H.E., 2016. Sequence-based methods for detecting and evaluating the human gut mycobiome. Lett. Appl. Microbiol. 62, 209–215.

Sutcliffe, S., De marzo, A.M., Sfanos, K.S., Laurence, M., 2014. MSMB variation and prostate cancer risk: clues towards a possible fungal etiology. Prostate 74, 569–578.

Szeimies, R., Torezan, L., Niwa, A., Valente, N., Unger, P., Kohl, E., et al., 2012. Clinical, histopathological and immunohistochemical assessment of human skin field cancerization before and after photodynamic therapy. Br. J. Dermatol. 167, 150–159.

Takenaka, K., Kise, Y., Miki, H., 2007. GSK3β positively regulates Hedgehog signaling through Sufu in mammalian cells. Biochem. Biophys. Res. Commun. 353, 501–508.

Teni, T.R., Sampat, M.B., Sheth, N.A., 1992. Inhibin (10.7 kD prostatic peptide) in normal, hyperplastic, and malignant human endometria: an immunohistochemical study. J. Pathol. 168, 35–40.

Theelen, B., Cafarchia, C., Gaitanis, G., Bassukas, I.D., Boekhout, T., Dawson jr, T.L., 2018. *Malassezia* ecology, pathophysiology, and treatment. Med. Mycol. 56, 10–25.

Thieu, K., Ruiz, M.E., Owens, D.M., 2013. Cells of origin and tumor-initiating cells for nonmelanoma skin cancers. Cancer Lett. 338, 82–88.

Urosevic, M., Maier, T., Benninghoff, B., Slade, H., Burg, G., Dummer, R., 2003. Mechanisms underlying imiquimod-induced regression of basal cell carcinoma in vivo. Arch. Dermatol. 139, 1325–1332.

Villalobos, A., 2016. Tumors of the skin in dogs, The Merck Veterinary Manual, eleventh ed. Merck.

Vlachos, C., Schulte, B.M., Magiatis, P., Adema, G.J., Gaitanis, G., 2012. *Malassezia*-derived indoles activate the aryl hydrocarbon receptor and inhibit Toll-like receptor-induced maturation in monocyte-derived dendritic cells. Br. J. Dermatol. 167, 496–505.

von Bomhard, W., Goldschmidt, M.H., Shofer, F.S., Perl, L., Rosenthal, K.L., Mauldin, E.A., 2007. Cutaneous neoplasms in pet rabbits: a retrospective study. Vet. Pathol. 44, 579–588.

Vos, M., Plasmeijer, E.I., Van bemmel, B.C., Van Der bij, W., Klaver, N.S., Erasmus, M.E., et al., 2018. Azathioprine to mycophenolate mofetil transition and risk of squamous cell carcinoma after lung transplantation. J. Heart Lung Transplant. 37, 853–859.

Weiber, H., Andersson, C., Murne, A., Rannevik, G., Lindstrom, C., Lilja, H., et al., 1990. Beta microseminoprotein is not a prostate-specific protein. Its identification in mucous glands and secretions. Am. J. Pathol. 137, 593–603.

Weiner, A., Kuo, G., Lee, C., Rosenblatt, J., Choo, Q., Houghton, M., et al., 1990. Detection of hepatitis C viral sequences in non-A, non-B hepatitis. Lancet 335, 1–3.

Wille, G., Mayser, P., Thoma, W., Monsees, T., Baumgart, A., Schmitz, H.-J., et al., 2001. Malassezin—a novel agonist of the arylhydrocarbon receptor from the yeast *Malassezia furfur*. Bioorg. Med. Chem. 9, 955–960.

Williams, R., 1908. The microbic theory of cancer. The Natural History of Cancer. William Wood and Company, New York.

Wood, D.L., Lachner, N., Tan, J.-M., Tang, S., Angel, N., Laino, A., et al., 2018. A natural history of actinic keratosis and cutaneous squamous cell carcinoma microbiomes. mBio 9, e01432–18.

Xu, X., Valtonen-Andre, C., Savblom, C., Hallden, C., Lilja, H., Klein, R.J., 2010. Polymorphisms at the microseminoprotein-beta locus associated with physiologic variation in beta-microseminoprotein and prostate-specific antigen levels. Cancer Epidemiol. Biomarkers. Prev. 19, 2035–2042.

Xuan, J.W., Chin, J.L., Guo, Y., Chambers, A.F., Finkelman, M.A., Clarke, M.W., 1995. Alternative splicing of PSP94 (prostatic secretory protein of 94 amino acids) mRNA in prostate tissue. Oncogene 11, 1041–1047.

Zuther, K., Mayser, P., Hettwer, U., Wu, W., Spiteller, P., Kindler, B.L., et al., 2008. The tryptophan aminotransferase Tam1 catalyses the single biosynthetic step for tryptophan-dependent pigment synthesis in *Ustilago maydis*. Mol. Microbiol. 68, 152–172.

Zwald, F.O.R., Brown, M., 2011. Skin cancer in solid organ transplant recipients: advances in therapy and management: part II. Management of skin cancer in solid organ transplant recipients. J. Am. Acad. Dermatol. 65, 263–279.

CHAPTER

A PREREQUISITE FOR HEALTH: PROBIOTICS

11

Rodnei Dennis Rossoni[1], Felipe de Camargo Ribeiro[1], Patrícia Pimentel de Barros[1], Eleftherios Mylonakis[2] and Juliana Campos Junqueira[1]

[1]Department of Biosciences and Oral Diagnosis, Institute of Science and Technology, São Paulo State University (UNESP), São José dos Campos, São Paulo, Brazil [2]Infectious Diseases Division, Alpert Medical School & Brown University, Providence, RI, United States

INTRODUCTION: DEFINITIONS AND TERMINOLOGY

Probiotics are living microorganisms that can confer health benefits when administered in adequate doses (Guarner et al., 2012). The most commonly used microorganisms are *Lactobacillus* spp. and *Bifidobacterium* spp., and many strains of these genders have showed benefits to human health in controlled studies performed around the world (Hill et al., 2014; Sanders et al., 2018). Moreover, other isolates of Gram-positive bacteria (such as *Enterococcus faecium*, *Streptococcus thermophilus*, *Propionibacterium acidipropionici*, and *Bacillus* spp.), Gram-negative bacteria (such as *Escherichia coli*), and yeasts (*Saccharomyces* spp.) have been commonly used as probiotics in commercial products (Sanchez et al., 2017; Sanders et al., 2018). These probiotics have different functions in the host and they may result in numerous advantages for the human health. Their potential effects include suppression of pathogen colonization, reduction of gastrointestinal infections, control of serum cholesterol, stimulation of immune system, improvement of lactose digestion, increase in the synthesis of vitamins, and anticarcinogenic activity (Guarner et al., 2012; Schachtsiek et al., 2004; Galdeano et al., 2007; Oelschlaeger, 2010; Markowiak and Slizewska, 2017).

Prebiotics are ingredients and substances that can promote the growth of specific bacteria in the gut. A more recent definition was established during the 2017 Meeting of the International Scientific Association of Probiotics and Prebiotics, determining prebiotics as substrates "selectively utilized by host microorganisms that confer a health benefit" (Gibson et al., 2017). The substances recognized for their propensity to influence gastrointestinal health comprise certain nondigestible oligosaccharides (NDOs), soluble fermentable fibers, and human milk oligosaccharides (Delcour et al., 2016; De Leoz et al., 2015). NDOs are low-molecular-weight carbohydrates intermediate between simple sugars and polysaccharides such as fructooligosaccharides and galactooligosaccharides (Charbonneau et al., 2016; Vandeputte et al., 2017; Cerdo et al., 2017). Prebiotics usually enhance the activity of *Lactobacillus* spp. and *Bifidobacterium* spp., but they can also be metabolized by other beneficial bacteria, including *Roseburia* spp., *Eubacterium* spp., or *Faecalibacterium* spp. (Boets et al., 2017; Canani et al., 2011; Gibson et al., 2017). The presence of prebiotics (inulin or oligofructose) in the diet may lead to numerous health benefits, such as the reduction of the

Microbiomics. DOI: https://doi.org/10.1016/B978-0-12-816664-2.00011-6

blood low-density lipoprotein level, stimulation of the immunological system, increase in calcium absorbability, maintenance of intestinal pH value, inhibition of the pathobiome of the gut, and regulation of the vaginal mycobiome to subpathogenic levels (Charbonneau et al., 2016; Vandeputte et al., 2017; Cerdo et al., 2017; Gibson et al., 2017).

The combination of *probiotics* and *prebiotics* is defined as *synbiotics*. The stimulation of *probiotics* with *prebiotics* results in the modulation of metabolic activity with the maintenance of the intestinal biostructure, development of beneficial microbiota, and inhibition of potential pathogens present in the gastrointestinal tract (GIT) (De Vrese et al., 2001; Markowiak and Slizewska, 2017). Therefore *synbiotics* have both *probiotic* and *prebiotic* properties, and an appropriate combination of these components in a single product should ensure a greater activity compared to the use of either alone (Markowiak and Slizewska, 2017).

In the last years, new terms such as *paraprobiotics* and *postbiotics* have emerged based on the concept that bacterial viability is not essential for probiotics to exert their benefits and effects on human health (Taverniti and Guglielmetti, 2011; Tsilingiri and Rescigno, 2013; Aguilar-Toalá et al., 2018). *Paraprobiotics*, also known as "inactivated probiotics," refer to nonviable microbial cells, that when administered in sufficient amounts, confer benefits to consumers (Taverniti and Guglielmetti, 2011; Tsilingiri and Rescigno, 2013; Aguilar-Toalá et al., 2018). Bacterial cell inactivation may be achieved by physical (mechanical disruption, heat treatment, UV irradiation, freeze–drying, or sonication) or chemical (acid deactivation) methods, which are capable of affecting the microbial cell structures or their physiological functions (Aguilar-Toalá et al., 2018). *Paraprobiotics* retain their immunomodulatory activity beyond their cell viability (Kanauchi et al., 2019). The major role of *paraprobiotics* on the modulation of host immune response seems to be associated with the structure components of the dead cells, mainly the cell wall constituents. Nonviable *probiotic* cells have advantages over live *probiotics*: the most important is the reduced risk of microbial translocation, potentially leading to infection or enhanced inflammatory responses in individuals with compromised immune system. Besides being safer, *paraprobiotics* provide more stable products with longer shelf lives compared to *probiotics* (Taverniti and Guglielmetti, 2011; Nakamura et al., 2016).

Postbiotics, also known as metabiotics, biogenics, or simply metabolites/CFSs (cell-free supernatants), refer to soluble factors secreted by live bacteria or released after bacterial lysis (Cicenia et al., 2014; Konstantinov et al., 2013; Tsilingiri and Rescigno, 2013; Aguilar-Toalá et al., 2018). These bioproducts include short-chain fatty acids (SCFAs), enzymes, antimicrobial peptides (AMPs), teichoic acids, peptidoglycan-derived muropeptides, endo- and exopolysaccharides, cell surface proteins, vitamins, plasmalogens, and organic acids (Konstantinov et al., 2013; Oberg et al., 2018; Tsilingiri and Rescigno, 2013). *Postbiotics* have drawn attention because of their clear chemical structure, safe dose parameters, long shelf life, and presence of various molecules with antimicrobial, antiinflammatory, immunomodulatory, antiobesogenic, antihypertensive, hypocholesterolemic, antiproliferative, and antioxidant activities. These properties indicate that *postbiotics* may contribute to the improvement of host health by improving specific physiological functions, although the exact mechanisms have not been entirely elucidated (Taverniti and Guglielmetti, 2011; Tsilingiri and Rescigno, 2013; Aguilar-Toalá et al., 2018).

Since *probiotics*, *prebiotics*, *synbiotics*, *paraprobiotics*, and *postbiotics* have local and systemic effects on the host's microbiome and immune system (Fig. 11.1), the purpose of this chapter is to describe their action mechanisms and clinical applications in the control of human infectious diseases.

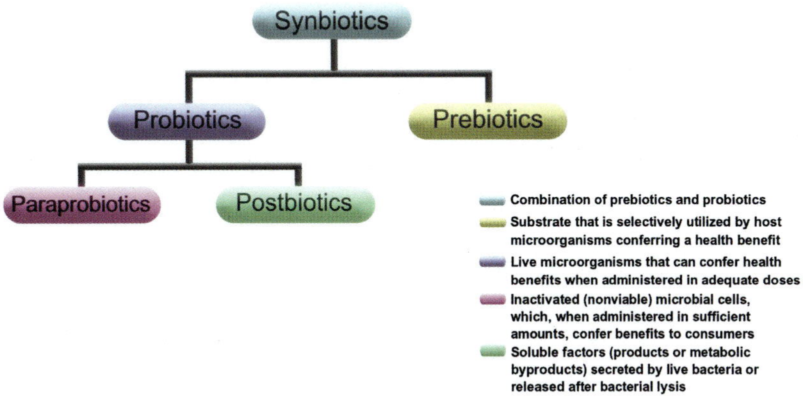

FIGURE 11.1

Schematic representation of terminology referring to probiotics.

MECHANISMS OF ACTION OF PROBIOTICS AGAINST PATHOGENS

There are different products and supplements containing probiotic strains currently on the market, and a common question is, how do probiotics work? There is no simple answer. Unlike most pharmacological agents that have their mechanisms of action (MoA) previously established by animal and human assays, the probiotics act against pathogens by diverse, complex, and heterogeneous pathways not yet elucidated (Reid, 2016). The MoA of some probiotics have been widely investigated and seem to have a focal role in the control of infectious diseases, including the competitive exclusion of pathogens by blocking binding sites, the production of bioactive compounds, and the modulation of the immune system.

COMPETITIVE EXCLUSION OF PATHOGENS BY BLOCKING BINDING SITES

The first step for a probiotic strain to exert its beneficial effects is associated with the capacity of adherence and colonization in host tissues (Fuochi et al., 2015; Grompone et al., 2012; Munoz-Quezada et al., 2013; Zhang et al., 2018). This action mechanism involves interspecies interactions that result in competitive exclusion of pathogens by blocking binding sites (Garmasheva and Kovalenko, 2005). To colonize the host tissue a probiotic strain has specific adhesion receptors to the mucosal cells, including mucus-binding proteins, S-layer proteins, lipoteichoic acid or exopolysaccharides, and nonspecific adhesion mechanisms as the electrostatic and hydrophobic interactions (Vandenplas et al., 2015; Zhang et al., 2018).

It is known that certain *Lactobacillus* strains present great ability to adhere to the host cells and enact a persistent colonization in specific sites, protecting the intestinal, oral, and vaginal mucosa against pathogens (Sengupta et al., 2013). Therefore the genetic basis of different phenotypic characteristics of *Lactobacillus* strains still need to be explored in depth in order to optimize industrial-level manufacturing approaches and clinical treatments (Shi et al., 2019). In an in vitro and in vivo study with *Lactobacillus* strains isolated from fermented dairy products, Shi et al. (2019) characterized the whole genome sequence of two strains of *Lactobacillus plantarum* (CGMCC12436 and CCFM605). Although these strains shared the majority of their genomic content, some genes were specific to each. Interestingly, the *cpsD* gene cluster was identified only in CGMCC12436, the strain that showed the longer colonization period in the mice gut. The *cpsD* encodes proteins involved in the biosynthesis of capsular polysaccharides, which are attached in the bacterial surface and contribute to the adhesion and colonization processes (Shi et al., 2019). Therefore *cpsD* seems to be implicated in the probiotic colonization, instigating further investigations.

After adherence, most bacterial and fungal pathogens form biofilms on the host tissues. Microbial biofilms are three-dimensional communities embedded in a matrix of extracellular polymeric substances (Darrene and Cecile, 2016; Tsui et al., 2016; Rossoni et al., 2018b). Since the biofilms show more resistance to antimicrobial compounds and immune cells than planktonic cells, the use of probiotics and their products on bacterial and fungal biofilms has also been widely investigated (Alcazar-Fuoli and Mellado, 2014).

Petrova et al. (2016) analyzed the effects of *Lactobacillus rhamnosus* on bacterial biofilms formed by various gastrointestinal and urogenital pathogens. In this study the genome sequence of *L. rhamnosus* GG (LGG) (ATCC 53103) was screened for the presence of lectin-like proteins. Two genomic regions (*LGGRS02780* and *LGGRS02750*) encoding two putative cell wall proteins were identified and two gene sequences were annotated as *llp1* and *llp2*, encoding the putative lectin-like protein 1 and 2 (Llp1 and Llp2), respectively. Both proteins showed inhibitory activity against biofilm formation by *Salmonella typhimurium* and *E. coli*, with Lip2 being more active than LIp1. These effects were associated with the specificity of LIp1 and LIp2 for complex glycans (mannan and to D-mannose), facilitating the adhesion of LGG to gastrointestinal and vaginal mucosal cells. Rossoni et al. (2018a) evaluated the activity of *Lactobacillus paracasei* strain 28.4, *L. rhamnosus* strain 5.2, and *Lactobacillus fermentum* strain 20.4 against *Candida albicans* biofilms (Fig. 11.2). All these probiotic strains were able to significantly reduce the number of viable cells and the total biomass of fungal biofilms. The inhibitory activity of *Lactobacillus* strains on *C. albicans* biofilms was associated with the downregulated expression of *C. albicans* biofilm-specific genes (*ALS3*, *HWP1*, *EFG1*, and *CPH1*).

PRODUCTION OF BIOACTIVE COMPOUNDS

Another MoA for probiotics is the production of metabolites (including, but not limited to, lactic acid, hydrogen peroxide, and bacteriocins) with a broad antimicrobial activity, including activity against fungi (Fig. 11.3), bacteria, and viruses (Zhang et al., 2018; Chew et al., 2015; Yang et al., 2015). Recently, many groups have been investigated for the CFS from centrifuged liquid cultures of probiotic strains, seeking to identify the presence of bioactive substances. Koohestani et al. (2018) studied the antibacterial and antibiofilm properties of CFS of *Lactobacillus acidophilus* LA5 and *Lactobacillus casei* 431 against *Staphylococcus aureus*. Antibacterial activity of both *Lactobacillus* strains was measured by the agar spot method (Natarajan Devi Avaiyarasi et al., 2016), while the antibiofilm activity of CFS was determined in polystyrene and glass models containing 2-day-old biofilm of *S. aureus*. The authors found an inhibition zone of *L. acidophilus* (50.2 mm) greater than *L. casei* (37.0 mm). In addition, the CFSs significantly reduced the *S. aureus* biofilm on both tested models. Rossoni et al. (2018b) evaluated whether the supernatants

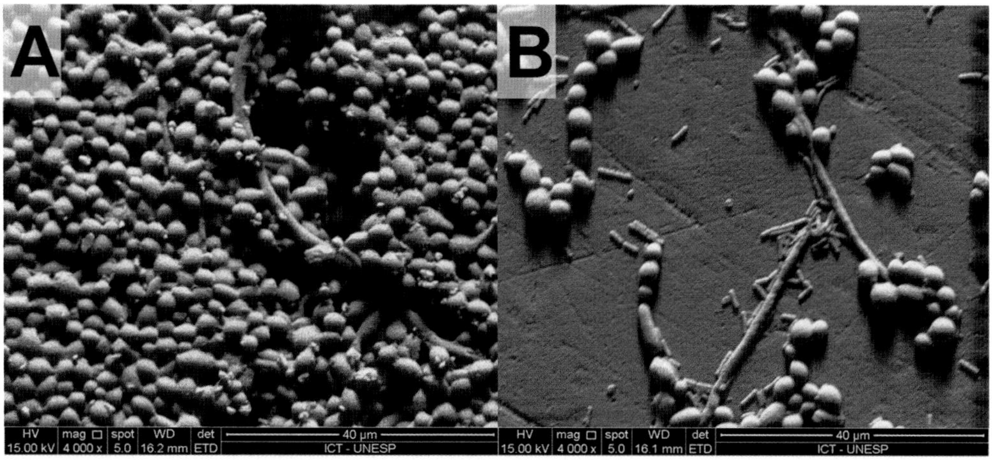

FIGURE 11.2

Antibiofilm activity of the probiotic strain *Lactobacillus fermentum* 20.3 on *Candida albicans*. Scanning electron microscopy of biofilms formed in vitro: (A) single biofilm of *C. albicans* ATCC 18804 and (B) mixed biofilm of *C. albicans* ATCC 18804 and *L. fermentum* 20.3.

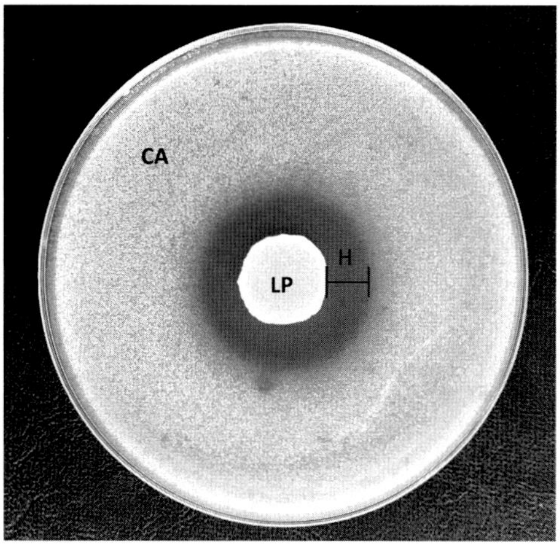

FIGURE 11.3

Agar diffusion test: zone of inhibition of CA ATCC 18804 (CA) indicative of its susceptibility to LP 28.4 (LP). *CA*, *Candida albicans*; *H*, inhibition halo; *LP*, *Lactobacillus paracasei*.

of *Lactobacillus* strains (*L. paracasei* 25.4, *L. fermentum* 20.4, *L. paracasei* 20.3, and *L. paracasei* 11.6) isolated from caries-free subjects could inhibit *Streptococcus mutans* biofilm, one of the most important bacteria for dental caries. The *S. mutans* biofilms were formed on hydroxyapatite disks, and after incubation times of 24 and 48 h, a strong inhibitory activity of *Lactobacillus* supernatant on *S. mutans* cells was observed for both time points tested. In addition, this reduction was not correlated with the decreasing of pH value in the medium, indicating that these *Lactobacillus* strains release bioactive substances that can inhibit *S. mutans* growth and biofilm formation.

The production of lactic acid has been considered an important factor for the antimicrobial properties of probiotic strains (Alakomi et al., 2000). According to Juturu and Wu (2016), the lactic acid is produced by many bacterial genera as either primary or secondary fermentation product. However, the term "lactic acid bacteria (LAB)" is reserved specifically for the genera belonging to the order Lactobacillales, which includes *Lactobacillus, Pediococcus, Aerococcus, Carnobacterium, Enterococcus, Tetragenococcus, Vagococcus, Leuconostoc, Oenococcus, Weissella, Streptococcus,* and *Lactococcus*.

O'Hanlon et al. (2011) evaluated the effects of lactic acid and hydrogen peroxide (H_2O_2) on a broad range of bacteria associated with bacterial vaginosis (BV). Cultures of seventeen species of bacteria (*Gardnerella vaginalis, Prevotella bivia, Prevotella corporis, Anaerococcus prevotii, Fusobacterium nucleatum, Porphyromonas levii, Bacteroides ureolyticus, Peptostreptococcus anaerobius, Anaerococcus tetradius, Atopobium vaginae, Megasphaera elsdenii, Propionibacterium acnes, Ureaplasma urealyticum, Mobiluncus curtisii, Mobiluncus mulieris, Mycoplasma hominis,* and *Micromonas micros*) were exposed to lactic acid or hydrogen peroxide in physiological concentrations. After 2 h the remaining viable bacteria were enumerated by growth on agar media plates. As a result, hydrogen peroxide was proven unable to inactivate the bacteria tested. On the other hand the lactic acid dramatically reduced the viability of all bacteria associated with BV; *M. mulieris* cells in particular were reduced below detectability threshold. These findings suggest that H_2O_2 production by *Lactobacillus* spp. is an implausible mechanism for suppressing these bacteria, while lactic acid production plays an important role in the treatment of BV by probiotics.

Probiotics that produce lactic acid also seem to have some potential protective effects against some viruses, such as human immunodeficiency virus (HIV) and herpes simplex virus (HSV) (Tachedjian et al., 2017). Aldunate et al. (2013) demonstrated that physiological concentrations of lactic acid have broad-spectrum virucidal activity on HIV, suggesting that women colonized vaginally with *Lactobacillus*-rich microbiome enjoy a protective factor in the sexual transmission of HIV. It has also been suggested that women with vaginal microbiome dominated by *Lactobacillus* strains are less likely to be infected with HSV-2 (Borgdorff et al., 2014) as these strains may inhibit the viral entry and replication or, alternatively, the lactic acid produced can directly affect the viral particles (Conti et al., 2009).

In addition, probiotics can secrete bacteriocins capable of acting locally and inhibiting the pathogens occupying the same habitat (Mokoena, 2017). Bacteriocins are a heterogeneous family of small peptides with potent antimicrobial activity that are produced by many bacterial species, including *Lactobacillus* strains (Cotter et al., 2005). Corr et al. (2007) evaluated the protective activity of *Lactobacillus salivarius* UCC118, a bacteriocin-producing strain, against infection by the invasive foodborne pathogen *Listeria monocytogenes* in a mouse model. Lactobacilli were administered orally by a pipette for 3 or 6 days before the oral infection with *L. monocytogenes*. The infection in liver and spleen was analyzed by CFU counts and bioluminescence assay. The probiotic administration was able to increase the resistance to infection and to reduce the numbers of *L. monocytogenes* bacteria in both liver and spleen. The authors confirmed that the antimicrobial effect was a result of direct antagonism between *L. salivarius* and the pathogen, mediated by the bacteriocin Abp118.

Another probiotic bacteriocin widely studied is the reuterin, which is produced by some strains of *Lactobacillus reuteri* during anaerobic fermentation of glycerol and shows activity against food-borne pathogens and spoilage microorganisms (Talarico et al., 1988; Mu et al., 2018; Urrutia-Baca et al., 2018; Ortiz-Rivera et al., 2017). Interestingly, this peptide is soluble in water, resistant to heat and some enzymes, and maintains its antimicrobial activity at low pH (Rasch, 2002; Rasch et al., 2007; Langa et al., 2014). Recently, Vimont et al. (2019) investigated the antifungal activity of reuterin against food isolates of yeasts and its potential application in biopreservation of yogurt. The authors quantified the minimum inhibitory concentration and the minimum fungicidal concentration (MFC) of reuterin against a representative panel of the most abundant fungal genera associated with food contamination (*Candida*, *Penicillium*, *Aspergillus*, *Aureobasidium*, *Eurotium*, *Paecilomyces*, *Kluyveromyces*, *Lachancea*, 15 *Rhodotorula*, *Saccharomyces*, *Schyzosaccharomyces*, *Torulospora*, *Wickerhamomyces*, and *Zygosaccharomyces*). They found high antifungal activity of reuterin that inhibited the growth of these microorganisms at a concentration of 11 mM or less. Moreover, reuterin showed a potent fungicidal activity (99.9% reduction for all tested microorganisms) at concentrations equal or below 15.6 mM as indicated by MFC. In another study, Ortiz-Rivera et al. (2017) found that reuterin produced by *L. reuteri* ATCC 53608 has significant antimicrobial activity against important human pathogens, such as *S. aureus*, *Salmonella enterica*, and *L. monocytogenes*. These studies support that reuterin has a high potential to be used as an antimicrobial food preservative, as well as in medical and veterinary contexts.

MODULATION OF IMMUNE SYSTEM

One of the major benefits of probiotics in immunological terms is the modulation of the host immune responses via interacting with the gastrointestinal mucosa (Zhang et al., 2018; Kim and Mylonakis, 2012). Once administered, oral probiotic bacteria interact with the intestinal epithelial cells or immune cells of the lamina propria, through toll-like receptors (TLRs), and induce the production of different cytokines and chemokines (Maldonado Galdeano et al., 2019). Probiotics interact with different cells of the immune system [enterocytes, dendritic cells (DCs), Th1, Th2, and T regulatory cells] in the intestine, modulating the immune response toward a pro- or antiinflammatory action (Galdeano and Perdigon, 2004). De Oliveira et al. (2017) found that certain strains of *L. rhamnosus*, *L. acidophilus*, and *L. paracasei* induced the expression of tumor necrosis factor-α and interleukin (IL)-1β, IL-6, and IL-17 in mouse macrophages and consequently protected *Galleria mellonella* model from infection by *C. albicans*.

Most of the immune mechanisms involve tissue-specific regulation of gene expression, particularly in the intestine and the liver (Fontana et al., 2013; Plaza-Diaz et al., 2014). In this context the modulation of gene expression by probiotics is an important issue that needs to be addressed (Plaza-Diaz et al., 2014). It is known that the mucins expressed at the surface of epithelial cells are essential to protect the human body from infections. The *MUC1* gene in humans encodes such a protein and the expression of mucin genes (MUC) can be modulated by probiotics (Plaza-Diaz et al., 2014). Likewise, TLR and nucleotide-binding oligomerization domain-like receptor genes as well as proinflammatory transcription factors, cytokines, and apoptosis-related enzyme genes can also be affected by bacteria (Plaza-Diaz et al., 2014; Mattar et al., 2002). Matsubara et al. (2017) verified the effects of some *Lactobacillus* strains (*L. rhamnosus* LR32, *L. casei* L324, and *L. acidophilus* NCFM) on human macrophages challenged with *C. albicans*. All the probiotic strains were able to downregulate the transcription of pattern-recognition receptors genes, such as *CLEC7A* and *TRL-2*, resulting in the decreased expression of dectin-1 on macrophages pretreated with probiotics.

Weiss et al. (2010) investigated whether *L. acidophilus* NCFM has the ability to induce antiviral defense gene expression (IFN-β, TLR-3, IL-12, and IL-10) in murine bone marrow–derived DCs. The gene expression profile of TLR-3 and IFN-β are key players involved in viral defense. *L. acidophilus* NCFM induced a stronger expression of IFN-β, IL-12, and IL-10 compared with the synthetic double-stranded RNA ligand poly I:C, whereas the levels of expressed TLR-3 were similar. Whole genome microarray gene expression analysis revealed that other genes related to viral defense were significantly upregulated. These results reveal that certain lactobacilli trigger the expression of viral defense genes in DCs dependent on IFN-β.

BIOENGINEERING FOR ENHANCING THE FUNCTIONAL PROPERTIES OF PROBIOTICS STRAINS

The increasing knowledge on the MoA of probiotic bacteria and their interactions with the local microbiota has provided a basis for the design of bioengineered probiotic strains. Nowadays, some researchers have worked in the optimization of existing probiotic strains, seeking improvement through alterations or acquisition of phenotypic traits able to positively affect their probiotic properties (Douillard and De vos, 2019; Mathipa and Thantsha, 2017). Thus specific genes can be manipulated to generate a range of phenotypes for the same probiotic strain. The bioengineering of probiotic strains can improve specific properties of theirs, applicable in the colonization of the gut, such as mucosal adherence, acid stress resistance, production of bioactive compounds, and modulation of the immune system. In addition, bioengineering offers the possibility of deleting undesirable, counterproductive, or problematic genes that may be detected in some probiotic strains, such as antibiotic-resistant genes found in various *Lactobacillus* spp. that can be laterally transferred among bacteria in a given microbiome (Douillard and De vos, 2019).

In this context, studies based on genomics, transcriptomics, and metabolomics providing valuable tools in this area are termed "probiogenomics." Various genomic tools are currently available and have been employed for LAB, including transformation methods, integration/homologous recombination techniques, and CRISPR–Cas-based editing (Borner et al., 2019).

Using bioengineered probiotics, various studies were conducted during the last decade seeking to improve different functional properties of the conventional probiotic strains (Mathipa and Thantsha, 2017). Volzing et al. (2013) engineered a *Lactobacillus lactis* strain featuring inducible the expression and secretion of AMPs with high activity against Gram-negative pathogens, especially *E. coli* and *Salmonella* spp. Bermudez-Humaran et al. (2015) engineered another recombinant *L. lactis* strain to express elafin, a protease inhibitor expressed in the intestinal epithelium that contributes to the reduction of inflammation. Then, either wild-type or recombinant *L. lactis* was used to treat mice with induced colonic inflammation. The recombinant *L. lactis* secreting the elafin caused a significant reduction in all inflammation parameters (colonic thickness, elastase activities, and granulocyte infiltration) compared to the wild-type treatment.

Koo et al. (2012) developed a recombinant *L. paracasei* strain harboring the *L. monocytogenes* adhesion protein in order to contain *L. monocytogenes* infection by enacting competitive exclusion for binding sites. They observed that the wild-type probiotic strain effected no significant reduction in the adhesion of the *L. monocytogenes* to the cell monolayer, while the recombinant probiotic strain achieved a 60% reduction of adhesion. Chu et al. (2005) cloned and expressed the K99 fimbriae from enterotoxinogenic *E. coli* (ETEC) into the probiotic *L. acidophilus* and found that the recombinant strain was able to reduce the attachment of ETEC to porcine intestinal brush border in a dose-dependent manner. These studies indicated that the reduction of the adherence of the pathogen by the recombinant probiotic prevents the actual binding of the pathogen, and consequently, decreases the development of infections.

Therefore bioengineering technology has a great potential for the development of novel strategies for the prevention and treatment of several infectious diseases (Mathipa and Thantsha, 2017).

CLINICAL APPLICATIONS

Due to their diverse MoA, probiotics present several clinical applications, including reduction of the prevalence of antibiotic-associated diarrhea, antagonistic activities against several pathogens, enhancement/regeneration of compromised gut epithelial barrier, reduction of allergic sensitization, and antiinflammatory properties (Dietrich et al., 2014; Kozakova et al., 2016; Lee et al., 2016). The use of probiotics for bacterial infections of the GIT and oral cavity, and for oral and vaginal candidiasis, will be developed further herein.

BACTERIAL INFECTIONS OF THE GASTROINTESTINAL TRACT

The human GIT is home to countless species of microorganisms that established a symbiotic relationship with the host and form the GIT microbiome. The human intestinal microbiome plays an important role in the development of the gut immune system, in metabolism, nutrition absorption, production of SCFAs, and essential vitamins, in developing resistance to pathogenic microorganisms, and additionally it modulates a normal immunological response. The use of probiotics in GIT can enhance these characteristics by maintaining a balance among intestinal microbiota and improving the bowel functional regularity (Javanmard et al., 2018). In addition, certain probiotic strains can increase the bioavailability of nutrients necessary for the human host; such are the cases of the synthesis of growth factors (spermidin) and of the production of B-complex vitamins (B2, B9, and B12). Other beneficial intervention of such strains is their implication in digestive processes such as the degradation of lactose. (De Vrese et al., 2001; Crittenden et al., 2003; Turpin et al., 2010).

When the intestinal microbiome gets off balance, several intestinal infections may occur and one of the main GIT diseases is the infectious diarrhea caused by rotavirus, which affects adults and—mainly—children. Some studies report the use of probiotics for the treatment of such infections, in which the results indicate a reduction in the duration of diarrhea and fever. (Allen et al., 2010; Szajewska et al., 2011; Salari et al., 2012). Another important GIT disease is the *Clostridioides (Clostridium) difficile* infection (CDI), a multifactorial disease caused by a complex interaction of factors (Ziakas and Mylonakis, 2014; Allen et al., 2013). The treatments for CDI remain largely ineffective, with high relapse rates (Goldstein et al., 2017) and improvements over existing practice or novel approaches are required. In a recent study, Monteiro et al. (2019) studied the effects of *L. plantarum* ATCC 8014 and its supernatant against *Clostridium butyricum* ATCC 860, *C. difficile* ATCC 9689, and *Clostridium perfringens* ATCC 12924. Using agar spot assay, inhibition of gas production, and coaggregation test, the authors verified that *L. plantarum* was able to inhibit the three isolates and this inhibition was maintained even when only the cell-free probiotic supernatant was used, suggesting that this probiotic strain produces antimicrobial molecules.

Dudzicz et al. (2018) studied the efficacy of *L. plantarum* 299v (LP299v) strain, belonging to the bacteria that naturally occur on the surface of human intestinal mucosa, in the prophylaxis from CDI during antibiotic therapy in patients treated with immunosuppressive drugs in nephrology and transplantation ward. Before the implementation of LP299v, the authors found that 18 patients had CDI; however, when LP299v was used as a prophylaxis, the cases of CDI were down to 2 per year. These promising results are in agreement with Yang and Yang (2018), who verified that a probiotic strain of *Bifidobacterium breve* was able to modify the expression of the virulence genes *tcdA* and *tcdB*, two core elements of *C. difficile* infection, which control the production of toxins.

Another relevant bowel disease is the ulcerative colitis (UC), a chronic inflammatory disorder of the GIT with unknown etiology, which features mucosal inflammation, diarrhea, abdominal pain, weight loss, ulcerations of intestinal mucosa and presents periods of activity and remission (Derikx et al., 2016). The use of probiotics for treating the UC is still not cleared due to the limitations of the respective studies. Interestingly, some studies report an improvement in the clinical condition of the patients treated with probiotics (Bibiloni et al., 2005; Soo et al., 2008). Tursi et al. (2004) studied a combination of VSL3 (a probiotic cocktail containing: *L. casei*, *L. plantarum*, *L. acidophilus*, *Lactobacillus delbrueckii* subsp. *bulgaricus*, *Bifidobacterium longum*, *B. breve*, *Bifidobacterium infantis*, and *Streptococcus salivarius* subsp. *thermophilus*) with balsalazide in the remission of UC. The combination of VSL3 and balsalazide presented higher remission rate of UC when compared to balsalazide alone and showed improvement in rectal bleeding, reduction of stool frequency, and also improved the well-being of the patients.

The numerous direct or indirect benefits of the use of probiotics may render their employment for the prevention and treatment of infectious and inflammatory diseases of the GIT very promising and inaugurate a dynamic market in the therapeutics segment.

ORAL INFECTIONS

The use of probiotics also shows promise for the containment of bacterial infections of the oral cavity. Probiotics have emerged as both preventive and therapeutic approaches for oral infections associated with microbiomic imbalance, such as dental caries, periodontal diseases, and halitosis (Meurman and Stamatova, 2018; Bustamante et al., 2019).

Dental caries, the primary cause of oral pain and tooth loss, is an infectious disease caused by cariogenic bacteria (such as *S. mutans)* that ferment carbohydrates and produce organic acids (Featherstone, 2004; Selwitz et al., 2007; Bustamante et al., 2019). Thus oral probiotics are likely to decrease the cases of dental caries by inhibiting the growth of cariogenic pathogens (Teughels et al., 2011). Nase et al. (2001) evaluated whether the consumption of milk containing LGG could interfere with the risk of caries in children from 1 to 6 years old. The children who consumed milk with LGG showed lower *S. mutans* counts and less dental caries compared to the control group, confirming that probiotics may have beneficial effects on children's dental health. In other clinical assays, strains of *Bifidobacterium* (Caglar et al., 2005), *L. reuteri* (Caglar et al., 2006), and *L. paracasei* (Holz et al., 2013) also decreased the number of cariogenic bacteria and prevented dental caries. In contrast, *Bifidobacterium animalis* administered in yogurt failed to reduce the salivary levels of *S. mutans* in children. A metaanalysis of 50 randomized and controlled clinical trials concluded that the evidence about the effects of *Lactobacillus* spp. on dental caries had at the time been insufficient for recommending probiotics as a therapeutic strategy (Gruner et al., 2016), and therefore further studies are still needed to evaluate their efficacy against caries (Zhang et al., 2018).

Chronic periodontitis is an inflammatory process that affects the soft and hard structures that support the teeth. Its etiology is multifactorial, involving host factors, environmental conditions, and the presence of a complex biofilm. The main bacteria associated with chronic periodontitis are *Porphyromonas gingivalis*, *Treponema denticola*, *Aggregatibacter actinomycetemcomitans*, *Prevotella intermedia*, and *Tannerella forsythia* (Cardoso et al., 2018; Larsen and Fiehn, 2017). Several studies indicate that probiotic therapy may prove beneficial for the prevention and treatment of gingivitis and periodontitis. *L. reuteri* has been found able to inhibit the gingival index, plaque index, probing pocket depth, and clinical attachment level, as well as to reduce the levels of periodontal pathogens (*A. actinomycetemcomitans*, *P. intermedia*, and *P. gingivalis*) (Vivekananda et al., 2010; Teughels et al., 2013; Vicario et al., 2013). Short-term daily consumption of probiotic lozenges (twice a day, for a 4-week period) with LGG and *Bifidobacterium lactis* BB-12 reduced significantly the gingival index in probiotic group compared to the placebo group. In addition, there was a significant decrease of *A. actinomycetemcomitans* and *P. gingivalis* in the saliva and biofilm of the patients treated with probiotics (Alanzi et al., 2018).

Another common oral disorder is halitosis, a multifactorial condition with oral and/or nonoral origins (Van den Broek et al., 2007; Oliveira-Neto et al., 2013). Halitosis is usually attributed to degradation of salivary and food proteins, generating amino acids that transform into volatile sulfur compounds (VSCs) (H_2S and CH_4S) by Gram-negative anaerobic bacteria (Oliveira-Neto et al., 2013). In a crossover trial, Iwamoto et al. (2010) found that *L. salivarius* WB21 alleviated oral malodor parameters by using organoleptic test score. Georgiou et al. (2018) performed a systematic review to study the effects of probiotics on the severity of halitosis, measured by VSC levels, organoleptic scores (ORN), or hydrogen sulfide, methyl mercaptan, and dimethyl sulfide levels. The study resulted in 1104 original research articles being initially identified of which 6 were selected as being eligible totaling results from 129 subjects. Among these six articles, five were reporting studies with randomized placebo-controlled clinical trials; two of these reported a significant reduction in ORN between probiotic and placebo groups, and two studies were based on total VSC levels. In conclusion, probiotics may have beneficial effects in treating halitosis. However, due to scarce data and the heterogeneity of the studies, the efficacy of probiotics needs to be investigated further.

Oropharyngeal candidiasis causes significant morbidity in patients immunocompromised due to AIDS, neutropenia, diabetes mellitus, or the use of immunosuppressive drugs (Verma et al., 2017). Oral candidiasis is one of the most common opportunistic infections of the oral cavity, presenting different clinical manifestations with diverse diagnostic and therapeutic approaches (Quindos et al., 2019). The main clinical phenotypes are pseudomembranous, erythematous, and hyperplastic candidiasis, prone to evolve to chronic forms with high rate of relapse (Millsop and Fazel, 2016; Quindos et al., 2019). Besides immunosuppression, oral candidiasis is associated with other predisposing factors, including—but not restricted to—the use of dentures, inhaled corticosteroids, reduced salivary flow, and administration of broad-spectrum antibiotics (Coronado-Castellote and Jimenez-Soriano, 2013; Costa et al., 2013). As in the oral bacterial infections, probiotic bacteria have been widely studied as an alternative method for the control of oral candidiasis.

Until now, studies related to the use of probiotics for the prevention of oral candidiasis were focused only on candidiasis of elderly denture wearers. Probiotics reduced the oral *Candida* spp. colonization in elderly patients (Hatakka et al., 2007; Mendonca et al., 2012; Kraft-Bodi et al., 2015; Ishikawa et al., 2015) and the prevalence of hyposalivation (feeling of dry mouth) (Hatakka et al., 2007) while they increased the anti-*Candida* IgA levels in saliva (Mendonca et al., 2012).

Miyazima et al. (2017) evaluated the effect of consumption of cheese supplemented with probiotics on the oral colonization of *Candida* in denture wearers. The study group consumed cheese supplemented with *L. acidophilus* NCFM or *L. rhamnosus* Lr-32 daily for 8 weeks and showed reduction in the *Candida* levels [≥ 3 (log CFU)/mL] compared to the control group (cheese without probiotics). Interestingly, the reduction in the levels of oral *Candida* occurred independently of the colonizing *Candida* species, participant age, and use of bi- or unimaxillary dentures. Such data suggest the use of probiotic bacteria in reducing the risk of oral candidiasis in highly susceptible populations.

VULVOVAGINAL CANDIDIASIS

Candida spp. are commensal fungi of the vaginal mucosa and may cause symptoms when the balance among the fungus, mucosa, and host defense mechanisms is disturbed, leading to the development of candidiasis (Thompson et al., 2010; Cassone and Cauda, 2012; Costa et al., 2013; Hebecker et al., 2014; Salvatori et al., 2016; Matsubara et al., 2016).

Vulvovaginal candidiasis (VVC) is a very common infection that affects both immunocompetent and immunocompromised women, presenting symptoms and signs of vaginal and vulval inflammation caused by various *Candida* species, mainly *C. albicans* (Dovnik et al., 2015; Makanjuola et al., 2018). It is known that most women present one or more episodes of VVC during their life. This infection affects women's quality of life, mental health, and sexual activity due to its intense symptoms: vulvar erythema, excoriation, pruritus, and a vaginal discharge that may be accompanied by a change in vaginal odor (Blostein et al., 2017). The treatment of VVC is by the use of topical or systemic antifungals, depending on the extent of infection; however, high rates of antifungal resistance have been reported (Wang et al., 2016; Khan et al., 2018). The current spectrum of antifungal resistance, combined with evidence that certain probiotic strains inhibit activity against *Candida* species (Santos et al., 2019), led to studies conceived to evaluate the clinical application of *Lactobacillus* for the VVC treatment (Xie et al., 2017; Buggio et al., 2019).

Pendharkar et al. (2015) evaluated that the antifungal efficacy of EcoVag, a commercially available probiotic product composed of *Lactobacillus gasseri* (DSM 14869) and *L. rhamnosus* (DSM 14870), was combined with fluconazole for the treatment of VVC. The group treated with the antifungal plus EcoVag and the group treated with fluconazole alone had a cure rate of 100%. However, after the end of treatment, 33% of the women treated with fluconazole alone had a relapse, requiring treatment anew, while the group treated with the combination regimen did not present relapse of any case. In a similar study, Davar et al. (2016) investigated the effect of Pro-Digest (*L. acidophilus*, *Bifidobacterium bifidum*, and *B. longum*) plus fluconazole in the treatment of recurrent VVC. The authors found that the recurrence rate of the group treated with fluconazole was 35.5% compared with 7.2% for the group treated with fluconazole plus Pro-Digest. The results of these studies suggest that probiotics may enhance the effect of fluconazole and provide a protective, sustainable solution after the end of antifungal therapy.

CONCLUSION

Given that the diverse MoA of probiotics, the availability of current bioengineering tools and the evidence for their effectiveness in clinical trials, exciting prospects emerge for the field in the context of global management of human health (Puebla-Barragan and Reid, 2019). For the near future, one of the most intriguing challenges for the probiotics to overcome will be the development of more effective therapeutic products in terms of functional properties and clinical performance, which will target specific human hosted (sub)microbiomes providing new perspectives for the treatment of infectious diseases.

ACKNOWLEDGMENT

The authors would like to thank the Fundação de Amparo à Pesquisa do Estado de São Paulo (FAPESP) (grants 2016/25544-1, 2017/19219-3, and 2017/02652-6).

REFERENCES

Aguilar-Toalá, J.E., Garcia-Varela, R., Garcia, H.S., Mata-Haro, V., González-Córdova, A.F., Vallejo-Cordoba, B., et al., 2018. Postbiotics: an evolving term within the functional foods field. Trends Food Sci. Technol. 75, 105–114.

Alakomi, H.L., Skytta, E., Saarela, M., Mattila-Sandholm, T., Latva-Kala, K., Helander, I.M., 2000. Lactic acid permeabilizes Gram-negative bacteria by disrupting the outer membrane. Appl. Environ. Microbiol. 66, 2001–2005.

Alanzi, A., Honkala, S., Honkala, E., Varghese, A., Tolvanen, M., Soderling, E., 2018. Effect of *Lactobacillus rhamnosus* and *Bifidobacterium lactis* on gingival health, dental plaque, and periodontopathogens in adolescents: a randomised placebo-controlled clinical trial. Benefic. Microbes 9, 593–602.

Alcazar-Fuoli, L., Mellado, E., 2014. Current status of antifungal resistance and its impact on clinical practice. Br. J. Haematol. 166, 471–484.

Aldunate, M., Tyssen, D., Johnson, A., Zakir, T., Sonza, S., Moench, T., et al., 2013. Vaginal concentrations of lactic acid potently inactivate HIV. J. Antimicrob. Chemother. 68, 2015–2025.

Allen, S.J., Martinez, E.G., Gregorio, G.V., Dans, L.F., 2010. Probiotics for treating acute infectious diarrhoea. Cochrane Database Syst. Rev. CD003048. Available from: https://doi.org/10.1002/14651858.CD003048.pub3.

Allen, S.J., Wareham, K., Wang, D., Bradley, C., Sewell, B., Hutchings, H., et al., 2013. A high-dose preparation of lactobacilli and bifidobacteria in the prevention of antibiotic-associated and *Clostridium difficile* diarrhoea in older people admitted to hospital: a multicentre, randomised, double-blind, placebo-controlled, parallel arm trial (PLACIDE). Health Technol. Assess. 17, 1–140.

Bermudez-Humaran, L.G., Motta, J.P., Aubry, C., Kharrat, P., Rous-Martin, L., Sallenave, J.M., et al., 2015. Serine protease inhibitors protect better than IL-10 and TGF-beta anti-inflammatory cytokines against mouse colitis when delivered by recombinant lactococci. Microb. Cell Fact. 14, 26.

Bibiloni, R., Fedorak, R.N., Tannock, G.W., Madsen, K.L., Gionchetti, P., Campieri, M., et al., 2005. VSL#3 probiotic-mixture induces remission in patients with active ulcerative colitis. Am. J. Gastroenterol. 100, 1539–1546.

Blostein, F., Levin-Sparenberg, E., Wagner, J., Foxman, B., 2017. Recurrent vulvovaginal candidiasis. Ann. Epidemiol. 27, 575–582e3.

Boets, E., Gomand, S.V., Deroover, L., Preston, T., Vermeulen, K., De preter, V., et al., 2017. Systemic availability and metabolism of colonic-derived short-chain fatty acids in healthy subjects: a stable isotope study. J. Physiol. 595, 541–555.

Borgdorff, H., Tsivtsivadze, E., Verhelst, R., Marzorati, M., Jurriaans, S., Ndayisaba, G.F., et al., 2014. *Lactobacillus*-dominated cervicovaginal microbiota associated with reduced HIV/STI prevalence and genital HIV viral load in African women. ISME J. 8, 1781–1793.

Borner, R.A., Kandasamy, V., Axelsen, A.M., Nielsen, A.T., Bosma, E.F., 2019. Genome editing of lactic acid bacteria: opportunities for food, feed, pharma and biotech. FEMS Microbiol. Lett. 366. Available from: https://doi.org/10.1093/femsle/fny291.

Buggio, L., Somigliana, E., Borghi, A., Vercellini, P., 2019. Probiotics and vaginal microecology: fact or fancy? BMC Womens Health 19, 25.

Bustamante, M., Oomah, B.D., Mosi-Roa, Y., Rubilar, M., Burgos-Diaz, C., 2019. Probiotics as an adjunct therapy for the treatment of halitosis, dental caries and periodontitis. Probiotics Antimicrob. Proteins . Available from: https://doi.org/10.1007/s12602-019-9521-4.

Caglar, E., Sandalli, N., Twetman, S., Kavaloglu, S., Ergeneli, S., Selvi, S., 2005. Effect of yogurt with *Bifidobacterium* DN-173 010 on salivary mutans streptococci and lactobacilli in young adults. Acta Odontol. Scand. 63, 317–320.

Caglar, E., Cildir, S.K., Ergeneli, S., Sandalli, N., Twetman, S., 2006. Salivary mutans streptococci and lactobacilli levels after ingestion of the probiotic bacterium *Lactobacillus reuteri* ATCC 55730 by straws or tablets. Acta Odontol. Scand. 64, 314–318.

Canani, R.B., Costanzo, M.D., Leone, L., Pedata, M., Meli, R., Calignano, A., 2011. Potential beneficial effects of butyrate in intestinal and extraintestinal diseases. World J. Gastroenterol. 17, 1519–1528.

Cardoso, E.M., Reis, C., Manzanares-Cespedes, M.C., 2018. Chronic periodontitis, inflammatory cytokines, and interrelationship with other chronic diseases. Postgrad. Med. 130, 98–104.

Cassone, A., Cauda, R., 2012. *Candida* and candidiasis in HIV-infected patients: where commensalism, opportunistic behavior and frank pathogenicity lose their borders. AIDS 26, 1457–1472.

Cerdo, T., Ruiz, A., Suarez, A., Campoy, C., 2017. Probiotic, prebiotic, and brain development. Nutrients 9. Available from: https://doi.org/10.3390/nu9111247.

Charbonneau, M.R., O'donnell, D., Blanton, L.V., Totten, S.M., Davis, J.C., Barratt, M.J., et al., 2016. Sialylated milk oligosaccharides promote microbiota-dependent growth in models of infant undernutrition. Cell 164, 859–871.

Chew, S.Y., Cheah, Y.K., Seow, H.F., Sandai, D., Than, L.T., 2015. Probiotic *Lactobacillus rhamnosus* GR-1 and *Lactobacillus reuteri* RC-14 exhibit strong antifungal effects against vulvovaginal candidiasis-causing *Candida* glabrata isolates. J. Appl. Microbiol. 118, 1180–1190.

Chu, H., Kang, S., Ha, S., Cho, K., Park, S.M., Han, K.H., et al., 2005. *Lactobacillus acidophilus* expressing recombinant K99 adhesive fimbriae has an inhibitory effect on adhesion of enterotoxigenic *Escherichia coli*. Microbiol. Immunol. 49, 941–948.

Cicenia, A., Scirocco, A., Carabotti, M., Pallotta, L., Marignani, M., Severi, C., 2014. Postbiotic activities of lactobacilli-derived factors. J. Clin. Gastroenterol. 48 (Suppl. 1), S18–S22.

Conti, C., Malacrino, C., Mastromarino, P., 2009. Inhibition of herpes simplex virus type 2 by vaginal lactobacilli. J. Physiol. Pharmacol. 60 (Suppl. 6), 19–26.

Coronado-Castellote, L., Jimenez-Soriano, Y., 2013. Clinical and microbiological diagnosis of oral candidiasis. J. Clin. Exp. Dent. 5, e279–e286.

Corr, S.C., Li, Y., Riedel, C.U., O'toole, P.W., Hill, C., Gahan, C.G., 2007. Bacteriocin production as a mechanism for the antiinfective activity of *Lactobacillus salivarius* UCC118. Proc. Natl. Acad. Sci. U.S.A. 104, 7617–7621.

Costa, A.C., Pereira, C.A., Junqueira, J.C., Jorge, A.O., 2013. Recent mouse and rat methods for the study of experimental oral candidiasis. Virulence 4, 391–399.

Cotter, P.D., Hill, C., Ross, R.P., 2005. Bacteriocins: developing innate immunity for food. Nat. Rev. Microbiol. 3, 777–788.

Crittenden, R.G., Martinez, N.R., Playne, M.J., 2003. Synthesis and utilisation of folate by yoghurt starter cultures and probiotic bacteria. Int. J. Food Microbiol. 80, 217–222.

Darrene, L.N., Cecile, B., 2016. Experimental models of oral biofilms developed on inert substrates: a review of the literature. Biomed. Res. Int. 2016, 7461047.

Davar, R., Nokhostin, F., Eftekhar, M., Sekhavat, L., Bashiri zadeh, M., Shamsi, F., 2016. Comparing the recurrence of vulvovaginal candidiasis in patients undergoing prophylactic treatment with probiotic and placebo during the 6 months. Probiotics Antimicrob. Proteins 8, 130–133.

Delcour, J.A., Aman, P., Courtin, C.M., Hamaker, B.R., Verbeke, K., 2016. Prebiotics, fermentable dietary fiber, and health claims. Adv. Nutr. 7, 1–4.

De leoz, M.L., Kalanetra, K.M., Bokulich, N.A., Strum, J.S., Underwood, M.A., German, J.B., et al., 2015. Human milk glycomics and gut microbial genomics in infant feces show a correlation between human milk oligosaccharides and gut microbiota: a proof-of-concept study. J. Proteome Res. 14, 491–502.

De oliveira, F.E., Rossoni, R.D., De barros, P.P., Begnini, B.E., Junqueira, J.C., Jorge, A.O.C., et al., 2017. Immunomodulatory effects and anti-*Candida* activity of lactobacilli in macrophages and in invertebrate model of *Galleria mellonella*. Microb. Pathog. 110, 603–611.

Derikx, L.A., Dieleman, L.A., Hoentjen, F., 2016. Probiotics and prebiotics in ulcerative colitis. Best Pract. Res. Clin. Gastroenterol. 30, 55–71.

De vrese, M., Stegelmann, A., Richter, B., Fenselau, S., Laue, C., Schrezenmeir, J., 2001. Probiotics--compensation for lactase insufficiency. Am. J. Clin. Nutr. 73, 421S–429S.

Dietrich, C.G., Kottmann, T., Alavi, M., 2014. Commercially available probiotic drinks containing *Lactobacillus casei* DN-114001 reduce antibiotic-associated diarrhea. World J. Gastroenterol. 20, 15837–15844.

Douillard, F.P., De vos, W.M., 2019. Biotechnology of health-promoting bacteria. Biotechnol. Adv. 37, 107369.

Dovnik, A., Golle, A., Novak, D., Arko, D., Takac, I., 2015. Treatment of vulvovaginal candidiasis: a review of the literature. Acta Dermatovenerol. Alp. Pannonica Adriat. 24, 5–7.

Dudzicz, S., Kujawa-Szewieczek, A., Kwiecien, K., Wiecek, A., Adamczak, M., 2018. *Lactobacillus plantarum* 299v reduces the incidence of *Clostridium difficile* infection in nephrology and transplantation ward-results of one year extended study. Nutrients 10. Available from: https://doi.org/10.3390/nu10111574.

Featherstone, J.D., 2004. The continuum of dental caries--evidence for a dynamic disease process. J. Dent. Res. 83 (Spec No C), C39–C42.

Fontana, L., Bermudez-Brito, M., Plaza-Diaz, J., Munoz-Quezada, S., Gil, A., 2013. Sources, isolation, characterisation and evaluation of probiotics. Br. J. Nutr. 109 (Suppl. 2), S35–S50.

Fuochi, V., Petronio, G.P., Lissandrello, E., Furneri, P.M., 2015. Evaluation of resistance to low pH and bile salts of human *Lactobacillus* spp. isolates. Int. J. Immunopathol. Pharmacol. 28, 426–433.

Galdeano, C.M., Perdigon, G., 2004. Role of viability of probiotic strains in their persistence in the gut and in mucosal immune stimulation. J. Appl. Microbiol. 97, 673–681.

Galdeano, C.M., De Moreno De leblanc, A., Vinderola, G., Bonet, M.E., Perdigon, G., 2007. Proposed model: mechanisms of immunomodulation induced by probiotic bacteria. Clin. Vaccine Immunol. 14, 485–492.

Garmasheva, I.L., Kovalenko, N.K., 2005. Adhesive properties of lactic acid bacteria and methods of their investigation. Mikrobiol. Z. 67, 68–84.

Georgiou, A.C., Laine, M.L., Deng, D.M., Brandt, B.W., Van loveren, C., Dereka, X., 2018. Efficacy of probiotics: clinical and microbial parameters of halitosis. J. Breath Res. 12, 046010.

Gibson, G.R., Hutkins, R., Sanders, M.E., Prescott, S.L., Reimer, R.A., Salminen, S.J., et al., 2017. Expert consensus document: the International Scientific Association for Probiotics and Prebiotics (ISAPP) consensus statement on the definition and scope of prebiotics. Nat. Rev. Gastroenterol. Hepatol. 14, 491–502.

Goldstein, E.J.C., Johnson, S.J., Maziade, P.J., Evans, C.T., Sniffen, J.C., Millette, M., et al., 2017. Probiotics and prevention of *Clostridium difficile* infection. Anaerobe 45, 114–119.

Grompone, G., Martorell, P., Llopis, S., Gonzalez, N., Genoves, S., Mulet, A.P., et al., 2012. Anti-inflammatory *Lactobacillus rhamnosus* CNCM I-3690 strain protects against oxidative stress and increases lifespan in *Caenorhabditis elegans*. PLoS One 7, e52493.

Gruner, D., Paris, S., Schwendicke, F., 2016. Probiotics for managing caries and periodontitis: systematic review and meta-analysis. J. Dent. 48, 16–25.

Guarner, F., Khan, A.G., Garisch, J., Eliakim, R., Gangl, A., Thomson, A., et al., 2012. World gastroenterology organisation global guidelines: probiotics and prebiotics October 2011. J. Clin. Gastroenterol. 46, 468–481.

Hatakka, K., Ahola, A.J., Yli-Knuuttila, H., Richardson, M., Poussa, T., Meurman, J.H., et al., 2007. Probiotics reduce the prevalence of oral *Candida* in the elderly--a randomized controlled trial. J. Dent. Res. 86, 125–130.

Hebecker, B., Naglik, J.R., Hube, B., Jacobsen, I.D., 2014. Pathogenicity mechanisms and host response during oral *Candida albicans* infections. Expert Rev. Anti-Infect. Ther. 12, 867–879.

Hill, C., Guarner, F., Reid, G., Gibson, G.R., Merenstein, D.J., Pot, B., et al., 2014. Expert consensus document. The International Scientific Association for Probiotics and Prebiotics consensus statement on the scope and appropriate use of the term probiotic. Nat. Rev. Gastroenterol. Hepatol. 11, 506–514.

Holz, C., Alexander, C., Balcke, C., More, M., Auinger, A., Bauer, M., et al., 2013. *Lactobacillus paracasei* DSMZ16671 reduces mutans streptococci: a short-term pilot study. Probiotics Antimicrob. Proteins 5, 259–263.

Ishikawa, K.H., Mayer, M.P., Miyazima, T.Y., Matsubara, V.H., Silva, E.G., Paula, C.R., et al., 2015. A multispecies probiotic reduces oral *Candida* colonization in denture wearers. J. Prosthodont. 24, 194–199.

Iwamoto, T., Suzuki, N., Tanabe, K., Takeshita, T., Hirofuji, T., 2010. Effects of probiotic *Lactobacillus salivarius* WB21 on halitosis and oral health: an open-label pilot trial. Oral Surg. Oral Med. Oral Pathol. Oral Radiol. Endodontol. 110, 201–208.

Javanmard, A., Ashtari, S., Sabet, B., Davoodi, S.H., Rostami-Nejad, M., Esmaeil akbari, M., et al., 2018. Probiotics and their role in gastrointestinal cancers prevention and treatment; an overview. Gastroenterol. Hepatol. Bed Bench 11, 284–295.

Juturu, V., Wu, J.C., 2016. Microbial production of lactic acid: the latest development. Crit. Rev. Biotechnol. 36, 967–977.

Kanauchi, M., Kondo, A., Asami, K., 2019. Eliminating lipopolysaccharide (LPS) using lactic acid bacteria (LAB) and a fraction of its LPS-elimination protein. Methods Mol. Biol. 1887, 167–174.

Khan, M., Ahmed, J., Gul, A., Ikram, A., Lalani, F.K., 2018. Antifungal susceptibility testing of vulvovaginal *Candida* species among women attending antenatal clinic in tertiary care hospitals of Peshawar. Infect. Drug Resist. 11, 447–456.

Kim, Y., Mylonakis, E., 2012. *Caenorhabditis elegans* immune conditioning with the probiotic bacterium *Lactobacillus acidophilus* strain NCFM enhances Gram-positive immune responses. Infect. Immun. 80, 2500–2508.

Konstantinov, S.R., Kuipers, E.J., Peppelenbosch, M.P., 2013. Functional genomic analyses of the gut microbiota for CRC screening. Nat. Rev. Gastroenterol. Hepatol. 10, 741–745.

Koo, O.K., Amalaradjou, M.A., Bhunia, A.K., 2012. Recombinant probiotic expressing *Listeria* adhesion protein attenuates *Listeria monocytogenes* virulence in vitro. PLoS One 7, e29277.

Koohestani, M., Moradi, M., Tajik, H., Badali, A., 2018. Effects of cell-free supernatant of *Lactobacillus acidophilus* LA5 and *Lactobacillus casei* 431 against planktonic form and biofilm of *Staphylococcus aureus*. Vet. Res. Forum 9, 301–306.

Kozakova, H., Schwarzer, M., Tuckova, L., Srutkova, D., Czarnowska, E., Rosiak, I., et al., 2016. Colonization of germ-free mice with a mixture of three *Lactobacillus* strains enhances the integrity of gut mucosa and ameliorates allergic sensitization. Cell. Mol. Immunol. 13, 251–262.

Kraft-Bodi, E., Jorgensen, M.R., Keller, M.K., Kragelund, C., Twetman, S., 2015. Effect of probiotic bacteria on oral *Candida* in frail elderly. J. Dent. Res. 94, 181S–186SS.

Langa, S., Martin-Cabrejas, I., Montiel, R., Landete, J.M., Medina, M., Arques, J.L., 2014. Short communication: combined antimicrobial activity of reuterin and diacetyl against foodborne pathogens. J. Dairy Sci. 97, 6116–6121.

Larsen, T., Fiehn, N.E., 2017. Dental biofilm infections - an update. APMIS 125, 376–384.

Lee, J., Yang, W., Hostetler, A., Schultz, N., Suckow, M.A., Stewart, K.L., et al., 2016. Characterization of the anti-inflammatory *Lactobacillus reuteri* BM36301 and its probiotic benefits on aged mice. BMC Microbiol. 16, 69.

Makanjuola, O., Bongomin, F., Fayemiwo, S.A., 2018. An update on the roles of non-albicans *Candida* species in vulvovaginitis. J. Fungi (Basel) 4. Available from: https://doi.org/10.3390/jof4040121.

Maldonado Galdeano, C., Cazorla, S.I., Lemme dumit, J.M., Velez, E., Perdigon, G., 2019. Beneficial effects of probiotic consumption on the immune system. Ann. Nutr. Metab. 74, 115–124.

Markowiak, P., Slizewska, K., 2017. Effects of probiotics, prebiotics, and synbiotics on human health. Nutrients 9. Available from: https://doi.org/10.3390/nu9091021.

Mathipa, M.G., Thantsha, M.S., 2017. Probiotic engineering: towards development of robust probiotic strains with enhanced functional properties and for targeted control of enteric pathogens. Gut Pathog. 9, 28.

Matsubara, V.H., Bandara, H.M., Mayer, M.P., Samaranayake, L.P., 2016. Probiotics as antifungals in mucosal candidiasis. Clin. Infect. Dis. 62, 1143–1153.

Matsubara, V.H., Ishikawa, K.H., Ando-Suguimoto, E.S., Bueno-Silva, B., Nakamae, A.E.M., Mayer, M.P.A., 2017. Probiotic bacteria alter pattern-recognition receptor expression and cytokine profile in a human macrophage model challenged with *Candida albicans* and lipopolysaccharide. Front. Microbiol. 8, 2280.

Mattar, A.F., Teitelbaum, D.H., Drongowski, R.A., Yongyi, F., Harmon, C.M., Coran, A.G., 2002. Probiotics up-regulate MUC-2 mucin gene expression in a Caco-2 cell-culture model. Pediatr. Surg. Int. 18, 586–590.

Mendonca, F.H., Santos, S.S., Faria ida, S., Goncalves e Silva, C.R., Jorge, A.O., Leao, M.V., 2012. Effects of probiotic bacteria on *Candida* presence and IgA anti-*Candida* in the oral cavity of elderly. Braz. Dent. J. 23, 534–538.

Meurman, J.H., Stamatova, I.V., 2018. Probiotics: evidence of oral health implications. Folia Med. (Plovdiv) 60, 21–29.

Millsop, J.W., Fazel, N., 2016. Oral candidiasis. Clin. Dermatol. 34, 487–494.

Miyazima, T.Y., Ishikawa, K.H., Mayer, M., Saad, S., Nakamae, A., 2017. Cheese supplemented with probiotics reduced the *Candida* levels in denture wearers-RCT. Oral Dis. 23, 919–925.

Mokoena, M.P., 2017. Lactic acid bacteria and their bacteriocins: classification, biosynthesis and applications against uropathogens: a mini-review. Molecules 22. Available from: https://doi.org/10.3390/molecules 22081255.

Monteiro, C., Do carmo, M.S., Melo, B.O., Alves, M.S., Dos santos, C.I., Monteiro, S.G., et al., 2019. In vitro antimicrobial activity and probiotic potential of *Bifidobacterium* and *Lactobacillus* against species of *Clostridium*. Nutrients 11. Available from: https://doi.org/10.3390/nu11020448.

Mu, Q., Tavella, V.J., Luo, X.M., 2018. Role of *Lactobacillus reuteri* in human health and diseases. Front. Microbiol. 9, 757.

Munoz-Quezada, S., Chenoll, E., Vieites, J.M., Genoves, S., Maldonado, J., Bermudez-Brito, M., et al., 2013. Isolation, identification and characterisation of three novel probiotic strains (*Lactobacillus paracasei* CNCM I-4034, *Bifidobacterium breve* CNCM I-4035 and *Lactobacillus rhamnosus* CNCM I-4036) from the faeces of exclusively breast-fed infants. Br. J. Nutr. 109 (Suppl. 2), S51–S62.

Nakamura, F., Ishida, Y., Aihara, K., Sawada, D., Ashida, N., Sugawara, T., et al., 2016. Effect of fragmented *Lactobacillus amylovorus* CP1563 on lipid metabolism in overweight and mildly obese individuals: a randomized controlled trial. Microb. Ecol. Health Dis. 27, 30312.

Nase, L., Hatakka, K., Savilahti, E., Saxelin, M., Ponka, A., Poussa, T., et al., 2001. Effect of long-term consumption of a probiotic bacterium, *Lactobacillus rhamnosus* GG, in milk on dental caries and caries risk in children. Caries Res. 35, 412–420.

Natarajan Devi Avaiyarasi, A.D.R., Venkatesh, P., Arul, V., 2016. In vitro selection, characterization and cytotoxic effect of bacteriocin of *Lactobacillus sakei* GM3 isolated from goat milk. Food Control 69, 124–133.

Oberg, T.S., Steele, J.L., Ingham, S.C., Smeianov, V.V., Briczinski, E.P., Abdalla, A., et al., 2018. Correction to: Intrinsic and inducible resistance to hydrogen peroxide in *Bifidobacterium* species. J. Ind. Microbiol. Biotechnol. 45, 765.

Oelschlaeger, T.A., 2010. Mechanisms of probiotic actions - a review. Int. J. Med. Microbiol. 300, 57–62.

O'Hanlon, D.E., Moench, T.R., Cone, R.A., 2011. In vaginal fluid, bacteria associated with bacterial vaginosis can be suppressed with lactic acid but not hydrogen peroxide. BMC Infect. Dis. 11, 200.

Oliveira-Neto, J.M., Sato, S., Pedrazzi, V., 2013. How to deal with morning bad breath: a randomized, crossover clinical trial. J. Indian Soc. Periodontol. 17, 757–761.

Ortiz-Rivera, Y., Sanchez-Vega, R., Gutierrez-Mendez, N., Leon-Felix, J., Acosta-Muniz, C., Sepulveda, D.R., 2017. Production of reuterin in a fermented milk product by *Lactobacillus reuteri*: inhibition of pathogens, spoilage microorganisms, and lactic acid bacteria. J. Dairy Sci. 100, 4258–4268.

Pendharkar, S., Brandsborg, E., Hammarstrom, L., Marcotte, H., Larsson, P.G., 2015. Vaginal colonisation by probiotic lactobacilli and clinical outcome in women conventionally treated for bacterial vaginosis and yeast infection. BMC Infect. Dis. 15, 255.

Petrova, M.I., Imholz, N.C., Verhoeven, T.L., Balzarini, J., Van damme, E.J., Schols, D., et al., 2016. Lectin-like molecules of *Lactobacillus rhamnosus* GG inhibit pathogenic *Escherichia coli* and *Salmonella* biofilm formation. PLoS One 11, e0161337.

Plaza-Diaz, J., Gomez-Llorente, C., Fontana, L., Gil, A., 2014. Modulation of immunity and inflammatory gene expression in the gut, in inflammatory diseases of the gut and in the liver by probiotics. World J. Gastroenterol. 20, 15632–15649.

Puebla-Barragan, S., Reid, G., 2019. Forty-five-year evolution of probiotic therapy. Microb. Cell 6, 184–196.

Quindos, G., Gil-Alonso, S., Marcos-Arias, C., Sevillano, E., Mateo, E., Jauregizar, N., et al., 2019. Therapeutic tools for oral candidiasis: current and new antifungal drugs. Med. Oral Patol. Oral Cir. Bucal 24, e172–e180.

Rasch, M., 2002. The influence of temperature, salt and pH on the inhibitory effect of reuterin on *Escherichia coli*. Int. J. Food Microbiol. 72, 225–231.

Rasch, M., Metris, A., Baranyi, J., Bjorn budde, B., 2007. The effect of reuterin on the lag time of single cells of *Listeria innocua* grown on a solid agar surface at different pH and NaCl concentrations. Int. J. Food Microbiol. 113, 35–40.

Reid, G., 2016. Probiotics: definition, scope and mechanisms of action. Best Pract. Res. Clin. Gastroenterol. 30, 17–25.

Rossoni, R.D., De barros, P.P., De alvarenga, J.A., Ribeiro, F.C., Velloso, M.D.S., Fuchs, B.B., et al., 2018a. Antifungal activity of clinical *Lactobacillus* strains against *Candida albicans* biofilms: identification of potential probiotic candidates to prevent oral candidiasis. Biofouling 34, 212–225.

Rossoni, R.D., Velloso, M.D.S., De barros, P.P., De alvarenga, J.A., Santos, J.D.D., Santos prado, A., et al., 2018b. Inhibitory effect of probiotic *Lactobacillus* supernatants from the oral cavity on *Streptococcus mutans* biofilms. Microb. Pathog. 123, 361–367.

Salari, P., Nikfar, S., Abdollahi, M., 2012. A meta-analysis and systematic review on the effect of probiotics in acute diarrhea. Inflamm. Allergy Drug Targets 11, 3–14.

Salvatori, O., Puri, S., Tati, S., Edgerton, M., 2016. Innate immunity and saliva in *Candida albicans*-mediated oral diseases. J. Dent. Res. 95, 365–371.

Sanchez, B., Delgado, S., Blanco-Miguez, A., Lourenco, A., Gueimonde, M., Margolles, A., 2017. Probiotics, gut microbiota, and their influence on host health and disease. Mol. Nutr. Food Res. 61. Available from: https://doi.org/10.1002/mnfr.201600240.

Sanders, M.E., Benson, A., Lebeer, S., Merenstein, D.J., Klaenhammer, T.R., 2018. Shared mechanisms among probiotic taxa: implications for general probiotic claims. Curr. Opin. Biotechnol. 49, 207–216.

Santos, R.B., Scorzoni, L., Namba, A.M., Rossoni, R.D., Jorge, A.O.C., Junqueira, J.C., 2019. *Lactobacillus* species increase the survival of *Galleria mellonella* infected with *Candida albicans* and non-albicans *Candida* clinical isolates. Med. Mycol. 57, 391–394.

Schachtsiek, M., Hammes, W.P., Hertel, C., 2004. Characterization of *Lactobacillus coryniformis* DSM 20001T surface protein Cpf mediating coaggregation with and aggregation among pathogens. Appl. Environ. Microbiol. 70, 7078–7085.

Selwitz, R.H., Ismail, A.I., Pitts, N.B., 2007. Dental caries. Lancet 369, 51–59.

Sengupta, R., Altermann, E., Anderson, R.C., Mcnabb, W.C., Moughan, P.J., Roy, N.C., 2013. The role of cell surface architecture of lactobacilli in host-microbe interactions in the gastrointestinal tract. Mediators Inflamm. 2013, 237921.

Shi, Y., Zhao, J., Kellingray, L., Zhang, H., Narbad, A., Zhai, Q., et al., 2019. In vitro and in vivo evaluation of *Lactobacillus* strains and comparative genomic analysis of *Lactobacillus plantarum* CGMCC12436 reveal candidates of colonise-related genes. Food Res. Int. 119, 813–821.

Soo, I., Madsen, K.L., Tejpar, Q., Sydora, B.C., Sherbaniuk, R., Cinque, B., et al., 2008. VSL#3 probiotic upregulates intestinal mucosal alkaline sphingomyelinase and reduces inflammation. Can. J. Gastroenterol. 22, 237–242.

Szajewska, H., Wanke, M., Patro, B., 2011. Meta-analysis: the effects of *Lactobacillus rhamnosus* GG supplementation for the prevention of healthcare-associated diarrhoea in children. Aliment. Pharmacol. Ther. 34, 1079–1087.

Tachedjian, G., Aldunate, M., Bradshaw, C.S., Cone, R.A., 2017. The role of lactic acid production by probiotic *Lactobacillus* species in vaginal health. Res. Microbiol. 168, 782–792.

Talarico, T.L., Casas, I.A., Chung, T.C., Dobrogosz, W.J., 1988. Production and isolation of reuterin, a growth inhibitor produced by *Lactobacillus reuteri*. Antimicrob. Agents Chemother. 32, 1854–1858.

Taverniti, V., Guglielmetti, S., 2011. The immunomodulatory properties of probiotic microorganisms beyond their viability (ghost probiotics: proposal of paraprobiotic concept). Genes Nutr. 6, 261–274.

Teughels, W., Loozen, G., Quirynen, M., 2011. Do probiotics offer opportunities to manipulate the periodontal oral microbiota? J. Clin. Periodontol. 38 (Suppl. 11), 159–177.

Teughels, W., Durukan, A., Ozcelik, O., Pauwels, M., Quirynen, M., Haytac, M.C., 2013. Clinical and microbiological effects of *Lactobacillus reuteri* probiotics in the treatment of chronic periodontitis: a randomized placebo-controlled study. J. Clin. Periodontol. 40, 1025–1035.

Thompson 3rd, G.R., Patel, P.K., Kirkpatrick, W.R., Westbrook, S.D., Berg, D., et al., 2010. Oropharyngeal candidiasis in the era of antiretroviral therapy. Oral Surg. Oral Med. Oral Pathol. Oral Radiol. Endodontol. 109, 488–495.

Tsilingiri, K., Rescigno, M., 2013. Postbiotics: what else? Benefic. Microbes 4, 101–107.

Tsui, C., Kong, E.F., Jabra-Rizk, M.A., 2016. Pathogenesis of *Candida albicans* biofilm. Pathog. Dis. 74, ftw018.

Turpin, W., Humblot, C., Thomas, M., Guyot, J.P., 2010. Lactobacilli as multifaceted probiotics with poorly disclosed molecular mechanisms. Int. J. Food Microbiol. 143, 87–102.

Tursi, A., Brandimarte, G., Giorgetti, G.M., Forti, G., Modeo, M.E., Gigliobianco, A., 2004. Low-dose balsalazide plus a high-potency probiotic preparation is more effective than balsalazide alone or mesalazine in the treatment of acute mild-to-moderate ulcerative colitis. Med. Sci. Monit. 10, PI126–PI131.

Urrutia-Baca, V.H., Escamilla-Garcia, E., de la Garza-Ramos, M.A., Tamez-Guerra, P., Gomez-Flores, R., Urbina-Rios, C.S., 2018. In vitro antimicrobial activity and downregulation of virulence gene expression on *Helicobacter pylori* by reuterin. Probiotics Antimicrob. Proteins 10, 168–175.

Van den Broek, A.M., Feenstra, L., De baat, C., 2007. A review of the current literature on aetiology and measurement methods of halitosis. J. Dent. 35, 627–635.

Vandenplas, Y., Huys, G., Daube, G., 2015. Probiotics: an update. J. Pediatr. (Rio J) 91, 6–21.

Vandeputte, D., Falony, G., Vieira-Silva, S., Wang, J., Sailer, M., Theis, S., et al., 2017. Prebiotic inulin-type fructans induce specific changes in the human gut microbiota. Gut 66, 1968–1974.

Verma, A., Gaffen, S.L., Swidergall, M., 2017. Innate immunity to mucosal *Candida* infections. J. Fungi (Basel) 3. Available from: https://doi.org/10.3390/jof3040060.

Vicario, M., Santos, A., Violant, D., Nart, J., Giner, L., 2013. Clinical changes in periodontal subjects with the probiotic *Lactobacillus reuteri* Prodentis: a preliminary randomized clinical trial. Acta Odontol. Scand. 71, 813–819.

Vimont, A., Fernandez, B., Ahmed, G., Fortin, H.P., Fliss, I., 2019. Quantitative antifungal activity of reuterin against food isolates of yeasts and moulds and its potential application in yogurt. Int. J. Food Microbiol. 289, 182–188.

Vivekananda, M.R., Vandana, K.L., Bhat, K.G., 2010. Effect of the probiotic *Lactobacilli reuteri* (Prodentis) in the management of periodontal disease: a preliminary randomized clinical trial. J. Oral Microbiol. 2. Available from: https://doi.org/10.3402/jom.v2i0.5344.

Volzing, K., Borrero, J., Sadowsky, M.J., Kaznessis, Y.N., 2013. Antimicrobial peptides targeting Gram-negative pathogens, produced and delivered by lactic acid bacteria. ACS Synth. Biol. 2, 643–650.

Wang, F.J., Zhang, D., Liu, Z.H., Wu, W.X., Bai, H.H., Dong, H.Y., 2016. Species distribution and in vitro antifungal susceptibility of vulvovaginal *Candida* Isolates in China. Chin. Med. J. (Engl.) 129, 1161–1165.

Weiss, G., Rasmussen, S., Zeuthen, L.H., Nielsen, B.N., Jarmer, H., Jespersen, L., et al., 2010. *Lactobacillus acidophilus* induces virus immune defence genes in murine dendritic cells by a Toll-like receptor-2-dependent mechanism. Immunology 131, 268–281.

Xie, H.Y., Feng, D., Wei, D.M., Mei, L., Chen, H., Wang, X., et al., 2017. Probiotics for vulvovaginal candidiasis in non-pregnant women. Cochrane Database Syst. Rev. 11, CD010496.

Yang, J., Yang, H., 2018. Effect of *Bifidobacterium breve* in Combination With Different Antibiotics on *Clostridium difficile*. Front Microbiol. 9, 2953.

Yang, X., Twitchell, E., Li, G., Wen, K., Weiss, M., Kocher, J., et al., 2015. High protective efficacy of rice bran against human rotavirus diarrhea via enhancing probiotic growth, gut barrier function, and innate immunity. Sci. Rep. 5, 15004.

Zhang, Z., Lv, J., Pan, L., Zhang, Y., 2018. Roles and applications of probiotic *Lactobacillus* strains. Appl. Microbiol. Biotechnol. 102, 8135–8143.

Ziakas, P.D., Mylonakis, E., 2014. ACP Journal Club. Probiotics did not prevent antibiotic-associated or *C. difficile* diarrhea in hospitalized older patients. Ann. Intern. Med. 160, JC6.

CHAPTER

MICROBIOMIC PROSPECTS IN FERMENTED FOOD AND BEVERAGE TECHNOLOGY

12

Paraskevi Bouki[1], Chrysanthi Mitsagga[1], Manousos E. Kambouris[2] and Ioannis Giavasis[1]

[1]Laboratory of Food Microbiology and Biotechnology, Department of Food Science and Nutrition, University of Thessaly, Karditsa, Greece [2]Golden Helix Foundation, London, United Kingdom

INTRODUCTION

Food fermentation is one of the oldest methods of food preservation and of worldwide application but is also a method for producing food with unique sensory characteristics, due to the growth of indigenous—or exogenous, in the case of added starter cultures—microbes that ferment substrate found within the food and transform the sugars, peptides, and lipids into organic acids, alcohol, volatile and aromatic compounds, polysaccharides, and other metabolites that affect the quality characteristics, the safety, and shelf-life of fermented foods and beverages (Hui et al., 2004). The word "fermentation" originally describes the anaerobic catabolism of sugars into metabolites of lower molecular weight (e.g., ethanol and lactate) in order to produce energy for microorganisms with fermentative metabolism. However, in the context of food production the term "fermentation" describes the spontaneous or controlled, anaerobic or aerobic, microbial process that converts any food component—predominantly carbohydrates—into new molecules, via microbial metabolism, and transforms plant and animal raw materials into completely different foods of high quality, palatability, shelf-life, and safety (Caplice and Fitzgerald, 1999). In addition, the extensive research on traditional types of fermented foods from many different communities, localities, and countries has revealed the highly versatile nature of the respective microbiomes and showed that most of them also exhibit significant probiotic properties and affect human health positively (Farnworth, 2003).

The use of industrially produced starter cultures in many modern fermented products enhances controllability and ensures a smooth, fast, and successful fermentation while facilitating the standardization of fermented products in terms of quality. Still, many fermented products are industrially produced by exploiting the natural, beneficial, autochthonous microbiota, as in the case of pickles, fermented olives, vinegar, soy sauce, and several alcoholic beverages (Steinkraus, 2004, 2018). To produce fermented products of distinct character reflecting local, distinct microbiomes that differ from the massively produced fermented foodstuff (the latter being of almost uniform taste and scent), many innovative entrepreneurs such as winemakers and cheese producers prefer natural, more traditional, and authentic fermentation(s) that result from autochthonous microorganisms, found in local raw materials, instead of commercially available starter cultures (Settanni and

Microbiomics. DOI: https://doi.org/10.1016/B978-0-12-816664-2.00012-8

Moschetti, 2014). This is evident most clearly in the recent trend for "natural wines," which is spreading in Europe and the United States (González and Dans, 2018).

Therefore the natural microbiome of fermented foods is crucial to their development and determines to a great extend the taste, scent, texture, and color, as well as the self-life and the safety of the final product. The above occurs through the transformation of the food substrate; the production of flavor, color, and textural compounds; and the interactions with other microorganisms. These interactions may include either stimulatory activity, for example, the production of vitamins and peptides by lactic acid bacteria (LAB) which are used up by other microorganisms, or inhibitory effects, such as the production of bacteriocins, lactate, acetate by LAB, or other bacteria (Jeevaratnam et al., 2005). Besides, even when starter cultures are used, these usually grow along with the indigenous microbiota (especially when using unpasteurized raw materials), which can be involved in different microbial interactions. Also, fermented foods are a rich source of probiotic microorganisms, and their consumption is believed to be beneficial for the health, as they may exert immunostimulatory, antimicrobial, anticancer, antiinflammatory, hypocholesterolemic, hypolipidemic, or hypoglycemic effects (Veiga et al., 2014). On the other hand, natural fermentations of foods with indigenous microbiota may incur some undesirable effects, such as the production of biogenic amines, as in wine or fermented meat, which may raise safety concerns (Ordóñez et al., 2016).

In recent years, molecular microbiology techniques such as high-throughput sequencing (HTS), metagenomics, and improved bioinformatics tools have resolved the microbiomes implicated in several naturally fermented foods (Ercolini, 2013). However, the exact roles and interactions of microbial species or communities in a food matrix during and after fermentation are complex and not fully understood (Wolfe and Dutton, 2015). In many cases, bacteria coexist with yeasts or molds in the active fermentative microbiome inside the food matrix and collaborate or compete for the utilization of available nutrients (Fig. 12.1). A complex microbiome is believed to enhance

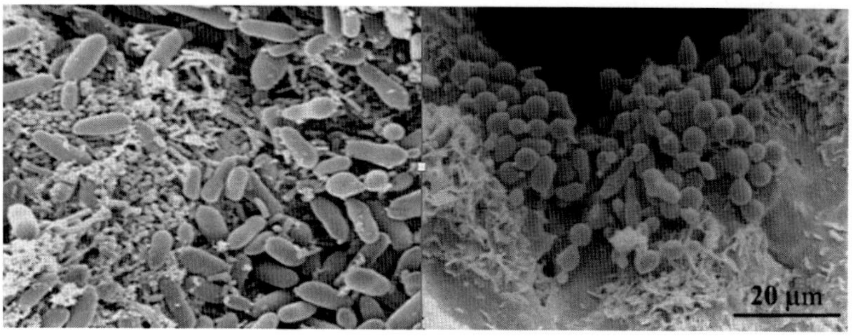

FIGURE 12.1

Scanning electron micrographs of microstructures and microbial communities of kefir grains (left) and black olives (right) fermented with mixed cultures of bacteria and yeasts.

Adapted from Garofalo, C., Osimani, A., Milanović, V., Aquilanti, L., De Filippis, F., Stellato, G., et al., 2015. Bacteria and yeast microbiota in milk kefir grains from different Italian regions. Food Microbiol. 49, 123–133; Nychas, G.J., Panagou, E.Z., Parker, M. L., Waldron, K.W., Tassou, C.C., 2002. Microbial colonization of naturally black olives during fermentation and associated biochemical activities in the cover brine. Lett. Appl. Microbiol. 34, 173–177.

product quality, due to a greater variety of flavor compounds or a better, accumulative, and diverse bioprotective effect against spoilage microorganisms (Wouters et al., 2002). Still, although the industrial food microbiology of standardized products and precisely controlled processes has been served well by simple molecular microbiology at its most evolved iterations, the current twist of the market toward novel foodstuff of unique "signature" requires genomics if not multiomics approaches. Similar products with special characteristics due to local conditions and indigenous species or strains of fermenting microbiota can be quality controlled and bulk produced without compromising their safety and quality characteristics, only by precise analysis of the implicated microbiomes and, subsequently, their metabolomes. This may amend seasonal fluctuations, contain microbial spoilage and pathogenicity, and stabilize the quality of commercially produced fermented food.

THE MICROBIOME OF NATURALLY FERMENTED DAIRY PRODUCTS

Fermented dairy foods such as yogurt, fermented or cultured milk, kefir, acidophilus milk, kumis, curd, buttermilk, and cheese have long been important in traditional and balanced dietary habits. They are usually fermented by different species of LAB, as is the case of yogurt, sour milk, feta cheese, cheddar cheese, or mozzarella. Alternatively, mixed fermentations by LAB and yeasts, as in kefir fermentation, or by LAB and molds, as in blue cheese fermentation may be implemented (Kumar et al., 2015). LAB are industrially important organisms recognized for their fermentative ability and comprise an ecologically diverse group of Gram-positive bacterial genera, namely, *Lactobacillus*, *Lactococcus*, *Leuconostoc*, *Pediococcus*, *Enterococcus*, *Tetragenococcus*, *Vagococcus*, *Carnobacterium*, and *Streptococcus*, which share some common characteristics, predominantly the formation of lactic acid from the metabolism of sugars, especially glucose and lactose (Zamfir et al., 2006). This diversity contributes to the taste, flavor, texture, color, and nutritional properties of the final products, which are highly appreciated for their organoleptic characteristics (Oberman and Libudzisz, 1998). LAB are generally recognized as safe for use in food and for direct consumption of live cells (Steele et al., 2013; Sharma et al., 2012; Carr et al., 2002; Holzapf et al., 2001).

LAB used in the dairy fermentations can roughly be divided into two groups: the mesophilic LAB, with optimum growth temperature between 20°C and 30°C, and the thermophilic ones, with optimum growth between 30°C and 45°C. Studies showed that the traditional fermented products from subtropical countries harbor mainly thermophilic LAB, whereas the products treated with mesophilic LAB originate from Western and Northern European countries (Gemechu, 2015). LAB are also divided by their metabolic products. Heterofermentative LAB (all *Leuconostoc* and some *Lactobacillus* species) produce CO_2, along with other metabolites such as acetate. Homofermentative LAB (all other LAB) do not produce CO_2 (Jay et al., 2008).

Starter LAB (SLAB, mainly *Lactococcus* and *Lactobacillus* species) are responsible for acid production and are involved in the initiation of the fermentation of milk. Non-SLAB (NSLAB) are primarily involved in the maturation process (e.g., during long-term cheese maturation). NSLAB include species of lactobacilli, pediococci, enterococci, and leuconostocs, which collectively contribute to the flavor of products such as cheese (Beresford et al., 2001).

Milk fermentation by LAB can be initiated either by inoculation of standardized starter cultures, selected for their acid production and flavor development capacity, or by indigenous LAB. The latter are used preferentially in some traditional products, acting as spontaneous starter cultures. When standard starter cultures are used, the autochthonous LAB naturally present in raw milk may still participate in the maturation of the products (Wouters et al., 2002). To acquire natural LAB, a common practice is to incubate raw milk or whey under controlled temperature and use it to inoculate milk (Cogan et al., 1997). This approach is still common in many Southern European countries and the resulting cheese is generally designated as "artisanal," meaning it being produced by farmers and shepherds (Cogan et al., 1997).

Dairy industries currently use starter cultures for rapid, controlled acidification, resulting in standardized organoleptic and physical properties of the dairy products, because the relatively small amounts of indigenous LAB in raw milk take effect rather slowly and may be outcompeted by other indigenous microbiota, such as members of Enterobacteriaceae or Bacillaceae family, leading to cheese spoilage and fouled fermentation (Trmčić et al., 2015; Montel et al., 2014; Dal Bello et al., 2010). The use of starter cultures is necessary when fermenting pasteurized milk, which is the industrial standard so as to conform to safety regulations Thus the dairy industry has introduced new fermentation techniques that utilize the fresh raw milk microbiota as an inoculum in addition to SLAB and NSLAB cultures, and this combination and microbial complexity are believed to improve the characteristic flavors of traditional cheeses (Wouters et al., 2002).

In artisanal cheese the indigenous microbiota of raw milk, as well as the local microenvironment of the cheese-making facilities and the processing equipment contribute to the organoleptic characteristics and the typical features of high-value traditional cheese varieties (Garabal, 2007). From 38 artisanal cheese, 4379 isolates were identified by solid culturing, microscopic and biochemical tests as *Lactococcus* (38%), *Enterococcus* (17%), *Streptococcus thermophilus* (14%), mesophilic *Lactobacillus* (12%), *Leuconostoc* (10%), and thermophilic *Lactobacillus* (9%) (Table 12.1). Considerable variation was found in the types of LAB present in all these artisanal products and in their ability to produce acids, EPS, bacteriocins, and proteinases, suggesting that each cheese is a unique ecosystem (Cogan et al., 1997).

More recently, the health benefits of many fermented dairy products have been the subject of intense investigation. Many LAB, especially *Lactobacillus* species, which can withstand the acidic environment of the stomach, the antibacterial activity of bile salts, and peptic enzymes in the gut and thus colonize the intestinal epithelium are known for their probiotic properties, which include hypolipidemic, hypocholesterolemic, antiinflammatory, antimicrobial, and anticancer effects (Parmjit, 2011). For instance, *Lactobacillus acidophilus*, *Lactobacillus casei* (Shirota), *Lactobacillus plantarum* have been studied and used industrially due to their probiotic properties (Marco et al., 2017). Also, *Lactobacillus* species such as *L. helveticus* are believed to produce bioactive peptides from milk casein and have showed antihypertensive effect, immunomodulatory activity, and anticancer properties (Nouaille et al., 2003). Besides, many LAB can improve the nutritional value of fermented milk products, either by producing nutritious metabolites, such as vitamins (e.g., B9, B12) and short chain fatty acids (Forssén et al., 2000; Buttriss, 1997; Rao et al., 1984) or by hydrolyzing lactose into glucose and galactose by β-galactosidase (lactase), which makes these products more suitable for people with lactose intolerance (Prentice, 2014; Otieno, 2010). The multiple health benefits, enhanced nutritious value, and the fact that fermented dairy foods are rated as "natural"—as opposed to "processed"—products have further increased their

Table 12.1 Identification and Quantification of Isolates From Different Artisanal Dairy Products.

Product	*Lactococcus*	*Leuconostoc*	*Enterococcus*	*Streptococcus thermophilus*	Atypical cocci	Mesophilic *Lactobacillus*	Thermophilic *Lactobacillus*	Atypical rods
Kefir/fermented milk	361	41						
Kasseri cheese	5		90		90		47	60
Feta cheese	100		28		3		24	4
Galotyri cheese							3	
Scottain nesto			2	93			54	
Fiore Sardo cheese	17		30					
Casuaxedu cheese	73		11			13	3	
Mozzarella, natural starter	57	1	1	41				8
Caciotta, natural starter				30				
Grana, natural starter								73
Pecorino, natural starter				12				61
Manchego cheese	169	101	59			31		
Gredos cheese	186	58	74			55		
Majorero cheese	47	11	74		10	107		
Fresh French cheese	105	19	12		6	1		2
Chevret cheese	12		3		1	1		
Tomme cheese	72	10	17	2	5	19	2	

Adapted from Cogan, T.M., Barbosa, M., Beuvier, E., Bianchi-Salvadori, B., Cocconcelli, P.S., Fernandes, I., et al., 1997. Characterization of the lactic acid bacteria in artisanal dairy products. J. Dairy Res. 64, 409–421.

popularity (Panesar, 2011; Kumar et al., 2015). Thus cultured dairy foods such as "Bioghurt," "Yakult," "Actimel," and "Acidophilus milk" have been marketed as therapeutic and dietetic products.

RESOLVING THE COMPOSITION OF THE MICROBIOMES

Studies on the microbial diversity of naturally fermented milk have relied for many years on culture-dependent methods allowing observation, isolation, and subsequently identification of specific isolates. However, difficulties in simulating the proper growth conditions and the uncultivable status of many microorganisms hindered detection of certain microbial taxa (Ercolini et al., 2001). Many PCR-based, sequencing-based, or other molecular analysis methods are now available for typing, identification, or even enumeration of microbial communities in food and environmental samples, along with advanced non-NAA approaches (Table 12.2). Some of them have already been applied in the investigation of fermented dairy food microbiomes, such as denaturing gradient gel electrophoresis (DGGE), single-strand conformational polymorphism, DNA microarrays, whole genome sequencing, and proteomics studies (Vinusha et al., 2018; Jany and Barbier, 2008; Pfeiler and Klaenhammer, 2007). Application of microbiological, molecular, and physiological techniques allowed the study of complex microbiomes involved in the production of fermented milk and cheese. For instance, the random amplification of polymorphic DNA (RAPD) technique was applied for the characterization of the lactobacilli down to strain level; the sequences of the major rDNA gene complex have been used for characterization and identification of diverse and multiple taxa (Baruzzi et al., 2000). These techniques, coupled to physiological and biochemical assays, have also revealed the presence of undesirable bacteria in cheese batches submitted to prolonged maturation in uncontrolled temperature and variable salt content (Baruzzi et al., 2000).

High-throughput pyrosequencing detected 135 genera in 85 samples (55 yogurts, 18 yak milks, 6 types of kumis, and 6 types of cheese) collected from China, Mongolia, and Russia (Zhong et al., 2016). The genera *Lactobacillus*, *Lactococcus*, *Streptococcus*, *Acetobacter*, *Acinetobacter*, *Leuconostoc*, and *Macrococcus* accounted for 96.63% of the bacterial sequences retrieved. Different types of fermented dairy products were characterized by distinct predominant LAB genera. *Lactococcus* (55.42%) and *Streptococcus* (31.90%) were predominant in cheese; *Lactobacillus* (82.75%) and *Streptococcus* (13.93%) in kumis; *Lactobacillus* (78.44%), *Streptococcus* (14.08%), and *Lactococcus* (5.30%) in fermented (sour) yak milk; and *Lactobacillus* (64.69%), *Lactococcus* (14.62%), *Streptococcus* (10.29%), and *Acetobacter* (4.78%) in yogurt. Intergenus competition seems to occur occasionally: the relative abundance of *Lactobacillus* was negatively correlated with *Streptococcus*, *Lactococcus*, *Enterococcus*, *Acinetobacter*, *Leuconostoc*, *Acetobacter*, and *Macrococcus* (Zhong et al., 2016).

Several *S. thermophilus* isolates from Italian DPO cheeses were typed by RAPD (Table 12.2) and attributed according to the milk origin (Andrighetto et al., 2002). DGGE of PCR-amplified bacterial 16S rDNA fragments (Table 12.2), as well as reverse transcriptase–PCR, revealed the shift in microbial populations during production and maturation of artisanal Ragusano-type cheese, produced in different areas of the Hyblean region in Sicily: in raw milk Leuconostoc, *Lactococcus lactis*, and Macrococcus caseolyticus were the dominant bacteria, while S. thermophilus took the lead during lactic fermentation, and *S. thermophilus* along with thermophilic *Lactobacillus delbrueckii* and *Lactobacillus fermentum* prevailed during maturation (Randazzo et al., 2002).

Table 12.2 Synopsis of Methods Used to Determine Fermenting and Spoilage Microbiota in Food and Beverage Microbiomes.

Nucleic Acid Analysis (NAA) Methods	Non-NAA Methods
RAPD **/ T**	MALDI-TOF MS **/ IQ**
qPCR **/ I T Q**	flow cytometry **/ ITQ**
Simplex, specific PCR **I T Q**	Microscopy **/ IQ**
Multiplex PCR **/ I T**	Culture (macromorphology, biochemical, and physiological tests) **/ I T Q**
PCR-TGGE **/ I T**	
RT-PCR **/ I T Q**	
PCR-RFLP/REA**/ I T**	
PCR-DGGE**/ I T**	
PCR-SSCP**/ I T**	
PCR-sequencing **I T Q**	
PFGE-RFLP/REA **/ I T**	
DNA microarrays **/ I T Q**	
Species-specific Hybridization **/ I Q**	
FISH **/ I T Q**	
NGS**/ I T Q**	
WGS**/ I T Q**	
HTS**/ I T**	

Red: PCR-based or, provisionally, PCR-dependent methods; Blue: sequencing-based methods; Black: non-amplification-based methods. The envisaged results per method may be of any combination of three basic categories, these being I: identification; T: typing; Q: quantification (including semiquantification).

WILD LACTOCOCCI

Lactococci isolated from artisanal manufacture of fermented dairy products without the application of industrially prepared starter cultures and from nondairy environments are generally referred to as "wild" lactococci. *Lactococcus* strains have been isolated from kefir, naturally fermented Mozzarella, feta, kasseri cheese, and other products (Table 12.1).

Studies showed that wild *L. lactis* strains differ in phenotype from respective industrial starter strains, as does *L. lactis* subsp. *lactis* which hydrolyzes arginine, metabolizes a number of sugars, and grows at 40°C and/or in the presence of 4% NaCl. *L. lactis* subsp. *cremoris* cannot grow under these conditions (Ayad et al., 1999; Cogan and Accolas, 1996). In addition, wild lactococci differ from reference strains by RAPD typing: in seasonal samples of raw milk from the Camembert cheese registered designation of origin area, strains of lactococci were isolated and while they had uniform L. lactis subsp. *lactis* phenotype, their RAPD signature identified them either as L. lactis subsp. *lactis* or as L. lactis subsp. *cremoris* (Corroler et al., 1998).

Amino-acid requirements of wild lactococcal strains are lower than these of the industrial strains that are better adapted to the milk environment. The highly diversified microbiomes in raw milk or dairy and nondairy environments harbor a larger pool of enzymes such as active amino-acid convertases, which produce amino acid–derived flavor components, and thus increase the flavor intensity and complexity of dairy products fermented by indigenous microbiomes, compared to respective products fermented by commercial starter cultures (Ayad et al., 2000; Ayad et al., 1999; Deguchi and Morishita, 1992).

Organic acids (e.g., lactate and acetate) and bacteriocins produced by LAB are generally considered as safe natural antimicrobials, contributing to the preservation and safety of dairy and other fermented foods (Zacharof and Lovitt, 2012). Bacteriocins are antimicrobial proteins or peptides, active against closely related bacterial species, and can be used to control the growth of spoilage and pathogenic microbiota, such as *Listeria monocytogenes* and *Clostridium* species (Cosentino et al., 2012; O'sullivan et al., 2002; De Vuyst and Vandamme, 1994; Klaenhammer, 1993). Most LAB bacteriocins are cationic, heat-tolerant, amphiphilic peptides of low molecular weight (<10 kDa).

Bacteriocin-producing lactococcal strains have been used successfully in starter cultures to improve the safety and quality of the cheese (Maisnier-Patin et al., 1992; Delves-Broughton et al., 1996). Out of 79 strains of wild lactococci isolated from artisanal production of dairy products and nondairy wild strains, 32 were antimicrobially active in agar-well diffusion assay against two target organisms, either *Micrococcus flavus* NIZO B423 or *L. lactis* subsp. *cremoris SK110*, producing bacteriocins and bacteriocin-like compounds (Ayad et al., 2002). The high percentage of bacteriocin-producing strains is probably due to the highly competitive and complex habitat, where bacteriocins provide a competitive advantage.

MESOPHILIC LACTOBACILLI

Mesophilic lactobacilli reside in raw milk and dairy products, although acidification of their environment allows more vigorous growth of the genus *Lactococcus* (Wouters et al., 2002). Mesophilic lactobacilli are a significant component of the microbiome of Kasseri, Feta, Serra da Estrela, Gredos, and Majorero cheese (Cogan et al., 1997), as they often function as secondary fermentation flora, during the maturation of different cheese varieties. Nonstarter lactobacilli originating from raw milk or the environment may outgrow inocula of the starter culture in cheese varieties made from pasteurized milk (Shakeel-Ur-Rehman et al., 2000).

NSLAB are important in cheese maturation, as they release proteolytic and lipolytic enzymes, thus producing small peptides and amino acids and contributing to the maturation of cheese—especially in varieties from raw milk (Williams and Banks, 1997; Fox, 1998). Selected adjunct NSLAB cultures using strains isolated from raw milk cheese can be added to accelerate maturation and to produce desirable flavor (McSweeney et al., 1993; Beresford et al., 2001; Law, 2010). NSLAB isolates from Cheddar were reported to belong to the species *Lactobacillus paracasei*, *L. plantarum*, *Lactobacillus rhamnosus*, and *Lactobacillus curvatus* and their quantitative representation varied by the day of fermentation and the age of cheese (Fitzsimons et al., 2001). In Spanish artisanal starter-free cheese varieties, such as hard, semihard, and blue-veined varieties made from ewes' and goats' milk, homofermentative NSLAB were classified as *L. plantarum*, *L. casei* subsp. *pseudoplantarum*, *L. curvatus*, and *L. casei* subsp. *casei* by phenotypical and biochemical characterization and grouping in clusters using the simple match coefficient (SMC) and the unweighted pair group algorithm with arithmetic averages (UPGMA) (Lopez and Mayo, 1997). Other strains of *L. curvatus*, *L. paracasei* subsp. *paracasei*, *L. plantarum*, and *Lactobacillus brevis* were identified in Pecorino Toscana cheese and 12% of the isolates from artisanal cheese varieties were characterized as mesophilic lactobacilli, by combining phenotypic and genotypic analyses (Bizzarro et al., 2000; Cogan et al., 1997).

Apparently, NSLAB population during maturation in Cheddar cheese may include up to 20 strains (Crow et al., 2001), and some strains may affect the development of others. For instance, the growth of a *L. rhamnosus* strain, added on purpose as an adjunct culture in a Cheddar cheese trial, could be delayed by the addition of a *L. casei* strain (Martley and Crow, 1993).

On the other hand, the presence of NSLAB in cheese may result in defects or spoilage, undesirable flavors, like those caused by *L. brevis* usually during the later stages of maturation (Peterson and Marshall, 1990), or synthesis of potentially toxic amines responsible for food poisoning, such as histamine, produced by the heterofermentative *Lactobacillus buchneri* in Swiss-type cheese (Linares et al., 2011; Taylor et al., 1982). Spoilage may be due to succession of microbial populations, especially in spontaneous, natural milk fermentation. For example, the lactate produced by NSLAB can be utilized by *Clostridium* species such as *C. tyrobutyricum* to produce butyric acid, which is perceived as off-flavor in yogurt, sour milk, and many types of cheese. CO_2 production by heterofermentative NSLAB can further induce the growth of anaerobic Clostridia and result in additional CO_2 formation by saccharolytic Clostridia, as in the case of the "late gas" spoilage during cheese maturation (Sheehan, 2007).

THERMOPHILIC LACTIC ACID BACTERIA

Thermophilic LAB are best known as starters for fermented milks that were originally produced from spontaneous acidification of raw milk by indigenous microbiota. Nowadays, the fermentation of milk into yoghurt is a standardized industrial process using starter cultures to develop the desired viscosity, acidity, and flavor. The typical starter culture used in yogurt production is a 1:1 mixture of *L. delbrueckii* subsp. *bulgaricus* and *Streptococcus salivarius* subsp. *thermophilus*, added in milk and incubated at 37°C–45°C. These two species act synergistically, as the production of formic acid and carbon dioxide from lactose by the latter stimulates the growth of the former (Driessen et al., 1982). As the pH drops lactobacilli become dominant in yogurt, due to their high tolerance in acidic environments, especially during yogurt preservation (Makino et al., 2006; Antunes et al., 2005).

The texture of yogurt is affected by the secreted polysaccharides, which include homopolysaccharides (HoPS) composed of a single type of monosaccharide and heteropolysaccharides (HePS), containing several types of monosaccharides; both are produced by *Lactobacillus* species, such as the abovementioned L. delbrueckii subsp. *bulgaricus* (Han et al., 2016; Lamothe et al., 2002).

The characteristic flavor of yogurt is the result of the formation of aromatic compounds from milk, such as acetaldehyde that produces a green apple or nutty flavor (Chaves et al., 2002), acetoin, diacetyl, 2,3-butanediol, and ethanol, resulting from LAB-induced lactose metabolism and subsequent citrate formation, as shown in Fig. 12.2 (Chen et al., 2017; Routray and Mishra, 2011, Cheng, 2010; Neves et al., 2005; Slocum et al., 1988). Other flavor compounds such as aromatic peptides, ammonia, amines, aldehydes, and indole are formed via proteolysis and amino-acid conversion by LAB, while lipolysis by LAB results in the release of aromatic short chain fatty acids and their further oxidation to aldehydes, ketones, and ketoacids (Chen et al., 2017; Cheng, 2010).

Vitamin synthesis is also of great interest. For instance, *S. thermophilus* produces folic acid (Vitamin B9) during milk fermentation for the production of yogurt; thus higher concentration of

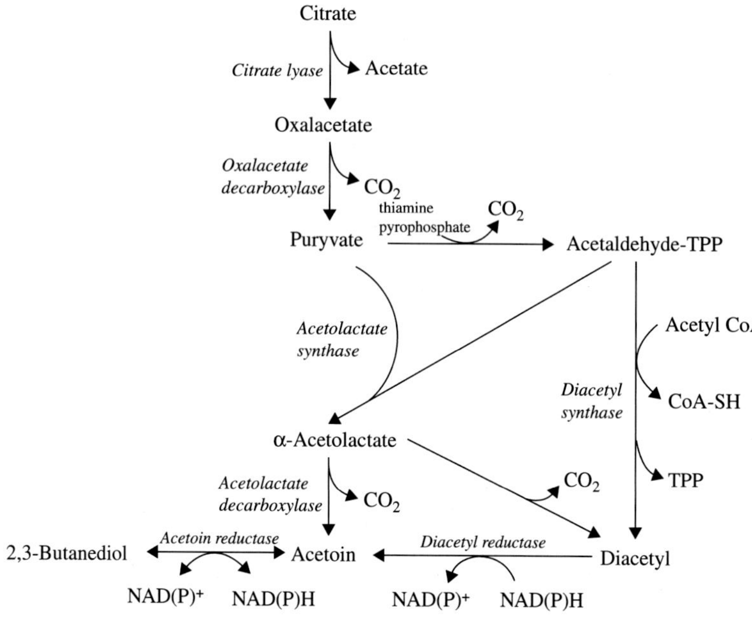

FIGURE 12.2

Metabolic pathways for the formation of acetaldehyde, diacetyl, acetoin, and 2,3-butanediol in yogurt fermentation.

Adapted from Cheng, H., 2010. Volatile flavor compounds in yogurt: a review. Crit. Rev. Food Sci. Nutr. 50 (10), 938–950.

folic acid is observed in yoghurt than in milk. An important goal in yogurt fermentation is to select strains balancing optimal organoleptic characteristics and increased content of folic acid (Wouters et al., 2002; Rao et al., 1984).

THE MICROBIOME OF NATURALLY FERMENTED MEAT PRODUCTS

Meat is an irreplaceable source of protein, fat, vitamins, minerals, and other nutrients, as well as a very favorable medium for growth and propagation of microbes. In meat fermentations, bacteria, yeasts, and molds coexist, forming a challenging microbiome that participates in the fermentation process. Microorganisms native of meat (or of the spices added into meat) such as LAB take part in the processes of fermentation and ripening of meat products (Cukon et al., 2012).

The fermentation process is controlled by the supplemented sodium chloride and the gradual drying process, which transform raw meat and fat into edible, tasty, ready-to-eat final product, safe for human consumption, and stable during preservation, with highly appreciated organoleptic characteristics (Cocolin et al., 2011).

The main source of the indigenous microbiome is the animal carcass, the implicated workforce, processing tools and equipment, as well as the production environments. It mainly consists of Gram-negative intestinal bacteria such as *Escherichia coli*, *Salmonella* spp., *Pseudomonas* spp., and of Gram-positive bacteria (mainly LAB), such as lactobacilli, pediococci, and enterococci, which take part in meat fermentation (Cukon et al., 2012). Apart from LAB, meat and meat products also contain other Gram-positive bacteria; both aerobic, such as *Bacillus*, *Staphylococcus*, and *Listeria*, and anaerobic, such as *Clostridium*, *Peptococcus*, and *Ruminococcus* (Holzapfel, 1998). Likewise, *Penicillium*, *Mucor*, *Cladosporium*, *Candida*, and *Rhodotorula* are the most frequent genera of fungi (Cukon et al., 2012).

The main LAB genera isolated from fermented dry sausages are *Lactobacillus*, *Pediococcus*, *Leuconostoc*, *Weissella*, and *Enterococcus* (Albano et al., 2009; Ammor and Mayo, 2007). The most frequently detected LAB species in dry sausages are *Lactobacillus sakei*, *L. curvatus*, and *L. plantarum* (Kozačinski et al., 2008; Papamanoli et al., 2003; Hammes et al., 1990). Dry salami is also characterized by the presence of some *Pediococcus* species, such as *P. acidilactici*, *P. pentosaceus*, and *P. dextrinicus* (Yuksekdag and Aslim, 2010) and *Weissella* species, such as *W. viridescens*, *W. hellenica*, and *W. paramesenteroides* (Papamanoli et al., 2003; Samelis et al., 1994). *Lactobacillus pentosus*, *L. paracasei*, *L. casei*, and *L. alimentarius* have been isolated more sparingly (Aymerich et al., 2006; Papamanoli et al., 2003). *Enterococcus faecium* and *Enterococcus faecalis* were the dominant species among enteroccocal isolates in slightly fermented Spanish sausages, as revealed by RAPD and sequencing of 16S rRNA and *sodA* genes (Martin et al., 2005).

The indigenous LAB found in fermented meats are adapted to the respective environment, which controls the ripening processes and inhibits the growth of haphazardly present microorganisms (Hugas and Monfort, 1997). The salting and dry-curing of fermented meat products cause a "microbial inversion" and the primary representatives of the initial microbiome of the raw materials (including many Gram-negative bacteria) are replaced by *lactobacilli*, *micrococci*, *enterococci* in the final dry-cured meat product (Mioković and Zdolec, 2004).

The population of LAB and Micrococcaceae (*Staphylococcus*, *Micrococcus*, and *Kocuria* species) in the raw materials is usually low ($\sim 10^3-10^5$ cfu/g), but it increases during fermentation and ripening for 3–4 weeks to $\sim 10^7$ cfu for staphylococci and 10^8 cfu/g or higher for lactobacilli in traditional sausages (Fonseca et al., 2013; Comi et al., 2005; Greco et al., 2005; Coppola et al., 2000). The growth and reproduction of the beneficial Gram-positive bacteria (LAB, staphylococci, micrococci) is enhanced by salts like sodium chloride, which acts as humectant, depriving microorganisms of water, and sodium nitrite and/or sodium nitrate, which have antimicrobial activity against both Gram-negative pathogens such as Enterobacteriaceae species, and Gram-positive ones like *Clostridium botulinum* (Cukon et al., 2012; González and Dıez, 2002), while leaving intact the population of LAB and Micrococcaceae (González and Dıez, 2002).

Furthermore, during spontaneous or controlled fermentation of meat LAB produce lactic acid by the fermentation of glycogen in meat, which leads to a significant decrease of pH value to 5.9–4.6 and they participate to the development and stability of the color and contribute to the overall flavor, through their antioxidant properties and the production of free amino acids (Leroy et al., 2006; Talon et al., 2002; Hammes and Knauf, 1994). Moreover, LAB are responsible for the nice, "tangy" flavor of sausages and produce small amounts of acetic acid, ethanol, acetoin, pyruvic acid, and carbon dioxide (Demeyer, 1982; Pearson and Dutson, 1986). Also, due to the increased acidity, muscle proteins coagulate, which leads to the decrease in their elasticity and hydration ability, further facilitating the dehydration process (Cukon et al., 2012).

Notably, several LAB produce bacteriocins in fermented meat, which act against other LAB, but also against the main pathogens usually present therein, such as *L. monocytogenes*, *Staphylococcus aureus*, *Clostridium perfringens*, *Escherichia* spp., and *Salmonella* spp. (Oliveira et al., 2018; Cukon et al., 2012). *L. curvatus*, *L. sakei*, *P. acidilactici*, and *P. pentosaceus* are some of the most common bacteriocinogenic agents of the native microbiome in raw-fermented sausages but are also available as bioprotective starter cultures and as biocontrol agents for meat products (Oliveira et al., 2018; Cukon et al., 2012; Yuksekdag and Aslim, 2010).

With regard to LAB population diversity in fermented meat, different kinds of sausages produced in Hungary, Italy, and Greece were analyzed and from 295 LAB isolates, clustering analysis yielded 5 different clusters for the 27 isolates of *L. plantarum*, 9 clusters for the 100 isolates of *L. curvatus* and 19 clusters for the 168 isolates of *L. sakei* (Rantsiou et al., 2005a,b). In terms of intraspecies diversity, from 250 LAB isolates collected from 21 different Spanish slightly fermented sausages, 144 different strains were isolated, of which 112 were *L. sakei*, 23 were *L. curvatus*, and 9 were *L. mesenteroides* after analysis by a combination of RAPD typing and plasmid profiling (Aymerich et al., 2006). In addition, 73 isolates of *L. sakei* sourced from different countries were determined using a PCR-based method that detects the presence of 60 chromosomal genes belonging to the collective gene pool of the species (Table 12.2). The identification of strain clusters of different genotypes in similar types of meat products suggests a complex microbiome where intraspecies diversity may be required for successful adaptation (Chaillou et al., 2009) or, alternatively, caused in its wake.

To explore microbial dynamics during fermented sausage manufacturing, culture-dependent and culture-independent methods are used, as the latter detect the DNA of both culturable and nonculturable microorganisms. The culture-independent method of choice in sausage fermentation is DGGE (Rantsiou and Cocolin, 2008). Furthermore, different methods for the identification of microbial species of interest for sausage fermentation have been developed, due to the increasing availability of the sequences of the 16S rRNA gene and the intergenic region between 16S rRNA and 23S rRNA genes (Table 12.2). For the identification of LAB and coagulase-negative cocci isolated from fermentation sausages, an impressive array of methods has been applied, including, but not restricted to, probes targeting rDNA, species-specific PCR primers, restriction fragment length polymorphism (RFLP), sequence analysis of the 16S rRNA gene coupled to HTS technology, multiplex PCR, temperature-gradient gel electrophoresis, and DGGE coupled with DNA sequencing (Stellato et al., 2015; Rantsiou and Cocolin, 2008). LAB strains with optimal adaptation in meat fermentation can be isolated, propagated, and used as functional starters (Talon et al., 2007; Rantsiou et al., 2006) as they ensure safety and stable organoleptic qualities during fermentation and ripening (Coppola et al., 1997).

Apart from LAB, the genera *Micrococcus* and *Staphylococcus* play significant roles in fermented meat products, as indigenous flora or starter cultures. They are more tolerant than LAB to the increased concentrations of salt of dry-cured meats, thus surviving and growing in environments with significantly lower water activity, as that of ripe fermented meats (Cukon et al., 2012). *Micrococcus* spp. are usually abundant in meat products of longer ripening (Frece et al., 2014). The staphylococci frequently represent the second largest population in dry fermented meat, varying from 10^5 to 10^7 cfu/g in traditional sausages (Iacumin et al., 2006; Blaiotta et al., 2004). Among coagulase-negative staphylococci (CNS), *Staphylococcus carnosus* and *Staphylococcus xylosus* are the most frequently isolated strains, while *Staphylococcus equorum* and *Staphylococcus saprophyticus* are also isolated from dry sausages (Leroy et al., 2010; Bonomo et al., 2009; Cocolin et al., 2001; Coppola et al., 2000).

Micrococcus and *Staphylococcus* spp. stabilize the red color in meat products, due to the production of nitrate reductases that reduce nitrates to nitrites, as well as nitrite reductases, which are both involved in reactions leading to nitric oxide (NO) formation; the latter leads to the formation of nitrosomyoglobin, the pigment that renders the characteristic red color in cured and fermented meat (Hammes, 2012). Since staphylococci are sensitive to acids, reduction of nitrates to nitrites takes place at an early state of fermentation, before LAB reach their peak population and increase acidity significantly (Cukon et al., 2012). In addition, *Micrococcus* and *Staphylococcus* species contribute to the development of flavor in fermented meat products, due to their proteolytic and lipolytic activity (Yamanaka et al., 2005). Their amino-acid and lipid catabolism during fermentation and ripening releases compounds of low molecular weight such as amino acids, peptides, free-fatty acids, aldehydes, amines, and esters (Frece et al., 2014; Leroy et al., 2006). In long-term fermented ham (420−600 days) the most profound flavor compounds were methyl-alcanoates, such as 2-methyl-ethyl-butanoate, ethyl hexanoate, and other esters (Ruiz et al., 1999). Another useful trait of micrococci and staphylococci is their ability to decompose hydrogen peroxide—which may be produced by indigenous lactobacilli and pediococci—thanks to their strong catalase activity, thus preventing the formation of grayish or greenish discoloration in fermented meat (Hammes and Knauf, 1994).

Despite being essential for meat fermentation, LAB (and to some extent staphylococci) can be harmful, due to biogenic amine formation, which constitutes a health risk. Free amino acids can be decarboxylated by LAB, resulting in the release of biogenic amines (Pereira et al., 2001; Rice and Koehler, 1976), a feature that is species- and strain dependent. Up to 48% of the LAB isolated from dry fermented sausages decarboxylated one or more amino acids (Latorre-Moratalla et al., 2010). *L. curvatus*, *E. faecium*, and *E. faecalis* strains produced tyramine and phenylethylamine, whereas all the *L. sakei* strains were nonaminogenic (Latorre-Moratalla et al., 2010; Aymerich et al., 2006; Hugas et al., 2003). In contrast, the percentage of aminogenic CNS is low, representing 6%–14% of the total isolates from fermented meat (Even et al., 2010; Latorre-Moratalla et al., 2010; Martín et al., 2006a,b). Biogenic amine production by staphylococci was related to *S. xylosus*, *Staphylococcus warneri*, *Staphylococcus epidermidis*, and *S. carnosus* which mainly produced phenylethylamine and, to a lesser extent, tyramine (Martín et al., 2006a,b).

Therefore through the exploitation of autochthonous microbiota from naturally fermented meat products and the selection of LAB and CNS by their acidification profile, the production of bacteriocin and aromatic compounds, their reducing capacity, and the absence of amino-decarboxylase or amino-oxidase activity, starter cultures can be produced upon specifications and applied in meat fermentations, thus ensuring product safety, palatability, and high and stable quality characteristics (Talon and Leroy, 2011).

Yeasts and molds are also important parts of autochthonous microbiota in fermented meat and may be added as starter cultures. *Penicillium* spp. are the most ubiquitous: They can grow at low temperature and low water activity (a_w); they have antioxidant and proteolytic effects and limited lipolytic activity, which contribute to the development of flavor and contribute to the formation and preservation of the red coloration. They also provide a typical external appearance to sausages and, when growing on the surface of fermented sausages, they create a favorable microclimate, which prevents excessive dehydration of their skin (Martinović and Vesković-Moračanin, 2006; Martín et al., 2006a; Ludemann et al., 2004; López-Díaz et al., 2001). The characteristic aroma of fermented sausages is mostly produced by white molds, such as *Penicillium camemberti* (Bruna et al., 2003) and *Penicillium nalgiovence* (Ludemann et al., 2004). Notably, although *P. nalgiovence* showed the highest proteolytic activity among several *Penicillium* isolates of Argentinian dry salami, there is strong interspecies variability in the proteolytic capacity which is favored in lower fermentation temperatures (14°C), in contrast to the fungal growth which is faster at 25°C (Ludemann et al., 2004). Regarding yeasts, the genus *Debaryomyces* (e.g., *D. hansenii*, which is also used as starter culture) is most commonly encountered in fermented sausages, but *Trichosporon*, *Candida*, and *Cryptococcus* species are also present (Coppola et al., 2000; Metaxopoulos et al., 1996).

Although yeasts and molds grow slowly compared to LAB and micrococcaceae, they contribute to the end sensory characteristics of fermented sausages by increasing the quantity of ammonia, producing free amino acids, aromatic compounds such as ethyl-2-methyl propanoate, ethyl-2-butanoate, and 2-methyl-butanoate, decreasing the quantity of lactic acid and exerting proteolytic activity (Flores et al., 2004; Santos et al., 2001). The aromatic compounds produced by indigenous fungi in naturally fermented ham contained more short chain aliphatic carboxylic acids and their esters, branched carbonyls and alcohols and some sulfur compounds than those produced by ham inoculated with *Penicillium chrysogenum* and *D. hansenii*. The latter resulted in increased concentrations of long chain aliphatic and branched hydrocarbons, furanones, esters of long chain carboxylic acids (Martín et al., 2006a). The yeast population in fermented sausages inoculated with *D. hansenii* has been also correlated with the reduction of undesirable volatile products of lipid oxidation such as aldehydes and hydrocarbons, which means they can prevent or delay rancidity (Flores et al., 2004). This antioxidant effect was also evident by the decrease in the development of yellow coloration on the superficial fat of fermented hams inoculated with *D. hansenii* (Martín et al., 2006a).

In terms of population dynamics, yeasts prevail on the surface of fermented sausages at the beginning of the fermentation, but after 2 weeks molds are represented equally on the surface and prevail toward the end of ripening (Toldrá, 2010). In fermented sausages, yeasts are also found internally as well as on the surface, while molds are not, due to lack of oxygen (Coppola et al., 2000). One drawback of the presence of molds on fermented salami and ham is the potential presence of mycotoxins, especially ochratoxin-A, but citrinin, cyclopiazonic acid, and aflatoxin B1 may also be present to levels close to detection limits as well (Pleadin et al., 2015; Markov et al., 2013; Rodríguez et al., 2012; Iacumin et al., 2009; Bailly et al., 2005). Ochratoxin-A (OTA), in particular, is commonly found on the surface of fermented meat in concentrations up to 10-fold the recommended safety levels for meat products in EU (Pleadin et al., 2015), although in fermented sausages OTA is only present on the skin which can be removed, or brushed and washed, leaving no molds and no detectable OTA on the outer surface of the sausages (Iacumin et al., 2009). *Penicillium nordicum* and *Penicillium* verrucosum are the main species that have been linked to OTA synthesis and their accumulation in fermented meats has been detected by real-time quantitative PCR (qPCR, Table 12.2) and targeting the *otanps*PN gene (Rodríguez et al., 2012).

THE MICROBIOME OF NATURALLY FERMENTED OLIVES AND PICKLES

Fermentation of vegetables (pickles and olives) is an ancient preservation method, as fresh vegetables are very susceptible to microbial spoilage (Wouters et al., 2013). Fermented vegetables and olives exploit the native microbiomes of the plants (Karasu et al., 2010).

TABLE OLIVES

Table olives are divided into three categories: (1) the green Spanish-style, unripe olives, which are debittered by alkaline oxidation; (2) the naturally debittered (in water/brine) black, ripe olives known as Greek-style olives; and (3) the California-style olives, ripe and debittered by alkaline oxidation (Bautista-Gallego et al., 2011). Generally, both the green and black olives are spontaneously fermented by a mixture of yeasts and LAB (Tofalo et al., 2012). Though the fermentation of green, Spanish-style olives is performed mainly by LAB, while black, ripe, Greek-style olive are mainly fermented by yeasts (Colmagro et al., 2001).

The microbiome in spontaneous fermentation is affected by the indigenous olive microbiota; salt concentration in brine (usually 8% w/v or higher); diffusion of nutrients from the drupe to the brine; water activity of olives; pH of olives; the levels of antimicrobial compounds, such as polyphenols, in olives; fermentation temperature; and oxygen availability (Medina-Pradas et al., 2017; Nychas et al., 2002). Enterobacteriaceae, LAB (and more specifically *Lactobacillus plantarum*, *L. casei*, *Leuconostoc mesenteroides*, *P. pentosaceus*, Propionibacteriaceae) and yeasts such as *Pichia membranaefaciens*, *Pichia fermentans*, *Saccharomyces cerevisiae*, *Candida oleophila*, *Candida silvae*, and *Cystofilobasidium capitatum* are the most common microorganisms implementing the fermentation (Bautista-Gallego et al., 2011; Garrido-Fernández et al., 1997).

Brine and olive samples from Spanish-style green olives of the cultivars Conservolea and Halkidiki and black olives of the cultivars Conservolea and Kalamata were used, from which a total of 145 LAB were recovered and identified. For species identification, sequence analysis of 16S rRNA gene was performed by PCR of the 16S rRNA gene followed by (1) DGGE in the case of black olives or by (2) restriction enzyme analysis of the amplicons (PCR-REA) in the cases of the brine and green olives. Pulsed field gel electrophoresis (PFGE) of A*pa*I macrorestriction fragments was used for distinguishing strains within the species (Table 12.2). A total of 71 different stains were recovered: 17 *L. mesenteroides*, 51 *L. plantarum*, 2 *L. paracasei*, and 1 *Leuconostoc pseudomesenteroides* (Doulgeraki et al., 2013). Randazzo et al. (2004) investigated the Lab of naturally fermented homemade green olives from different areas of Sicily by PCR-REA (Table 12.2). Thirteen heterofermentative isolates were *L. brevis*, 24 homofermentative strains were *L. casei*, and 11 isolates were *E. faecalis*.

Apart from LAB, which affect olive fermentation (more specifically the acidification rate and the organoleptic characteristics of the final products), yeasts also play an important role (Campus et al., 2018; Medina-Pradas et al., 2017). Yeasts consume lactic and acetic acids and result in a milder taste and less intense acidification, which might lead to shorter shelf-life if LAB are not dominant in the first stage of the fermentation (Arroyo-López et al., 2008). On the other hand, yeasts produce vitamins, amino acids, and purines that stimulate the growth of lactobacilli (Viljoen, 2006). In addition, yeasts produce killer factors, namely, proteins and glycoproteins, which, along with the accumulation of ethanol, prevent mold growth (Viljoen, 2006). Last but not least, yeast fermentation affects greatly the flavor of table olives due to the production of glycerol, ethanol, higher molecular weight alcohols, esters, and other volatiles (Arroyo-López et al., 2012, 2008).

The diversity of yeast species during fermentation is affected by the geographic area, cultivar, by the interactions with LAB and by parameters including, but not restricted to, pH, brine salinity, and temperature (Campus et al., 2018). Kalamata black olives were spontaneously fermented in plain 7% w/v NaCl brine, brine acidified with 0.5% v/v vinegar, and brine with 0.1% v/v lactic acid. Yeasts were evaluated at 4, 34, 90, 140, and 187 days of fermentation. Sequencing of the D1/D2 domain of 26S rRNA gene (Table 12.2) permitted the characterization of 260 isolates to species level. During fermentation *Candida boidinii*, *Aureobasidium pullulans*, *D. hansenii*, *Metschnikowia pulcherrima*, *Barnettozyma californica*, *Pichia guilliermondii*, *Pichia manshurica*, *Pichia membranifaciens*, *Zygoascus hellenicus* were identified. Until the 90th day of fermentation *A. pullulans*, *B. californica*, *Candida diddensiae*, *Candida naeodendra*, *C. oleophila*, *C. silvae*, *Citeromyces matritensis*, *Cryptococcus laurentii*, *Cystofilobasidium bisporidii*, *D. hansenii*, *M. pulcherrima*, *P. guilliermondii*, *Pichia kluyveri*, *Rhodosporidium diobovatum*, *Rhodotorula glutinis*, and *Rhodotorula mucilaginosa* were identified. At the end of fermentation, *P. manshurica*, *P. membranifaciens*, *S. cerevisiae*, and *Z. hellenicus* were identified. More specifically, at the fourth day of fermentation the dominant species in the control and in the vinegar treated samples was *A. pullulans*, while *C. naeodendra* was dominant in lactic acid treated samples. At the 90th day of fermentation in the control treatment the dominant microorganism was *C. boidinii*, while in samples treated with vinegar or lactic acid the dominant species was *C. molendinolei* (Bonatsou et al., 2018).

Several bacteria, yeasts, and molds are also responsible for the spoilage of table olives. Gas pocket (formed below the olive skin, or in the interior of olive) is the result of CO_2 accumulation by gas-generating Gram-negative bacteria and some yeasts, mostly *Aeromonas*, *Escherichia*, *Citrobacter*, *Klebsiella*, *Enterobacter*, *Pichia anomala*, *Saccharomyces kluyveri*, and *S. cerevisiae* (Medina-Pradas et al., 2017; Lanza, 2013). *Clostridium butyricum* and *Clostridium acetobutyricum* are responsible for butyric fermentation and release of butyrate odor. Zapateria is caused by

Propionibacterium pentosaceum and *Pratylenchus zeae*, as well as by H_2S-producing Clostridia, which form gas holes in the interior of olives, and produce several acids, such as malic, succinic, propionic or butyric, formic, valeric, and caproic acids, which create an odor perceived as spoilage (Medina-Pradas et al., 2017; Lanza, 2013). Yeasts or LAB—such as *L. plantarum*—can cause white spots on the surface of olives, while softening is due to pectinolytic enzymes of *S. cerevisiae*, *S. kluyveri*, *P. anomala*, *P. manshurica*, *Rhodotorula rubra*, *Penicillium* spp., *Fusarium* spp., *Mucor* spp., and *Alternaria* spp. (Medina-Pradas et al., 2017; Lanza, 2013; Golomb et al., 2013). The abnormal fermentation caused by these microorganisms is usually due to the presence of O_2 and inadequate salinity in brine, to high pH or slow rate of acidification, poor hygiene or poor quality of olive crops, or even to slow growth of beneficial LAB, occasionally due to bacteriophages (Medina-Pradas et al., 2017; Lanza, 2013).

PICKLES

Fermented pickles are a major category of fermented vegetables, bearing a fermentation microbiome that resembles, to some extent, to the one of table olives; their successful fermentation is also based on salt concentration and adequate acidification (Medina-Pradas et al., 2017). Salt concentration in the brine of pickles varies greatly in relation to the diameter of vegetables, from $\sim 1\%-4\%$ w/v in sauerkraut (fermented cabbage) to $12\%-15\%$ in fermented cucumbers, where salt concentration below 12% may be detrimental to the fermentation process and may lead to spoilage if no acidifiers (e.g., acetic acid, lactic acid) are added (Medina-Pradas et al., 2017; Viander et al., 2003; Battcock and Azam-Ali, 1998).

The amount of salt added in the brine dictates the rate of fermentation and the development of the microbiota. For example, in sauerkraut with 2% brine LAB grew faster than in 5% and 8% brine, but *E. coli* may also sustain dense populations ($> 10^5$ cfu/mL) for several days, while fungi are not favored in this environment; probably due to antagonism from bacteria. However, bacterial spoilage may be caused by Enterobacteriaceae under such conditions. On the contrary, at 8% salt concentration in the brine, LAB growth and sugar metabolism were significantly slower, thus the fermentation was proceeding best at 5% salt concentration, which allowed fast growth of LAB, high rate of sugar utilization, and the highest death rate of *E. coli* (Xiong et al., 2016).

Robust growth and prevalence of LAB in pickled vegetables are crucial, not only to ensure fast acidification to prevent spoilage but also to create the characteristic flavor of such products (Medina-Pradas et al., 2017). Apart from lactic acid and acetic acid that contribute to the flavor, several volatile compounds are produced during vegetable fermentation by LAB, such as ethanol, methanol, *n*-hexanol, acetaldehyde, methyl–ethyl–ketone, acetone, propionaldehyde amino acids, fatty acids, and nucleotides that contribute to the aroma of pickles (Yan and Xue, 2005; Cheigh et al., 1994).

The fermenting microbiota are predominantly specific LAB species. The microbiome of naturally fermented Chinese sauerkraut (made in brine of 6% NaCl and 2% sugar) consisted mainly of *E. faecalis*, *L. lactis* subsp. *lactis*, *L. mesenteroides* subsp. *mesenteroides*, *L. plantarum*, *L. casei*, and *Lactobacillus zeae*. The fermentation was initiated by *L. mesenteroides* subsp. *mesenteroides*, proceeded in the middle stage of fermentation with *E. faecalis*, *L. lactis*, *L. zeae*, while *L. plantarum* and *L. casei* were dominant at the end of the fermentation (Xiong et al., 2012).

The profile of the microbial community of industrial, naturally fermented sauerkraut, during a 14-day fermentation at 21°C until pH dropped below 3.6 was examined by 16S rDNA amplicon sequencing (Table 12.2). During the actual fermentation the phylogenetic diversity is greatly reduced, although there is high phylogenetic diversity within the industrial environment of sauerkraut fermentation, and moderate diversity of microbiomes in the raw materials. Thus only a few microbes (LAB) can fit into this ecological niche, and these can ensure a stable fermentation (Zabat et al., 2018). In line with previous findings, *Leuconostoc* spp. were dominant within the fermentation microbiome from day 2 till day 10, after which lactobacilli prevailed until the end of the fermentation process.

In pickled cucumber fermentation, LAB such as *L. mesenteroides*, *P. pentosaceus*, *L. brevis* and *L. plantarum* lead the acidification process in the beginning of the fermentation, while *L. plantarum* usually finishes the fermentation, due to its high acid tolerance, in comparison to pediococci and *Leuconostoc* spp. (Fleming, 1984). PCR has been used (Table 12.2) to investigate the succession of LAB and their antagonism in cucumber fermented in brine (Singh and Ramesh, 2008). *Lactobacillus and Pediococcus* were dominant toward the end of the fermentation, although at the initial stage of the fermentation *Leuconostoc* and *Lactobacillus* dominated. Apart from differences in pH tolerance, this succession of populations was also attributed to the production of mesentericin, pediocin, and plantaricin A, which were detected during fermentation (Singh and Ramesh, 2008).

Yeasts, such as *P. manshurica* and *Issatchenkia occidentalis* and bacteria, such as *L. buchneri*, *Clostridium bifermentans*, and *Enterobacter cloacae* are associated with alterations in industrial cucumber fermentation (Franco et al, 2012; Franco and Pérez-Díaz, 2013). The colony appearance and microscopic morphology (Table 12.2) of some of the major spoilage yeasts isolated from fermented cucumbers are shown in Fig. 12.3. Furthermore, culture-independent identification methods for species-level characterization of isolates from spoiled commercial fermented cucumber samples revealed that at pH below 3.4 the spoilage bacteria were mainly lactobacilli and *Acetobacter* species (*Leuconostoc rapi*, *L. buchneri*, *Lactobacillus namurensis*, *Lactobacillus acetotolerans*, *Lactobacillus panis*, *Acetobacter aceti*, *Acetobacter peroxydans*, and *Acetobacter pasterianus*), whereas at pH above 4, the spoilage microbiome included lactate-consuming bacteria such as *Propionibacterium*, *Pectinatus*, *Veillonella*, and *Dialister* (Medina et al., 2016; Breidt et al., 2013).

THE MICROBIOME OF NATURALLY FERMENTED WINE AND BEER

WINE

One of the most important fermented beverages is wine, one of the oldest alcoholic beverages known, and the result of the fermentation of the grape juice by yeasts (Sabate et al., 1998). Wine is considered to be a complex microbial ecosystem in which yeasts and LAB play important roles (Liu et al., 2017; Renouf et al., 2007). Yeasts enact the alcoholic fermentation, while LAB enact the malolactic fermentation (Renouf et al., 2007), thus being both responsible for the quality and the sensory properties of the wine which define its character (Liu et al., 2017). Most of the microbiota involved in the spontaneous fermentation of wine are native in the skin of grape berries but also reside at the winery equipment (Liu et al., 2017; Renouf et al., 2007). Indicative genera

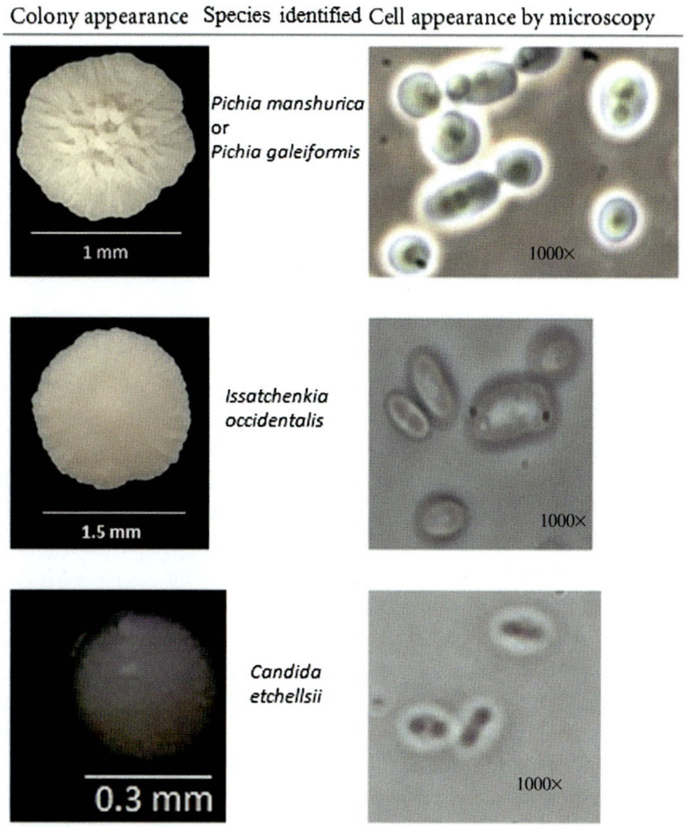

FIGURE 12.3

Macroscopic (Colony) and microscopic appearance of spoilage yeasts isolated from secondary fermentation of cucumber pickles.

Adapted from Franco, W., Pérez-Díaz, I.M., Johanningsmeier, S.D., McFeeters, R.F., 2012. Characteristics of spoilage-associated secondary cucumber fermentation. Appl. Environ. Microbiol. 78(4), 1273–1284.

include: yeasts such as *Kloeckera* and *Hanseniaspora*, LAB such as *Lactobacillus*, *Pediococcus*, *Oenococcus*, acetic acid bacteria such as *Gluconobacter* and *Acetobacter* and molds such as *Botrytis* and *Aspergillus*, although acetic acid bacteria and molds are spoilage microorganisms and do not productively participate in winemaking (Walker, 2014). Usually the population of the yeasts on the skin of the immature grape berries is very low ($10-10^3$ cfu/g) but it increases to 10^4-10^5 cfu/g, depending on environmental conditions, at harvest (Walker, 2014; Fleet, 2003). Yeasts usually found in the winery environment are *Schizosaccharomyces pombe*, *Zygosaccharomyces bailii* and *Zygosaccharomyces fermentati* (Fleet, 2000) although their contribution is questionable, as they may be inhibited by other wine yeasts (Fleet, 2003).

The fermentation of the wine is divided in two categories, the spontaneous and the "controlled." During the early stages of the spontaneous fermentation the dominant microorganisms are usually the indigenous yeast genera *Hanseniaspora* and *Candida*; and generally speaking non-*Saccharomyces* yeasts. Due to their sensitivity in ethanol levels over 5% v/v, their presence gradually declines and, eventually, indigenous strains of *S. cerevisiae* become the dominant fermenting agents and complete the process (Walker, 2014; Romano et al., 2003). Other yeast genera possibly present during fermentation are *Brettanomyces*, *Kluyveromyces*, *Schizosaccharomyces*, *Torulaspora*, *Zygosaccharomyces*, and *Saccharomycodes* (Romano et al., 2003). In the controlled fermentation, in order to avoid biochemical reactions by autochthonous microbiota which may lead to spoilage, winemakers inoculate the grape must with lyophilized starter cultures of *S. cerevisiae* about 10^6-10^7 cfu/g and add sulfite to reduce the growth of the autochthonous spoilage microorganisms. Although this practice helps the wineries to perform a controlled and reproducible process which favors standardization, it also results in the universal production of wines with similar organoleptic properties, irrespectively of differences in the local microbiota, vineyard microenvironment, and types of grape (Comitini et al., 2017; Romano et al., 2003).

The location of the wineries and the microbiota of the grapes determine to a great extend the composition of the natural microbiome of the wine, which is variable and undergoes changes during spontaneous wine fermentation. In the case of spontaneous fermentation of the wine Cabernet Sauvignon produced in Ningxia, China, must samples at different stages of fermentation were collected from three different wineries and 400 different yeast colonies were isolated and identified by sequencing the D1/D2 domain of the 26S rRNA gene (Table 12.2). Despite variations between the three wineries, nine dominant yeasts were identified: *Hanseniaspora uvarum*, *Hanseniaspora occidentalis*, *Metschnikowia pulcherimma*, *Candida zemplinina*, *Hanseniaspora vinae*, *Issatchenkia orientalis*, *Z. bailii*, *P. kluyveri*, and *S. cerevisiae*. Generally, non-*Saccharomyces* yeasts took part in the early phase of the fermentation, but *H. uvarum*, I. orientalis, and P. kluyveri tolerated high alcohol concentrations and remain active throughout the fermentation (Li et al., 2011).

In another study, the populations of were analyzed during the spontaneous fermentation of Catalanesca white grape, a variety from Campania, Italy, 18 different naturally occurring yeast species were identified with PCR-DGGE and partial sequence analysis of the 26S rRNA gene (Table 12.2). The dominant species were *Hanseniaspora* spp., *Issatchenkia* spp., *Candida* spp., and *S. cerevisiae*. The most interesting finding in this study was the isolation of four different *Issatchenkia* species that are rarely isolated during wine fermentation (Di Maro et al., 2007). In Austrian wines the yeasts identified by a combination of D1/D2 rRNA gene sequencing and RAPD fingerprinting (Table 12.2) included *Candida*, *Hanseniaspora*, *Issatchenkia*, *Kregervanrija*, *Lachancea*, *Metschnikowia*, *Pichia*, *Saccharomyces*, and *Zygoascus* (Lopandic et al., 2008). The microbiome of six different grape must varieties from Valle del Andarax (Spain) was characterized by PCR-RFLP (Table 12.2); the main indigenous genera, being *Candida*, *Hanseniaspora*, *Issatchenkia*, *Metschnikowia*, *Pichia*, and *Saccharomyces* (Clemente-Jimenez et al., 2004).

Wine is a substrate rich in sugars and other nutrients and thus susceptible to spoilage microorganisms, including fungi and bacteria, from the beginning of the fermentation until storage. Spoilage microorganisms produce several compounds that mar the organoleptic properties of wine. Wild, autochthonous yeasts participate in alcoholic fermentation, but they may also spoil the wine at different production stages, producing off-flavors, which include high concentrations of hydrogen sulfide and other sulfur volatiles, acetic acid and esters. For example, in bottled wines the most common spoilage genera are *Zygosaccharomyces*, *Dekkera*, *Saccharomyces*, and *Saccharomycodes* which synthesize, once beyond a certain concentration threshold (Schifferdecker et al., 2014) some of the said compounds (Fleet, 2003).

The "pharmaceutical" spoilage of wines is usually due to phenolic compounds such as 4-vinylphenol, 4-vinylguiacol, 4-ethylphenol, and 4-ethylguiacol, while the "mousy" or "wet horse" spoilage is caused by carbonyl nitrogenous compounds, such as 2-acetyl-3,4,5,6-tetrahydropyridine, 2-acetyl-1,2,5,6-tetrahydropyridine, and 2-ethyl-3,4,5,6-tetrahydropyridine, as a result of the activation of phenol reductases and decarboxylases (Godoy et al., 2009; Pretorius, 2000).

The most efficient way to avoid yeast spoilage is to follow good manufacturing and hygiene practices (Fleet, 2003) and to use engineered strains of *S. cerevisiae* or other species with killer activity (Du Toit and Pretorius, 2000). Autochthonous yeasts and bacteria have been used as biocontrol agents to contain the growth of *Brettanomyces bruxellensis*, a wine spoilage yeast. Cocultivating autochthonous strains of *S. cerevisiae* with non-*Saccharomyces* strains, or *S. cerevisiae* with *Oenococcus oeni* could be an effective way to control the growth of *B. bruxellensis* (Berbegal et al., 2017). Seven strains of *M. pulcherrima* were evaluated for suppressive activity against yeasts such as *Pichia*, *Candida*, *Brettanomyces*, and *Dekkera* and seemed effective against the last two (Oro et al., 2014).

Molds occurring during the cultivation of grapes may also spoil wine fermentation, especially if aerobic conditions occur during fermentation. Many genera that are responsible for the spoilage of grapes, such as *Botrytis*, *Uncinula*, *Alternaria*, *Plasmopara*, *Aspergillus*, *Penicillium*, *Rhizopus*, *Oidium*, and *Cladosporium*, are also found in spoiled wine (Fleet, 2003). Fungi do not only produce off-flavors during wine fermentation, but they can also synthesize mycotoxins such as ochratoxin-A, as mentioned previously, which is produced mainly by *Penicillium* and *Aspergillus* species and is nephrotoxic, hepatotoxic, teratogenic, and carcinogenic (Moreno-Arribas and Polo, 2009). Bacterial wine spoilage is usually due to *Lactobacillus*, *Pediococcus*, and some *Oenococcus* species, as well as to acetic acid bacteria, which produce compounds such as acetic acid, acetaldehyde, and ethyl-acetate, which have a solvent-like odor (Bartowsky, 2009). Sulfur dioxide, bacteriocins, and natural products such as lysozyme have been used to contain the growth of spoilage bacteria in wine (Bartowsky, 2009).

BEER

Beer is another important fermented beverage, believed to originate from the Middle East and Egypt since the third millennium BC (Pavsler and Buiatti, 2009a,b). Beer is made from malted grain, HoPS, water and fermentation yeasts (Rodhouse and Carbonero, 2019) and divided into two basic categories: (1) the top and (2) the bottom fermented beers. The former, or ale beers, are fermentation products mainly of *S. cerevisiae* at high temperatures, 18°C–24°C for 2–3 days (Briggs et al., 2004). The latter, or lager beers, are fermentation products of *Saccharomyces pastorianus* or *Saccharomyces carlsbergensis* at low temperatures of 8°C–14°C for 3–4 weeks. Ales usually have higher alcoholic content than lagers. Beers inoculated with starter cultures are common in large, industrial-scale production; there are also beers fermented spontaneously with indigenous microbiota at small production scale or in artisanal practice. Lambic sour beer is one of the oldest beers the traditional beer in Belgium, produced by spontaneous fermentation and its sour taste is due to the interactions among yeasts, LAB, and acetic acid bacteria (Spitaels et al., 2014). Samples collected in a 2-year period during and after fermentation from different fermentation batches and different brewers were subjected to MALDI-TOF mass spectrometry, 16s rDNA PCR-DGGE analysis. The results showed that in the first month of the fermentation the dominant microorganisms were Enterobacteriaceae, but were replaced by *Pediococcus damnosus* and *Saccharomyces* spp. after 2 months of fermentation and by *D. bruxellensis* after 6 months (Spitaels et al., 2014). This succession of microbial populations is due to the sensitivity of Enterobacteriaceae to ethanol, to the higher growth rate of *Saccharomyces* on glucose compared to *Dekkera*, and to the fact that after prolonged fermentation, *Dekkera* species are more tolerant to the exhaustion of carbon and nitrogen sources compared to *Saccharomyces* and demonstrate a late expression of ethanol stress tolerance genes (Schifferdecker et al., 2014).

The microbiome of the Belgian red-brown acidic ale from samples from three different breweries subjected to pyrosequencing of the 16s rDNA and of the internal transcribed spacer (ITS) region has shown the dominant fermentative microorganisms to be *P. damnosus*, *D. bruxellensis*, and *Acetobacter pasteurianus*. The main metabolites, determined by HPLC, were L-lactic acid, D-lactic acid, and ethanol and in lower concentration acetic acid (Snauwaert et al., 2016). Also, Takahashi et al. (2015) examined the microbial diversity of three types of lager-type beer (with 100% malt, 25% malt/75% barley, and 50% malt/50% barley) in a pilot-scale brewing process. The fermentation was performed by adding *S. pastorianus* starter culture for 23–24 days at 10°C, followed by a ripening period at 2°C. The microbial communities were investigated by next generation sequencing, quantitative PCR, fluorescent in situ hybridization, flow cytometry, and PCR-RFLP (Table 12.2). After boiling the malt or malt/barley wort the beer bacteriome during fermentation contained *Acinetobacter*, *Bacillus*, *Escherichia*, *Methylobacterium*, *Paenibacillus*, *Propionibacterium*, *Pseudomonas*, *Ralstonia*, the orders Actinomycetales and Rhizobiales, and unidentified bacteria. Heat-resistant *Bacillus* and *Paenibacillus* were detected in all types of beer during the early and middle stages of fermentation. The bacterial population, quantified by qPCR (Table 12.2), gradually decreased from an initial 10^7–10^9 cfu/mL count in the mash, to ~5×10^5–5×10^6 log cfu/mL before wort boiling, and ~ 10^3–10^4 cfu/mL after wort boiling and remained practically stable at that level during fermentation. They did not exceed 5×10^3 log cfu/mL even after 6 months of storage.

On the other hand, the fungal population, which consisted of up to 38 genera at the start of mashing, was gradually reduced to levels undetectable by flow cytometry even before boiling the wort, but showed an increase to 5×10^5–5×10^6 cfu/g after addition of *S. pastorianus*, the population of which increased slowly to ~7×10^6–10^7 log cfu/mL on day 12 of the fermentation and dominated the eukaryotic population of the fermented beer: >98% of all fungal sequences belonged to *S. pastorianus*. This proves that wild yeasts and molds did not survive the boiling process nor did they affect the fermentation. Nonetheless, *S. pastorianus* declined during ripening to $<10^5$ cfu/mL and was not detected in the beer after filtration (Takahashi et al., 2015).

Beer is a poor medium for microbial growth, since fermentation yeasts consume most nutrients during fermentation (Suzuki, 2011) and thus becomes hostile to spoilage microorganisms, due to the presence of ethanol, hop, organic acids, and carbon dioxide, in combination with the lack of oxygen (Sakamoto and Konings, 2003; Gaunzle et al., 2001). However, spoilage microorganisms do occur in beer. These include both bacteria such as *Lactobacillus*, *Pediococcus*, *Pectinatus*, and *Megasphaera* and wild yeasts, which affect the turbidity and acidity and produce unwanted odorants, such as diacetyl- and hydrogen sulfide (Sakamoto and Konings, 2003). Some of the most common beer spoilage species causing serious damage in breweries are *L. brevis*, *Lactobacillus lindneri*, *Lactobacillus brevisimilis*, *Lactobacillus malefermentans*, *Lactobacillus parabuchneri*, *Lactobacillus collinoides*, *L. paracasei*, *P. acidilactici*, *P. damnosus*, *P. dextrinicus*, *Pediococcus halophilus*, *Pediococcus inopinatus*, *Pediococcus parvulus*, *P. pentosaceus*, *Micrococcus kristinae*, *Pectinatus cerevisiiphilus*, *Pectinatus frisingensis*, *Megasphaera elsdenii*, and *Megasphaera cerevisiae* (Sakamoto and Konings, 2003).

Thermal pasteurization of beer can eradicate the spoilage LAB and fungi at relatively low temperatures, 60°C–65°C for a few minutes: $D_{60^\circ C}$ values of 0.02 and 0.04 min have been observed for *L. paracasei* var. *paracasei* and *Aspergillus niger*, respectively. On the other hand, an adverse effect on the organoleptic properties, is observed, due to loss of volatile flavor compounds (Reveron et al., 2005). As an alternative, high pressure pasteurization used at 300 MPa for 5 min is effective in destroying *L. plantarum* in beer, and this effect is enhanced at high alcohol concentrations (Gaunzle et al., 2001). Natural antimicrobials such as lysozyme can also help prevent beer spoilage. Samples of unpasteurized beers from four different breweries in the Netherlands, Belgium, and Italy were treated with lysozyme in order to restrict growth of spoilage bacteria. The results showed that 100 ppm lysozyme exhibited great inhibitory action against LAB such as *L. brevis* and *L. malefermentans* and significantly prolonged shelf-life of the beers, without having any adverse effect on flavor (Silvetti et al., 2010).

REFERENCES

Albano, H., Pinho, C., Leite, D., Barbosa, J., Silva, J., Carneiro, L., et al., 2009. Evaluation of a bacteriocin-producing strain of *Pediococcus acidilactici* as a biopreservative for "alheira", a fermented meat sausage. Food Control 20 (8), 764–770.

Ammor, M.S., Mayo, B., 2007. Selection criteria for lactic acid bacteria to be used as functional starter cultures in dry sausage production: an update. Meat Sci. 76, 138–146.

Andrighetto, C., Borney, F., Barmaz, A., Stefanon, B., Lombardi, A., 2002. Genetic diversity of *Streptococcus thermophilus* strains isolated from Italian traditional cheeses. Int. Dairy J. 12, 141–144.

Antunes, A.E., Cazetto, T.F., Bolini, H.M., 2005. Viability of probiotic micro-organisms during storage, postacidification and sensory analysis of fat-free yogurts with added whey protein concentrate. Int. J. Dairy Technol. 58, 169–173.

Arroyo-López, F.N., Querol, A., Bautista-Gallego, J., Garrido-Fernández, A., 2008. Role of yeasts in table olive production. Int. J. Food Microbiol. 128, 189–196.

Arroyo-López, F.N., Romero-Gil, V., Bautista-Gallego, J., Rodríguez-Gómez, F., Jiménez-Díaz, R., García-García, P., et al., 2012. Yeasts in table olive processing: desirable or spoilage microorganisms? Int. J. Food Microbiol. 160, 42–49.

Ayad, E.H., Verheul, A., de Jong, C., Wouters, J.T., Smit, G., 1999. Flavour forming abilities and amino acid requirements of *Lactococcus lactis* strains isolated from artisanal and non-dairy origin. Int. Dairy J. 9, 725–735.

Ayad, E.H., Verheul, A., Wouters, J.T., Smit, G., 2000. Application of wild starter cultures for flavour development in pilot plant cheese making. Int. Dairy J. 10, 169–179.

Ayad, E.H., Verheul, A., Wouters, J.T., Smit, G., 2002. Antimicrobial-producing wild lactococci isolated from artisanal and non-dairy origins. Int. Dairy J. 12, 145–150.

Aymerich, T., Martin, B., Garriga, M., Vidal-Carou, M.C., Bover-Cid, S., Hugas, M., 2006. Safety properties and molecular strain typing of lactic acid bacteria from slightly fermented sausages. J. Appl. Microbiol. 100, 40–49.

Bailly, J.D., Tabuc, C., Quérin, A., Guerre, P., 2005. Production and stability of patulin, ochratoxin A, citrinin, and cyclopiazonic acid on dry cured ham. J. Food Protect. 68, 1516–1520.

Bartowsky, E.J., 2009. Bacterial spoilage of wine and approaches to minimize it. Lett. Appl. Microbiol. 48, 149–156.

Baruzzi, F., Morea, M., Matarante, A., Cocconcelli, P.S., 2000. Changes in the *Lactobacillus* community during Ricotta forte cheese natural fermentation. J. Appl. Microbiol. 89, 807–814.

Battcock, M., Azam-Ali, S., 1998. Fermented Fruits and Vegetables: A Global Perspective (No. FAO ASB-134). FAO, Rome.

Bautista-Gallego, J., Rodríguez-Gómez, F., Barrio, E., Querol, A., Garrido-Fernández, A., Arroyo-López, F.N., 2011. Exploring the yeast biodiversity of green table olive industrial fermentations for technological applications. Int. J. Food Microbiol. 147, 89–96.

Berbegal, C., Garofalo, C., Russo, P., Pati, S., Capozzi, V., Spano, G., 2017. Use of autochthonous yeasts and bacteria in order to control *Brettanomyces bruxellensis* in wine. Fermentation 3, 65.

Beresford, T.P., Fitzsimons, N.A., Brennan, N.L., Cogan, T.M., 2001. Recent advances in cheese microbiology. Int. Dairy J. 11, 259–274.

Bizzarro, R., Tarelli, G.T., Giraffa, G., Neviani, E., 2000. Phenotypic and genotypic characterization of lactic acid bacteria isolated from Pecorino Toscano cheese. Ital. J. Food Sci. 12, 303–316.

Blaiotta, G., Pennacchia, C., Villani, F., Ricciardi, A., Tofalo, R., Parente, E., 2004. Diversity and dynamics of communities of coagulase-negative staphylococci in traditional fermented sausages. J. Appl. Microbiol. 97, 271–284.

Bonatsou, S., Paramithiotis, S., Panagou, E.Z., 2018. Evolution of yeast consortia during the fermentation of kalamata natural black olives upon two initial acidification treatments. Front. Microbiol. 8, 2673.

Bonomo, M.G., Ricciardi, A., Zotta, T., Sico, M.A., Salzano, G., 2009. Technological and safety characterization of coagulase-negative staphylococci from traditionally fermented sausages of Basilicata region (Southern Italy). Meat Sci. 83, 15–23.

Breidt, F., Medina, E., Wafa, D., Pérez-Díaz, I., Franco, W., Huang, H.Y., et al., 2013. Characterization of cucumber fermentation spoilage bacteria by enrichment culture and 16S rDNA cloning. J. Food Sci. 78 (3), M470–M476.

Briggs, D.E., Brookes, P.A., Stevens, R., Boulton, C.A., 2004. Brewing: Science and Practice. Elsevier.

Bruna, J.M., Hierro, E.M., de la Hoz, L., Mottram, D.S., Fernández, M., Ordóñez, J.A., 2003. Changes in selected biochemical and sensory parameters as affected by the superficial inoculation of *Penicillium camemberti* on dry fermented sausages. Int. J. Food Microbiol. 85, 111–125.

Buttriss, J., 1997. Nutritional properties of fermented milk products. Int. J. Dairy Technol. 50, 21–27.

Campus, M., Değirmencioğlu, N., Comunian, R., 2018. Technologies and trends to improve table olive quality and safety. Front. Microbiol. 9, 617.

Caplice, E., Fitzgerald, G.F., 1999. Food fermentations: role of microorganisms in food production and preservation. Int. J. Food Microbiol. 50, 131–149.

Carr, F.J., Chill, D., Maida, N., 2002. The lactic acid bacteria: a literature survey. Crit. Rev. Microbiol. 28, 281–370.

Chaillou, S., Daty, M., Baraige, F., Dudez, A.M., Anglade, P., Jones, R., et al., 2009. Intraspecies genomic diversity and natural population structure of the meat-borne lactic acid bacterium *Lactobacillus sakei*. Appl. Environ. Microbiol. 75, 970–980.

Chaves, A.C.S.D., Fernandez, M., Lerayer, A.L.S., Mierau, I., Kleerebezem, M., Hugenholtz, J., 2002. Metabolic engineering of acetaldehyde production by *Streptococcus thermophilus*. Appl. Environ. Microbiol. 68, 5656–5662.

Cheigh, H.S., Park, K.Y., Lee, C.Y., 1994. Biochemical, microbiological, and nutritional aspects of kimchi (Korean fermented vegetable products). Crit. Rev. Food Sci. Nutr. 34, 175–203.

Chen, C., Zhao, S., Hao, G., Yu, H., Tian, H., Zhao, G., 2017. Role of lactic acid bacteria on the yogurt flavour: a review. Int. J. Food Prop. 20 (Suppl. 1), S316–S330.

Cheng, H., 2010. Volatile flavor compounds in yogurt: a review. Crit. Rev. Food Sci. Nutr. 50 (10), 938–950.

Clemente-Jimenez, J.M., Mingorance-Cazorla, L., Martínez-Rodríguez, S., Las Heras-Vázquez, F.J., Rodríguez-Vico, F., 2004. Molecular characterization and oenological properties of wine yeasts isolated during spontaneous fermentation of six varieties of grape must. Food Microbiol. 21, 149–155.

Cocolin, L., Manzano, M., Cantoni, C., Comi, G., 2001. Denaturing gradient gel electrophoresis analysis of the 16S rRNA gene V1 region to monitor dynamic changes in the bacterial population during fermentation of Italian sausages. Appl. Environ. Microbiol. 67, 5113–5121.

Cocolin, L., Dolci, P., Rantsiou, K., 2011. Biodiversity and dynamics of meat fermentations: the contribution of molecular methods for a better comprehension of a complex ecosystem. Meat Sci. 89, 296–302.

Cogan, T.M., Accolas, J.P., 1996. Dairy Starter Cultures. VCH Publishers.

Cogan, T.M., Barbosa, M., Beuvier, E., Bianchi-Salvadori, B., Cocconcelli, P.S., Fernandes, I., et al., 1997. Characterization of the lactic acid bacteria in artisanal dairy products. J. Dairy Res. 64, 409–421.

Colmagro, S., Collins, G., Sedgley, M., 2001. Processing technology of the table olive. Hortic. Rev. 25, 235–260.

Comi, G., Urso, R., Iacumin, L., Rantsiou, K., Cattaneo, P., Cantoni, C., et al., 2005. Characterisation of naturally fermented sausages produced in the North East of Italy. Meat Sci. 69, 381–392.

Comitini, F., Capece, A., Ciani, M., Romano, P., 2017. New insights on the use of wine yeasts. Curr. Opin. Food Sci. 13, 44–49.

Coppola, R., Iorizzo, M., Saotta, R., Sorrentino, E., Grazia, L., 1997. Characterization of micrococci and staphylococci isolated from soppressata molisana, a Southern Italy fermented sausage. Food Microbiol. 14, 47–53.

Coppola, S., Mauriello, G., Aponte, M., Moschetti, G., Villani, F., 2000. Microbial succession during ripening of Naples-type salami, a southern Italian fermented sausage. Meat Sci. 56, 321–329.

Corroler, D., Mangin, I., Desmasures, N., Gueguen, M., 1998. An ecological study of lactococci isolated from raw milk in the Camembert cheese registered designation of origin area. Appl. Environ. Microbiol. 64, 4729–4735.

Cosentino, S., Fadda, M.E., Deplano, M., Melis, R., Pomata, R., Pisano, M.B., 2012. Antilisterial activity of nisin-like bacteriocin-producing *Lactococcus lactis* subsp. lactis isolated from traditional Sardinian dairy products. BioMed Res. Int. 2012. Available from: https://doi.org/10.1155/2012/376428.

Crow, V., Curry, B., Hayes, M., 2001. The ecology of non-starter lactic acid bacteria (NSLAB) and their use as adjuncts in New Zealand Cheddar. Int. Dairy J. 11, 275–283.

Cukon, N., Fleck, Ž.C., Bratulić, M., Kozačinski, L., Njari, B., 2012. Diversity of microflora in meat and meat products. Meso 14.

Dal Bello, B., Rantsiou, K., Bellio, A., Zeppa, G., Ambrosoli, R., Civera, T., et al., 2010. Microbial ecology of artisanal products from North West of Italy and antimicrobial activity of the autochthonous populations. LWT-Food Sci. Technol. 43, 1151–1159.

De Vuyst, L., Vandamme, E.J., 1994. Nisin, a lantibiotic produced by *Lactococcus lactis* subsp. lactis: properties, biosynthesis, fermentation and applications. Bacteriocins of Lactic Acid Bacteria. Springer, Boston, MA, pp. 151–221.

Deguchi, Y., Morishita, T., 1992. Nutritional requirements in multiple auxotrophic lactic acid bacteria: genetic lesions affecting amino acid biosynthetic pathways in *Lactococcus lactis*, *Enterococcus faecium*, and *Pediococcus acidilactici*. Biosci. Biotechnol. Biochem. 56, 913–918.

Delves-Broughton, J., Blackburn, P., Evans, R.J., Hugenholtz, J., 1996. Applications of the bacteriocin, nisin. Antonie Van Leeuwenhoek 69, 193–202.

Demeyer, D.I., 1982. Stoichiometry of dry sausage fermentation. Antonie Van Leeuwenhoek 48, 414–416.

Di Maro, E., Ercolini, D., Coppola, S., 2007. Yeast dynamics during spontaneous wine fermentation of the Catalanesca grape. Int. J. Food Microbiol. 117, 201–210.

Doulgeraki, A.I., Pramateftaki, P., Argyri, A.A., Nychas, G.J.E., Tassou, C.C., Panagou, E.Z., 2013. Molecular characterization of lactic acid bacteria isolated from industrially fermented Greek table olives. LWT-Food Sci. Technol. 50, 353–356.

Driessen, F.M., Kingma, F., Stadhouders, J., 1982. Evidence that *Lactobacillus bulgaricus* in yogurt is stimulated by carbon dioxide produced by *Streptococcus thermophilus*. Netherl. Milk Dairy J. (Netherlands) 36, 135–144.

Du Toit, M., Pretorius, I.S., 2000. Microbial spoilage and preservation of wine: using weapons from nature's own arsenal—a review. S. Afr. J. Enol. Vitic. 21, 74–96.

Ercolini, D., 2013. High-throughput sequencing and metagenomics: moving forward in the culture-independent analysis of food microbial ecology. Appl. Environ. Microbiol. 79, 3148–3155.

Ercolini, D., Moschetti, G., Blaiotta, G., Coppola, S., 2001. The potential of a polyphasic PCR-DGGE approach in evaluating microbial diversity of natural whey cultures for water-buffalo mozzarella cheese production: bias of culture-dependent and culture-independent analyses. Syst. App. Microbiol. 24, 610–617.

Even, S., Leroy, S., Charlier, C., Zakour, N.B., Chacornac, J.P., Lebert, I., et al., 2010. Low occurrence of safety hazards in coagulase negative staphylococci isolated from fermented foodstuffs. Int. J. Food Microbiol. 139, 87–95.

Farnworth, E.R.T. (Ed.), 2003. Handbook of Fermented Functional Foods. CRC Press.

Fitzsimons, N.A., Cogan, T.M., Condon, S., Beresford, T., 2001. Spatial and temporal distribution of non-starter lactic acid bacteria in Cheddar cheese. J. Appl. Microbiol. 90, 600–608.

Fleet, G.H., 2000. Schizosaccharomyces. In: Robinson, R.K., Batt, C.A., Patel, P.D. (Eds.), Encyclopedia of Food Microbiology, vol. 3. Academic Press, London.

Fleet, G.H., 2003. Yeast interactions and wine flavour. Int. J. Food Microbiol. 86, 11–22.

Fleming, H.P., 1984. Developments in cucumber fermentation. J. Chem. Technol. Biotechnol. 34, 241–252.

Flores, M., Durá, M.A., Marco, A., Toldrá, F., 2004. Effect of *Debaryomyces* spp. on aroma formation and sensory quality of dry-fermented sausages. Meat Sci. 68, 439–446.

Fonseca, S., Cachaldora, A., Gómez, M., Franco, I., Carballo, J., 2013. Monitoring the bacterial population dynamics during the ripening of Galician chorizo, a traditional dry fermented Spanish sausage. Food Microbiol. 33, 77–84.

Forssén, K.M., Jägerstad, M.I., Wigertz, K., Witthöft, C.M., 2000. Folates and dairy products: a critical update. J. Am. Coll. Nutr. 19 (Suppl. 2), 100S–110S.

Fox, P.F., 1998. Developments in biochemistry of cheese ripening. In: Proceedings of Dairy Science and Technology, 25th Int. Dairy Congress, Aarhus, Denmark, September 21–24, 1998, pp. 11–37.

Franco, W., Pérez-Díaz, I.M., 2013. Microbial interactions associated with secondary cucumber fermentation. J. Appl. Microbiol. 114, 161–172.

Franco, W., Pérez-Díaz, I.M., Johanningsmeier, S.D., McFeeters, R.F., 2012. Characteristics of spoilage-associated secondary cucumber fermentation. Appl. Environ. Microbiol. 78 (4), 1273–1284.

Frece, J., Cvrtila, J., Topić, I., Delaš, F., Markov, K., 2014. Lactococcus lactis ssp. lactis as Potential Functional Starter Culture. Food Technol Biotechnol. 52, 489–494.

Garabal, J.I., 2007. Biodiversity and the survival of autochthonous fermented products. Int. J. Microbiol. 1.

Garofalo, C., Osimani, A., Milanović, V., Aquilanti, L., De Filippis, F., Stellato, G., et al., 2015. Bacteria and yeast microbiota in milk kefir grains from different Italian regions. Food Microbiol. 49, 123–133.

Garrido-Fernández, A., Fernández Díez, M.J., Adams, M.R., 1997. Table Olives: Production and Processing. Chapman & Hall, London.
Gaunzle, M.G., Ulmer, H.M., Vogel, R.F., 2001. High pressure inactivation of *Lactobacillus plantarum* in a model beer system. J. Food Sci. 66, 1174–1181.
Gemechu, T., 2015. Review on lactic acid bacteria function in milk fermentation and preservation. Afr. J. Food Sci. 9, 170–175.
Godoy, L., Garrido, D., Martínez, C., Saavedra, J., Combina, M., Ganga, M.A., 2009. Study of the coumarate decarboxylase and vinylphenol reductase activities of *Dekkera bruxellensis* (anamorph *Brettanomyces bruxellensis*) isolates. Lett. Appl. Microbiol. 48, 452–457.
Golomb, B.L., Morales, V., Jung, A., Yau, B., Boundy-Mills, K.L., Marco, M.L., 2013. Effects of pectinolytic yeast on the microbial composition and spoilage of olive fermentations. Food Microbiol. 33, 97–106.
González, B., Dıez, V., 2002. The effect of nitrite and starter culture on microbiological quality of "chorizo"—a Spanish dry cured sausage. Meat Sci. 60, 295–298.
González, P.A., Dans, E.P., 2018. The 'terrorist' social movement: the reawakening of wine culture in Spain. J. Rural Stud. 61, 184–196.
Greco, M., Mazzette, R., De Santis, E.P.L., Corona, A., Cosseddu, A.M., 2005. Evolution and identification of lactic acid bacteria isolated during the ripening of Sardinian sausages. Meat Sci. 69, 733–739.
Hammes, W.P., 2012. Metabolism of nitrate in fermented meats: the characteristic feature of a specific group of fermented foods. Food Microbiol. 29, 151–156.
Hammes, W.P., Knauf, H.J., 1994. Starters in the processing of meat products. Meat Sci. 36, 155–168.
Hammes, W.P., Bantleon, A., Min, S., 1990. Lactic acid bacteria in meat fermentation. FEMS Microbiol. Rev. 7, 165–173.
Han, X., Yang, Z., Jing, X., Yu, P., Zhang, Y., Yi, H., et al., 2016. Improvement of the texture of yogurt by use of exopolysaccharide producing lactic acid bacteria. Bio. Med Res. Int. 2016.
Holzapfel, W.H., 1998. The Gram-positive bacteria associated with meat and meat products, In: A. Davies (Ed.) The microbiology of meat and poultry. Blackie Academic & Professional London, New York, 35–84.
Holzapfel, W.H., Haberer, P., Geisen, R., Björkroth, J., Schillinger, U., 2001. Taxonomy and important features of probiotic microorganisms in food and nutrition. Am. J. Clin. Nutr. 73, 365s–373s.
Hugas, M., Monfort, J.M., 1997. Bacterial starter cultures for meat fermentation. Food Chem. 59 (4), 547–554.
Hugas, M., Garriga, M., Aymerich, M.T., 2003. Functionality of enterococci in meat products. Int. J. Food Microbiol. 88, 223–233.
Iacumin, L., Comi, G., Cantoni, C., Cocolin, L., 2006. Ecology and dynamics of coagulase-negative cocci isolated from naturally fermented Italian sausages. Syst. Appl. Microbiol. 29, 480–486.
Iacumin, L., Chiesa, L., Boscolo, D., Manzano, M., Cantoni, C., Orlic, S., et al., 2009. Molds and ochratoxin A on surfaces of artisanal and industrial dry sausages. Food Microbiol. 26, 65–70.
Jany, J.L., Barbier, G., 2008. Culture-independent methods for identifying microbial communities in cheese. Food Microbiol. 25, 839–848.
Jay, J.M., Loessner, M.J., Golden, D.A., 2008. Modern Food Microbiology. Springer Science Business Media.
Jeevaratnam, K., Jamuna, M., Bawa, A.S., 2005. Biological preservation of foods—bacteriocins of lactic acid bacteria. Indian J. Biotechnol. 4, 446–454.
Karasu, N., Şimşek, Ö., Çon, A.H., 2010. Technological and probiotic characteristics of *Lactobacillus plantarum* strains isolated from traditionally produced fermented vegetables. Ann. Microbiol. 60, 227–234.
Klaenhammer, T.R., 1993. Genetics of bacteriocins produced by lactic acid bacteria. FEMS Microbiol. Rev. 12, 39–85.

Kozačinski, L., Drosinos, E., Čaklovica, F., Cocolin, L., Gasparik-Reichardt, J., Vesković, S., 2008. Investigation of microbial association of traditionally fermented sausages. Food Technol. Biotechnol. 46, 93–106.

Kumar, B.V., Vijayendra, S.V.N., Reddy, O.V.S., 2015. Trends in dairy and non-dairy probiotic products—a review. J. Food Sci. Technol. 52, 6112–6124.

Lamothe, G., Jolly, L., Mollet, B., Stingele, F., 2002. Genetic and biochemical characterization of exopolysaccharide biosynthesis by *Lactobacillus delbrueckii* subsp. bulgaricus. Arch. Microbiol. 178, 218–228.

Lanza, B., 2013. Abnormal fermentations in table-olive processing: microbial origin and sensory evaluation. Front. Microbiol. 4, 91.

Latorre-Moratalla, M.L., Bover-Cid, S., Talon, R., Aymerich, T., Garriga, M., Zanardi, E., et al., 2010. Distribution of aminogenic activity among potential autochthonous starter cultures for dry fermented sausages. J. Food Protect. 73, 524–528.

Law, B., 2010. Cheese adjunct cultures. Aust. J. Dairy Technol. 65, 45.

Leroy, F., Verluyten, J., De Vuyst, L., 2006. Functional meat starter cultures for improved sausage fermentation. Int. J. Food Microbiol. 106, 270–285.

Leroy, S., Giammarinaro, P., Chacornac, J.P., Lebert, I., Talon, R., 2010. Biodiversity of indigenous staphylococci of naturally fermented dry sausages and manufacturing environments of small-scale processing units. Food Microbiol. 27, 294–301.

Li, E., Liu, A., Xue, B., Liu, Y., 2011. Yeast species associated with spontaneous wine fermentation of Cabernet Sauvignon from Ningxia, China. World J. Microbiol. Biotechnol. 27, 2475–2482.

Linares, D.M., Martín, M., Ladero, V., Alvarez, M.A., Fernández, M., 2011. Biogenic amines in dairy products. Crit. Rev. Food Sci. Nut. 51, 691–703.

Liu, Y., Rousseaux, S., Tourdot-Maréchal, R., Sadoudi, M., Gougeon, R., Schmitt-Kopplin, P., et al., 2017. Wine microbiome: a dynamic world of microbial interactions. Crit. Rev. Food Sci. Nutr. 57, 856–873.

Lopandic, K., Tiefenbrunner, W., Gangl, H., Mandl, K., Berger, S., Leitner, G., et al., 2008. Molecular profiling of yeasts isolated during spontaneous fermentations of Austrian wines. FEMS Yeast Res. 8, 1063–1075.

Lopez, S., Mayo, B., 1997. Identification and characterization of homofermentative mesophilic *Lactobacillus* strains isolated from artisan starter-free cheeses. Lett. Appl. Microbiol. 25, 233–238.

López-Díaz, T.M., Santos, J.A., García-López, M.L., Otero, A., 2001. Surface mycoflora of a Spanish fermented meat sausage and toxigenicity of *Penicillium* isolates. Int. J. Food Microbiol. 68, 69–74.

Ludemann, V., Pose, G., Pollio, M.L., Segura, J., 2004. Determination of growth characteristics and lipolytic and proteolytic activities of *Penicillium* strains isolated from Argentinean salami. Int. J. Food Microbiol. 96, 13–18.

Maisnier-Patin, S., Deschamps, N., Tatini, S.R., Richard, J., 1992. Inhibition of *Listeria monocytogenes* in Camembert cheese made with a nisin-producing starter. Le Lait 72, 249–263.

Makino, S., Ikegami, S., Kano, H., Sashihara, T., Sugano, H., Horiuchi, H., et al., 2006. Immunomodulatory effects of polysaccharides produced by *Lactobacillus delbrueckii* ssp. bulgaricus OLL1073R-1. J. Dairy Sci. 89, 2873–2881.

Marco, M.L., Heeney, D., Binda, S., Cifelli, C.J., Cotter, P.D., Foligné, B., et al., 2017. Health benefits of fermented foods: microbiota and beyond. Curr. Opin. Biotechnol. 44, 94–102.

Markov, K., Pleadin, J., Bevardi, M., Vahčić, N., Sokolić-Mihalak, D., Frece, J., 2013. Natural occurrence of aflatoxin B1, ochratoxin A and citrinin in Croatian fermented meat products. Food Control 34, 312–317.

Martin, B., Garriga, M., Hugas, M., Aymerich, T., 2005. Genetic diversity and safety aspects of enterococci from slightly fermented sausages. J. Appl. Microbiol. 98, 1177–1190.

Martín, A., Córdoba, J.J., Aranda, E., Córdoba, M.G., Asensio, M.A., 2006a. Contribution of a selected fungal population to the volatile compounds on dry-cured ham. Int. J. Food Microbiol. 110, 8–18.

Martín, B., Garriga, M., Hugas, M., Bover-Cid, S., Veciana-Nogués, M.T., Aymerich, T., 2006b. Molecular, technological and safety characterization of Gram-positive catalase-positive cocci from slightly fermented sausages. Int. J. Food Microbiol. 107, 148–158.

Martinović, A., Vesković-Moračanin, S., 2006. Primena starter kultura u industriji mesa. Tehnologija mesa 47, 226–230.

Martley, F.G., Crow, V.L., 1993. Interactions between non-starter microorganisms during cheese manufacture and repening. Int. Dairy J. 3, 461–483.

McSweeney, P.L.H., Fox, P.F., Lucey, J.A., Jordan, K.N., Cogan, T.M., 1993. Contribution of the indigenous microflora to the maturation of Cheddar cheese. Int. Dairy J. 3, 613–634.

Medina, E., Pérez-Díaz, I.M., Breidt, F., Hayes, J., Franco, W., Butz, N., et al., 2016. Bacterial ecology of fermented cucumber rising ph spoilage as determined by nonculture-based methods. J. Food Sci. 81, M121–M129.

Medina-Pradas, E., Perez-Dıaz, I.M., Garrido-Fernandez, A., Arroyo-Lopez, F.N., 2017. Review of vegetable fermentations with particular emphasis on processing modifications, microbial ecology, and spoilage. In: Bevilacqua, A., Corbo, M.R., Sinigaglia, M. (Eds.), The Microbiological Quality of Food: Foodborne Spoilers. Woodhead Publishing, Cambridge.

Metaxopoulos, J., Stravropoulos, S., Kakouri, A., Samelis, J., 1996. Yeasts isolated from traditional Greek salami. Ital. J. Food Sci. 1, 25–32.

Mioković, B., Zdolec, N., 2004. Značenjehalo filnihbakterija u preradi mesa i ribe. Meso: prvihrvatskičasopis o mesu 6, 36–42.

Montel, M.C., Buchin, S., Mallet, A., Delbes-Paus, C., Vuitton, D.A., Desmasures, N., et al., 2014. Traditional cheeses: rich and diverse microbiota with associated benefits. Int. J. Food Microbiol. 177, 136–154.

Moreno-Arribas, M.V., Polo, C., 2009. In: Moreno-Arribas, M.V., Polo, C. (Eds.), Wine Chemistry and Biochemistry. Springer, New York.

Neves, A.R., Pool, W.A., Kok, J., Kuipers, O.P., Santos, H., 2005. Overview on sugar metabolism and its control in *Lactococcus lactis*—the input from in vivo NMR. FEMS Microbiol. Rev. 29, 531–554.

Nouaille, S., Ribeiro, L.A., Miyoshi, A., Pontes, D., Le Loir, Y., Oliveira, S.C., et al., 2003. Heterologous protein production and delivery systems for *Lactococcus lactis*. Genet. Mol. Res. 2, 102–111.

Nychas, G.J., Panagou, E.Z., Parker, M.L., Waldron, K.W., Tassou, C.C., 2002. Microbial colonization of naturally black olives during fermentation and associated biochemical activities in the cover brine. Lett. Appl. Microbiol. 34, 173–177.

O'sullivan, L., Ross, R.P., Hill, C., 2002. Potential of bacteriocin-producing lactic acid bacteria for improvements in food safety and quality. Biochimie 84, 593–604.

Oberman, H., Libudzisz, Z., 1998. Fermented milks. In: Wood, B.J. (Ed.), Microbiology of Fermented Foods, second ed. Blackie Academic & Professional, London, pp. 308–350.

Oliveira, M., Ferreira, V., Magalhães, R., Teixeira, P., 2018. Biocontrol strategies for Mediterranean-style fermented sausages. Food Res. Int. 103, 438–449.

Ordóñez, J.L., Troncoso, A.M., García-Parrilla, M.D.C., Callejón, R.M., 2016. Recent trends in the determination of biogenic amines in fermented beverages—a review. Analyt. Chim. Acta 939, 10–25.

Oro, L., Ciani, M., Comitini, F., 2014. Antimicrobial activity of *Metschnikowia pulcherrima* on wine yeasts. J. Appl. Microbiol. 116, 1209–1217.

Otieno, D.O., 2010. Synthesis of β-galactooligosaccharides from lactose using microbial β-galactosidases. Compr. Rev. Food Sci. Food 9, 471–482.

Panesar, P.S., 2011. Fermented dairy products: starter cultures and potential nutritional benefits. Food Nutr. Sci. 2, 47.

Papamanoli, E., Tzanetakis, N., Litopoulou-Tzanetaki, E., Kotzekidou, P., 2003. Characterization of lactic acid bacteria isolated from a Greek dry-fermented sausage in respect of their technological and probiotic properties. Meat Sci. 65, 859–867.

Parmjit, S., 2011. Fermented dairy products: starter cultures and potential nutritional benefits. Food Nutr. Sci. 2011.

Pavsler, A., Buiatti, S., 2009b. Non-lager beer. In: Preedy, V.R. (Ed.), Beer in Health and Disease Prevention. Academic Press, pp. 17–30.

Pavsler, A., Buiatti, S., 2009a. Lager beer. In: Preedy, V.R. (Ed.), Beer in Health and Disease Prevention. Academic Press, pp. 31–43.

Pearson, A.M., Dutson, T.R., 1986. Meat and Poultry Microbiology. AVI Publishing Co.

Pereira, C.I., Crespo, M.B., San Romao, M.V., 2001. Evidence for proteolytic activity and biogenic amines production in *Lactobacillus curvatus* and *L. homohiochii*. Int. J. Food Microbiol. 68, 211–216.

Peterson, S.D., Marshall, R.T., 1990. Nonstarter lactobacilli in Cheddar cheese: a review. J. Dairy Sci. 73 (6), 1395–1410.

Pfeiler, E.A., Klaenhammer, T.R., 2007. The genomics of lactic acid bacteria. Trends Microbiol. 15 (12), 546–553.

Pleadin, J., Staver, M.M., Vahčić, N., Kovačević, D., Milone, S., Saftić, L., et al., 2015. Survey of aflatoxin B1 and ochratoxin A occurrence in traditional meat products coming from Croatian households and markets. Food Control 52, 71–77.

Prentice, A.M., 2014. Dairy products in global public health. Am. J. Clin. Nutr. 99, 1212S–1216S.

Pretorius, I.S., 2000. Tailoring wine yeast for the new millennium: novel approaches to the ancient art of winemaking. Yeast 16, 675–729.

Randazzo, C.L., Torriani, S., Akkermans, A.D., de Vos, W.M., Vaughan, E.E., 2002. Diversity, dynamics, and activity of bacterial communities during production of an artisanal Sicilian cheese as evaluated by 16S rRNA analysis. Appl. Environ. Microbiol. 68, 1882–1892.

Randazzo, C.L., Restuccia, C., Romano, A.D., Caggia, C., 2004. *Lactobacillus casei*, dominant species in naturally fermented Sicilian green olives. Int. J. Food Microbiol. 90, 9–14.

Rantsiou, K., Cocolin, L., 2008. Fermented meat products. Molecular Techniques in the Microbial Ecology of Fermented Foods. Springer, New York, pp. 91–118.

Rantsiou, K., Drosinos, E.H., Gialitaki, M., Urso, R., Krommer, J., Gasparik-Reichardt, J., et al., 2005a. Molecular characterization of *Lactobacillus* species isolated from naturally fermented sausages produced in Greece, Hungary and Italy. Food Microbiol. 22, 19–28.

Rantsiou, K., Urso, R., Iacumin, L., Cantoni, C., Cattaneo, P., Comi, G., et al., 2005b. Culture-dependent and-independent methods to investigate the microbial ecology of Italian fermented sausages. Appl. Environ. Microbiol. 71, 1977–1986.

Rantsiou, K., Drosinos, E.H., Gialitaki, M., Metaxopoulos, I., Comi, G., Cocolin, L., 2006. Use of molecular tools to characterize *Lactobacillus* spp. isolated from Greek traditional fermented sausages. Int. J. Food Microbiol. 112, 215–222.

Rao, D.R., Reddy, A.V., Pulusani, S.R., Cornwell, P.E., 1984. Biosynthesis and utilization of folic acid and vitamin B12 by lactic cultures in skim milk. J. Dairy Sci. 67, 1169–1174.

Renouf, V., Claisse, O., Lonvaud-Funel, A., 2007. Inventory and monitoring of wine microbial consortia. Appl. Microbiol. Biotechnol. 75, 149–164.

Reveron, I.M., Barreiro, J.A., Sandoval, A.J., 2005. Thermal death characteristics of *Lactobacillus paracasei* and *Aspergillus niger* in Pilsen beer. J. Food Sci. Eng. 66, 239–243.

Rice, S.L., Koehler, P.E., 1976. Tyrosine and histidine decarboxylase activities of *Pediococcus cerevisiae* and *Lactobacillus* species and the production of tyramine in fermented sausages. J. Milk. Food Technol. 39, 166–169.

Rodhouse, L., Carbonero, F., 2019. Overview of craft brewing specificities and potentially associated microbiota. Crit. Rev. Food Sci. Nutr. 59, 462–473.

Rodríguez, A., Rodríguez, M., Martín, A., Delgado, J., Córdoba, J.J., 2012. Presence of ochratoxin A on the surface of dry-cured Iberian ham after initial fungal growth in the drying stage. Meat Sci. 92, 728–734.

Romano, P., Fiore, C., Paraggio, M., Caruso, M., Capece, A., 2003. Function of yeast species and strains in wine flavour. Int. J. Food Microbiol. 86, 169–180.

Routray, W., Mishra, H.N., 2011. Scientific and technical aspects of yogurt aroma and taste: a review. Compr. Rev. Food Sci. F. 10, 208–220.

Ruiz, J., Ventanas, J., Cava, R., Andrés, A., García, C., 1999. Volatile compounds of dry-cured Iberian ham as affected by the length of the curing process. Meat Sci. 52, 19–27.

Sabate, J., Cano, J., Querol, A., Guillamón, J., 1998. Diversity of Saccharomyces strains in wine fermentations: analysis for two consecutive years. Lett. Appl. Microbiol. 26, 452–455.

Sakamoto, K., Konings, W.N., 2003. Beer spoilage bacteria and hop resistance. Int. J. Food Microbiol. 89, 105–124.

Samelis, J., Maurogenakis, F., Metaxopoulos, J., 1994. Characterisation of lactic acid bacteria isolated from naturally fermented Greek dry salami. Int. J. Food Microbiol. 23 (2), 179–196.

Santos, N.N., Santos-Mendonça, R.C., Sanz, Y., Bolumar, T., Aristoy, M.C., Toldrá, F., 2001. Hydrolysis of pork muscle sarcoplasmic proteins by *Debaryomyces hansenii*. Int. J. Food Microbiol. 68, 199–206.

Schifferdecker, A.J., Dashko, S., Ishchuk, O.P., Piškur, J., 2014. The wine and beer yeast *Dekkera bruxellensis*. Yeast 31, 323–332.

Settanni, L., Moschetti, G., 2014. New trends in technology and identity of traditional dairy and fermented meat production processes: preservation of typicality and hygiene. Trends Food. Sci. Technol. 37, 51–58.

Shakeel-Ur-Rehman, Fox, P.F., McSweeney, P.L., 2000. Methods used to study non-starter microorganisms in cheese: a review. Int. J. Dairy Technol. 53, 113–119.

Sharma, R., Sanodiya, B.S., Bagrodia, D., Pandey, M., Sharma, A., Bisen, P.S., 2012. Efficacy and potential of lactic acid bacteria modulating human health. Int. J. Pharma Bio Sci. 3, 935–948.

Sheehan, J.J., 2007. What causes the development of gas during ripening? In: McSweeney, P.L.H. (Ed.), Cheese Problems Solved. Woodhead Publishing, Cambridge, p. 131.

Silvetti, T., Brasca, M., Lodi, R., Vanoni, L., Chiolerio, F., De Groot, M., et al., 2010. Effects of lysozyme on the microbiological stability and organoleptic properties of unpasteurized beer. J. Inst. Brew. 116, 33–40.

Singh, A.K., Ramesh, A., 2008. Succession of dominant and antagonistic lactic acid bacteria in fermented cucumber: insights from a PCR-based approach. Food Microbiol. 25, 278–287.

Slocum, S.A., Jasinski, E.M., Kilara, A., 1988. Processing variables affecting proteolysis in yogurt during incubation. J. Dairy Sci. 71, 596–603.

Snauwaert, I., Roels, S.P., Van Nieuwerburg, F., Van Landschoot, A., De Vuyst, L., Vandamme, P., 2016. Microbial diversity and metabolite composition of Belgian red-brown acidic ales. Int. J. Food Microbiol. 221, 1–11.

Spitaels, F., Wieme, A.D., Janssens, M., Aerts, M., Daniel, H.M., Van Landschoot, A., et al., 2014. The microbial diversity of traditional spontaneously fermented lambic beer. PLoS One 9, e95384.

Steele, J., Broadbent, J., Kok, J., 2013. Perspectives on the contribution of lactic acid bacteria to cheese flavor development. Curr. Opin. Biotechnol. 24, 135–141.

Steinkraus, K. (Ed.), 2004. Industrialization of Indigenous Fermented Foods, Revised and Expanded. CRC Press.

Steinkraus, K. (Ed.), 2018. Handbook of Indigenous Fermented Foods, Revised and Expanded. CRC Press.

Stellato, G., De Filippis, F., La Storia, A., Ercolini, D., 2015. Coexistence of lactic acid bacteria and potential spoilage microbiota in a dairy processing environment. Appl. Environ. Microbiol. 81, 7893–7904.

Suzuki, K., 2011. 125th anniversary review: microbiological instability of beer caused by spoilage bacteria. J. Inst. Brew. 117, 131–155.

Takahashi, M., Kita, Y., Kusaka, K., Mizuno, A., Goto-Yamamoto, N., 2015. Evaluation of microbial diversity in the pilot-scale beer brewing process by culture-dependent and culture-independent method. J. Appl. Microbiol. 118, 454–469.

Talon, R., Leroy, S., 2011. Diversity and safety hazards of bacteria involved in meat fermentations. Meat Sci. 89, 303–309.

Talon, R., Leroy-Sétrin, S., Fadda, S., 2002. Bacterial starters involved in the quality of fermented meat products. Res. Adv. Qual. Meat Meat Prod. 175–191.

Talon, R., Lebert, I., Lebert, A., Leroy, S., Garriga, M., Aymerich, T., et al., 2007. Traditional dry fermented sausages produced in small-scale processing units in Mediterranean countries and Slovakia. 1: Microbial ecosystems of processing environments. Meat Sci. 77, 570–579.

Taylor, S.L., Keefe, T.J., Windham, E.S., Howell, J.F., 1982. Outbreak of histamine poisoning associated with consumption of Swiss cheese. J. Food Prot. 45, 455–457.

Tofalo, R., Schirone, M., Perpetuini, G., Angelozzi, G., Suzzi, G., Corsetti, A., 2012. Microbiological and chemical profiles of naturally fermented table olives and brines from different Italian cultivars. Antonie Van Leeuwenhoek 102, 121–131.

Toldrá, F. (Ed.), 2010. Handbook of Meat Processing. John Wiley & Sons.

Trmčić, A., Martin, N.H., Boor, K.J., Wiedmann, M., 2015. A standard bacterial isolate set for research on contemporary dairy spoilage. J. Dairy Sci. 98, 5806–5817.

Veiga, P., Pons, N., Agrawal, A., Oozeer, R., Guyonnet, D., Brazeilles, R., et al., 2014. Changes of the human gut microbiome induced by a fermented milk product. Sci. Rep. 4, 6328.

Viander, B., Mäki, M., Palva, A., 2003. Impact of low salt concentration, salt quality on natural large-scale sauerkraut fermentation. Food Microbiol. 20, 391–395.

Viljoen, B.C., 2006. Yeast ecological interactions. Yeast–yeast, yeast bacteria, yeast–fungi interactions and yeasts as biocontrol agents. In: Querol, A., Fleet, H. (Eds.), Yeasts in Food and Beverages. Springer-Verlag, Berlin, pp. 83–110.

Vinusha, K.S., Deepika, K., Johnson, T.S., Agrawal, G.K., Rakwal, R., 2018. Proteomic studies on lactic acid bacteria: a review. Biochem. Biophys. Rep. 14, 140–148.

Walker, G.M., 2014. Wines: microbiology of winemaking. In: Robinson, R.K. (Ed.), Encyclopedia of Food Microbiology. Academic Press, pp. 787–792.

Williams, A.G., Banks, J.M., 1997. Proteolytic and other hydrolytic enzyme activities in non-starter lactic acid bacteria (NSLAB) isolated from Cheddar cheese manufactured in the United Kingdom. Int. Dairy J. 7, 763–774.

Wolfe, B.E., Dutton, R.J., 2015. Fermented foods as experimentally tractable microbial ecosystems. Cell 161, 49–55.

Wouters, D., Grosu-Tudor, S., Zamfir, M., De Vuyst, L., 2013. Applicability of *Lactobacillus plantarum* IMDO 788 as a starter culture to control vegetable fermentations. J. Sci. Food Agric. 93, 3352–3361.

Wouters, J.T., Ayad, E.H., Hugenholtz, J., Smit, G., 2002. Microbes from raw milk for fermented dairy products. Int. Dairy J. 12, 91–109.

Xiong, T., Guan, Q., Song, S., Hao, M., Xie, M., 2012. Dynamic changes of lactic acid bacteria flora during Chinese sauerkraut fermentation. Food Control 26, 178–181.

Xiong, T., Li, J., Liang, F., Wang, Y., Guan, Q., Xie, M., 2016. Effects of salt concentration on Chinese sauerkraut fermentation. LWT-Food Sci. Technol. 69, 169–174.

Yamanaka, H., Akimoto, M., Sameshima, T., Arihara, K., Itoh, M., 2005. Effects of bacterial strains on the development of the ripening flavor of cured pork loins. Anim. Sci. J. 76, 499–506.

Yan, P.M., Xue, W.T., 2005. Relation between lactic acid bacteria the flavor of fermented vegetable. Chin. Condiment 2.

Yuksekdag, Z.N., Aslim, B., 2010. Assessment of potential probiotic-and starter properties of *Pediococcus* spp. isolated from Turkish-type fermented sausages (sucuk). J. Microbiol. Biotechnol. 20, 161–168.

Zabat, M., Sano, W., Wurster, J., Cabral, D., Belenky, P., 2018. Microbial community analysis of sauerkraut fermentation reveals a stable and rapidly established community. Foods 7, 77.

Zacharof, M.P., Lovitt, R.W., 2012. Bacteriocins produced by lactic acid bacteria a review article. APCBEE Procedia 2, 50–56.

Zamfir, M., Vancanneyt, M., Makras, L., Vaningelgem, F., Lefebvre, K., Pot, B., et al., 2006. Biodiversity of lactic acid bacteria in Romanian dairy products. Syst. Appl. Microbiol. 29, 487–495.

Zhong, Z., Hou, Q., Kwok, L., Yu, Z., Zheng, Y., Sun, Z., et al., 2016. Bacterial microbiota compositions of naturally fermented milk are shaped by both geographic origin and sample type. J. Dairy Sci. 99, 7832–7841.

FURTHER READING

Albano, H., Henriques, I., Correia, A., Hogg, T., Teixeira, P., 2008. Characterization of microbial population of 'Alheira' (a traditional Portuguese fermented sausage) by PCR-DGGE and traditional cultural microbiological methods. J. Appl. Microbiol. 105, 2187–2194.

Alegría, Á., Delgado, S., Roces, C., López, B., Mayo, B., 2010. Bacteriocins produced by wild *Lactococcus lactis* strains isolated from traditional, starter-free cheeses made of raw milk. Int. J. Food Microbiol. 143, 61–66.

Dall'Asta, C., Galaverna, G., Bertuzzi, T., Moseriti, A., Pietri, A., Dossena, A., et al., 2010. Occurrence of ochratoxin A in raw ham muscle, salami and dry-cured ham from pigs fed with contaminated diet. Food Chem. 120, 978–983.

Hui, Y.H., Meunier-Goddik, L., Josephsen, J., Nip, W.K., Stanfield, P.S. (Eds.), 2004. Handbook of Food and Beverage Fermentation Technology, vol. 134. CRC Press.

Fermentation and fermented dairy products. In: Jay, J.M. (Ed.), Modern Food Microbiology. Springer, New York.

Kumar, B., Balgir, P.P., Kaur, B., Garg, N., 2011. Cloning and expression of bacteriocins of *Pediococcus* spp.: a review. Arch. Clin. Microbiol. 2, 4.

Morot-Bizot, S.C., Leroy, S., Talon, R., 2006. Staphylococcal community of a small unit manufacturing traditional dry fermented sausages. Int. J. Food Microbiol. 108, 210–217.

CHAPTER

LEGACY AND INNOVATIVE TREATMENT: PROJECTED MODALITIES FOR ANTIMICROBIAL INTERVENTION 13

Mohammad Al Sorkhy and Rose Ghemrawi

College of Pharmacy, Al Ain University, Abu Dhabi, United Arab Emirates

INTRODUCTION

For centuries, saving humanity from the clutches of infectious diseases was considered a miraculous feat that required outrageous treatments, including bowel emptying, consumption of questionable chemical concoctions, ice baths, life-threatening diets, and bloodlettings. In actuality, these "treatments" probably caused more harm than good as they wore out the patients' bodies and weakened their immune system; it is believed that George Washington died due to severe blood loss via bloodletting that was supposed to treat a streptococcal throat infection, something that can be treated easily using antibiotics nowadays.

Even after the development of the germ theory, which widely increased the knowledge of microorganisms that cause infectious diseases, very little could be done to properly treat patients until the emergence of antimicrobial agents in the 1940s which revolutionized medicine.

Doctors were amazed that a material that was so harmless to the human body could be used to fight off the microorganisms causing these diseases. As a result, millions of lives that could have been taken by infectious diseases were saved by the use of antibiotics, instilling a hope that infectious diseases could be eradicated once and for all. Unfortunately, this was not so. The consequences of the overuse of antibiotics began to manifest; microorganisms had begun to develop resistance to antibiotics that rendered the use of antibiotics futile, something that the World Health Organization has considered the greatest threat to human health.

This chapter discusses the discovery of antimicrobial drugs; how they have become the backbone of healthcare due to their ability to treat viral, bacterial, fungal, and parasitic infections; the evolution of antimicrobial resistance; and how this may be overcome.

A BRIEF HISTORY OF THE ANTIMICROBIAL STRUGGLE

THE HISTORY OF CHEMOTHERAPY ORIGINATED WITH PAUL EHRLICH

In the 1900s the leading causes of death were infectious diseases. Due to this, the focus of microbiologists was dedicated to boosting the immune system in an attempt to control and possibly cure

Microbiomics. DOI: https://doi.org/10.1016/B978-0-12-816664-2.00013-X

infectious diseases. A scientist named Paul Ehrlich, who knew that certain dyes could be used to stain specific types of bacteria, thereby identifying them, hypothesized that certain chemicals might be toxic to specific types of bacteria, allowing us to target those bacteria alone and kill them without harming the infected human body, thus introducing the "selective toxicity" concept.

After the synthesis of many arsenic–phenol compounds, Sahachiro Hata, a colleague of Paul Ehrlich, isolated a compound that could successfully treat syphilis in animals and humans without harming them. This compound was appropriately named Salvarsan as it offered salvation from syphilis and contained arsenic.

Throughout the following decades, any dye made for industrial use was also tested for antimicrobial properties. This paid off in 1932 when a German bacteriologist named Gerhard Domagk discovered a compound in a red dye that was able to inhibit the effects of Gram-positive bacterial species such as staphylococci and streptococci. This dye was named Prontosil, and its discovery is considered to have started modern-day chemotherapy.

FLEMING'S OBSERVATION OF THE PENICILLIN EFFECT USHERED IN THE ERA OF ANTIBIOTICS

In 1928 microbiologist Alexander Fleming, who had previously theorized the existence of antibiotics and their functions, noticed that a blue-green mold had begun to grow on one of the staphylococci nutrient agar plates, and that no bacteria would grow near it. He examined this mold and found it to be a species of *Penicillium*, a type of fungus able to kill bacteria by a substance it produced. Fleming failed to identify and analyze this bactericidal substance but nonetheless named it "penicillin."

In 1939, during World War II, a pathologist named Howard Florey and a biochemist named Ernst Chain led a research group at Oxford University and found that penicillin could be used to treat a number of diseases caused by bacterial infections such as gonorrhea, meningitis, tetanus, and diphtheria thanks to its bactericidal properties. However, due to the war, penicillin could not be mass-produced in England. Instead, a group of American companies managed to find ways to accomplish this, making penicillin commercially available.

THE CURRENT ANTIBACTERIAL ARSENAL

Antibiotics can be classified based on their way to inhibit bacterial growth; herein we discuss the four major groups of antibacterial agents.

METABOLIC ANTAGONISTS

In 1935 the active form of Prontosil was identified as a chemical called sulfanilamide, a substance that was highly effective against Gram-positive bacterial species. Sulfanilamide soon became useful to soldiers in World War II for treating infected wounds.

Sulfanilamide became the first broad-spectrum synthetic agent of a group called *sulfonamides*. This group of drugs has a bacteriostatic effect, disrupting the metabolism of Gram-positive and

Gram-negative bacterial cells by preventing the synthesis of folic acid—an important growth factor used in the synthesis of nucleic acids. Folic acid is produced by enzymatic joining of three components, one of which is *para-aminobenzoic acid (PABA)*—a molecule similar in structure to a sulfonamide called sulfamethoxazole (SMZ). SMZ competes with PABA for the bacterial enzyme's active site. This *competitive inhibition ultimately* prevents the synthesis of nucleic acids thereby preventing DNA replication. However, mutations in some bacterial species have now emerged, allowing them to absorb folic acid from extracellular sources and making them resistant to sulfonamides (Griffith et al., 2018).

Sulfonamides have therapeutic uses in chronic obstructive pulmonary disease, ulcer, asthma, arthritis, and cancer (Jain et al., 2013). Recently, some sulfonamides, such as acetazolamide, ethoxzolamide, dichlorophenamide, dorzolamide, sulthiame, and 4-(2-hydroxymethyl-4-nitrophenyl-sulphonamido)ethylbenzenesulphonamide, were found to be effective inhibitors of species belonging to the genus *Clostridium*, known to cause serious human disease, such as tetanus, botulism, gas gangrene, and bacterial corneal keratitis (Vullo et al., 2018).

Advanced forms of sulfonamides consist of a mixture of two drugs, such as the cotrimoxazole (Bactrim), a combination of trimethoprim and SMZ, used to treat infections in the urinary tract, lungs, and ears. Combinations of drugs are an example of *drug synergism*, meaning that these drugs when administered together are more effective than the sum of their individual efficiencies. The combination of drugs is important because much lower doses are needed to effectively counter the susceptible microbiota, decreasing the possibility of resistance development.

NUCLEIC ACIDS INHIBITORS

Quinolones are a group of synthetic bactericidal drugs that block DNA synthesis in both Gram-positive and Gram-negative bacterial cells.

It is one of the most commonly prescribed classes of antibacterials worldwide and is used to treat a variety of bacterial infections in humans. The cellular targets for quinolones are the bacterial type II topoisomerases, gyrase, and topoisomerase IV. They act by converting their targets, gyrase, and topoisomerase IV into toxic moieties that fragment the bacterial chromosome (Aldred et al., 2014).

Fluoroquinolones, such as levofloxacin (Levaquin) and ciprofloxacin (Cipro), are commonly prescribed antibiotics that are used to treat a number of infections such as intestinal infections, urinary tract infections, gonorrhea, and chlamydial infection. The obvious result of overprescription of fluoroquinolones is the development of worldwide resistance of microbiota against this class of drugs; thus new fluoroquinolones are under development such as avarofloxacin (Taneja and Kaur, 2016).

A semisynthetic bacteriostatic drug, called Rifampin, works to impede the synthesis of bacterial RNA; it inhibits transcription by binding to the prokaryotic DNA-dependent RNA-polymerase and blocking the RNA strand initialization stage (Artsimovitch et al., 2003). Rifamycin was isolated from the Gram-positive bacterium, *Amycolatopsis mediterranei* (*Streptomyces mediterranei*). It is prescribed with isoniazid and ethambutol against *Mycobacterium tuberculosis*. It is also prescribed against the meningitis-causing bacterial genera *Neisseria* and *Haemophilus*. The combination of rifampin with colistin and meropenem/doripenem has demonstrated synergistic effects against multidrug-resistant (MDR) *Pseudomonas* spp., *Acinetobacter* spp., and carbapenemase-producing *Enterobacteria* (Tangden, 2014).

CELL WALL SYNTHESIS INHIBITORS

Penicillin has saved millions of lives since its commercial production and has become the go-to drug in eradicating many infections due to its high chemotherapeutic index. All forms of penicillin have the same beta-lactam nucleus and only differ in the side chains attached to it.

Penicillin kills bacteria by obstructing the cross-linking between peptidoglycan layers during the formation of the bacterial cell wall (see next). This severely weakens the cell wall, leading to the internal osmotic pressure, cell's swelling, and bursting. Penicillin effectively behaves as bactericidal when bacterial cells are multiplying rapidly but has a mere bacteriostatic effect when cells multiply slowly or do not multiply at all.

Natural penicillins

Penicillin G is the most common penicillin prescribed by physicians. It is primarily delivered intravenously as it is sensitive to acidity and therefore cannot be administered orally. Penicillin V, another natural penicillin, is less sensitive to acidity and therefore is administered orally.

These penicillins tend to have a narrow therapeutic spectrum and are only useful against some species of Gram-positive bacteria. Many bacterial species have become resistant to natural penicillins by producing the enzyme *beta-lactamase*, which inactivates beta-lactam antibiotics. Through hydrolysis, the beta-lactamase enzyme breaks the beta-lactam ring, deactivating the molecule's antibacterial properties and rendering natural penicillins such as penicillin G harmless by converting them into penicilloic acid.

Semisynthetic penicillins

In the 1950s scientists manufactured semisynthetic penicillins by attaching different functional groups to the beta-lactam nucleus, thus widening the therapeutic spectrum of the penicillins and making them resistant to beta-lactamase. Ampicillin and amoxicillin are examples of semisynthetic penicillins. These penicillins are administered orally and are absorbed by the intestine, as they are not sensitive to stomach acids. The other semisynthetic penicillin, carbenicillin, has an even broader therapeutic spectrum; it is effective against even more Gram-negative bacterial species and is commonly used to treat urinary tract infections. Despite these modifications, resistance was still able to develop as in methicillin-resistant *Staphylococcus aureus* (MRSA).

Certain penicillins such as amoxicillin can be combined with a beta-lactamase inhibitor such as clavulanic acid to allow them to perform their function without being destroyed by bacterial beta-lactamase.

Other beta-lactam antibiotics

Bacterial cells are surrounded by a cell wall made of peptidoglycan, which consists of long carbohydrate polymers and peptides. The peptide chains cross-link the carbohydrate chains, thus strengthening the cell wall. A cardinal link is the one formed between the D-alanyl-alanine portion of the peptide chain and the glycine residues in the presence of penicillin-binding proteins (PBPs).

Beta-lactam ring mimics the D-alanyl-D-alanine portion of the peptide chain. PBP interacts with beta-lactam ring instead of interacting with the D-alanyl-D-alanine, leading to the disruption of peptidoglycan layer and the eventual lysis of bacterium (Kapoor et al., 2017).

Cephalosporins have beta-lactam rings and carry out their functions much in the same way as penicillins. Cephalosporins are isolated from the fungus *Cephalosporium acremonium* and chemically resemble penicillins. This resemblance makes them a suitable alternative when target bacterial cells have developed penicillin resistance or when the patient is allergic to penicillin. Their broader bactericidal spectrum against Gram-negative bacteria makes them a more-than-worthy alternative to penicillin.

As with penicillins, many drugs are produced by combination of cephalosporins with beta-lactamase inhibitors, such as ceftolozane/tazobactam (Zerbaxa) and ceftazidime/avibactam (Avycaz) (Taneja and Kaur, 2016).

Carbapenems are another beta-lactam-containing group of antibiotics that has a very broad therapeutic spectrum. They are synthesized by a bacterium called *Streptomyces cattleya* and are highly resistant to beta-lactamase enzymes, making them a very useful "last resort" antibiotic when infectious agents such as *Escherichia coli* and *Klebsiella pneumoniae* show resistance to other antibiotics. Of the early carbapenems evaluated, thienamycin demonstrated the greatest antimicrobial activity and became the parent compound for all subsequent carbapenems. To date, more than 80 compounds with mostly improved antimicrobial properties are described in the literature (Papp-Wallace et al., 2011). However, the problem with carbapenems like imipenem is that they are readily degraded in the kidneys and cleared before acting on the disease. To remedy this, imipenem is usually prescribed in combination with cilastatin, the latter preventing the degradation of the antibiotic and giving it enough time to perform its antimicrobial function. The imipenem/cilastatin combination is called Primaxin and is effective against 98% of Gram-positive and Gram-negative bacteria.

The medical community is worried because of the bacterial development of New Delhi metallo-beta-lactamase (NDM-1). This enzyme is coded by bla—NDM-1 or NDM-1 gene carried on plasmids. NDM-1 hydrolyzes the beta-lactams, particularly carbapenem. It makes bacterial pathogens that are normally isolated from healthcare and community infections, such as those in the Enterobacteriaceae family, resistant to antibiotics. There are no new antibiotic developments to treat the *carbapenem-resistant Enterobacteriaceae (CRE)*.

Vancomycin is an antibiotic that kills bacterial cells by disrupting their cell wall. The sugar component of the peptidoglycan consists of alternating residues of β-(1,4) linked *N*-acetylglucosamine and *N*-acetylmuramic acid. Vancomycin prevents the synthesis and polymerization of these sugar residues causing leakage of intracellular components and bacterial cell death.

Oral vancomycin has low systemic absorption and is only effective for treating intestinal infections. Therefore it is indicated for the treatment of *Clostridium difficile*-associated diarrhea, pseudomembranous colitis, and staphylococcal enterocolitis.

Vancomycin is used for the treatment of infections caused by Gram-positive bacteria, when administered via intravenous injection. It is used most often to treat staphylococcal diseases where penicillin would be redundant either due to penicillin allergy or to resistance against multiple types of antibiotics, including methicillin. Vancomycin has actually been found to be vital in the treatment of MRSA, and, similar to carbapenems, thus it is considered a "last resort" drug.

Unfortunately, certain bacterial species have developed resistance to it. Nephrotoxicity and ototoxicity have been associated with the excessive use of vancomycin (Monteiro et al., 2018).

Bacitracin and polymyxin B are antibiotics that comprise polypeptides. Bacitracin kills bacteria by impeding the transport of precursors of bacterial cell walls through the cell membrane. It was also found to induce hydrolytic degradation of nucleic acids (Ciesiołka et al., 2014). Bacitracin is incredibly toxic when administered systematically and may damage the kidneys; therefore the only safe way to use this antibiotic is to apply it as an ointment to treat skin infections caused by Gram-positive bacteria. When bacitracin is combined with polymyxin B and neomycin, they form the drug named Neosporin.

Polymyxin B kills Gram-negative bacteria, especially bacilli, by increasing the permeability of the cell membrane. The cationic polypeptides of polymyxin B interact with anionic lipopolysaccharide molecules in the outer membrane of Gram-negative bacteria. This leads to calcium (Ca^{2+}) and magnesium (Mg^{2+}) displacement, thus deranging the cell membrane, causing leakage of cell contents, and ultimately cell death.

Polymyxin B is used to treat superficial infections, abrasions, and burns and, in combination with bacitracin and Gramicidin, forms an antibacterial ointment called Polysporin. Polymyxin B is increasingly used for the treatment of MDR pathogens, and it used in combination with meropenem against carbapenemase-producing *K. pneumoniae* (Sharma et al., 2016).

PROTEIN SYNTHESIS INHIBITORS

Some types of antibiotics tend to tamper with bacteria's ribosome by acting on its subunits, the small 30S subunit and the large 50S subunit. These types of antibiotics tend to be produced by species of the genus *Streptomyces*.

Aminoglycosides are broad spectrum, commonly prescribed for children, primarily for infections caused by Gram-negative pathogens. They cause misreading and premature termination of the translation of mRNA by interacting with the 16S r-RNA of the 30S ribosomal subunit near the A site. Thus they insert a bias to termination missense and thwart the entire protein production of the bacterial cell.

Aminoglycosides include gentamicin, amikacin, tobramycin, neomycin, and streptomycin. They are polar drugs, with poor gastrointestinal absorption, so intravenous or intramuscular administration is needed. Streptomycin, the first aminoglycoside isolated from *Streptomyces griseus* in 1943, was considered a magnificent discovery as it could treat tuberculosis along with other diseases caused by Gram-negative bacteria. Gentamicin is the most commonly used antibiotic in United Kingdom neonatal units. It is used to treat urinary tract infections caused by Gram-negative bacteria. The aforementioned neomycin is an aminoglycoside discovered in Waksman's lab. Tobramycin is an aminoglycoside that, when in an aerosolized form, is used to treat cystic fibrosis under the drug name Tobi (Germovsek et al., 2017).

Tetracyclines act upon the conserved sequences of the 16S r-RNA of the 30S ribosomal subunit to prevent binding of t-RNA to the A site. Tetracyclines can be natural as are chlortetracyclines or semisynthetic as are minocycline and doxycycline. As indicated in their name, tetracyclines have four benzene rings in their structure. The fact that tetracyclines can be administered orally led to misuse between the 1950s and 1960s.

The less serious side effects are discoloration of the teeth to a yellow-gray color and the more serious the stunted bone growth in children. Despite the side effects, tetracyclines are still prescribed because of their activity against Gram-positive and negative pathogens, and the availability of intravenous and oral formulations for most members of the class. They are the go-to drug to treat rickettsial and chlamydial diseases, and due to their ability to act against many Gram-negative bacteria, they are used to treat primary atypical pneumonia, pneumococcal pneumonia, syphilis, and gonorrhea.

Tigecycline is an antibiotic from a class that is similar to tetracyclines called glycylcyclines, it became useful in treating MRSA infections, but a few years after its development, some Enterobacteriaceae had developed resistance against it. New synthetic derivatives were created with improved in vitro potency and in vivo efficacy, which can be used against current and emerging MDR pathogens, including CRE, MDR *Acinetobacter* species, and *Pseudomonas aeruginosa* (Grossman, 2016).

Chloramphenicol is a potent inhibitor of protein synthesis and acts by binding reversibly to the 50S subunit of the bacterial ribosome. It is the drug of choice when treating typhoid fever and is considered an alternative to tetracyclines to treat epidemic typhus and Rocky Mountain spotted fever, due to its capability to inhibit many Gram-positive and Gram-negative species. Chloramphenicol has good oral bioavailability and excellent tissue penetration. However, it causes serious side effects such as accumulation in the blood of newborns leading to the breakdown of the cardiovascular system, known as *gray syndrome* (Cassir et al., 2014). Because of this, it can only be used in life-threatening situations.

The activity of chloramphenicol against ESKAPE pathogens (*Enterococcus faecium*, *S. aureus*, *K. pneumoniae*, *Acinetobacter baumannii*, *P. aeruginosa*, *Enterobacter* spp.) is good for Gram-positives, but less so for Gram-negatives. However, synergy is observed when combined with colistin. The risk–benefit related to chloramphenicol toxicity needs to be reexamined in light of the emerging problem of multidrug-resistant pathogens (Čivljak et al., 2014).

Macrolides are a type of bacteriostatic antibiotics, which interfere with protein synthesis by binding to bacterial 50S ribosomal subunit, thus leading to a premature detachment of incomplete peptide chains. They are composed of a macrocyclic lactone of different ring sizes, to which one or more deoxy sugar or amino sugar residues are attached. Erythromycin was the first macrolide to be discovered and is still widely used as an alternative to penicillin in fighting Gram-positive bacteria and species of *Neisseria* and *Chlamydia* affecting the eyes of newborns.

Some of the best antibiotics in the world are semisynthetic macrolides (the second-generation of macrolides), such as clarithromycin (Biaxin) and azithromycin (Zithromax), which have an even broader therapeutic spectrum than erythromycin. In order to address increasing antibiotic resistance, a third generation of macrolides was developed, more specifically, the ketolide telithromycin. However, it exhibited rare but serious irreversible hepatotoxicity named "Ketek effects" (Dinos, 2017).

Lincosamides are mostly active against Gram-positive pathogens and against selected Gram-negative anaerobes and protozoa. The majority of the semisynthetic derivatives of this class originate from lincomycin, naturally produced by *Streptomyces lincolnensis*. Of a large number of lincosamide derivatives reported to date, clindamycin, the chlorinated analog of lincomycin, is the only semisynthetic lincosamide used in clinics, thus alleviating the need for the development of novel lincosamides (Matzov et al., 2017).

Clindamycin is a bacteriostatic drug used against aerobic Gram-positive cocci and anaerobic Gram-negative bacilli. The antimicrobial spectrum of clindamycin includes staphylococci, group A and B streptococci, *Streptococcus pneumoniae*, most anaerobic bacteria, and *Chlamydia trachomatis*, several protozoa, such as *Plasmodium* spp. and *Toxoplasma* spp. However, it shows no activity against most aerobic Gram-negative bacilli, *Nocardia* spp., *Mycobacterium* spp., as well as *Enterococcus faecalis* and *E. faecium*. Clindamycin can only be used to treat serious infections and that when resistance to antibiotics is met, as it practically exterminates the intestinal microbime and thus promotes colonization by *C. difficile* to grow to dysbiosis, incurring *pseudomembranous colitis*.

Streptogramins include pristinamycin, virginiamycin, mikamycin, and quinupristin–dalfopristin and consist of two structurally different components, A and B. The A components are polyunsaturated macrolactones, such as pristinamycin IIA, virginiamycin M, mikamycin A, or dalfopristin. The B components are cyclic hexadepsipeptides, such as pristinamycin IB, virginiamycin S, mikamycin B, or quinupristin (Giguère, 2013).

Streptogramins cannot cross the outer membrane of most Gram-negative bacteria and are primarily effective against Gram-positive bacteria. Synthetic versions are prescribed in combination with another cyclic peptide called quinupristin–dalfopristin (Synercid) to effectively kill many Gram-positive bacterial species, including *S. aureus*, and pathogens affecting the respiratory system.

Oxazolidinones selectively inhibit bacterial protein synthesis by binding to 23S r-RNA of the 50S subunit and interacting with peptidyl-t-RNA.

Linezolid (Zyvox) falls under this class; it is active against Gram-positive bacteria. Linezolid is indicated for adults in the treatment of community-acquired pneumonia, nosocomial pneumonia, and skin structure infections (Pandit et al., 2012). However, like some aforementioned antibiotics, oxazolidinones due to their toxic nature should only be used when all else fail.

NONBACTERIAL MICROBES

Based on the Centers for Disease Control and Prevention (CDC) report, about 60% of microbial diseases are of nonbacterial nature; this number is even higher in the developing world. We have plenty of available drugs that allow us to fight bacterial infections but very few that can battle viral, fungal, and parasitic diseases.

The fact that antibacterial drugs cannot be used effectively to fight fungi and viruses results to another problem. Similarly to animals and humans, fungi are eukaryotes and therefore have similar cellular structures, knowing that viruses are intracellular parasites that use the host's cellular machinery to survive and multiply, this makes it difficult to target viruses and fungi since this may be harmful to hosts.

ANTIVIRAL DRUGS

Since the 1960s antiviral drugs are used to interfere with viral development. Some of these can be naturally isolated from animals, plants, bacteria, or fungi, while others must be chemically synthesized, either by design or randomly. Viruses are continually developing new resistant strains, due to their high spontaneous mutation rate, which make continuous research for new antiviral compounds a grim necessity (Villa et al., 2017).

Antivirals display a variety of mechanisms of action, including

- preventing viruses from entering cells—a method used to treat human immunodeficiency virus (HIV),
- preventing viruses from replicating (targeting DNA replication) by blocking a specific enzyme or a particular step in the viral replication cycle—a method used to treat herpesviruses and HIV,
- thwarting the production of new viruses at their spontaneous assembly step, and
- enhancing the animal immune system.

Almost all approved antiviral drugs target viral proteins; some examples are as follows:

- Herpesvirus drugs target the viral DNA polymerases.
- The five licensed hepatitis B virus inhibitors are polymerase inhibitors of nucleotide nature.
- Influenza can be inhibited by two classes of inhibitors, viral neuraminidase inhibitors and M2 channel blockers.
- Two drugs are already approved for hepatitis C that inhibits the viral NS3 protease, and many more are in development, targeting the viral NS5B RNA polymerase and its modulator, the NS5A protein.
- More than two dozen HIV type-1 (HIV-1) drugs have been approved which target the viral reverse transcriptase, protease, or integrase; for example, maraviroc, an inhibitor blocking HIV-1 entry is a host-targeting antiviral that has been successfully developed (Lin and Gallay, 2013).

ANTIFUNGAL AGENTS

Invasive fungal infections contribute to more than one million deaths annually and continue to be a serious danger to human health (Brown et al., 2012). Invasive candidiasis and disseminated cryptococcosis contribute to around 70% of all the systematic fungal infections (Diekema et al., 2012). More than 80% of the deaths are caused by species belonging to *Candida*, *Aspergillus*, *Cryptococcus*, *Pneumocystis*, and *Rhizopus* (Denning and Hope, 2010). However, new species belonging to *Zygomycetes*, *Fusarium*, or *Scedosporium* have increasingly reported as causative agents of invasive systemic mycoses (Castelli et al., 2014).

Development of antifungal drugs is a challenging process since fungi are eukaryotes and many key therapeutic targets are found in humans with a significant host toxicity margin (Roemer and Krysan, 2014).

Antifungal drugs categorized into seven major groups based on their therapeutic targets, which are discussed in the following sections.

Inhibitors of ergosterol biosynthesis

Ergosterol is the major constituent of the fungal cell membrane and contributes in many cellular activities such as fluidity and integrity of the membrane and the proper function of membrane enzymes.

Azoles inhibit the cytochrome P450-dependent enzyme 14a-lanosterol demethylase (CYP51) that converts lanosterol to ergosterol in the cell membrane, thus inhibiting fungal growth and replication (Carrillo-Munoz et al., 2006). Azoles are the antifungal drug of choice in clinical practice due to their broad spectrum and low toxicity range (Sheehan et al., 1999).

Fungal membrane disrupters

Macrocyclic organic compounds bind to the lipid bilayer forming complexes with ergosterol, altering the latter's distribution and thus producing pores; Nystatin and Amphotericin B are the most widely used in this category (Hossain and Ghannoum, 2000).

Fungal cell wall inhibitors

Members of this category are either glucan synthesis inhibitors (such as echinocandins) or chitin synthesis inhibitors (such as nikkomycin). These drugs are not widely used due to their poor absorption in the gastrointestinal tract and their hydrolytic liability (Lorand and Kocsis, 2007).

Sphingolipids biosynthesis inhibitors

Sphingolipids are major parts of membranes in eukaryotic cells, and some of them play an important role in fungal pathogenesis. Recent studies have demonstrated that inhibiting enzymes involved in sphingolipid biosynthesis could undermine the virulence of fungal pathogens. Aureobasidin A is an example of this group (Rollin-Pinheiro et al., 2016).

Nucleic acid synthesis inhibitors

The fungistatic flucytosine (5-fluorocytosine) is a fluorinated pyrimidine analog, which interferes with pyrimidine metabolism. It is converted by cytosine deaminase to 5-fluorouracil that is transformed by UMP pyrophosphorylase into 5-fluorouridine monophosphate. The latter is phosphorylated and incorporated in the RNA, replacing UTP and inhibiting protein synthesis. It is considered a safe drug since there is little or no cytosine deaminase activity in mammalian cells (Zhao et al., 2010).

Protein synthesis inhibitors

Protein synthesis inhibitor target different stages of the process, for example, tavaborole inhibits the leucyl-t-RNA synthetase, while sordarin inhibits translation elongation factor 2 (Gupta and Versteeg, 2016).

Microtubules inhibitors

Microtubules are highly organized structures, forming the cellular skeleton in all eukaryotic cells. Antifungal agents, belonging to this group such as griseofulvin or vinblastine, bind to tubulin, interfering with fungal microtubule assembly and inhibit mitosis (Gauwerky et al., 2009).

ANTIPROTIST AGENTS

As protists are also eukaryotes, drugs attempting to eradicate them must be targeted to metabolic pathways and structures unique to protists to prevent any harm to the host.

Malaria is a common but serious protistan disease caused by members of the genus *Plasmodium*; its mode of action is by infecting and destroying red blood cells of the afflicted individual. Aminoquinolines such as quinines, which are toxic to malarial parasites, are used to treat in this disease by invading the infected red blood cells and destroying the causative parasite. Quinine, which was discovered in the 1500 s, is considered to be one of the first antimicrobial agents and was primarily used to treat malaria until quinine resistance emerged.

New synthetic and more effective versions of aminoquinolines, such as chloroquine, mefloquine, and primaquine, have been developed and are now used to treat malaria. Chloroquine is the primary drug of choice to treat malaria caused by *Plasmodium*, whereas mefloquine is the go-to drug to treat malaria caused specifically by *Plasmodium falciparum*. Despite the efficacy of these drugs, resistance developed nonetheless, necessitating the production of *artemisinin*—a drug effective against strains of *P. falciparum* is resistant to multiple drugs. Its mode of action is the release of free radicals which kill malarial microorganisms. However, as expected, resistance to *artemisinin* has emerged in South East Asia and may spread to other parts of Asia and Africa.

The opportunist free-living protists such as *Acanthamoeba* spp. and *Balamuthia mandrillaris* have become a serious threat to human life; they can transform into metabolically inactive cyst forms, which are not susceptible to available drugs (Siddiqui et al., 2013). Currently, broad-spectrum antiprotist agents, acting against multiple cyst-forming protists, are under development.

ANTIHELMINTHIC DRUGS

Helminths are multicellular, eukaryotic parasites. While most antimicrobial drugs operate better on dividing cells of the targeted agents, antihelminthic agents work in the opposite way—they specifically attack the nondividing cells.

Praziquantel affects flukes and tapeworms by altering the permeability of their plasma membranes and thus marring the electrolyte balance, resulting in muscle contraction and paralysis.

Mebendazole prevents glucose and nutrient uptake by larval and adult forms of parasites living in a host's intestines. This halts the parasite's ability to synthesize ATP, eventually necrosing it.

Avermectins like ivermectin work against various types of nematodes by affecting their nervous system and causing muscle paralysis.

Recently, seed extracts from pumpkin (*Cucurbita pepo* L.) were found to have anthelmintic activity on two model nematodes: *Caenorhabditis elegans* (*C. elegans*) and *Heligmosoides bakeri* (*H. bakeri*). The anthelmintic action of the extracts was observed on egg hatching, larval development, and adult worms' motility. However, no significant effect of the tested extracts was observable regarding *C. elegans* integrity or motility. In addition, *C. pepo* seed extract administrated to mice was effective in reducing both the fecal egg counts and the number of adult stages of *H. bakeri*. Thus pumpkin seed extracts may be used to control gastrointestinal nematode infections. This relatively inexpensive alternative to the currently available chemotherapeutics is considered as a novel drug candidate (Grzybek et al., 2016).

ANTIBIOTIC RESISTANCE

Antibiotic resistance is reported to occur when a drug loses its ability to effectively inhibit microbial growth. Resistant microbes continue to replicate even in the presence of previously therapeutic levels of the antibiotics.

In 1945, only 5 years after the introduction of sulfonamides and penicillin, resistance was reported. Resistance to tetracycline, streptomycin, and chloramphenicol was found in the 1950s. Methicillin was introduced in 1959, and MRSA was identified in 1961. Linezolid and daptomycin were introduced in the 2000s, and respective resistance was reported within 5 years. Currently, the most notorious resistant bacterium is *S. aureus.* The most antibiotic-resistant microorganisms, including *E. faecium*, *S. aureus*, *K. pneumoniae*, *A. baumannii*, *P. aeruginosa*, and *Enterobacter* species, known collectively as "ESKAPE" microorganisms, have caused significant morbidity and mortality (Boucher et al., 2009). A bacterium resistant to more than one antibiotic is said to be MDR. Many of them have been reported in orthopedic implant-associated infections, including MRSA, vancomycin-resistant *S. aureus*, MDR *Acinetobacter*, extended spectrum beta-lactamase producing Enterobacteriaceae, and MDR *P. aeruginosa*.

Resistance may be brought about in two ways:

1. *Mutations*: Mutations occurring in the bacterial genome may incur antibiotic resistance. In general, resistance arising due to mutational changes is diverse and varies in complexity. Mutations resulting in antimicrobial resistance can alter the antibiotic action via
   a. modifications of the molecular target of the antibiotic, leading to decreased affinity for the drug;
   b. a decrease in the drug uptake;
   c. activation of efflux mechanisms; the production of complex bacterial mechanisms capable to expel a toxic compound out of the cell can also result in antimicrobial resistance. Many classes of efflux pumps have been characterized in both Gram-negative and Gram-positive pathogens; and
   d. changes in important metabolic pathways via modulation of regulatory networks.

   In 2007 scientists observed a patient who had been infected with a strain of *S. aureus* for 30 months. During that period of time, the strain of *S. aureus* presented 18 sequential mutations allowing it to develop resistance to many antibiotics including vancomycin, meaning that the patient could not be treated with any of those antibiotics, and therefore was never treated.
2. *Horizontal gene transfer*: Bacterial cells often become resistant to antibiotics by acquiring resistance genes from other bacterial cells that had already become resistant. This method is called *horizontal gene transfer* and may occur via the following:
   a. Transformation: Bacteria take up naked DNA from their environment.
   b. Conjugation: Bacteria directly transfer DNA to each other.
   c. Transduction: Bacteriophages (bacterial viruses) move genes from one cell to another.

Bacteria that acquire resistance genes via mutations or gene transfer become resistant to more than one family of antibiotics. Fig. 13.1 summarizes some bacterial resistance mechanisms.

Bacterial cells resist antibiotics in two ways, described as "offensive" and "defensive" methods. The former focuses on modifying the antimicrobial agent, while the latter modifies the targeted cellular elements and pathways, making them resistant. Table 13.1 summarizes some of these mechanisms.

OFFENSIVE RESISTANCE STRATEGIES

- *Antibiotic hydrolysis*: Enzymatic hydrolysis of the antibiotics using enzymes, such as beta-lactamase, an enzyme produced by penicillin-resistant species of bacteria. Beta-lactamase works

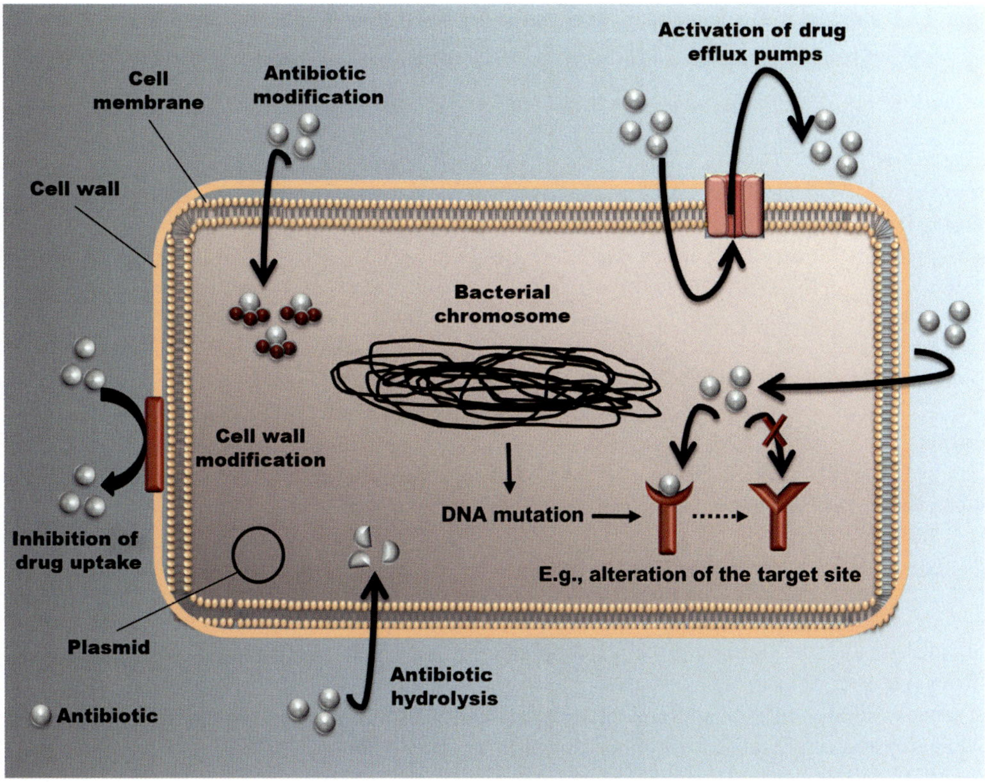

FIGURE 13.1 Bacterial antibiotic resistance mechanisms.

Mechanisms of antibiotic resistance include expression of efflux pumps on the cell membrane, degradation of antibiotics by enzymatic action, inactivation of the antibiotic by inducing covalent modification, resistance mutations modifying target proteins, and cell wall modification leading to the inhibition of drug uptake. In addition, overproduction of the target moieties may saturate the antibiotic molecules, especially the ones with fast clearance.

by breaking the beta-lactam ring of penicillin and cephalosporin, preventing them from inhibiting cell wall synthesis.

- *Chemical alterations of the antibiotic*: Enzymatic inactivation of antibiotics. Streptomycin-resistant bacterial cells can do this by phosphorylation, acetylation, or adenylation of the antibiotic, hindering its binding to ribosomes.

DEFENSIVE RESISTANCE STRATEGIES

- *Cell wall and membrane modifications*: This prevents antibiotics from entering the cells. Penicillin-resistant *Pseudomonas* modifies its membrane so penicillin cannot cross it to enter the cells. *E. coli* and *S. aureus* pump out tetracyclines as soon as they enter the cells; both

Table 13.1 Bacterial Mechanisms of Resistance Toward Specific Antibiotic Classes.

Antibiotic	Mechanism of Action	Mechanism of Resistance
Beta-lactams	Interference with cell wall synthesis	Altered cell permeability. Reduction of affinity of the target to the antibiotic
Glycopeptides		Antibiotic hydrolysis
Macrolides Chloramphenicol	Inhibition of protein synthesis (binding to 50S ribosomal subunit)	Reduction of affinity of the target to the antibiotic
Aminoglycosides	Inhibition of protein synthesis (binding to 30S ribosomal subunit)	Antibiotic inactivation via antibiotic's acetylation, phosphorylation or adenylation
Tetracyclines		Modified cell permeability Active efflux from cells
Fluoroquinolones	Inhibition of DNA synthesis	Alteration of antibiotic target Altered cell permeability
Rifampin	Inhibition of RNA synthesis	Alteration in antibiotic target Antibiotic inactivation
Sulfonamides	Inhibition of metabolic pathway	Metabolic bypass of inhibited reaction Overproduction of antibiotic target
Polymyxins	Disruption of bacterial membrane	Altered cell permeability

membrane and cytoplasmic proteins participate in this efflux mechanism to prevent the drug from acting on ribosomes.

- *Target modification*: Some strains of streptomycin-resistant bacteria can alter the structure of their ribosomes, preventing antibiotics from binding to them and therefore allowing protein synthesis in the cell to go on unhindered. Some types of modifications involve modifying RNA polymerase and enzymes participating in DNA synthesis.
- *Overproduction of antibiotic target*: Increasing the levels of drug target means higher antibiotic concentrations are required to neutralize the excess target so as to prevent bacterial growth; this spoils greatly the antibiotic effectiveness.

NEW APPROACHES OF ANTIMICROBIAL DISCOVERY

The massive appearance of drug-resistant bacteria and the lack of other therapeutic alternatives terrify public health professionals and put them most of the time in an uncomfortable position when it comes to prescribing antibiotics. The widespread use of antibiotics has paved the way for bacteria to mutate and develop resistance to these drugs. As a result, a small number of bacterial organisms decipher how to resist the drug's bactericidal effects; this led to the emergence of a population of antibiotic-resistant strains able to survive all the known antibiotic arsenals; these are called the "Superbugs."

Some of the most well-known superbugs include MRSA, *C. difficile*, CRE, MDR tuberculosis, antibiotic-resistant gonorrhea.

This resistance resulted in a huge increase in the economic burden for both hospitals and patients due to the prolonged hospital stay and the concomitant higher costs required, and in increasingly limited and more expensive antibiotic choices for infection control.

The consequences of antibiotic resistance are very serious and could have a significant impact on morbidity and mortality (Friedman et al., 2016). According to the CDC, it is estimated that nearly two million individuals develop bacterial infections caused by antibiotic-resistant bacteria every year, and about 23,000 of them die mainly because of these infections (Lushniak, 2014).

THE TWO-COMPONENT SYSTEM

To overcome this resistance a new component in the bacterial physiology must be targeted to stop or prevent their growth despite the presence of bacterial resistance. The two-component system (TCS) illustrated in Fig. 13.2 is an attractive antibacterial target in almost all bacteria except mycoplasma and is crucial for bacterial signal transduction pathway to adjust to any environmental

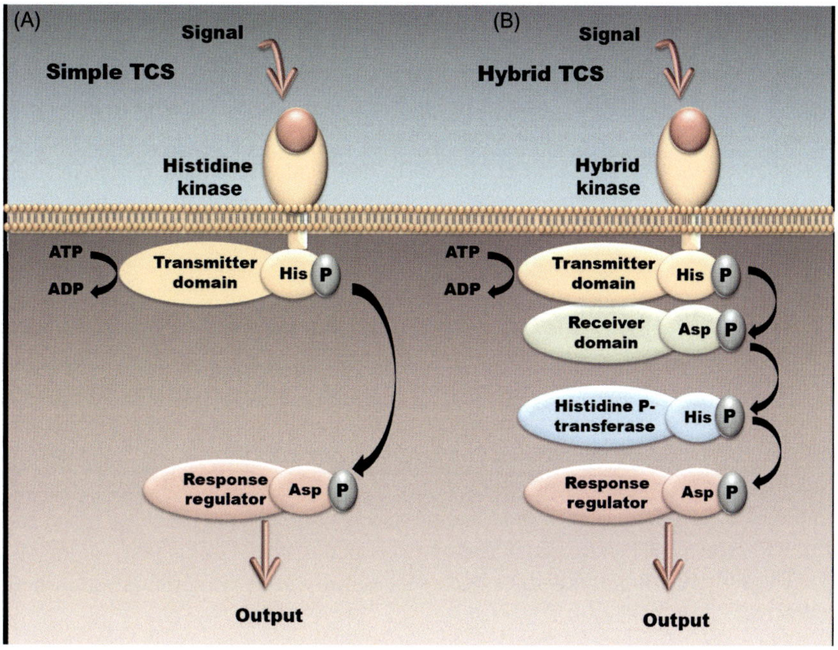

FIGURE 13.2 Schematics of the two types of TCSs.

(A) A simple TCS signaling comprises of a single sensor kinase and a single response regulator. The sensor kinase responds to its signal by autophosphorylating the conserved histidine residue in its transmitter domain. The phosphate group is then transferred to a conserved aspartate residue within the response regulator. (B) A hybrid-type TCS signaling contains, besides the sensor kinase and the response regulator, an intermediate receiver domain and a phosphotransferase protein containing a histidine. Arrows indicate transfer of phosphate groups during phosphorylation events. *TCS*, Two-component system.

changes (Bijlsma and Groisman, 2003). In addition, the two-component regulatory system seems to contribute in the tuning of the bacterial virulence and its resistance to antibiotics (Gooderham and Hancock, 2009).

A typical bacterial pathogen contains various TCSs that are essential for its outgrowth, survival, and virulence (Hancock and Perego, 2004). The standard TCS contains an inner membrane-spanning histidine kinase (HK) receptor, acting as a sensor, that receives specific signals and a cytoplasmic cognate response regulator (RR), which translates these signals into a desired output through RR phosphorylation (Capra and Laub, 2012). In response to an extracellular environmental signal, the HK binds ATPs process at the conserved histidine residue to start the autophosphorylation and subsequently transfers the phosphate group to a conserved aspartate residue on the RR. The phosphorylated RR modifies its conformation and triggers the signal transduction pathway all the way either to DNA (thus affecting the reactive transcription process) or to a hub of protein--protein interactions (Gilmour et al., 2005).

HK inhibitor forestalls the autophosphorylation in both Gram-positive and Gram-negative bacteria. Bacterial HK inhibition results in necrosis in many pathogenic species, including MRSA (Chekabab et al., 2014).

TCS is the focus of the new antibiotic discovery research, which continues to increase; for example, studying the crystal structures of both HKs and RRs is under process. Moreover, work is underway on developing novel regimens suppressing or circumventing or in any other way neutralizing the mechanisms of antibiotic resistance. Such regimens are formed by coupling an antibiotic compound with one or more existing or novel adjuvants such as beta-lactamase inhibitors (Shlaes, 2013), efflux pump inhibitors (Abuzaid et al., 2012), and outer membrane permeabilizers (Urakawa et al., 2010).

BETA-LACTAMASE INHIBITORS

The combination of a beta-lactamase inhibitor with a beta-lactam formulation can increase the efficacy and therapeutic spectrum of the antibiotic. Examples of beta-lactamase inhibitors are FPI-1465, metallopolymer, NagZ inhibitor, cobaltocenium containing polymers, avibactam, clavulanic acid, sulbactam, and tazobactam. Studies constantly attempt to find novel and more potent beta-lactamase inhibitors.

An example of the combination of beta-lactam and beta-lactamase inhibitor is the administration of penicillin with the beta-lactamase inhibitors clavulanic acid, sulbactam, or tazobactam (Gill et al., 2015). These three compounds have been successfully used in combination in both parenteral and oral regimens.

EFFLUX PUMP INHIBITORS

Efflux pump overexpression is an important mechanism of microbial resistance that results in antibiotics being expelled from the microbial cells. Pump inhibition is a strategy that could reestablish the potency of current antibiotics against some resistant bacteria and perhaps drive the development of new antibiotics. Many pump inhibitors are used such as boronic acid derivatives, capsaicin, homoisoflavonoid, and phenylalanine–arginine β-naphthylamide (Gill et al., 2015).

OUTER MEMBRANE PERMEABILIZERS

Due to the permeability barrier provided by the outer membrane, Gram-negative bacteria are intrinsically resistant to most antibiotics. Thus permeabilizers represent an efficient method by which the activity of some antibiotics, severely limited by the presence of the outer membrane, can be increased. These compounds are typically cationic and amphiphilic or chelators, which can be developed from peptides, peptide-like compounds, polymers, or lipids, such as, antimicrobial peptides and cholic acid (Gill et al., 2015).

CONCLUSION REMARKS

The present scene of antimicrobials, resistance, and therapy is definitely not restricted to clinical microbiology as it was in the beginning of the antibiotic era. The problem nowadays is a sophisticated one, requiring concerted efforts of microbiologists, healthcare specialists, educational institutions, policy makers, agricultural and pharmaceutical industry bodies, and the public to deal with the problem at hand. In reality, this is everyone's concern, because, in the end, there is always a chance for anyone at some stage to be infected with an antibiotic-resistant pathogen.

Thus to combat microbial infections effectively, the life span of our current repertoire of antimicrobial agents should be extended. Here are some suggestions:

- Minimizing the quantity of antibiotics used in agriculture, especially in livestock, by providing stringent controls on animals' hygiene and optimizing nutrition to enhance natural immunity.
- Reducing the number of antibiotic prescriptions for nonmicrobial diseases.
- Boosting scientific research for discovery of new classes of antimicrobial agents. Some examples are as follows:
  - New nonantibiotic biocides, including disinfectants, preservatives, antiseptics, pesticides, herbicides, fungicides, and insecticides, are consistently finding their way to the market, and their use is widening in various clinical applications.
  - Bacteriophage therapy shows significant potential to be used as a viable strategy to restrain and cure infectious diseases. Phages receive increased attention due to their special characteristics, such as widespread distribution, self-replication, and a lack of effects on the normal microflora of treated animals. A number of commercially produced phage products have been approved to as biocontrol agents in the field of poultry raising, cattle breeding, and food preservation.

Finally, we must adapt new methodologies in targeting pathogens to reduce the indiscriminate use of antimicrobial agents. Exploring the microbiome for potential mechanisms of antibiotic resistance may contribute to design the necessary measures to preserve the efficacy of our current antimicrobial arsenal.

REFERENCES

Abuzaid, A., Hamouda, A., Amyes, S.G., 2012. *Klebsiella pneumoniae* susceptibility to biocides and its association with cepA, qacDE and qacE efflux pump genes and antibiotic resistance. J. Hosp. Infect. 81 (2), 87–91.

Aldred, K.J., Kerns, R.J., Osheroff, N., 2014. Mechanism of quinolone action and resistance. Biochemistry 53 (10), 1565–1574.

Artsimovitch, I., Chu, C., Lynch, A.S., Landick, R., 2003. A new class of bacterial RNA polymerase inhibitor affects nucleotide addition. Science 302, 650–654.

Bijlsma, J.J., Groisman, E.A., 2003. Making informed decisions: regulatory interactions between two-component systems. Trends Microbiol. 11 (8), 359–366.

Boucher, H.W., Talbot, G.H., Bradley, J.S., et al., 2009. Bad bugs, no drugs: no ESKAPE! An update from the Infectious Diseases Society of America. Clin. Infect. Dis. 48, 1–12.

Brown, G.D., Denning, D.W., Gow, N.A.R., Levitz, S.M., Netea, M.G., White, T.C., 2012. Hidden killers: human fungal infections. Sci. Transl. Med. 4 (165), 165rv13.

Capra, E.J., Laub, M.T., 2012. The evolution of two-component signal transduction systems. Annu. Rev. Microbiol. 66, 325–347.

Carrillo-Munoz, A.J., Giusiano, S., Ezkurra, P.A., Quindos, G., 2006. Antifungal agents: mode of action in yeast cells. Rev. Esp. Quimioter. 19 (2), 130–139.

Cassir, N., Rolain, J.M., BrouqUI, P., 2014. A new strategy to fight antimicrobial resistance: the revival of old antibiotics. Front. Microbiol. 5, 551.

Castelli, M.V., Butassi, E., Monteiro, M.C., Svetaz, L.A., Vicente, F., Zacchino, S.A., 2014. Novel antifungal agents: a patent review (2011-present). Expert Opin. Ther. Pat. 24 (3), 323–338.

Chekabab, S.M., Jubelin, G., Dozois, C.M., Harel, J., 2014. PhoB activates *Escherichia coli* O157:H7 virulence factors in response to inorganic phosphate limitation. PLoS One 9 (4), e94285.

Ciesiołka, J., Jeżowska-Bojczuk, M., Wrzesiński, J., Stokowa-Sołtys, K., Nagaj, J., Kasprowicz, A., et al., 2014. Antibiotic bacitracin induces hydrolytic degradation of nucleic acids. Biochim. Biophys. Acta 1840 (6), 1782–1789.

Čivljak, R., Giannella, M., Di Bella, S., Petrosillo, N., 2014. Could chloramphenicol be used against ESKAPE pathogens? A review of in vitro data in the literature from the 21st century. Expert Rev. Anti-Infect. Ther. 12 (2), 249–264.

Denning, D.W., Hope, W.W., 2010. Therapy for fungal diseases: opportunities and priorities. Trends Microbiol. 18 (5), 195–204.

Diekema, D., Arbefeville, S., Boyken, L., Kroeger, J., Pfaller, M., 2012. The changing epidemiology of healthcare-associated candidemia over three decades. Diagn. Microbiol. Infect. Dis. 73 (1), 45–48.

Dinos, G.P., 2017. The macrolide antibiotic renaissance. Br. J. Pharmacol. 174 (18), 2967–2983.

Friedman, N.D., Temkin, E., Carmeli, Y., 2016. The negative impact of antibiotic resistance. Clin. Microbiol. Infect. 22 (5), 416–422.

Gauwerky, K., Borelli, C., Korting, H.C., 2009. Targeting virulence: a new paradigm for antifungals. Drug Discov. Today 4 (3–4), 214–222.

Germovsek, E., Barker, C.I., Sharland, M., 2017. What do I need to know about aminoglycoside antibiotics? Arch. Dis. Child. Educ. Pract. Ed. 102 (2), 89–93.

Giguère, S., 2013. Lincosamides, pleuromutilins, and streptogramins. In: Giguère, S., Prescott, J.F., Dowling, P.M. (Eds.), Antimicrobial Therapy in Veterinary Medicine, fifth ed. Wiley, Hoboken, NJ, pp. 199–210.

Gill, E.E., Franco, O.L., Hancock, R.E., 2015. Antibiotic adjuvants: diverse strategies for controlling drug-resistant pathogens. Chem. Biol. Drug. Des. 85 (1), 56–78.

Gilmour, R., Foster, J.E., Sheng, Q., Mcclain, J.R., Riley, A., Sun, P.M., et al., 2005. New class of competitive inhibitor of bacterial histidine kinases. J. Bacteriol. 187 (23), 8196–8200.

Gooderham, W.J., Hancock, R.E., 2009. Regulation of virulence and antibiotic resistance by two-component regulatory systems in *Pseudomonas aeruginosa*. FEMS Microbiol. Rev. 33 (2), 279–294.

Griffith, E.C., Wallace, M.J., Wu, Y., Kumar, G., Gajewski, S., Jackson, P., et al., 2018. The structural and functional basis for recurring sulfa drug resistance mutations in *Staphylococcus aureus* dihydropteroate synthase. Front. Microbiol. 17 (9), 1369.

Grossman, T.H., 2016. Tetracycline antibiotics and resistance. Cold Spring Harb. Perspect. Med. 6 (4), a025387.

Grzybek, M., Kukula-Koch, W., Strachecka, A., Jaworska, A., Phiri, A.M., Paleolog, J., et al., 2016. Evaluation of anthelmintic activity and composition of pumpkin (*Cucurbita pepo* L.) seed extracts—in vitro and in vivo studies. Int. J. Mol. Sci. 9, 1456.

Gupta, A.K., Versteeg, S.G., 2016. Tavaborole – a treatment for onychomycosis of the toenails. Expert Rev. Clin. Pharmacol. 9 (9), 1145–1152.

Hancock, L.E., Perego, M., 2004. Systematic inactivation and phenotypic characterization of two-component signal transduction systems of *Enterococcus faecalis* V583. J. Bacteriol. 186 (23), 7951–7958.

Hossain, M.A., Ghannoum, M.A., 2000. New investigational antifungal agents for treating invasive fungal infections. Expert Opin. Investig. Drugs 9 (8), 1797–1813.

Jain, P., Saravanan, C., Singh, S.K., 2013. Sulphonamides: deserving class as MMP inhibitors? Eur. J. Med. Chem. 60, 89–100.

Kapoor, G., Saigal, S., Elongavan, A., 2017. Action and resistance mechanisms of antibiotics: a guide for clinicians. J. Anaesthesiol. Clin. Pharmacol. 33 (3), 300–305.

Lin, K., Gallay, P., 2013. Curing a viral infection by targeting the host: the example of cyclophilin inhibitors. Antiviral. Res. 99 (1), 68–77.

Lorand, T., Kocsis, B., 2007. Recent advances in antifungal agents. Mini Rev. Med. Chem. 7 (9), 900–911.

Lushniak, B.D., 2014. Surgeon general's perspectives. Public Health Rep. 129 (4), 314–316.

Matzov, D., Eyal, Z., Benhamou, R.I., Shalev-Benami, M., Halfon, Y., Krupkin, M., et al., 2017. Structural insights of lincosamides targeting the ribosome of *Staphylococcus aureus*. Nucleic Acids Res. 45 (17), 10284–10292.

Monteiro, J.F., Hahn, S.R., Gonçalves, J., Fresco, P., 2018. Vancomycin therapeutic drug monitoring and population pharmacokinetic models in special patient subpopulations. Pharmacol. Res. Perspect. 6 (4), e00420.

Pandit, N., Singla, R.K., Shrivastava, B., 2012. Current updates on oxazolidinone and its significance. Int. J. Med. Chem. 2012, 159285.

Papp-Wallace, K.M., Endimiani, A., Taracila, M.A., Bonomo, R.A., 2011. Carbapenems: past, present, and future. Antimicrob. Agents Chemother. 55 (11), 4943–4960.

Roemer, T., Krysan, D.J., 2014. Antifungal drug development: challenges, unmet clinical needs, and new approaches. Cold Spring Harb. Perspect. Med. 4 (5), a019703.

Rollin-Pinheiro, R., Singh, A., Barreto-Bergter, E., Del Poeta, M., 2016. Sphingolipids as targets for treatment of fungal infections. Future Med. Chem. 8 (12), 1469–1484.

Sharma, R., Patel, S., Abboud, C., Diep, J., Ly, N.S., Pogue, J.M., et al., 2016. Polymyxin B in combination with meropenem against carbapenemase-producing *Klebsiella pneumoniae*: pharmacodynamics and morphological changes. Int. J. Antimicrob. Agents 49 (2), 224–232.

Sheehan, D.J., Hitchcock, C.A., Sibley, C.M., 1999. Current and emerging azole antifungal agents. Clin. Microbiol. Rev. 12 (1), 40–79.

Shlaes, D.M., 2013. New b-lactam-b-lactamase inhibitor combinations in clinical development. Ann. N.Y. Acad. Sci. 1277, 105–114.

Siddiqui, R., Aqeel, Y., Khan, N.A., 2013. Killing the dead: chemotherapeutic strategies against free-living cyst-forming protists (*Acanthamoeba* sp. and *Balamuthia mandrillaris*). J. Eukaryot. Microbiol. 60 (3), 291–297.

Taneja, N., Kaur, H., 2016. Insights into newer antimicrobial agents against Gram-negative bacteria. Microbiol. Insights 9, 9–19.

Tangden, T., 2014. Combination antibiotic therapy for multidrug-resistant Gram-negative bacteria. Ups. J. Med. Sci. 119 (2), 149–153.

Urakawa, H., Yamada, K., Komagoe, K., Ando, S., Oku, H., Katsu, T., et al., 2010. Structure-activity relationships of bacterial outer-membrane permeabilizers based on polymyxin B heptapeptides. Bioorg. Med. Chem. Lett. 20 (5), 1771–1775.

Villa, T.G., Feijoo-Siota, L., Rama, J.L.R., Ageitos, J.M., 2017. Antivirals against animal viruses. Biochem. Pharmacol. 133, 97–116.

Vullo, D., Kumar, R.S., Scozzafava, A., Ferry, J.G., Supurana, C.T., 2018. Sulphonamide inhibition studies of the β-carbonic anhydrase from the bacterial pathogen *Clostridium perfringens*. J. Enzyme Inhib. Med. Chem. 33 (1), 31–36.

Zhao, C., Huang, T., Chen, W., Deng, Z., 2010. Enhancement of the diversity of polyoxins by a thymine-7-hydroxylase homolog outside the polyoxin biosynthesis gene cluster. Appl. Environ. Microbiol. 76 (21), 7343–7347.

FURTHER READING

Gould, I.M., Bal, A.M., 2013. New antibiotic agents in the pipeline and how they can overcome microbial resistance. Virulence 4 (2), 185–191.

Podgornaia, A.I., Laub, M.T., 2013. Determinants of specificity in two-component signal transduction. Curr. Opin. Microbiol. 16, 156–162.

Rossolini, G.M., Arena, F., Pecile, P., Pollini, S., 2014. Update on the antibiotic resistance crisis. Clin Opin Pharmacol 18, 56–60.

Rudkin, J.K., Mcloughlin, R.M., Preston, A., Massey, R.C., 2017. Bacterial toxins: offensive, defensive, or something else altogether? PLoS Pathog. 13 (9), e1006452.

Sharan, A., Soni, P., Nongpiur, R.C., Singla-Pareek, S.L., Pareek, A., 2017. Mapping the 'two-component system' network in rice. Sci. Rep. 7 (1), 9287.

Wiskirchen, D.E., Crandon, J.L., Nicolau, D.P., 2013. Impact of various conditions on the efficacy of dual carbapenem therapy against KPC-producing *Klebsiella pneumoniae*. Int. J. Antimicrob. Agents 41 (6), 582–585.

Wright, G.D., 2014. Something old, something new: revisiting natural products in antibiotic drug discovery. Can. J. Microbiol. 60 (3), 147–154.

CHAPTER

ELECTROMAGNETISM AND THE MICROBIOME(S) 14

Stavroula Siamoglou[1], Ilias Boltsis[2], Constantinos A. Chassomeris[1] and Manousos E. Kambouris[3]

[1]Department of Pharmacy, University of Patras, Patras, Greece [2]Department of Cell Biology, Erasmus Medical Centre, Rotterdam, The Netherlands [3]Golden Helix Foundation, London, United Kingdom

INTRODUCTION

The antibiotic effect is long established for alternate current (AC) fields (Doevenspeck, 1961). Direct current (DC) pulsed fields were proven antimicrobial in the 1990s (Grahl and Märkl, 1996), while DC static fields produced inconclusive results (Kermanshahi and Sailani, 2005), as did magnetic and electromagnetic fields (EMF) (Hristov and Perez, 2011; Fojt et al., 2009). The therapeutic effect of conductive electrostimulation (ES) is well-documented for quite some time. There are many protocols, differing in the type of current, AC, DC, or other (Daeschlein et al., 2007; Merriman et al., 2004; Petrofsky et al., 2008) or the electrodes used (Kloth, 2005) and, of course, dosage (Kloth, 2009) and every other conceivable factors and effects (Kloth, 2005). A number of studies describe positive (Berovic et al., 2008) and another, considerable number, negative effects on microbial growth (Kermanshahi and Sailani, 2005; Petrofsky et al., 2008; Daeschlein et al., 2007; Merriman et al., 2004). Generally, microbes tend to act galvanophilic in culture and eventually display galvanotaxis (Fig. 14.1), while the effect is reversed in traumas (Kambouris et al., 2014).

HISTORY AND LORE

Microbiota of both economic/environmental interest, but of clinical as well, have been exposed occasionally to a number of electromagnetic modalities to determine stimulatory or inhibitive effect and the respective transitional dynamics, if any, and then to plot dose–response curves. The whole concept is loosely based on the principles of electrotherapy, a recurring approach with a notorious past.

In ancient Greece, *Asklepeia* were built preferably near waterfalls to capitalize on the ionizing effect of high-pressure spraying of the waterfall; electric eels were also kept to treat by slight electrocution numerous ailments. This empiricism goes all the way to the 17th century with the charged golden leaves used to resolve smallpox lesions (Robertson, 1925; Mercola and Kirsch, 1995; Vodovnik and Karba, 1992; Poltawski and Watson, 2009; Kloth, 2005). The reinvigorating result of the electromagnetic energy within and after the Enlightenment is coined by the lore of

Microbiomics. DOI: https://doi.org/10.1016/B978-0-12-816664-2.00014-1

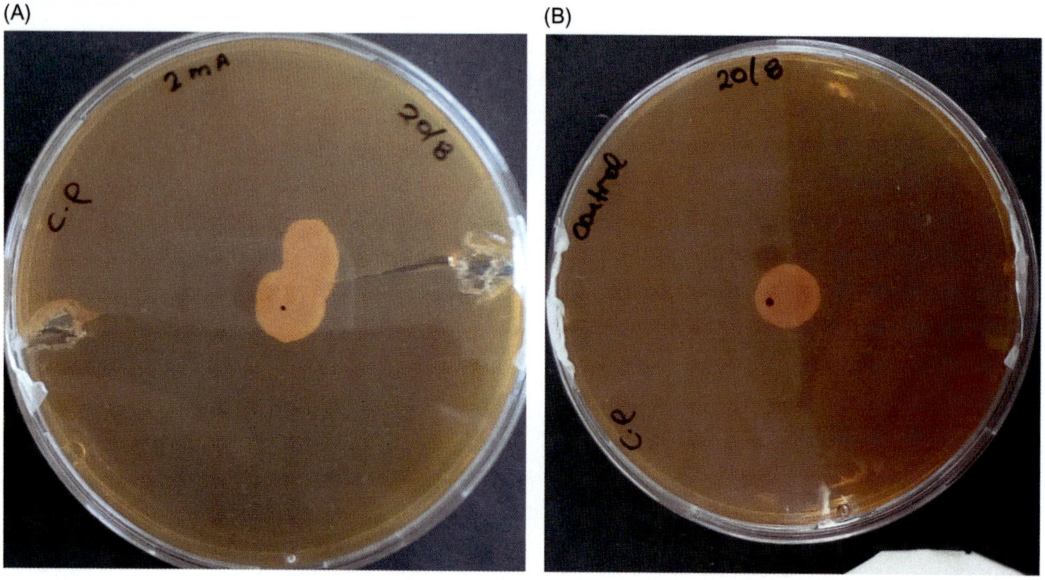

FIGURE 14.1

The asymmetric growth of the treated sample in a conductive format with pin electrodes inserted horizontally on axis (A), compared to the untreated control (B), indicates a galvanotactic organism (yeast).

Photo S. Kritikou, with permission.

Frankenstein, where an assortment of body parts of cadavers was reinvigorated as a functional body, returned by the energy of lightning bolts (Kambouris et al., 2014).

At the same timeline, Feng Shui principles were applied in China, mostly (although not exclusively) on the basis of magnetic flux and its differentiations by terrain, orientation, altitude, and environment. Similar concepts under the term chorothesia (positioning in space) were followed in ancient Greece but were discontinued after the advent of Christianity.

Therapeutic currents were routine practice in the 19th century, mainly for orthopedic and psychic ailments (Harishorne, 1841; Mercola and Kirsch, 1995; Kloth, 2005; Lente, 1850), only to be slighted and discontinued due to the Fleiner Results in 1910, released by a commission of the Carnage Foundation formed to review complaints reporting individual incidents of malpractice (Mercola and Kirsch, 1995; Kambouris et al., 2014).

ELECTRONS AND MICROBES: THE FORMAL MEETING

The interaction of electrical modalities with microbiota surfaced in the context of wound healing (Poltawski and Watson, 2009). After electrical modalities started going beyond the well-established orthopedics and neurologic applications anew, wound healing was an advantageous field, where treatment not only improved healing rates (Kambouris et al., 2014), but also decontaminated the wound site (Ramadhinara and Poulas, 2013; Rowley et al., 1974); a fact most beneficial for the healing process (Kloth, 2005).

The new era was officially inaugurated with the concept of "*electroceuticals*" (Famm et al., 2013), meaning to coin a term respective to "*pharmaceuticals*" (Kambouris et al., 2017; Kambouris et al., 2014). The idea was to declare the therapeutic potential in a systematic and self-contained manner but also the extremely variable character of electric modalities, not unlike the vast number of chemical ones, and, in the near future, of *Biopharmaceuticals*, i.e. preparations of live organisms (Adhya et al., 2014; Golkar et al., 2014; Krylov, 2014; Desnues et al., 2012; Taylor et al., 2014; Parratt and Laine, 2016). Of course, this new era was already present with many discrete and individual, isolated efforts since mid-1990s (Kloth, 2005; Poltawski and Watson, 2009) massing up to critical so as to achieve recognition by health insurance providers as valid approaches (Healthcare, 2008).

The next step has been the introduction of the terms "*elektrodynamics*" and "*elektrokinetics*" (Kambouris et al., 2014, 2017), respective to "pharmacodynamics" and "pharmacokinetics" and intended to differentiate the dissipation, reception, and focus of the provided energy from the quantitative and qualitative characteristics of the energy proper. The last theoretical iteration is the present proposal for the concept of (counterpart to pharmacogenomics). It is proposed so as to denote a defined, though holistic approach to the genomic elements controlling, or implicated in, the cellular elements which define the susceptibility to electroceuticals; though, the basics of the concept had been determined earlier, when ES and electroregulation had been associated with different expression rates of specific genes (Gao et al., 2005; Zhao et al., 2006; Meng et al., 2011).

An important contributor to any antimicrobial *in vivo* effect should be indirect; namely, the acceleration and bolstering of the immune system of the patient (Petrofsky et al., 2008; Asadi and Torkaman, 2014). To this effect, no or limited direct action onto the pathogen is required. Any destabilizing effect on the pathogen may slow the latter's growth rate, its production of biotoxins or of virulence factors (such as environment-regulating exoenzymes), or infringe with the accurate regulation of protective mechanisms, as the production of slime, of biofilm-forming extracellular matrix (Costerton et al., 1994; Caubet et al., 2004) or of iron intake moieties. Erroneous adaptation signals might also have a role: conceivably, blastic cells might start sporulation, thus decelerating the dynamics of the infection, or, in cases where sporulation might have been advantageous, it might be interrupted. Infringement with the germination process may suppress an infection event at its conception in prophylactic formats or lead to an early germination event and exposure to immune mechanisms before the full potential of the pathogen is mustered. A deregulation of germination may lead to necrosis due to nutritional shortage and biochemical incompatibility and may be a promising approach for more effective sanitization procedures in high-risk areas.

FORMATS, CONDITIONS, AND EFFECTS

The bioresponsive use of electromagnetism is achieved by the application of a respective field, which causes either the initiation of a current (inductive coupling) or simply polarization. Alternatively it can be achieved by actually routing current through the target moiety, thus forming part of an electric circuit (conductive coupling) as in Fig. 14.2. The current sparse use is partly the result of an unsorted multitude of forms [low-intensity DC, low-intensity pulsed DC, high-voltage (HV) pulsed DC, AC, pulsed electric fields, different magnetic fields, microcurrents] available as

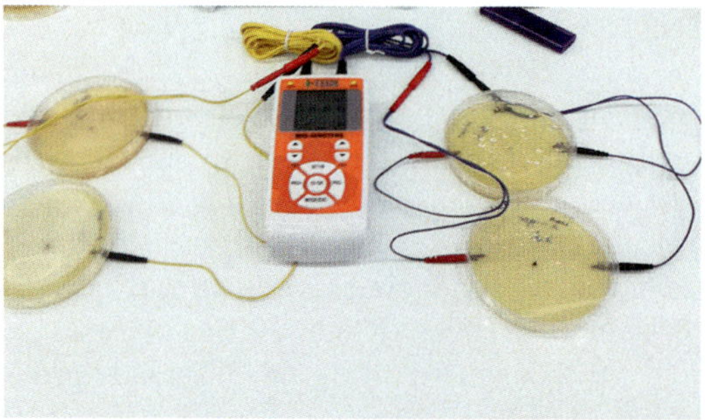

FIGURE 14.2

Low-throughput conductive electroculturomics format using standard orthopedic instrumentation and improvised culture bed (laterally drilled Petri dishes for horizontal insertion of pin electrodes).

Photo A. Milioni, with permission.

wanton protocols, without any notion of cohesion, standardization, and exhaustive testing on suitability so as to discriminate categories assignable to applications, as happens with conventional pharmaceuticals. Despite the many dedicated formats, there are few main categories of expanded electroceuticals (the "expanded" clause so as to include magnetism-based applications). These are (1) magnetic fields, (2) electric fields, (3) EMF/radiation, and (4) electric currents.

Their many forms and formats exercise either upregulation–induction or downregulation–suppression. The former applies in a multitude of primary and secondary sector productive industries, producing edible biomass (Jamil et al., 2012); biomass for biofuel (Hunt et al., 2009); or processing raw produce to foodstuff (Neelakantan et al., 1999); additionally, different biotechnological schemes may profit from this approach (de Souza and de Oliveira Magalhães, 2010). To the abovementioned ones, some new direct health sector concepts should be added. Such are the probiotics (Borchert et al., 2008) and the biopharmaceuticals, especially the live agent formulations, including, but not restricted to, phage therapy (Abedon et al., 2011), virophages (Desnues et al., 2012; Taylor et al., 2014), endomopathogenic fungi (Jaworska et al., 2016), and oncolytic bacteria (Din et al., 2016; Anderson et al., 2006) and viruses (Chiocca and Rabkin, 2014). But probably the most important application will be, in the near future, the stimulation of whole microbiomes either for rejecting aggressive strains (Kambouris et al., 2018b) or reshaping a microenvironment, possibly in biodegradation/bioremediation/biorestoration endeavors (Vidali, 2001; Olszanowski and Piechowiak, 2006; Gao et al., 2005; Mateescu et al., 2011).

Just as numerous are the suppression applications—but, in some cases more obvious. Health applications (for humans, animals, and plants) are the foremost for suppressive effects on bacteria, fungi, and viruses, including decontamination of trauma/injuries. But this is not all; a whole series of other applications follow, especially for bacteria and fungi: from the bacterial biofilms, which infest equipment and instruments (Grissom, 1995) to fungi, implicated in degradation of insulating plastic coatings (Mateescu et al., 2011), colors of artifacts (Arabatzis et al., 2008), and lining of lenses of telescopes (Velegraki et al., 2007) or to microbial foodstuff degradation and befouling (Miedaner and Geiger, 2015).

MAGNETIC FIELDS

Magnetic fields are produced by the motion of electrical charge, be that spontaneous (natural magnetism/ferrite magnets) or due to electric currents (electromagnetism). The biological effects of magnetism (*magnetoresponse*) comprise a series of wildly different interactions caused by natural magnetic fields and technidal ones—the latter defined as "resulting by any artificial product or action or lack of it" (see Chapter 15). The respective studies are problematic due to the fact that magnetic effects are observed for a huge range of magnetic flux densities ("flux density" being the word for "intensity" in magnetic colloquialism), which cover more than 10 orders of magnitude (Pazur et al., 2007).

The categorization of magnetoresponses of biological entities starts with static, low flux density fields, similar to that of the geomagnetic field (25–75 μT). The latter, although weak, is appreciable by living organisms, and of major importance in the case of magnetotactic bacteria, which contain ferromagnetic moieties used as compass needles (Pazur et al., 2007; Hergt et al., 2005; Grissom, 1995; Kirschvink et al., 1992). Thus fields similar to the geomagnetic one ("weak fields" 0–110 μT) could possibly affect magnetoresponsive organisms and processes and were studied as a result (Pazur et al., 2007; Grissom, 1995).

Higher flux fields were studied to establish effects of the electrical appliances, which emerged massively in the 1950s and were fed by just as massive power distribution networks. Thus research was tuned to 50–60 Hz alternating magnetic fields caused by electrification and was focused on negative effects and multicellular organisms, especially humans; and in the context of cancer before anything else (Kirschvink et al., 1992; Grissom, 1995; Pazur et al., 2007).

But massive studies of magnetoreception in other fields, static or not, moderate (100 μT to 1 mT), strong (1 mT to 1 T) or very strong (at the region of one to tens of Tesla), came later: Technidal magnetic fields were diligently studied during the 1970s and 1980s (Werner et al., 1978; Haberkorn and Michel-Beyerle, 1979); and despite the gradual assembly of a corpus on its microbiological applications (Hergt et al., 2005; Kirschvink et al., 1992), magnetism proper gained a foothold in diagnostics mostly, with MRI (magnetic resonance imaging) and NMR (nuclear magnetic resonance) formats (Gao et al., 2005) and was subsequently introduced in orthopaedics (Kubota et al., 1995).

Both diagnostics and therapeutics posed the question of interaction with what is currently known as appendage microbiomes (Gao et al., 2005; Brkovic et al., 2015; Kambouris et al., 2018b) and their possible stimulation to aggressive forms (Gao et al., 2005; Brkovic et al., 2015; Kambouris et al., 2018b). Starting from there and then moving to stand-alone applications of non-medical/biotechnological/environmental context (Gao et al., 2005; Fojt et al., 2009), a number of experiments produced a wealth of observations. Effects on DNA synthesis and transcription have been studied in the meantime (Repacholi and Greenebaum, 1999; Blank and Goodman, 2008).

Static magnetic fields

A focal observation has been the inhibition of *Streptococcus mutans* and *Staphylococcus aureus* when inoculated in the presence of moderate static magnetic fields (30, 60, 80, and 100 mT) in anaerobic conditions. The inhibition was a function of the time of the exposure and of the intensity of the field. In aerobic conditions, though, there were no inhibition to report; simultaneously, the samples of *Escherichia coli* treated identically showed no inhibition at all (Kohno et al., 2000), implying Gram-dependence (Soghomonyan et al., 2016). The latter contradicts with the observation that both clinical and reference strains of the same bacterium in liquid cultures displayed similar inhibition patterns of up to 40% (observed at maximum exposure time and field intensity) when exposed to static magnetic fields of 2–20 mT for up to 90 min by 15-min steps (Mousavian-Roshanzamir and Makhdoumi-Kakhki, 2017).

Though, the exposure of a diverse range of bacteria and one yeast to heterogeneous static field of ~60 mT revealed, in subsequent optical density (OD) counts, that the magnetic inhibition had been not only species-dependent and inversely proportional to the microbial density, but also of purely kinetic and not dynamic nature: even the most susceptible microbiote (the yeast), after a slow start, had been able to catch up with the growth of untreated standards within 48 h (Brkovic et al., 2015)

These results may be compared with other studies. Very strong magnetic fields (5.2 − 6.1 T) had been able to delay cell death in stationary cultures of *Bacillus subtilis* (Nakamura et al., 1997). Moderate static magnetic fields of 4.5 mT produced by permanent magnets during incubation increased the resistance of *E. coli* to antibiotics (Stansell et al., 2001), and strong ones (5.2–6.1 T) promoted its survival rate (Horiuchi et al., 2001) while increasing the detectable titer of S factor, encoded by the rpoS gene (Horiuchi et al., 2001). Similarly, *Shewanella oneidensis*, a Gram-negative, facultative anaerobe bacterium, exposed at very strong static fields (14.1 T) while at its exponential phase showed no appreciable change in growth parameters; still, a mixed pattern of upregulation and downregulation of a number of genes (Gao et al., 2005) indicated actual magneto-sensitivity and a profound genomic mobilization needed for compensation of the stimulus so as to achieve stability in growth. Although differences between Horiuchi et al. (2001) and Gao et al. (2005) in affecting growth were mostly attributed by the latter group to the growth phase of the exposed inocula, it is possible that species and the intensity/flux might both have been factors as well; as far as the latter is concerned, application at double or triple the base value might have proven no longer stimulative.

In fungi, weak magnetic fields may be stimulatory or inhibitory: fields of 0.1 − 1 mT stimulated, both in liquid and solid media, the growth of *Pseudomonas fluorescens*, *Staphylococcus albus*, and *Aspergillus niger* (Makarevich, 1999). Though similar static fields of 0.1, 0.5, and 1 mT caused 10% decrease to the colony growth of *Alternaria alternata*, *Fusarium oxysporum*, and *Curvularia inaequalis* (Nagy, 2005). Interestingly, static heterogeneous fields of 1 mT displayed very different effects (inhibitory—stimulatory/null) to a selection of molds depending on species, and on magnetic polarity (Abbas, 2013).

With regard to strong fields, *A. niger* spores inoculated on Czapek-Dox medium plates and exposed to static fields of 0.5 or 0.62 T for 7 days produced an atypical growth of the fungus, characterized by appreciably lower number of colonies, which were swollen and even bombastic (Mateescu et al., 2011). On the other hand, faster growth and enhanced pathogenicity were observed in the cases of *Beauveria bassiana* and *Isaria fumosorosea* when grown under static magnetic fields produced by permanent magnets and when grown on solid cultures made with water treated by magnets (Jaworska et al., 2016). Last but not least, *Saccharomyces cerevisiae* grown in a static magnetic field of 220 mT presented a clear increase in biomass compared to untreated controls (Muniz et al., 2007).

Moving to Protozoa, three species of *Acanthamoeba* displayed a growth decrease at modest static fields of 71 and 106 mT (Berk et al., 1997).

Alternating magnetic fields

Considering the more exhaustively studied alternating magnetic fields, fields of 10 mT oscillating at 50 Hz, the most biologically relevant frequency (Grissom, 1995), caused no appreciable changes in the morphology of sphere- and rod-shaped bacteria exposed to them for 1 h (Fojt et al., 2009), contrary to the effect of the same conditions applied for less than half the time (24 min) on the viability of the same bacteria expressed in terms of viable counts and OD counts. The observed decrease in viability differed by Gram status and shape: Gram-positive spherical bacteria show 20% decrease compared to respective controls, while Gram rod-like were 50%–100% more susceptible (Fojt et al., 2009) in a time- and intensity-dependent manner (Strašák et al., 2002, 2005; Fojt et al., 2004). In contrast, magnetic square wave signals (0.05 – 1 mT, 50 Hz) had no effect on the growth of *E. coli* (Del Re et al., 2004) which, though, showed complex dose-dependent increase with exposure to very specific intensities up to 22 mT at 16 and 50 Hz. (Aarholt et al., 1981).

In lower oscillation frequencies, both increased and diminished growth rates for *B. subtilis*, *Candida albicans*, *Halobacterium*, *Salmonella typhimurium*, and Staphylococci were observed (Moore, 1979), depending on AC frequencies of 0–0.3 Hz and magnetic flux densities of 5–90 mT; weaker AC fields 0.8–2.5 mT, but of higher frequency, 0.8 and 1 kHz, increased the growth of *B. subtilis* (Ramon et al., 1987).

Alternate 50 Hz sinusoid current was used to produce 5, 15, 25, and 100 mT alternating fields into which spawn of *Pleurotus* fungus were exposed in Petri dishes for 2, 5, or 15 min. Different combinations of intensity and time achieved optimization of different growth parameters, but 15 min exposure to 15 mT resulted in the widest range of optimization for different parameters, while time of exposure and intensity demonstrably appeared as changeable terms of a steady product, with increase in one compensating the other's diminution (Jamil et al., 2012).

Alternating fields of 1.8 mT, 72 Hz caused increased cell division rates to the protozoan species *Paramecium tetraurelia*, which, however, did not occur in the presence of a Ca^{2+} blocker (Dihel et al., 1985), demonstrating that the latter was a limiting factor.

In viral applications the effect of exposure of RNA coliphage MS2 inoculated into bacterial cultures exposed to 0.5 mT at 60 Hz had shown a significant delay in phage yield, 45–65 min after inoculation, compared with control cultures but no alteration of the final phage concentration—thus substantiating a purely kinetic change. At 2.5 mT though, impeded phage replication and increased phage yield were observed (Staczek et al., 1998). At roughly the same time, exposure of Akata cells, a human lymphoid cell line latently infected by the Epstein-Barr Virus (EBV) genome, to a 50 Hz EMF resulted in an increased number of cells expressing the virus early antigens indicating that DNA can be modulated by a magnetic field (Grimaldi et al., 1997). The confirmation came by the application of 1 mT, 50 Hz sine waves, on Kaposi's sarcoma (KS)-associated herpesvirus (KSHV) in BCBL-1 cell line: exposure did not affect the growth and viability of the cells. In the presence of tetradecanoylphorbol acetate, which induces lytic cycle from latent proviral forms (Pantry and Medveczky, 2009), total KSHV DNA content was found higher in extremely low frequency (ELF)-EMF exposed cells than in control BCBL-1 cells after 72 h of exposure, at the earliest. Viral progeny produced under ELF-EMF exposure consisted mainly of defective viral particles (Pica et al., 2006).

Mechanism of action

To propose any mechanism through which magnetic fields affect cells (magnetoresponsiveness) one should consider the receiving entities (magnetoreceptors) within the cell. As the affecting fields cover a huge dynamic range (Pazur et al., 2007), it is certain that more than one moiety should be involved (Pazur et al., 2007). The common sense centers onto macromolecules containing iron, which can be attracted by a powerful magnetic field, at the order of two-digit Tesla (Gao et al., 2005). Might it be so, such molecules should accumulate at specific positions along the magnetic dynamic lines and oriented accordingly; thus a shortage will ensue at every nonaligned position within the cell. This shortage should in any case trigger the production of more iron-containing molecules by inducing the respective genes, in order to compensate for the loss. If the magnetic stimulus stops, an increased quantity of such macromolecules may instigate faster growth; if not, the new ferromagnetic molecules will be drawn out of position as well and the cell will be short of their presence. Other iron-containing molecules will be downregulated as the ever limited quantities of iron will be directed to priority products, which will be accumulated and practically neutralized. This vicious circle will exhaust the cell in its effort to compensate the functionality loss by upregulating the production. This explains the observation of Gao et al. (2005) that under exposure some genes coding iron-containing proteins were upregulated while others were downregulated at the same time.

At the low end of flux density values, weak magnetic fields similar to geomagnetism affect cells through magnetosomes, the ferromagnetic particles responsible for magnetosensitivity in geomagnetic-like fields. Many eukaryotes, even multicellular ones do so, but the prime example is the magnetotactic bacteria (Pazur et al., 2007; Grissom, 1995; Hergt et al., 2005). Magnetosomes, moving in response to earth-strength ELF fields, are capable of opening transmembrane ion channels (Fojt et al., 2004), in a fashion similar to those predicted by ionic resonance models. Hence, the presence of trace levels of biogenic magnetite in virtually all human tissues examined suggests that similar biophysical processes may explain a variety of weak field ELF bioeffects (Kirschvink et al., 1992). Thus, although the bilayered biologic membranes are transparent to magnetic fields, they are affected indirectly in terms of permeability through the effect on ion channels and other ion-dependent moieties.

In-between, in the range of moderate fields, the main effect of magnetic fields to microbiota seems to occur through radical pairs (Haberkorn and Michel-Beyerle, 1979; Werner et al., 1978; Grissom, 1995; Fojt et al., 2004; Hunt et al., 2009). Proteins are in some cases magnetosensitive moieties, and enzymes are known to be affected by magnetism. Magnetosensitive molecules and respective physically distinct mechanisms likely to mediate magnetoreception are detected within the entire cell volume: cell membranes do not bar magnetic fields (Pazur et al., 2007; Hunt et al., 2009), thus DNA, highly charged by itself, has been proposed as a target (Pazur et al., 2007; Hunt et al., 2009). Even if one overlooks the, just as charged and much more agile, RNAs, it is obvious that the transcription and translation machineries are susceptible to magnetic stimuli (Pazur et al., 2007), which means that episomization of plasmids and transposition and incorporation/excision of proviruses are affected as well (Pazur et al., 2007; Hunt et al., 2009; Torgomyan and Trchounian, 2013; Soghomonyan et al., 2016). Indeed, cell-free systems of protein biosynthesis are receptive to magnetic fields (Pazur et al., 2007).

Similarly, enzymes are by definition susceptible due to their intermediate forms when processing the substrate, whence they are charged (Grissom, 1995; Pazur et al., 2007). Obviously this remains the case for enzymes handling or containing highly energized moieties, such as various ions (Pazur et al., 2007). It is thus easily understandable that the photosynthetic apparatus of microbiota, concentrated, charged, and positioned on the cell membrane and not in deep-seated chloroplasts, was among the first magnetoreceptors to be detected and studied (Grissom, 1995; Pazur et al., 2007; Werner et al., 1978; Haberkorn and Michel-Beyerle, 1979), with possibly far-reaching results: the colonization of different environments and the restoration of current, befouled ones will require massive production of oxygen, possibly available only through the resilient prokaryotic photoautotrophs.

The previous observations and argumentation explain why even very similar experiments in the field present extreme variability in results and an obvious lack of cohesion. The magnetic field amplitude is *not* the only parameter involved to determine the outcome in magnetobiological experiments; spatial variations, such as field gradient and symmetry, might also be of relevance (Engström et al., 2002).

The effect of magnetism is materialized through many different moieties; thus subtle differences in the system cell - field may produce important divergences. As the cell changes through stages and forms, the number and distribution of magnetoreceptor moieties change as well, altering the pattern of responsiveness in terms of intensity and frequency of the magnetic stimuli. The treatment-response curves are complex and present narrow *maxima* of magnetic flux and/or frequency—where applicable—depending on a host of parameters (Pazur et al., 2007). The magnetic signature, being a function of position, motion, quantity, and type of charges and charged moieties, guarantees that computations to define it will always be inaccurate and deductive at best - and irrelevant at worst.

ELECTRIC FIELDS

The use of electric fields is perhaps the most ubiquitous method of ES, electrosuppression, and electroeradication, especially in industry (Mattar et al., 2015; Delsart et al., 2015; Marsellés-Fontanet et al., 2009; Coustets et al., 2015; Singh et al., 2012).

The electric fields are produced by static charges, which may alternate in polarity (*alternating* fields) or simply change their accumulation rate (*pulsed* fields). Thus electrically charged moieties of the cells, which are insulated from the actual affecting charges, are attracted or repulsed toward the compatible charged edges of the field, changing their position and attitude (McGillivray and Gow, 1986). Steadily or occasionally charged moieties tend to move accordingly (*electrophoresis*), thus creating an electric current (*inductive coupling*), which appears through the insulating material. This interaction may attract a whole cell, especially when bearing itself a charge, as do the bacteria that exhibit a net negative charge and migrate toward anodes (Mainelis et al., 2002; Olszanowski and Piechowiak, 2006). It may also dislodge charged moieties from their position in the cell creating a cell polarity (*dielectrophoresis*). Similar effects may occur intramolecularly, with moieties induced and oriented by the field. This effect alters the disposition and distribution of different molecules and organelles, thus changing the functional polarity and the operational routine of the cells (McGillivray and Gow, 1986), especially considering the ion-dependent channels. This effect exerts shear and tear on the cytoskeleton and the membrane. Should a dimension of alternating—or even pulsating—be added to it, the resulting mechanical stress may dissociate the cytoskeleton-associated proteins of the ion channels and organelles. Thus, alternating and pulsating fields are more likely to exert antibiotic effect for a given energy consumption than static fields, explaining results such as the better efficacy of bipolar pulses to bioelectric effect (BEE) schemes compared to monopolar ones (Novickij et al., 2015). Moreover, the deregulation of channel function may indirectly cause even harsher antimicrobial effects by enhancing the action of (bio)chemical antimicrobials, or, more precisely, by comprimizing the influx–efflux mechanisms that constitute part of the antibiotic resistance (Munita and Arias, 2016).

It has been noted that electric fields and currents can influence the organization of biological membranes and affect metabolic and developmental processes of both prokaryotic and eukaryotic cells. Planktonic cells of *E. coli*, *Pseudomonas aeruginosa*, *Proteus mirabilis*, and *C. albicans* can be killed by alternating electric fields of 5 V/cm or whereabouts (Davis et al., 1989; Costerton et al., 1994), whereas low-intensity alternating electric fields of 1.5–20 V/cm can completely override the inherent resistance of biofilm bacteria to biocides and antibiotics (Costerton et al., 1994).

High-frequency alternating electric fields affect dividing bacteria during cytokinesis due to the non-homogeneous innate electric fields generated near the bridge separating the daughter cells. These non-homogeneous fields exert unidirectional dielectrophoresis forces on charged and polar particles and molecules and thus may result in their movement toward the furrow (Giladi et al., 2008). This falls in line with the differential observation that pulsed electric field (PEF) pulses of millisecond duration could be efficient for enzyme release (*electroextraction*) from bacterial cells in exponential growth phase, while inefficient when applied to cells in stationary phase (Coustets et al., 2015).

PEF treatment with 100 pulses at 45 kV/cm for gels incorporating high cell densities (10^6–10^7 CFU/mL) showed the highest log reduction onto *S. cerevisiae* (~6.5 log 10 CFU/mL) and the lowest onto *Staphylococcus epidermidis* (~0.5 log 10 CFU/mL), indicating higher resistance of the smaller, more compact and simpler bacterial cells compared to complex, more elaborate and bigger yeast ones (Griffiths et al., 2012). In these conditions, there was no saturation of the reduction effect as the number of pulses increased; but under similar conditions at lower field strength (25 kV/cm), 10 pulses was the maximum effective treatment for *S. cerevisiae* (Barbosa-Canovas et al., 1999), which meant a plateau of the antifungal dynamics at this specific field strength. With higher field strengths (45 kV/cm), all cells seem to exhibit signs of susceptibility (Griffiths et al., 2012).

Very low-intensity constant fields (5 V/m) applied on mixtures and pure strains of *Pseudomonas*, *Bacillus*, and other bacterial preparations, placed in clay soil contaminated with fuel oil hydrocarbon, increased the rate of biodegradation of the contaminants (Olszanowski and Piechowiak, 2006).

Treatment of *E. coli* by a field of 30 V/m at 18 GHz, which is more of electromagnetic radiation/wave than of EMF, resulted in poration, distinct but reversible morphological alterations and limited bactericidal effect of slightly over 10% in viable counts (Shamis et al., 2011). The same treatment applied on four Gram-positive cocci, (*S. epidermidis*, *Planococcus maritimus*, and two strains of *S. aureus*) produced poration resulting to different degrees of transient permeabilization, but no morphological alterations and only slightly increased lethality, of approximately 15% in viable counts (Nguyen et al., 2015). Treatment of a selection of bacteria (*Branhamella catarrhalis*, *Kocuria rosea*, and *Streptomyces griseus*) and *S. cerevisiae* yeast under the same conditions produced clear indication that the degree of permeabilization depends on the composition specifics of the cell envelope as are the membrane lipid composition, the cell microenvironment, and the presence of charged phospholipid head groups; moreover, prolonged multiple EMF exposures using two strains of *S. aureus* showed increase in permeabilization and overall decline in the viable cell numbers over repeated exposures (Nguyen et al., 2016).

A *S. cerevisiae* population was reduced by 2.1 log 10 units following exposure to a 30 kV/cm field of 20–60 kHz (Gevekeand and Brunkhorst, 2003). Square pulses of nanosecond-order duration induce apoptosis in yeasts, depending on energy parameters such as field strength, pulse period, and number of pulses (Simonis et al., 2017). On the other hand, PEF of up to 6 kV/cm, delivered for 1,000 pulses of 100 μ/s each with 100 ms interruptions to inoculum suspensions, stimulate *S. cerevisiae* to better fermentation kinetics (Mattar et al., 2015)

In the case of mycelial fungi, endogenous electric fields most probably are capable of mobilizing and localizing charged particles such as proteins and organelles in the cytoplasm by electrophoresis or electroosmosis, in order to achieve apical growth. Exogenous electrical fields may therefore polarize mycelial fungi either by electrophoresing the proteins and membrane-bound vesicles in the cytoplasm that are required for growth or by influencing the distribution of key morphogenetic proteins in the cytoplasm. The membrane proteins have been shown to become localized at a particular end of a cell if the cell is exposed to an exogenous electrical field (McGillivray and Gow, 1986). Conidiospores of *Neurospora crassa* were embedded in agarose and exposed to electrical fields between 0 and 40 V/cm either until germination or just after germination. *Aspergillus nidulans*, *Achlyabisexualis*, *Trichoderma harzianum*, and *Mucor mucedo* were also used. Germination timing and polarity and polarity of branching and of hyphal growth were assessed as functions of field strength and duration of treatment. Maximal polarization of germ tubes of *N. crassa* occurred at fields of 30–40 V/cm; germination was also more synchronous in treated than in untreated spores. Contrarily, fields of a similar strength had little effect on the polarity of germ tube formation in *A. nidulans*. Hyphae became aligned perpendicularly to the field as they grew longer and as the field strength increased. *N. crassa* and *A. bisexualis* grew and formed branches toward the anode (anodotropic) while *A. nidulans* and *M. mucedo* toward the cathode (cathodotropic). Galvanotropism of hyphae and branches of *T. harzianum* were in opposite directions. The mixed response of various fungi in applied fields, according to the authors, does not agree with the model of a common endogenous electrical polarity, although positive electrical current has been found to enter the growing hyphae in all five of them (McGillivray and Gow, 1986).

Pulses lasting a few microseconds appeared to be the optimal for amoebae for a given energy exposure. A strong field applied for short, cumulative pulse duration affects viability more than a weak field with a long cumulative pulsation (Coustets et al., 2015).

The simplicity of the virion, with no active metabolism and functionality, means that deleterious effects are caused by structural damage to the macromolecular components and to their disposition (and integrity) in 3D, mostly by dielectrophoresis. RNA is by definition more unstable than DNA, although the two-strand conformation affords increased stability. The lack of homeostasis means inability of the virion to self-cure, contrary to damaged cells (Somolinos et al., 2008).

Results are highly divergent: The rotavirus was found to be resistant to PEF treatment of 20–29 kV/cm, with no appreciable reductions in virus titer observed (Khadre and Yousef, 2002). On the other hand, human norovirus, prevalent cause of acute nonbacterial gastroenteritis, transported by fresh produce such as leafy greens due to little or no processing (especially thermal) after harvest, can be deactivated by PEF processing, using square wave 3 μs pulses of 29.5 kV and

8.8 A. Reductions of 1.2 and 2.4 log10 in murine norovirus-1 (the human norovirus surrogate) and Tulane virus respectively were thus achievable (Predmore, 2015).

Swine vesicular disease virus (SVDV) in deionized water or in bovine serum was inactivated effectively by 60–120 pulses of an electric field of 30 kV/cm, applied in exponential-decay waveform and resulting in empty virions of SVDV with intact capsid but no detectable RNA within. Equine herpesvirus-1 in minimum essential medium was inactivated by a minimum of 45 pulses causing observable damage to the envelope and to the contained DNA (Mizuno et al., 1990).

E. coli bacteriophages, M13mp18 and λ phage, were successfully inactivated by PEF, depending on treatment time under peak voltage of 5 or 7 kV; the survival ratios after 12 min PEF treatment were 10^{-4}–10^{-5}. *E. coli* proved to be more tolerant to PEF than the *E. coli* λ phage by two orders of magnitude—contrary to heat treatment results. As, in the case of λ phage, no degradation of total protein nor of DNA was observed, the inactivation was rather not due to the degradation of either, but to the destruction of phage particle structure (Tanino et al., 2013).

Mechanism of action

The concept of electric fields use to treat microbes applies a voltage to the carrier/container of microbes, which may induce current by the motion of charged particles (electrophoresis); fluorescently labeled proteins of opposite charge distributed themselves in opposite directions (McGillivray and Gow, 1986). Alternatively, the voltage may orient noncharged but polarized molecules and organelles, which bear charges onto different moieties but their electric loads add up to zero, and direct them to positions/attitudes of balance (dielectrophoresis). The change of the field polarity changes the distribution of charged and the orientation of polarized moieties and macromolecules, an event prone to disrupt the architecture of the cell and its functional orientation.

The most applicable form is the pulsed electric fields (PEF) which, if powerful enough (up to 1,000 V/cm; Giladi et al., 2008) may cause electroporation to cell membranes and alter their structural integrity (Novickij et al., 2015). Electroporation or electropermeabilization is the increase of the cell membrane permeability by means of application of pulsed electric field of controlled duration and intensity. Such an event allows insertion of molecules into the cell, which otherwise may not have been possible (Costerton et al., 1994; Novickij et al., 2015), thus inducing transformation; but also promotes conjugation or even fusion with other cells (Mattar et al., 2015) and the effect may range from temporary permeabilization of cell membrane(s) to irreversible electroporation resulting in cell wounding and perhaps necrosis (Novickij et al., 2015).

Irreversible permeabilization by PEF treatment (electroeradication) is proposed as an industrial tool for bacterial eradication in food and environmental contexts (Coustets et al., 2015), whence the leakage of plasma, and possibly organelles as well, may necrotize the cell or severely decrease its vigor and viability. This explains the observation that under the same field treatments (8 kV/cm), higher percentages of cells intake visual markers and are thus proven to be permeabilized, than losing their viability in viable counts (Novickij et al., 2015). The molecular pathways for electroeradication are complex, as field strength, pulse duration, pulse frequency, and postpulse incubation temperature must achieve a fragile balance so as to make the approach cost-effective.

Less aggressive treatment may reversibly permeabilize/solubilize membranes, in order to extract fully developed proteins from cellular translation systems such as bacterial hosts producing heterologous proteins in industrial context (electroextraction); when the pulse duration was several ms long, this resulted in a cytoplasmic soluble protein slow leakage (Coustets et al., 2015).

The procedure can be used in reverse, to insert antibiotics, biocides, and other toxic compounds to the cell (Novickij et al., 2015; Mosqueda-Melgar et al., 2008). The molecular alterations required for the two applications (electroextraction, electroeradication) were not the same (Coustets et al., 2015).

Lower field intensities may affect the targeted cells in different ways: When a cell is exposed to an electric field, the voltage across its membranes changes; this induced transmembrane voltage superimposes on the physiological transmembrane voltage of a cell and may alter the gating properties of ion channels (Simonis et al., 2017). This effect may induce the cellular activity as well, in contrary to amenities causing electroporation.

Moreover, high-frequency low-intensity fields in bacteria change the 3D disposition and clustering of water molecules, thus changing the hydration of macromolecules; excise DNA inserts such as proviruses and episomes; and affect membrane function by altering protein structure, infringing in enzyme recognition and cross-reaction with the substrates and, last but not least, modifying channel traffic by either pseudosignaling or by jamming legitimate signals (Hunt et al., 2009; Torgomyan and Trchounian, 2013; Soghomonyan et al., 2016). Frequency-dependent inhibition of bacterial growth relates to the suggested effect of alternating electric fields on the enzyme–substrate reaction equilibrium (Giladi et al., 2008), while external fields can influence proteins such as glycerol-3-phosphate dehydrogenase and FtsK, which are present in *P. aeruginosa* and *S. aureus*; affect the α-helix content and orientation of membrane proteins in eukaryotic cells; and impact on the electrophoretic mobility of bacterial membrane proteins (Costerton et al., 1994), and especially that of membrane-associated proteins, which carry exposed charged groups (McGillivray and Gow, 1986).

ELECTROMAGNETIC FIELDS

The realm of EMF, defined by the concomitant presence of electric and magnetic components, with ratios between 0.1 and 10, is able to cause effects to some distance and with typical EMF oscillation frequency of 100 kHz or more (Hunt et al., 2009).

The millimetric wavelength (MMW) irradiation, which is EMF oscillating at extremely high frequencies (of 30–300 GHz) and bound to produce effects at a distance, caused various and sometimes opposite effects on different bacteria. The kind of effect depends on parameters of the (1) EM amenity, (2) microbial growth system/culture, and (3) wider environment. In the first category, frequency, intensity, coherence, exposure duration and repetition, and mediated or direct effect are the basic factors. In the second category, bacterial species, growth phase (i.e., logarithmic/exponential or stationary or latent), composition of nutrient media, inoculum size and dilution, genetic features, peculiarities of metabolism, and membrane structures are the key features. The third category includes factors such as anaerobic or aerobic conditions of culture, incubation temperature, other interacting (micro)organisms and macromolecules. Thus, inhibitory effects depending on frequency of coherent MMW and exposure time are more prominent under direct irradiation of cells on solid medium than in aqueous liquid medium or suspension, and differ between Gram-negative and Gram-positive bacteria—the former being more resistant—and among different bacterial species: in addition to *E. coli*, *Enterococcus hirae*, and *L. acidophilus* were sensitive to MMW but to a lesser degree. Stimulant effect can also be observed in noise (*E. coli*, *Azotobacter* spp.) and in coherent (*S. aureus*, *Rhodobacter sphaeroides*) MMW amenities (Soghomonyan et al., 2016).

The cyanobacterium *Spirulina platensis* showed 50% growth increase after exposure for 30 min at 2.2 mW/cm and 41 GHz (Pakhomov et al., 1998). The archaebacterium *Methanosarcina barkeri* after three daily repeated 31.5 GHz treatment cycles of 2 h each presented 50% increase in methane production, higher cell numbers, increase in specific growth rate (manifest as a significant reduction in the lag phase), and a 20% increase in cell size (Banik et al., 2005). *S. cerevisiae* exposed in the frequency region of 41.83–41.96 GHz presented increase in growth rates up to 15% or decrease by 29%—depending on frequency (Grundler et al., 1977).

Different mechanisms implement the effect of MMW to the cell(s), depending on the frequency: changes in the conformation of the water molecules are a major intermediate in effecting cellular responses of different kinds (Hunt et al., 2009; Soghomonyan et al., 2016). The effective rearrangement of electron subsystem levels in DNA molecules might be induced by EMF of 41.3 and 51.8 GHz, while EMF of 51.8, 53, 70.6, and 73 GHz altered ion (H^+ and K^+) transport processes and enzymatic activities of plasma membranes in bacteria (Soghomonyan et al., 2016; Hunt et al., 2009). FOF1-ATPase can be among the primary cellular targets for MMW (Soghomonyan et al., 2016), while MMW could mimic—or suppress—control signals produced by cells through ion channels, which drive cellular responses (Soghomonyan et al., 2016; Hunt et al., 2009). It is of interest that, on top of chemical/ion-based signals, bacteria and the other cells might communicate with each other through EMF (Hunt et al., 2009). This function might be jammed by exogenous MMW (Soghomonyan et al., 2016).

High-power pulsed EMF (PEMF) in the form of 3.3 T, 0.19 kV/cm submicrosecond (450 ns) pulses were tested to permeabilize and necrotize pathogenic yeast *C. albicans* either as a standalone amenity or coupled to square wave, 100 μs PEF pulses of 8–17 kV/cm intensity. PEMF-only treatment produced limited loss of viability, approximately 20%, and negligible permeabilization, compared to PEF-only treatment, which scored over 55% loss of viability and 80% or whereabouts permeabilization, both in a dose-dependent way; the combination treatment did not improve scores any further once PEF started becoming efficient in terms of decreasing viability and was shown rather counterproductive, if not antagonistic, in terms of permeabilization (Novickij et al., 2018).

CURRENTS

Electric fields can induce currents to their target and mobilize charged entities, molecules or macromolecules and organelles (Tracy, 1932; Spilker and Gottstein, 1891); this was the quintessence of Franklinism, the first generation of ES introduced by Benjamin Franklin around CE 1750 (Kambouris et al., 2014). Actual current transfer to a living entity (conductive coupling), is another issue altogether, and marked the second generation of electrotherapy, introduced around CE 1800 by Galvin; it employed DC and was designated galvanism (Kambouris et al., 2014). Such application entails pin or dressing electrodes in contact with the target and cabled to a generator, which poses issues of biocompatibility, safety, operational efficiency, and contamination/infection of the target site, especially when wounds, burns, and ulcers are the object of the process, whence rash, itch, and pain are prominent.

Despite such pitfalls, the most effective ES protocols are still the conductive ones; thus the effect on microorganisms was naturally a matter of interest, not only for therapeutic concerns but also for other applications as well: the latter include nontherapeutic sterilization, a most important issue in medical amenities such as catheters (Davis et al., 1991); bioengineering, food and drug production/processing/conservation exemplified by heifer and juice decontamination (Unknown, 1987; Tracy, 1932 respectively); biotechnology applications, microbial biodegradation or decay of usable amenities and of organic litter and sewage (Ranalli et al., 2002), biorestoration, bioaugmentation of soil, and so on.

Conductive coupling may be used for inducing and for suppressing microbiota, but it is mostly used to the latter effect. It comprises mainly four different modalities, under more than four names: (1) alternating current (AC), (2) HV pulsed current (HVPC), (3) low-intensity DC, and (4) low-voltage PC (LVPC) as proposed in the most meaningful classification scheme (Daeschlein et al., 2007). Of these, only HVPC uses strong currents, up to 0.8 A (Kloth, 2005); the usual aim being to suppress and destroy microbiota, high amperage may though prove destructive for the entities sought to be protected from the microbiota (ohmic damage). The term “low electric currents,” denoting milliamperes and used up to the 21st century (Ranalli et al., 2002; Valle et al., 2007), has been supplemented by the prefix “micro", which covers the microamperage range (Denegar et al., 2015).

There has been vigorous research, mainly focused on the type of current, which may be AC, DC, HVPC, LVPC, or other (Daeschlein et al., 2007; Merriman et al., 2004; Maadi et al., 2010; Petrofsky et al., 2008); on the type, design, material, and polarity of the electrodes used (Barranco et al., 1974; Ong et al., 1994; Laatsch-Lybeck et al., 1995; Kloth, 2005; Valle et al., 2007; Kumagai et al., 2007) and, of course, on dosage (Kloth, 2009; Tracy, 1932). The Gram status of a bacterium might well be of some importance: in one study *E. coli* had been treated by graphite electrodes delivering 40 mA and inhibition of enzymatic activities and growth, plus a reduction in ATP content were observed; *Bacillus cereus* under the same conditions was stimulated in terms of growth, ATP content, and some enzyme activities (Valle et al., 2007); previously, 100 μA of DC delivered to cell cultures via a silver wire anode had a bacteriostatic effect on Gram-positive bacteria, whereas the same current amplitude and polarity produced a bactericidal effect on Gram-negative bacilli (Kloth, 2005). The matter clearly remains open and undecided (Hristov and Perez, 2011), but the bulk of the published work is performed using contact electrodes onto unicellular or acellular microbiota; mostly bacteria (Stone, 1909; Thornton, 1912; Davis et al., 1991; Ong et al., 1994; Daeschlein et al., 2007; Valle et al., 2007; Maadi et al., 2010; Hunckler and De Mel, 2017), but occasionally some yeasts as well (Tracy, 1932; Deitch et al., 1983; Davis et al., 1991; Ranalli et al., 2002; Berovic et al., 2008; Nakanishi et al., 1998), and viruses (Unknown, 1987; Kumagai et al., 2007, 2011; Kolsek, 2012).

Stimulatory effects of 0.1−0.3 mA DC have been recorded on bacterial and yeast growth since the early 20th century (Stone, 1909; Tracy, 1932). *S. cerevisiae* presented increased growth and EtOH production under 100 mA AC and 10 mA DC (Nakanishi et al., 1998) and *B. cereus* was stimulated by 40 mA DC, whereas *E. coli* was inhibited under the same conditions (Valle et al., 2007). The effect is reversed in traumas as shown with *S. aureus*−infected wounds (Carley and Wainapel, 1985; Spadaro et al., 1986); similar results were obtained with *Pseudomonas* and *Proteus* (Mercola and Kirsch, 1995; Wolcott et al., 1969). Still, full-blown infections are considered contraindications for trauma treatment, as the dynamics of ES onto appendage microbiomes remain unexplored (Kambouris et al., 2014, 2017, 2018b).

There are three main limitations to the use of DC currents, including weak currents, for the treatment of infection. The first is that such currents may stimulate nerves and muscles, causing pain and muscular contractions in the patient. The second relates to the spread of the currents in the body, which can be regarded as a volume conductor. Thus, unless the lesion is superficial or unless there is a conductor leading from the surface to a deeply situated lesion, a current density of sufficient intensity at the target can be obtained only when the density near the electrodes is of a damaging and cross-stimulating magnitude. The third limitation is that DC currents cannot be generated by insulated electrodes and are therefore always associated with electrolysis, metal ions, free radicals, etc. (Giladi et al., 2008).

The observation that *in vivo* microbial inhibition in wounds is preferably accomplished by negative DC of 200 μA—1 mA, while wound healing by similar positive DC (Mercola and Kirsch, 1995; Vodovnik and Karba, 1992; Wolcott et al., 1969) explains in part one advantage of AC current formats in wound healing. Still, AC currents, being the quintessence of Faradism, that is the third generation of ES c.1830 (Kambouris et al., 2014), are mostly ineffective in antimicrobial applications, (Kloth, 2005; Rowley et al., 1974; Merriman et al., 2004; Maadi et al., 2010) at least on a comparative basis, since they occasionally display adequate inhibitory potential against yeasts (Tracy, 1932) and bacteria (Thornton, 1912).

The concept of LVPC uses DC of low voltage, delivered intermittently with consecutive, programmable, and identical pulses, a repetitive context that does not change polarity. The concept has different spin-offs in pulse form but has proven inadequate (Merriman et al., 2004) despite some focused efforts to exonerate it (Daeschlein et al., 2007).

On the contrary, HVPC had been a thoroughly studied and used method (Merriman et al., 2004; Szuminsky et al., 1994; Kloth, 2005; Daeschlein et al., 2007; Hunckler and De Mel, 2017) as it employs high voltages; additionally, HVPC devices also allow selection of polarity and variation in pulse rates (Daeschlein et al., 2007). The operative idea behind HVPC is that the intermittent nature will save the host of the damage caused by HV, while the - less tolerant - microbiota should not survive. The high-power bursts are indeed effective (Szuminsky et al., 1994; Merriman et al., 2004), but their comparative effectiveness over DC remains controversial (Guffey and Asmussen, 1989; Kloth, 2005), a fact which led to the microcurrent electrotherapy (MET). The most novel spin-offs of MET have been the wireless microcurrent stimulation (WMCS)/non-contact current transfer (NCCT) concepts visited in following paragraphs.

The bulk of DC conductive modalities have been tested with bacteria through a period of half a century. Indicative research proved clearance of human wounds initially colonized with *Pseudomonas* and/or *Proteus* after several days of treatment with microampere levels of cathodal DC (Wolcott et al., 1969). Subsequently, cultures of *E. coli* were found to be practically unaffected by AC but obviously inhibited by DC (Rowley, 1972). DC of 0.4, 4.0, 40, and 400 μA delivered by stainless steel, platinum, gold, and silver electrodes caused a time- and intensity-depended bactericidal effect (Barranco et al., 1974) to *S. aureus*; an observation later confirmed and extended to HVPC (Guffey and Asmussen, 1989). Concomitantly, bacteriostatic effect was demonstrated *in vivo* with cathodal DC of 1.0 mA for 72 h to rabbit cutaneous wounds infected with *P. aeruginosa* (Rowley et al., 1974).

DC delivered by gold, carbon, and platinum electrodes to synthetic urine was shown to effectively reduce or eliminate Gram-positive and Gram-negative bacteria, and *C. albicans* inocula, depending on the electrode material (Davis et al., 1991). AC and DC (1.5, 3.5, 5.5, and 10 V) delivered to *P. aeruginosa* immediately after inoculation compared to delivery 19 h after inoculation produced different patterns of inhibition by voltage, substrate, and timing of exposure, but not by polarity, while AC exposure produced minimal inhibitive effect (Maadi et al., 2010).

Currents were explored as antiviral amenities rather early: HVPC, under the then-current term, High Voltage ElectroStimulation (HV-ES) was used to delivered 25 pulses/s of 700 V peak for four sessions of 30 s and Low Voltage ElectroStimulation (LV-ES) to deliver 14.3 pulses/s of 94 V peak for one session of 60 s to heifers. The idea had been to inactivate FMDV (foot and mouth disease virus) by causing a rapid depletion of ATP and glycogen in skeletal muscles with consequent accumulation of lactic acid and a drop of pH, which inactivated the FMDV (Unknown, 1987).

A little later, percutaneous electrical nerve stimulation treatment was shown to assist in herpes zoster virus (HZV) infection by decreasing pain levels during acute HZ (Ahmed et al., 1998). The results were confirmed relatively recently with a retrospective research showing that TENS (transcutaneous electrical nerve stimulation) indeed decreases pain levels in acute HZ and effectively inhibits PHN (post-herpetic neuralgia), a condition afflicting 20% of patients receiving conventional antiviral therapy instead of TENS. Interestingly, patients receiving both TENS and antiviral drugs presented no amelioration on PHN incidence, compared to antiviral drugs-only group, thus begging a hypothesis that antiviral drugs reduce the effect of TENS (Kolsek, 2012).

More recent research focused on the use of an indium-tin oxide flat electrode to stimulate HIV-1-infected cells (Kumagai et al., 2007) and free HIV-1 virions adsorbed on its surface (Kumagai et al., 2011) to determine possible inactivation. Indeed, the former study argues in favor of reactive oxygen species (ROS) induced by electrical stimulation as a key factor for decreased HIV-1 virulence upon host cells treated immediately after infection. The second study used DC for different exposure times and presented full inactivation of the adsorbed HIV-1 virions after 7 min 30 s treatment, which is within safety limits for host cell damage.

THE NEW GENERATION OF ELECTROSTIMULATION: WMCS-NCCT

In every application, low energy formats are desirable, as they entail lower financial cost and lower environmental burden. In health applications involving currents, the lower risk of ohmic damage and greater similarity to innate bioelectricity, as are the cutaneous bioelectric currents and the current of injury (Kloth, 2005), make low energy formats even more attractive (Kambouris et al., 2014). Thus microcurrent-level ES, also known as MET (Mercola and Kirsch, 1995), has been a field of considerable promise; its intersection with Non-Contact Current Transfer (NCCT) spawns the wireless microcurrent stimulation (WMCS) approach (Mercola and Kirsch, 1995). The WMCS turns atmospheric gases (oxygen and nitrogen) to hydrated ions and blows them onto the target; the latter usually is a wounded tissue. The circuit closes by flexible, wrappable return electrode placed so as to ensure that the resulting microcurrent is channeled through the intended target. The characteristics of this microcurrent are very similar to innate currents (Poltawski and Watson, 2009; Kambouris et al., 2014) and thus targeted manipulation of the cellular physiology is achievable. The basic application is thus wound healing, comprising burns, ulcers, and mechanical and chemical wounds. The observed sanitization of wound sites (Ramadhinara and Poulas, 2013; Lagoumintzis et al., 2014) may be due rather to the stimulation of the immunoresponses than to direct antimicrobial effect.

This is supported by the fact that *in vivo* microbial inhibition in wound sites seems to be attainable by a wide range of electrodes, the metallic ones (platinum, stainless steel, copper, and especially silver) being more effective in delivering preferably negative, continuous μA-level currents of 5 μA or more for 20 min or more, whence bacteriostatic (and sometimes bactericidal) effects are seen (Kloth, 2005; Mercola and Kirsch, 1995). Lower amperage has been used successfully, but only combined with silver electrodes (Asadi and Torkaman, 2014), an observation indicating the latter as the source supreme, although not unique, of antimicrobial efficacy in such cases. This observation begs the question of what happens within the 0.5–4 μA range, which is used by WMCS (Kambouris et al., 2014, 2017).

A substantiated answer on whether the NCCT has a positive or negative effect on microbial growth, and the subsequent determination of the involved kinetics and dynamics, will define the applicability of antimicrobial electrotherapies based on this modality either as alternatives to chemical antibiotics, thus sparing the patient the latter and the accompanying reverse effects or, to the contrary, as complementary and even synergistic co-factors in mixed regimens. On the other hand, combined protocols might be developed for resolving dysbiosis events, for example, by inducing benevolent microbiota (including but not restricted to probiotics) while preventing the proliferation of pathogens, thus expediting the therapy even more than in the current, stand-alone regimens. Identical considerations and prospects apply for every and any industry and application based on, and/or plighted by, microbiota.

Mechanism of action

There are intrinsic questions concerning the consistency of the outcome of ES treatment of a certain microbiote (Kambouris et al., 2017). Differences are expected and not only among taxa, where genera and species might present different maxima for current dose–response curves describing both stimulation and suppression (Ranalli et al., 2002; Griffiths et al., 2012). Reasons for dose–response variation include, without being restricted to, the following: (1) different forms and phases of a microbiote, with sessile and planktonic forms being the prime example (Costerton et al., 1994), and the phases onto the microbial growth curve the next (Ranalli et al., 2002; Giladi et al., 2008; Coustets et al., 2015); (2) different substrates/microenvironments (Maadi et al., 2010; Davis et al., 1989, 1992), and (3) inoculum density (Davis et al., 1989, 1992).

All the above apply even more in clinical, *in situ* contexts and other realistic sets of conditions, where the ambient conditions affect severely the actual charge dosage (Griffiths et al., 2012; Davis et al., 1989) and the resulting microenvironment (Costerton et al., 1994; Caubet et al., 2004; Petrofsky et al., 2008). Both electrokinetic and electrodynamic factors, such as polarity, type of current, and time of exposure, might prove important and are prone to be affected by the physiologic particularities of different taxa. Determining the intriguing multiparametric balances so as to project plausible effects and results in various but standardized settings requires high-throughput, low-cost experimental approaches. In such a context, comparative electroculturomics (see Chapter 8), using proper standards and high-throughput culture platforms adapted from current plasticware (as in Fig. 14.3), is likely to prove the best solution for determining various electroresponses (electrotoxicity, electrotolerance, or electrophilia) of different microbiota, in different culture formats (Kambouris et al., 2017, 2018a).

Still, a valid assumption is that the end result of electroceuticals is a matter of accumulated energy/charge and of rate of delivery, as noted long ago: "The fungicidal action of alternating current depends upon a definite quantity of electricity applied at, or above, a certain minimum current density" (Tracy, 1932). Thus intensity and duration of exposure should be viewed as interrelated and amenable to optimization (Karba et al., 1991; Griffiths et al., 2012), but their relationship is not a precise product, as anticipated, which implies additional factors at work (Tracy, 1932).

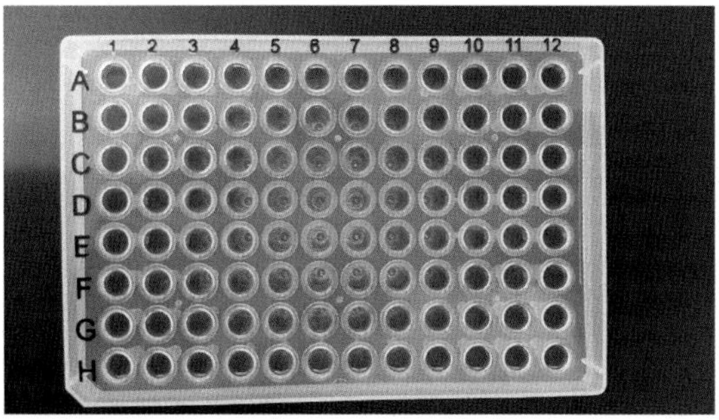

FIGURE 14.3

The miniaturization of culturomic platforms from Petri dishes to multiwell plates is prone to kick-start the electroculturomics. Multiwell plates currently started incorporating novel ideas, practices, and options, such as gated walls with microfluidics networks, separate, add-on bottom plates amenable to requirements of conductivity and built-in circuitry. Thus this new generation of plasticware is bound to accommodate massive, affordable formats of comparative electroculturomics offering automation-friendly solutions that achieve major savings in substrates, consumables, storage space, and incubation infrastructure while being adequately informative.

A peak inductive effect will be achieved in terms of duration and intensity, beyond which the treatment will become inhibitive even in cases of manifest electrophilia (Hristov and Perez, 2011; Berovic et al., 2008) as noticed in fermentors (Hristov and Perez, 2011). It is interesting though that, at least in some cases, exposure–inhibition curves are recurrent (Maadi et al., 2010), insinuating the implication of different mechanisms and/or receptors.

Galvanotaxis is the attraction of positively or negatively charged cells toward an electric field attribute of opposite polarity. Cells involved in wound healing migrate toward the anode or cathode of an electric field according to their specific type (Kloth, 2005). Negative polarity sports lower antibacterial performance but supplements it with additional, *in vivo*-induced indirect antibacterial effects such as increased blood flow, debridement and recruitment of neutrophils and macrophages (Hunckler and De Mel, 2017; Daeschlein et al., 2007). High-intensity pulsed electric currents (generating electric fields of 1,000 V/cm) cause electroporation, which means massive changes in the integrity and permeability of the membrane. Relatively low-intensity DC and low-frequency alternating electric fields, when applied through conductive electrodes, can lead to electrolysis, temperature increase within the wound, and to damages to the internal cell membranes resulting to an irreversible loss of the semipermeable barrier function (Ranalli et al., 2002; Hülsheger et al., 1983). Other possible lethal effects include, but are not restricted to, alteration of the pH, alterations in bacterial biofilm structure and the production of toxic derivatives and free radicals (Hunckler and De Mel, 2017; Giladi et al., 2008; Davis et al., 1992). ROS induced by electrical stimulation play a role in inhibition of HIV-1 infection (Kumagai et al., 2007).

Bacterial cells depend, as do all living cells, on physical phenomena such as membrane potentials for their basic metabolic activity. Delicate cellular electrical equilibria may be disturbed by the application of even weak electrical stimuli. Electric fields and currents can influence (1) the organization of biological membranes (by affecting the α-helix content and orientation of membrane proteins in eukaryotic cells and the electrophoretic mobilities of bacterial membrane proteins) and (2) a number of metabolic and developmental processes within both prokaryotic and eukaryotic cells (Costerton et al., 1994).

On the other hand the more specific and cell-type-sensitive, frequency-dependent inhibition of bacterial growth relates to the suggested effect of alternating electric fields on the enzyme–substrate reaction equilibrium. Glycerol-3-phosphate dehydrogenase and FtsK are both proteins of *P. aeruginosa* and *S. aureus* expected to be influenced by external electric fields (Giladi et al., 2008).

Radio-frequency alternating currents most probably target, among other moieties, the exopolysaccharide (EPS) matrix produced by sessile bacteria, which contains many types of charged particles and molecular chains with polar subsystems. Such a structure is susceptible to the influence of EMF; an RFC (radio-frequency current, whence no creation nor transportation of ions takes place), will induce vibration to polar molecules, charged particles, and polar parts of large molecular chains. The effect was exhibited by a mixed biofilm containing *Klebsiella pneumoniae*, *P. fluorescens*, and *P. aeruginosa* grown upon a wire electrode to a thickness of 50 μm in 3 days. When AC of 50 mA with an oscillating polarity (square wave) from 0.016 to 5 Hz was applied, the biofilm proper expanded or contracted accordingly; it expanded by approximately 4% when the wire was cathodic but was reduced to 74% of the original thickness when the wire was anodic, and the effect was partially replicated without current when buffers of pH 3 and pH 10 were alternated (Stoodley and Lappin-Scott, 1997). A molecular structure that is subject to an imposed vibration is prone to have its fluidity increased and its structure weakened. This could increase the exchanges between the bacterial cells in the biofilm and the surrounding liquid. The use of ultrasounds at frequencies between 70 kHz and 10 MHz to vibrate a biofilm produces synergy with antibiotics, a phenomenon very similar to the BEE (Caubet et al., 2004).

The fungicidal effect of AC on yeast cells depends upon current intensity as well as upon the total charge conducted. Higher lethality scores are attainable by longer exposure, which could have been substituted by increasing amperage had lethality been a unique function of total charge only. Since this does not happen, more factors are at work. Alternating current of 60 Hz passing through yeast cell suspensions in grape juice, at nonlethal temperatures of 42°C, succeeded in producing an evident lethal effect. Consequently, this latter is attributable to factors other than temperature (Tracy, 1932).

Though, AC treatment of 10–100 mA applied to yeast cultures caused alterations in the morphology and integrity of the cells; namely, the loss of cell organization and of primary turgidity. The membrane system was ruptured, leading to clear signs of loss of cytoplasmic and nuclear material from the cell envelope (Ranalli et al., 2002). Less potent treatment resulted in lighter membrane damage, incurring change in the chemiosmosis (Pethig, 1985), which caused altered electrochemical membrane potential and consequently affected the membrane transport mechanisms, the energy transfer process (through the inhibition of specific enzymatic complexes) and the membrane permeability (Rols et al., 1990; Ranalli et al., 2002).

The effects of different conductive modalities are partially understood and should be viewed through more holistic lenses; different experiments show different—and well-substantiated—results on the effect of a certain modality. For example, the research on the efficacy of LVPC has been contested (Merriman et al., 2004 vs Daeschlein et al., 2007). The former observed no effect and the latter considerable effects. The same studies also draw different conclusions on the subject of polarity and its effect; the former dismiss it altogether, the latter saw significant impact, although lower than the effect of common chemical antimicrobials for local decontamination.

A reconciliatory and synthetic approach might point to the fact that the former study used many different modalities and polarity was found not to be an issue in any of them; still, LVPC showed no effect whatsoever in this study and thus polarity cannot be assessed as a factor. At the same time the latter study used only LVPC, despite admitting that HVPC was at the time the most extensively tested modality. Once the discrepancy in the effect of LVPC is resolved, the issue of polarity will become important. Thus contradictory results on electrode material, polarity and modality might be interrelated: ES works without electrochemically active electrodes (Costerton et al., 1994); carbon and graphite electrodes were found least effective (Davis et al., 1991; Valle et al., 2007) but still effective nonetheless (Ranalli et al., 2002), despite producing no biologically active moieties when charged. Consequently, electrode material may start becoming important at certain optimal polarities and intensities (Ong et al., 1994; Valle et al., 2007; Barranco et al., 1974; Kloth, 2005; Davis et al., 1991). Nonoptimal intensities may fail to produce electrochemical reactions, or at least at adequate efficiency and rate, and products may not migrate far enough. On the other hand, nonoptimal polarities may be failing to produce drastic moieties at all, let alone mobilize them over significant distances.

The drastic antimicrobial effect of NCCT-type ES on microbial pathogens may be mediated by different, interacting or independent pathways, (Kambouris et al., 2014; Poltawski and Watson, 2009; Asadi and Torkaman, 2014; Sultana et al., 2015), and of course prokaryotes may—or may not—present marked differences compared to viruses and eukaryotes (Armstrong and Bezanilla, 1973; Kolsek, 2012; Karba et al., 1991; Kalinowski et al., 2004). The very low intensity of the current might not be enough for massive depolarization and consequent electrocution effects, but extensive deregulation due to ion spraying might cause an electrolyte imbalance and destabilize the microbial cell homeostasis. While the miniscule intensity of the conducted current may be too weak to disrupt the integrity of the cell membrane, at least in short-duration stimuli, it may well infringe with the transportation and distribution of charged moieties (Berovic et al., 2008), produce toxic molecules/radicals (Sultana et al., 2015), electrolyze surface molecules (Asadi and Torkaman, 2014), alter hydrophobicity, cell shape, and surface molecules (Luo et al., 2005), or compromise signaling (Zhao et al., 2006).

Similarly, yeast cells treated with AC proved to be more susceptible when at a higher concentration, in the logarithmic rather than in the lag growth phase (Ranalli et al., 2002); the authors suppose that the causality should be sought after in the concentration or in the absolute number of cells, but another aspect might be the altered physiology and electrical properties of vigorously dividing mitotic cells (Giladi et al., 2008). Higher susceptibility to PEF challenge is observed is such cases (Coustets et al., 2015) and it might be attributable to the engagement of the anabolism in the production of duplicates of all cellular components and thus its unavailability for repairs of damage caused by the treatment (Somolinos et al., 2008).

ELECTRORESISTANCE AND ELECTROSTIMULATION INTERACTION WITH ANTIBIOTICS

The interaction between electromagnetism and antibiotics might revolutionize the treatment of infectious diseases. Establishing either synergistic or antagonistic effect with current antibiotics in *in vivo/in situ* formats will allow, respectively, either alternate regimens, with drastic decrease of the pharmaceutical expenditure and of the expansion of microbiological resistance to antibiotics, or combined pharmaceutical–electroceutical protocols for drastic increase of the scope and effectiveness of the treatment. In the longer term the widespread use of different electroceuticals might be found to be an ubiquitous therapy applicable to enhance chemically focused schemes either expediting the recovery timeframe or decreasing the need for expensive to develop, produce, test, use, and dispose of chemical compounds, alleviating thus the pressure on the Research & Development centers and on the industry to continuously and promptly develop new compounds, especially when the antibiotics bottleneck is in full fledge (Kambouris et al., 2018b).

Regarding magnetic amenities, *E. coli*, *Pseudomonas*, and *Enterobacter* display, in a zero-magnetic field, modified resistance to various antibiotics (Pazur et al., 2007) as they become disorganized, having evolved in concert with the geomagnetic field. Static magnetic fields of some tens of mT produced by permanent magnets increased the resistance of exposed *E. coli* to piperacillin antibiotic (Stansell et al., 2001), but this might have been a result of induced growth before the addition of the antibiotic rather than a resistance mechanism triggered by, or dependent on, the magnetic field. After all, these results contradicted two important previous observations: (1) static magnetic field exposure of 30–120 min at 0.5–4.0 T had no significant influence on the growth of *E. coli* and *S. aureus*, nor were there any effects on their susceptibility to several antibiotics (Grosman et al., 1992); and (2) the observation that sessile *P. aeruginosa* displayed much higher susceptibility to gentamicin when subjected to static magnetic fields of 0.5–2 mT (Benson et al., 1994). The three abovementioned studies though use quite different intensities of static fields, which cover three orders of magnitude, and this might explain the different observations as the magnetoreceptors may differ not only among species but also within the very same cell for significantly different field dynamics (Pazur et al., 2007; Hunt et al., 2009). Long-term exposure (24 h) to AC magnetic field of 2 mT and 50 Hz did not influence the susceptibility of *E. coli* and *P. aeruginosa* to kanamycin, amikacin, ampicillin, cefazolin, ceftazidime, ceftriaxone, moxalactam, and levofloxacin, nor their growth rate (Segatore et al., 2012).

Similarly, although ES by DC at ~15 mA achieved by itself minimal effect on sessile cells of *P. aeruginosa*, its combination with tobramycin in moderate concentrations proved extremely effective, requiring a mere quadruplication of the dose to clear sessile cells compared to clearing their planktonic form, instead of the three orders of magnitude increase ostensibly required when no ES was applied (Costerton et al., 1994). Actually, the ES by itself achieved significant antibacterial effect in 24 h reducing the viable cells 500-fold, but in the next 24 h this effect had dropped to zero, meaning either a robust compensatory increased growth of the surviving cells, or a moderate compensatory growth followed by development of electroresistance. Similarly, tobramycin alone scored decently in absolute terms in 48 h, reducing effectively the viable population 100-fold, but the survival fraction and/or the recovery of the sessile cells canceled its effects. The demonstrated synergy of DC and antibiotics, with results far beyond an additive context, is defined as the BEE (Costerton et al., 1994) and extends to similar results shown for the use of other biocides instead of antibiotics (Blenkinsopp et al., 1992). Similar effects were observed for a range of bacteria and yeast fungi (Wellman et al., 1996; Khoury et al., 1992).

Later on, the BEE was shown to work with AC currents as well; high-frequency alternating currents (10 MHz) were reported not only to enhance the susceptibility of sessile *E. coli* to antibiotics (gentamicin and oxytetracycline), producing thus synergistic bactericidal effect, but also to decrease the number of bacteria in biofilms by 60%, even in the absence of an antibiotic (Caubet et al., 2004).

This observation clearly contrasts the previous ones, where stand-alone antimicrobial activity of the electric amenity had been minimal in residual terms (Costerton et al., 1994) and may insinuate that AC neutralize wholly or partially the compensatory growth and/or resistance mechanisms—at least in biofilm context. This work potentially disproves all the electrochemical theories explaining the BEE through either electrochemical reactions altering the pH, or producing toxic moieties (like free radicals–ROS) and ions at the medium and by the electrode, or through heat effects and electroporation.

The next step was to induce a BEE by AC field without a concomitant current. Fully insulated electrodes were used to create a field at 2–4 V/cm and 10 MHz, which enhanced the efficacy of chloramphenicol against planktonic cells of *S. aureus* and *P. aeruginosa*; the enhancement, nevertheless, was of additive, not synergistic, nature and value and dependent on amplitude and frequency (Giladi et al., 2008). *In vivo* though, the net effect would depend on the combined effects of the EM amenity to the immune system and to the infectious agent (Giladi et al., 2008); should mixed or combined regimens are used, the interplay between the amenities (EM and the selected drugs, in terms of altered pharmacodynamics and possibly pharmacokinetics) should be taken into consideration.

The combination of the antifungal drugs with Pulsing Electric Fields (PEF) in one more spin-off of the BEE may result in a more potent clinical method as an antifungal therapy to treat the skin infections, which are caused by drug-resistant pathogens such as—but not restricted to—*Candida lusitaniae*. Compared to single-pulse electroporation, the bipolar pulses on average result in a better inhibitory effect for the same energy delivered. The electric field strength and the total pulse energy remain the most important parameters (Novickij et al., 2015).

Possible development of resistance toward electroceuticals is a vital issue. In counterpain applications the desensitization is well established (Kambouris et al., 2014; Kloth, 2009; Meng et al., 2011), but it functions though multicellular and highly complex networks (Kambouris et al., 2014; Kloth, 2009; Meng et al., 2011). Bacteria did not apparently develop resistance to iontophoresis (Davis et al., 1991) nor to electric field exposure (Giladi et al., 2008) but genetic factors such as modified proteins, less sensitive to ionic changes are entirely possible to occur (Kambouris et al., 2014; Doevenspeck, 1961; Poltawski and Watson, 2009; Armstrong and Bezanilla, 1973; McCaig et al., 2005; Bezanilla, 2008).

The resistance mechanisms against antibiotics operate on the toxic moiety proper or on its access to the target entities; the latter is a heterogeneous, diverse defensive approach (Blair et al., 2015). Translation of such innate countermeasures to electroceuticals is not straightforward as they are an external factor not sharing coevolution with its targets, as has happened with antibiotics (Blair et al., 2015; Kambouris et al., 2014; Doevenspeck, 1961; Kloth, 2005; Asadi and Torkaman, 2014). Thus, evolved, responsive mechanisms of developing and dispersing resistance most probably are not in place yet. Resistance phenotypes will be developed; spontaneous mutations do produce defective or damaged organelles, macromolecules and moieties (Andersson and Hughes, 2010), which may respond less to electroceutical challenges and stimuli. Such mechanisms can be recognized at the thick Gram-positive cell wall, which alleviates HV electric current pulses (Kloth, 2005; Ranalli et al., 2002; Hülsheger et al., 1983), at the charged EPS matrix of the biofilms, with many types of charged particles and molecular chains incorporating polar subsystems, which blocks chloride species and other electroproduced biocidal moieties to protect sessile cells from iontophoretic killing (Costerton et al., 1994; Caubet et al., 2004) and at the increased growth, which compensates for high rates of electrodependent decease (Costerton et al., 1994; Ranalli et al., 2002).

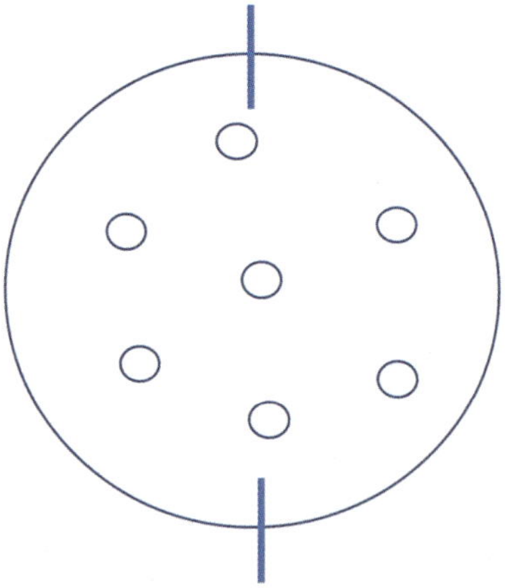

FIGURE 14.4

The concept of electroantibiogram (based on the disk-diffusion antibiotic susceptibility test) with conductive coupling. The pair of inserted electrodes is represented by the two line segments. Multiple diffusion disks of the same drug are used to establish possible positional differentiation within the Petri dish and allow statistic tests run on the inhibition zone diameter.

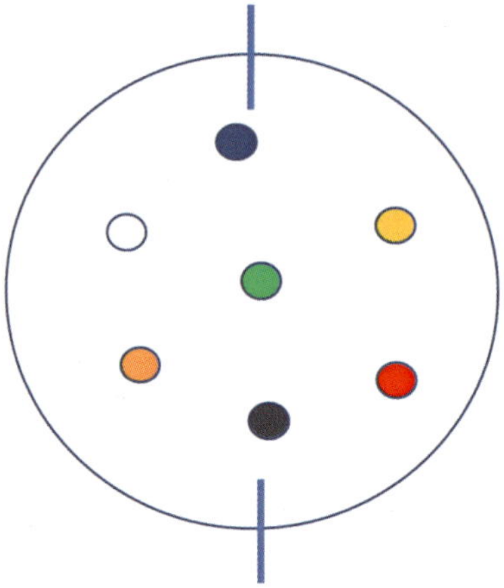

FIGURE 14.5

The concept of electroantibiogram (based on the disk-diffusion antibiotic susceptibility test) with conductive coupling. The pair of inserted electrodes is represented by the two line segments. Diffusion disks of an array of drugs are used, preferably with the same positional pattern as that of to an untreated antibiogram performed simultaneously (standard), to establish differences of sensitivity per drug under conductive electrostimulation of the microbiote.

The abovementioned mechanisms do not seem to be highly communicable, especially in a lateral manner, as happens with resistance factors against biochemical compounds or with genes encoding catabolic enzymes active against pharmaceutical moieties. Davis et al. (1991) specifically reported that during their iontophoretic experiments no resistance had been observed in tested bacteria.

Viral transduction, on the other hand, might transplant electroresistant genetic information from random survivors or from dedicated source organisms, as are the exoelectrogenic bacteria (Logan, 2009) to larger and diverse microbiomes. Consecutive transposition, transformation, and conjugation will occur, but most probably after a comfortable time lag (Logan, 2009; Shi et al., 2016; Blair et al., 2015; Stokes and Gillings, 2011; Brauner et al., 2016).

The synergy or antagonism of ES with antibiotics, biocides, and chemical sterilizers is a matter of conjecture in generalized applications. Different ES settings might differentiate or even reverse the effect of different chemical compounds; thus electroantibiograms, especially expanding on the disk-diffusion antibiotic susceptibility test (Figs. 14.4 and 14.5) as it is less amenable to positional bias and noise or incompatibilities due to the EM constituent, are prone to evolve to an indispensable assay, which shall provide definite answers in a most controversial context of overlapping and conflicting interactions.

Upregulation of cell functions might result in increased drug uptake and increased susceptibility. But it may also result in increased anabolism and thus restoration of cell damage and reinstating of functionality; or in acceleration of the catabolism of the drug or any other neutralizing mechanism, thus culminating in higher tolerance. On the other hand, cellular functions necessary for the uptake and activity of the drug, biocide, or sterilizer may be hampered by a suppressive ES, thus rendering the affected cell more tolerant to a certain drug (Brauner et al., 2016).

REFERENCES

Aarholt, E., Flinn, E.A., Smith, C.W., 1981. Effects of low-frequency magnetic fields on bacterial growth rate. Phys. Med. Biol. 26 (4), 613–621.

Abbas, Z., 2013. Effect of Magnetic Field on Growth and Biochemical Indices of Some Fungi (M.Sc. dissertation). Al-Nahrain University, Iraq.

Abedon, S.T., Kuhl, S.J., Blasdel, B.G., Kutter, E.M., 2011. Phage treatment of human infection. Bacteriophage 1 (2), 66–85.

Adhya, S., Merril, C.R., Biswas, B., 2014. Therapeutic and prophylactic applications of bacteriophage components in modern medicine. Cold Spring Harb. Perspect. Med. 4 (1), a012518.

Ahmed, H.E., Craig, W.F., White, P.F., Ghoname, E.S.A., Hamza, M.A., Gajraj, N.M., et al., 1998. Percutaneous electrical nerve stimulation: an alternative to antiviral drugs for acute herpes zoster. Anesth. Analg. 87 (4), 911–914.

Anderson, J.C., Clarke, E.J., Arkin, A.P., Voigt, C.A., 2006. Environmentally controlled invasion of cancer cells by engineered bacteria. J. Mol. Biol. 355 (4), 619–627.

Andersson, D.I., Hughes, D., 2010. Antibiotic resistance and its cost: is it possible to reverse resistance? Nat. Rev. Microbiol. 8 (4), 260–271.

Arabatzis, M., Dousi, A., Velegraki, A., 2008. Investigation and treatment of fungally deteriorated icons. In: Mikrobiokosmos 1st Annual Conference December 12–14, 2008, Athens.

Armstrong, C.M., Bezanilla, F., 1973. Currents related to movement of the gating particles of the sodium channels. Nature 242 (5398), 459–461.

Asadi, M.R., Torkaman, G., 2014. Bacterial inhibition by electrical stimulation. Adv. Wound Care 3 (2), 91–97.

Banik, S., Bandyopadhyay, S., Ganguly, S., Dan, D., 2005. Effect of microwave irradiated *Methanosarcina barkeri* DSM-804 on biomethanation. Bioresour. Technol. 97, 819–823.

Barbosa-Canovas, G.V., Gongora-Nieto, M.M., Pothakamury, U.R., Swanson, B.G., 1999. Preservation of Foods With Pulsed Electric Fields. Academic Press, Elsevier, London.

Barranco, S.D., Spadaro, J.A., Berger, T.J., Becker, R.O., 1974. In vitro effect of weak direct current on *Staphylococcus aureus*. Clin. Orthop. 100, 250–255.

Benson, D.E., Grissom, C.B., Burns, G.L., Mohammad, S.F., 1994. Magnetic field enhancement of antibiotic activity in biofilm forming *Pseudomonas aeruginosa*. J. Am. Soc. Artif. Internal Organs 40 (3), M371.

Berk, S.G., Srikanth, S., Mahajan, S.M., Ventrice, C.A., 1997. Static uniform magnetic fields and amoebae. Bioelectromagnetics 18 (1), 81–84.

Berovic, M., Potocnik, M., Strus, J., 2008. The influence of galvanic field on *Saccharomyces cerevisiae* in grape must fermentation. Vitis-Geilweilerhof 47 (2), 117–122.

Bezanilla, F., 2008. How membrane proteins sense voltage. Nat. Rev. Mol. Cell Biol. 9 (4), 323–332.

Blair, J.M., Webber, M.A., Baylay, A.J., Ogbolu, D.O., Piddock, L.J., 2015. Molecular mechanisms of antibiotic resistance. Nat. Rev. Microbiol. 13 (1), 42–51.

Blank, M., Goodman, R., 2008. Effect of static magnetic field on *E. coli* cells and individual rotations of ion–protein complexes. J. Cell Physiol. 214 (1), 20–26.

Blenkinsopp, S.A., Khoury, A.E., Costerton, J.W., 1992. Electrical enhancement of biocide efficacy against *Pseudomonas aeruginosa* biofilms. Appl. Environ. Microbiol. 58 (11), 3770–3773.

Borchert, D., Sheridan, L., Papatsoris, A., Faruquz, Z., Barua, J.M., Junaid, I., et al., 2008. Prevention and treatment of urinary tract infection with probiotics: review and research perspective. Indian J. Urol. 24 (2), 139–144.

Brauner, A., Fridman, O., Gefen, O., Balaban, N.Q., 2016. Distinguishing between resistance, tolerance and persistence to antibiotic treatment. Nat. Rev. Microbiol. 14 (5), 320–330.

Brkovic, S., Postic, S., Ilic, D., 2015. Influence of the magnetic field on microorganisms in the oral cavity. J. Appl. Oral Sci. 23 (2), 179–186.

Carley, P.J., Wainapel, S.F., 1985. Electrotherapy for acceleration of wound healing: low intensity direct current. Arch. Phys. Med. Rehabil. 66 (7), 443–446.

Caubet, R., Pedarros-Caubet, F., Chu, M., Freye, E., de Belem Rodrigues, M., Moreau, J.M., et al., 2004. A radio frequency electric current enhances antibiotic efficacy against bacterial biofilms. Antimicrob. Agents Chemother. 48 (12), 4662–4664.

Chiocca, E., Rabkin, S., 2014. Oncolytic viruses and their application to cancer immunotherapy. Cancer Immunol. Res. 2 (4), 295–300.

Costerton, J.W., Ellis, B., Lam, K., Johnson, F., Khoury, A.E., 1994. Mechanism of electrical enhancement of efficacy of antibiotics in killing biofilm bacteria. Antimicrob. Agents Chemother. 38 (12), 2803–2809.

Coustets, M, Ganeva, V., Galutzov, B., Teissie, J., 2015. Millisecond duration pulses for flow-through electro-induced protein extraction from *E. coli* and associated eradication. Bioelectrochemistry 103, 82–91.

Daeschlein, G., Assadian, O., Kloth, L.C., Meinl, C., Ney, F., Kramer, A., 2007. Antibacterial activity of positive and negative polarity low-voltage pulsed current (LVPC) on six typical Gram-positive and Gram-negative bacterial pathogens of chronic wounds. Wound Repair Regen. 15 (3), 399–403.

Davis, C.P., Weinberg, S., Anderson, M.D., Rao, G.M., Warren, M.M., 1989. Effects of microamperage, medium, and bacterial concentration on iontophoretic killing of bacteria in fluid. Antimicrob. Agents Chemother 33 (4), 442–447.

Davis, C.P., Wagle, N., Anderson, M.D., Warren, MM, 1991. Bacterial and fungal killing by iontophoresis with long-lived electrodes. Antimicrob. Agents Chemother. 35 (10), 2131–2134.

Davis, C.P., Wagle, N., Anderson, M.D., Warren, M.M., 1992. Iontophoresis generates an antimicrobial effect that remains after iontophoresis ceases. Antimicrob. Agents Chemother 36 (11), 2552–2555.

Deitch, E., Marino, A., Malakanok, V., Albright, J., 1983. Electrical augmentation of the antibacterial activity of silver nylon. In: Proceedings of the 3rd Annual BRAGS Oct 2-5; San Francisco.

Del Re, B., Bersani, F., Agostini, C., Mesirca, P., Giorgi, G., 2004. Various effects on transposition activity and survival of *Escherichia coli* cells due to different ELF-MF signals. Radiat. Environ. Biophys. 43 (4), 265–270.

Delsart, C., Grimi, N., Boussetta, N., Miot Sertier, C., Ghidossi, R., Vorobiev, E., et al., 2015. Impact of pulsed-electric field and high-voltage electrical discharges on red wine microbial stabilization and quality characteristics. J. Appl. Microbiol. 120 (1), 152–164.

Denegar, C.R., Saliba, E., Saliba, S., 2015. Microcurrent Electrical nerve stimulation for wound healing, Therapeutic Modalities for Musculoskeletal Injuries, fourth ed. Human Kinetics, Champaign.

Desnues, C., Boyer, M., Raoult, D., 2012. Sputnik, a virophage infecting the viral domain of life. Adv. Virus Res. 82, 63–89.

de Souza, P.M., de Oliveira Magalhães, P., 2010. Application of microbial α-amylase in industry - a review. Braz. J. Microbiol. 41 (4), 850–861.

Dihel, L.E., Smith-Sonneborn, J., Middaugh, C.R., 1985. Effects of extremely low frequency electromagnetic field on the cell division rate and plasma membrane of *Paramecium tetraurelia*. Biolelectromagnetics 6 (1), 61–71.

Din, M.O., Danino, T., Prindle, A., Skalak, M., Selimkhanov, J., Allen, K., et al., 2016. Synchronized cycles of bacterial lysis for in vivo delivery. Nature 536 (7614), 81.

Doevenspeck, H., 1961. Influencing cells and cell walls by electrostatic impulses. Fleischwirtschaft 13 (12), 968–987.

Engström, S., Markov, M.S., McLean, M.J., Holcomb, R.R., Markov, J.M., 2002. Effects of non-uniform static magnetic fields on the rate of myosin phosphorylation. Bioelectromagnetics 23 (6), 475–479.

Famm, K., Litt, B., Tracey, K.J., Boyden, E.S., Slaoui, M., 2013. Drug discovery: a jump-start for electroceuticals. Nature 496 (7444), 159.

Fojt, L., Strašák, L., Vetterl, V., Šmarda, J., 2004. Comparison of the low-frequency magnetic field effects on bacteria *Escherichia coli, Leclercia adecarboxylata* and *Staphylococcus aureus*. Bioelectrochemistry 63 (1–2), 337–341.

Fojt, L., Klapetek, P., Strasák, L., Vetterl, V., 2009. 50 Hz magnetic field effect on the morphology of bacteria. Micron 40 (8), 918–922.

Gao, W., Liu, Y., Zhou, J., Pan, H., 2005. Effects of a strong static magnetic field on bacterium *Shewanella oneidensis*: an assessment by using whole genome microarray. Bioelectromagnetics 26 (7), 558–563.

Geveke, D.J., Brunkhorst, C., 2003. Inactivation of *Saccharomyces cerevisiae* with radio frequency electric fields. J Food Prot. 66 (9), 1712–1715.

Giladi, M., Porat, Y., Blatt, A., Wasserman, Y., Kirson, E.D., Dekel, E., et al., 2008. Microbial growth inhibition by alternating electric fields. Antimicrob. Agents Chemother. 52 (10), 3517–3522.

Golkar, Z., Bagasra, O., Pace, D.G., 2014. Bacteriophage therapy: a potential solution for the antibiotic resistance crisis. J. Infect. Develop. Countries 8 (2), 129–136.

Grahl, T., Märkl, H., 1996. Killing of microorganisms by pulsed electric fields. Appl. Microbiol. Biotechnol. 45 (1-2), 148–157.

Griffiths, S., Maclean, M., Anderson, J.G., MacGregor, S.J., Grant, M.H., 2012. Inactivation of microorganisms within collagen gel biomatrices using pulsed electric field treatment. J. Mater. Sci. Mater. Med. 23 (2), 507–515.

Grimaldi, S., Pasquali, E., Barbatano, L., Lisi, A., Santoro, N., Serafino, A., et al., 1997. Exposure to a 50 Hz electromagnetic field induces activation of the Epstein-Barr virus genome in latently infected human lymphoid cells. J. Environ. Pathol. Toxicol. Oncol. 16 (2–3), 205–207.

Grissom, C.B., 1995. Magnetic field effects in biology: a survey of possible mechanisms with emphasis on radical-pair recombination. Chem. Rev. 95 (1), 3–24.

Grosman, Z., Kolár, M., Tesaríková, E., 1992. Effects of static magnetic field on some pathogenic microorganisms. Acta Univ. Palacki Olomuc Fac. Med. 134, 7–9.

Grundler, W., Keilmann, F., Fröhlich, H., 1977. Resonant growth rate response of yeast cells irradiated by weak microwaves. Phys. Lett. A 62 (6), 463–466.

Guffey, J., Asmussen, M., 1989. In vitro bactericidal effects of high voltage pulsed current versus direct current against *Staphylococcus aureus*. J. Clin. Electrophysiol. 1, 5–9.

Haberkorn, R., Michel-Beyerle, M.E., 1979. On the mechanism of magnetic field effects in bacterial photosynthesis. Biophys. J. 26 (3), 489–498.

Harishorne, E., 1841. On the causes and treatment of pseudarthrosis and especially that form of it sometimes called supernumerary joint. Am. J. Med. 1, 121–156.

Healthcare, C., 2008. Coverage position: electrical stimulation for wound healing. Available at: <https://my.cigna.com/teamsite/health/provider/medical/procedural/coverage_positions/medical/mm_0351_coveragepositioncriteria_electrical_stimulation_for_wound_healing.pdf> (accessed 29.03.14.).

Hergt, R., Hiergeist, R., Zeisberger, M., Schüler, D., Heyen, U., Hilger, I., et al., 2005. Magnetic properties of bacterial magnetosomes as potential diagnostic and therapeutic tools. J. Magn. Magn. Mater. 293 (1), 80–86.

Horiuchi, S., Ishizaki, Y., Okuno, K., Ano, T., Shoda, M., 2001. Drastic high magnetic field effect on suppression of *Escherichia coli* death. Bioelectrochemistry 53 (2), 149–153.

Hristov, J., Perez, V., 2011. Critical analysis of data concerning *Saccharomyces cerevisiae* free-cell proliferations and fermentations assisted by magnetic and electromagnetic fields. Int. Rev. Chem. Eng. 3 (1), 3–20.

Hülsheger, H., Potel, J., Niemann, E.G., 1983. Electric field effects on bacteria and yeast cells. Radiat. Environ. Biophys. 22 (2), 149–162.

Hunckler, J., de Mel, A., 2017. A current affair: electrotherapy in wound healing. J Multidiscip. Healthc 10, 179–194.

Hunt, R., Zavalin, A., Bhatnagar, A., Chinnasamy, S., Das, K., 2009. Electromagnetic biostimulation of living cultures for biotechnology, biofuel and bioenergy applications. Int. J. Mol. Sci. 10 (10), 4515–4558.

Jamil, Y., ul Haq, Z., Iqbal, M., Perveen, T., Amin, N., 2012. Enhancement in growth and yield of mushroom using magnetic field treatment. Int. Agrophys. 26 (4), 375–380.

Jaworska, M., Domanski, J., Tomasik, P., Znoj, K., 2016. Stimulation of pathogenicity and growth of entomopathogenic fungi with static magnetic field. J. Plant Dis. Prot. 123 (6), 295–300.

Kalinowski, D.P., Edsberg, L.E., Hewson, R.A., Johnson, R.H., Brogan, M.S., 2004. Low-voltage direct current as a fungicidal agent for treating onychomycosis. J. Am. Podiatr. Med. Assoc. 94 (6), 565–572.

Kambouris, M.E., Zagoriti, Z., Lagoumintzis, G., Poulas, K., 2014. From therapeutic electrotherapy to electroceuticals: formats, applications and prospects of electrostimulation. Ann. Res. Rev. Biol. 4 (20), 3054–3070.

Kambouris, M.E., Markogiannakis, A., Arabatzis, M., Manoussopoulos, Y., Kantzanou, M., Velegraki, A., 2017. Wireless Electrostimulation: a new approach in combating infection? Fut. Microbiol. 12 (3), 255–265.

Kambouris, M.E., Pavlidis, C., Skoufas, E., Arabatzis, M., Kantzanou, M., Velegraki, A., et al., 2018a. Culturomics: a new kid on the block of phenomics and pharmacomicrobiomics for personalized medicine. OMICS-JIB 22 (2), 108–118.

Kambouris, M.E., Gaitanis, G., Manoussopoulos, Y., Arabatzis, M., Kantzanou, M., Kostis, G.D., et al., 2018b. Humanome versus microbiome: games of dominance and pan-biosurveillance in the omics universe. OMICS-JIB 22 (8), 528–538.

Karba, R., Gubina, M., Vodovnik, L., 1991. Growth Inhibition in *Candida albicans* due to low intensity constant direct current. Electromagn. Biol. Med. 10 (1–2), 1–15.

Kermanshahi, R.K., Sailani, M.R., 2005. Effect of static electric field treatment on multiple antibiotic-resistant pathogenic strains of *Escherichia coli* and *Staphylococcus aureus*. J. Microbiol. Immunol. Infect. 38 (6), 394.

Khadre, M.A., Yousef, A.E., 2002. Susceptibility of human rotavirus to ozone, high pressure, and pulsed electric field. J. Food Prot. 65 (9), 1441–1446.

Khoury, A.E., Lam, K., Ellis, B., Costerton, J.W., 1992. Prevention and control of bacterial infections associated with medical devices. ASAIO J. 38 (3), M174–M178.

Kirschvink, J.L., Kobayashi-Kirschvink, A., Diaz-Ricci, J.C., Kirschvink, S.J., 1992. Magnetite in human tissues: a mechanism for the biological effects of weak ELF magnetic fields. Bioelectromagnetics 13 (S1), 101–113.

Kloth, L.C., 2005. Electrical stimulation for wound healing: a review of evidence from in vitro studies, animal experiments, and clinical trials. Int. J. Low Extrem. Wounds 4 (1), 23–44.

Kloth, L.C., 2009. Wound healing with conductive electrical stimulation- it's the dosage that counts. J. Wound Technol. 6, 28–35.

Kohno, M., Yamazaki, M., Kimura, I., Wada, M., 2000. Effect of static magnetic fields on bacteria: *Streptococcus mutans, Staphylococcus aureus*, and *Escherichia coli*. Pathophysiology 7 (2), 143–148.

Kolsek, M., 2012. TENS – an alternative to antiviral drugs for acute herpes zoster treatment and postherpetic neuralgia prevention. Swiss Med. Wkly. 141, w13229.

Krylov, V.N., 2014. Bacteriophages of *Pseudomonas aeruginosa*: long-term prospects for use in phage therapy. Adv. Virus Res. 88, 227–278.

Kubota, K., Yoshimura, N., Yokota, M., Fitzsimmons, R.J., Wikesjö, U.M., 1995. Overview of effects of electrical stimulation on osteogenesis and alveolar bone. J. Periodon. 66 (1), 2–6.

Kumagai, E., Tominaga, M., Nagaishi, S., Harada, S., 2007. Effect of electrical stimulation on human immunodeficiency virus type-1 infectivity. Appl. Microbiol. Biotechnol. 77 (4), 947–953.

Kumagai, E., Tominaga, M., Harada, S., 2011. Sensitivity to electrical stimulation of human immunodeficiency virus type 1 and MAGIC-5 cells. AMB Express 1 (1), 23.

Laatsch-Lybeck, L.J., Ong, P.C., Kloth, L.C., 1995. In vitro effects of two silver electrodes on select wound pathogens. J. Clin. Electrophysiol. 7, 10–15.

Lagoumintzis, G., Sideris, S.G., Kambouris, M.E., Poulas, K., Koutsojannis, C., Rennekampff, H.-O., 2014. Wireless Micro Current Stimulation technology improves firework burn healing: clinical applications of WMCS technology. In: 4th International Conference on Wireless Mobile Communication and Healthcare (Mobihealth), Athens 3-5 Nov, pp. 172–175.

Lente, F., 1850. Cases of united fractures treated by electricity. N.Y. State J. Med. 5, 5117–5118.

Logan, B.E., 2009. Exoelectrogenic bacteria that power microbial fuel cells. Nat. Rev. Microbiol. 7 (5), 375–381.

Luo, Q., Wag, H., Zhang, X., Qian, Y., 2005. Effect of direct electric current on cell surface properties of phenol-degrading bacteria. Appl. Environ. Microbiol. 71 (1), 423–427.

Maadi, H., Haghi, M., Delshad, R., Kangarloo, H., Mohammadnezhady, M.A., Hemmatyar, G.R., 2010. Effect of alternating and direct currents on *Pseudomonas aeruginosa* growth in vitro. Afr. J. Biotechnol. 9 (38), 6373–6379.

Mainelis, G., Górny, R.L., Reponen, T., Trunov, M., Grinshpun, S.A., Baron, P., et al., 2002. Effect of electrical charges and fields on injury and viability of airborne bacteria. Biotechnol. Bioeng. 79 (2), 229–241.

Makarevich, A.V., 1999. Effect of magnetic fields of magnetoplastics on the growth of microorganisms. Biofizika 44 (1), 70–74.

Marsellés-Fontanet, À.R., Puig, A., Olmos, P., Mínguez-Sanz, S., Martín-Belloso, O., 2009. Optimising the inactivation of grape juice spoilage organisms by pulse electric fields. Int. J. Food Microbiol. 130 (3), 159–165.

Mateescu, C., Buruntea, N., Stancu, N., 2011. Investigation of *Aspergillus niger* growth and activity in a static magnetic flux density field. Rom. Biotechnol. Lett. 16 (4), 6364–6368.

Mattar, J.R., Turk, M.F., Nonus, M., Lebovka, N.I., El Zakhem, H., Vorobiev, E., 2015. *S. cerevisiae* fermentation activity after moderate pulsed electric field pre-treatments. Bioelectrochemistry 103, 92–97.

McCaig, C.D., Rajnicek, A.M., Song, B., Zhao, M., 2005. Controlling cell behavior electrically: current views and future potential. Physiol. Rev. 85 (3), 943–978.

McGillivray, AM, Gow, NA, 1986. Applied electrical fields polarize the growth of mycelial fungi. J. Gen. Microbiol. 132 (9), 2515–2525.

Meng, S., Rouabhia, M., Zhang, Z., 2011. Electrical stimulation in tissue regeneration. In: Gargiulo, G. McEwan, A. (Ed.), Applied Biomedical Engineering, IntechOpen, pp. 38–63.

Mercola, J.M., Kirsch, D.L., 1995. The basis for micro current electrical therapy in conventional medical practice. J. Adv. Med. 8 (2), 107–120.

Merriman, H.L., Hegyi, C.A., Albright-Overton, C.R., Carlos, J., Putnam, R.W., Mulcare, J.A., 2004. A comparison of four electrical stimulation types on *Staphylococcus aureus* growth in vitro. J. Rehabil. Res. Develop. 41 (2), 139–146.

Miedaner, T., Geiger, H.H., 2015. Biology, genetics, and management of ergot (*Claviceps* spp.) in rye, sorghum, and pearl millet. Toxins (Basel) 7 (3), 659–678.

Mizuno, A., Inoue, T., Yamaguchi, S., Sakamoto, K.I., Saeki, T., Matsumoto, Y., et al., 1990. Inactivation of viruses using pulsed high electric field. In: Industry Applications Society Annual Meeting, Conference Record of the 1990 IEEE, pp. 713–719.

Moore, R.L., 1979. Biological effects of magnetic fields: studies with microorganisms. Can. J. Microbiol. 25 (10), 1145–1151.

Mousavian-Roshanzamir, S., Makhdoumi-Kakhki, A., 2017. The inhibitory effects of static magnetic field on *Escherichia coli* from two different sources at short exposure time. Rep. Biochem. Mol. Biol. 5 (2), 112.

Mosqueda-Melgar, J., Raybaudi-Massilia, R.M., Martin-Belloso, O., 2008. Combination of high-intensity pulsed electric fields with natural antimicrobials to inactivate pathogenic microorganisms and extend the shelf-life of melon and watermelon juices. Food Microbiol. 25 (3), 479–491.

Munita, J.,M., Arias, C.A., 2016. Mechanisms of antibiotic resistance. Microbiol. Spectr. 4 (2).

Muniz, J.B., Marcelino, M., Motta, M.D., Schuler, A., Motta, M.A.D., 2007. Influence of static magnetic fields on *S. cerevisiae* biomass growth. Braz. Arch. Biol. Technol. 50 (3), 515–520.

Nagy, P., 2005. The effect of low inductivity static magnetic field on some plant pathogen fungi. J. Centr. Eur. Agricult. 6 (2), 167–171.

Nakamura, K., Okuno, K., Ano, T., Shoda, M., 1997. Effect of high magnetic field on growth of *Bacillus subtilis* measured in a newly developed superconducting magnetic biosystem. Bioelectrochem. Bioenerg. 43 (1), 123–128.

Nakanishi, K., Tokuda, H., Soga, T., Yoshinaga, T., Takeda, M., 1998. Effect of electric current on growth and alcohol production by yeast cells. J. Ferment. Bioeng. 85 (2), 250–253.

Neelakantan, S., Mohanty, A.K., Kaushik, J.K., 1999. Production and use of microbial enzymes for dairy processing. Curr. Sci. 77 (1), 143–148.

Nguyen, T.H.P., Shamis, Y., Croft, R.J., Wood, A., McIntosh, R.L., Crawford, R.J., et al., 2015. 18 GHz electromagnetic field induces permeability of Gram–positive cocci. Sci. Rep. 5, 10980.

Nguyen, T.H., Pham, V.T., Nguyen, S.H., Baulin, V., Croft, R.J., Phillips, B., et al., 2016. The bioeffects resulting from prokaryotic cells and yeast being exposed to an 18 GHz electromagnetic field. PLoS One 11 (7), e0158135.

Novickij, V., Grainys, A., Lastauskienė, E., Kananavičiūtė, R., Pamedytytė, D., Zinkevičienė, A., et al., 2015. Growth inhibition and membrane permeabilization of *Candida lusitaniae* using varied pulse shape electroporation. BioMed Res. Intl. 2015, 457896.

Novickij, V., Lastauskienė, E., Švedienė, J., Grainys, A., Staigvila, G., Paškevičius, A., et al., 2018. Membrane permeabilization of pathogenic yeast in alternating sub-microsecond electromagnetic fields in combination with conventional electroporation. Membr. Biol. 251 (2), 189–195.

Olszanowski, A., Piechowiak, K., 2006. The use of an electric field to enhance bacterial movement and hydrocarbon biodegradation in soils. Polish J. Environ. Stud. 15 (2), 303–309.

Ong, P.C., Laatsch, L.J., Kloth, L.C., 1994. Antibacterial effects of a silver electrode carrying microamperage direct current in vitro. J. Clin. Electrophysiol. 6, 14–18.

Pakhomov, A.G., Akyel, Y., Pakhomova, O.N., Stuck, B.E., Murphy, M.R., 1998. Current state and implications of research on biological effects of millimeter waves: a review of the literature. Bioelectromagnetics 19, 393–413.

Pantry, S.N., Medveczky, P.G., 2009. Epigenetic regulation of Kaposi's sarcoma-associated herpes virus replication. Semin. Cancer Biol. 19 (3), 153–157.

Parratt, S., Laine, A.-L., 2016. The role of hyperparasitism in microbial pathogen ecology and evolution. ISME J. 10, 1815–1822.

Pazur, A., Schimek, C., Galland, P., 2007. Magnetoreception in microorganisms and fungi. CEJB 2 (4), 597–659.

Pethig, R., 1985. Dielectric and electrical properties of biological materials. J. Bioelectric 4 (2), vii–ix.

Petrofsky, J., Laymon, M., Chung, W., Collins, K., Yang, T.N., 2008. Effect of electrical stimulation on bacterial growth. J. Orthop. Neurolog. Surg. 31, 43–49.

Pica, F., Serafino, A., Divizia, M., Donia, D., Fraschetti, M., Sinibaldi-Salimei, P., et al., 2006. Effect of extremely low frequency electromagnetic fields (ELF-EMF) on Kaposi's sarcoma-associated herpes virus in BCBL-1 cells. Bioelectromagnetics 27 (3), 226–232.

Poltawski, L., Watson, T., 2009. Bioelectricity and microcurrent therapy for tissue healing – a narrative review. Phys. Ther. Rev. 14 (2), 104–114.

Predmore, A.N., 2015. Various Non-Thermal Technologies and Their Effectiveness Against Human Norovirus Surrogates (Doctoral dissertation). The Ohio State University.

Ramadhinara, A., Poulas, K., 2013. Use of wireless microcurrent stimulation for the treatment of diabetes-related wounds: 2 case reports. Adv. Skin Wound Care 26 (1), 1–4.

Ramon, C., Martin, J.T., Powell, M.R., 1987. Low-level, magnetic-field-induced growth modification on *Bacillus subtilis*. Biolelectromagnetics 8 (3), 275–282.

Ranalli, G., Iorizzo, M., Lustrato, G., Zanardini, E., Grazia, L., 2002. Effects of low electric treatment on yeast microflora. J. Appl. Microbiol. 93 (5), 877–883.

Repacholi, M.H., Greenebaum, B., 1999. Interaction of static and extremely low frequency electric and magnetic fields with living systems: health effects and research needs. Bioelectromagnetics 20, 133–160.

Robertson, W.S., 1925. Digby's receipts. Ann. Med. Hist. 7 (3), 216–219.

Rols, M.P., Dahhou, F., Mishra, K.P., Teissié, J., 1990. Control of electric field induced cell membrane permeabilization by membrane order. Biochemistry 29 (12), 2960–2966.

Rowley, B.A., 1972. Electrical current effects on *E. coli* growth rates. Proc. Soc. Exp. Biol. Med. 139 (3), 929–934.

Rowley, B.A., McKenna, J.M., Chase, G.R., Wolcott, L.E., 1974. The influence of electrical current on an infecting microorganism in wounds. Ann. N.Y. Acad. Sci. 238 (1), 543–551.

Segatore, B., Setacci, D., Bennato, F., Cardigno, R., Amicosante, G., Iorio, R., 2012. Evaluations of the effects of extremely low-frequency electromagnetic fields on growth and antibiotic susceptibility of *Escherichia coli* and *Pseudomonas aeruginosa*. Int. J. Microbiol. 2012, 587293.

Shamis, Y., Taube, A., Mitik-Dineva, N., Croft, R., Crawford, R.J., Ivanova, E.P., 2011. A study of the specific electromagnetic effects of microwave radiation on *Escherichia coli*. Appl. Environ. Microbiol. 77 (9), 3017–3023.

Shi, L., Dong, H., Reguera, G., Beyenal, H., Lu, A., Liu, J., 2016. Extracellular electron transfer mechanisms between microorganisms and minerals. Nat. Rev. Microbiol. 14 (10), 651–662.

Simonis, P., Kersulis, S., Stankevich, V., Kaseta, V., Lastauskiene, E., Stirke, A., 2017. Caspase dependent apoptosis induced in yeast cells by nanosecond pulsed electric fields. Bioelectrochemistry 115, 19–25.

Singh, P.K., Kumar, S., Kumar, P., Bhat, Z.F., 2012. Pulsed light and pulsed electric field-emerging non-thermal decontamination of meat. Am. J. Food Technol. 7 (9), 506–516.

Soghomonyan, D., Trchounian, K., Trchounian, A., 2016. Millimeter waves or extremely high frequency electromagnetic fields in the environment: what are their effects on bacteria? Appl. Microbiol. Biotechnol. 100 (11), 4761–4771.

Somolinos, M., García, D., Condón, S., Mañas, P., Pagán, R., 2008. Biosynthetic requirements for the repair of sublethally injured *Saccharomyces cerevisiae* cells after pulsed electric fields. J. Appl. Microbiol. 105 (1), 166–174.

Spadaro, J.A., Chase, S.E., Webster, D.A., 1986. Bacterial inhibition by electrical activation of percutaneous silver implants. J. Biomed. Mater. Res. 20 (5), 565–577.

Spilker, W., Gottstein, A., 1891. Ueber die Vernichtung von Mikroorganismen durch die Induktions elektricitat. Centralblf Bakt. Parasitenk. 9, 77–88.

Staczek, J., Marino, A.A., Gilleland, L.B., Pizarro, A., Gilleland Jr, H.E., 1998. Low-frequency electromagnetic fields alter the replication cycle of MS2 bacteriophage. Curr. Microbiol. 36 (5), 298–301.

Stansell, M.J., Winters, W.D., Doe, R.H., Dart, B.K., 2001. Increased antibiotic resistance of *E. coli* exposed to static magnetic fields. Bioelectromagnetics 22 (2), 129–137.

Stokes, H.W., Gillings, M.R., 2011. Gene flow, mobile genetic elements and the recruitment of antibiotic resistance genes into Gram-negative pathogens. FEMS Microbiol. Rev. 35 (5), 790–819.

Stone, G.E., 1909. Influence of electricity on micro-organisms. Bot. Gaz. 48 (5), 359–379.

Stoodley, P., Lappin-Scott, H.M., 1997. Influence of electrical fields and pH on biofilm structure as related to the bioelectric effect. Antimicrob. Agents Chemother. 41 (9), 1876–1879.

Strašák, L., Vetterl, V., Šmarda, J., 2002. Effects of low-frequency magnetic fields on bacteria *Escherichia coli*. Bioelectrochemistry 55 (1–2), 161–164.

Strašák, L., Vetterl, V., Fojt, L., 2005. Effects of 50 Hz magnetic fields on the viability of different bacterial strains. Electromagn. Biol. Med. 24 (3), 293–300.

Sultana, S.T., Atci, E., Babauta, J.T., Falghoush, A.M., Snekvik, K.R., Call, D.R., et al., 2015. Electrochemical scaffold generates localized, low concentration of hydrogen peroxide that inhibits bacterial pathogens and biofilms. Sci. Rep. 5, 14908.

Szuminsky, N.J., Albers, A.C., Unger, P., Eddy, J.G., 1994. Effect of narrow, pulsed high voltages on bacterial viability. Phys. Ther. 74 (7), 660–667.

Tanino, T., Yoshida, T., Sakai, K., Ohshima, T., 2013. Inactivation of Escherichia coli phage by pulsed electric field treatment and analysis of inactivation mechanism. J. Phys. Conf. Ser. 418 (1), 012108.

Taylor, B.P., Cortez, M.H., Weitz, J.S., 2014. The virus of my virus is my friend: ecological effects of virophage with alternative modes of coinfection. J. Theor. Biol. 354, 124–136.

Thornton, W.M., 1912. The electrical conductivity of bacteria, and the rate of sterilisation of bacteria by electric currents. Proc. Roy. Soc. Lond. (B) 85 (580), 331–344.

Torgomyan, H., Trchounian, A., 2013. Bactericidal effects of low-intensity extremely high frequency electromagnetic field: an overview with phenomenon, mechanisms, targets and consequences. Crit. Rev. Microbiol. 39 (1), 102–111.

Tracy, R.L., 1932. Lethal effect of alternating current on yeast cells. J. Bacteriol. 24 (6), 423–438.

Unknown, 1987. Investigation on the Possible Effect of Electrical Stimulation on pH and Survival of Foot-and-Mouth Disease Virus in Meat and Offals from Experimentally Infected Animals. Report EUR 10048 EN.

Valle, A., Zanardini, E., Abbruscato, P., Argenzio, P., Lustrato, G., Ranalli, G., et al., 2007. Effects of low electric current (LEC) treatment on pure bacterial cultures. J. Appl. Microbiol. 103 (5), 1376–1385.

Velegraki, A., Arabatzis M., Goudis C., Hantzios P., Arsenis G., et al., 2007. Investigation and treatment of fungal contaminations of public buildings. The case of "Aristarchos" telescope of the National Observatory of Athens 14-11-2007.

Vidali, M., 2001. Bioremediation. An overview. Pure Appl. Chem. 73 (7), 1163–1172.
Vodovnik, L., Karba, R., 1992. Treatment of chronic wounds by means of electric and electromagnetic fields part 1 literature review. Med. Biol. Eng. Comput. 30 (3), 257–266.
Wellman, N., Fortun, S.M., McLeod, B.R., 1996. Bacterial biofilms and the bioelectric effect. Antimicrob. Agents Chemother. 40 (9), 2012–2014.
Werner, H.J., Schulten, K., Weller, A., 1978. Electron transfer and spin exchange contributing to the magnetic field dependence of the primary photochemical reaction of bacterial photosynthesis. Biochim. Biophys. Acta 502 (2), 255–268.
Wolcott, L.E., Wheeler, P.C., Hardwicke, H.M., Rowley, B.A., 1969. Accelerated healing of skin ulcers by electrotherapy: preliminary clinical results. South Med. J. 62 (7), 795–801.
Zhao, M., Song, B., Pu, J., Wada, T., Reid, B., Tai, G., et al., 2006. Electrical signals control wound healing through phosphatidylinositol-3-OH kinase-gamma and PTEN. Nature 442 (7101), 457–460.

CHAPTER

MICROBIOMICS: A FOCAL POINT IN GCBR AND BIOSECURITY 15

Manousos E. Kambouris[1], Konstantinos Grivas[2], Basilis Papathanasiou[3], Dimitris Glistras[4] and Maria Kantzanou[5]

[1]*Golden Helix Foundation, London, United Kingdom* [2]*Hellenic Military Academy, Kitsi, Greece* [3]*Department of Turkish and Contemporary Asian Studies, National and Kapodistrian University of Athens, Athens, Greece* [4]*Department of History and Archaeology, National and Kapodistrian University of Athens, Athens, Greece* [5]*Department of Hygiene, Epidemiology & Medical Statistics, National Retrovirus Reference Center, School of Medicine, National and Kapodistrian University of Athens, Athens, Greece*

INTRODUCTION

Degrading human immunity and diminishing availability of healthcare, combined with novel pathogens emerging by 3T (trade, transport and travel), by biotechnological advances and due to the rupture of the host-species barrier, challenge current global infectious diseases' management.

The international status of public health seems bleak, despite ongoing successes in the scientific, the conceptual, and the therapeutic fields (WHO, 2015a; Kambouris et al., 2017). Limited accessibility and extremely low affordability in global terms cause an actual unavailability of health services and products for large parts of the global population: even within the first world, affordability has become a limiting factor for healthcare, including the purchase of medicaments.

This actual—or "functional"—unavailability is further exacerbated in the Third World by issues of accessibility, be that social or—more often—geographic. In both cases, though, the healthcare expenditure remains huge—a clear sign of inflated pricing and questionable marketing policies, as expensive tests and substances replace existing, massively available inexpensive ones with good diagnostic/therapeutic records (WHO, 2015b).

The ageing first world population witnesses marked deterioration of health standards: a drop in immunocompetence is attributable to unhealthy living environment and everyday habits. Though, the increased life expectancy achieved by better standards of life and a blooming medical science and healthcare sector played a substantial part.

Moreover, an unprecedented emergence of pathogens is recorded, attributable to (1) "(sub)speciation," meaning newly discovered or mutated/evolved pathogens with new qualities and niches; (2) "local emergence," meaning pathogens appearing in different localities/environments than their original habitats, mostly due to travel and migration infections; and (3) "rehosting," meaning existing pathogens finding new species or anatomical sites/biocompartments to colonize, especially among the immunodeficient host (Woolhouse, 2002).

Microbiomics. DOI: https://doi.org/10.1016/B978-0-12-816664-2.00015-3

Bioterrorism or industrial-scale accidents in biotechnology units further bleaken the prospects, especially as a predisposing factor for multiple, simultaneous, and even interrelated outbreaks. New microbiotes might reshape massively existing microbiomes of different sizes, habitats, and functions; might participate in genetic code exchanges by means of the local exogenomes, the latter defined as "the DNA available for lateral/horizontal gene transfer (LGT/HGT)" especially through transformation and transduction.

EMERGENCE OF NEW, AGGRESSIVE, AND BETTER ADAPTED PATHOGENS

The fast development of resistance, and the ability to disseminate it within a single generation through mobile DNA elements (Stokes and Gillings, 2011) implementing HGT, is a factor that definitively stalls and perhaps actually forestalls antimicrobial efforts, at least by stretching antibiotic resources toward faster obsolescence. On top of this fact, a wider basis of pathogens has come into play, collectively named "emerging pathogens" (Morens and Fauci, 2013). Basically, they originate from three distinct classes of events: (1) breach of host-species barrier (Fan and Moon, 2017), which is breach of immunological and microecological limitations; (2) high-volume transportation-travel, where the latter includes refugee flows and any kind of migration (Doganay and Demiraslan, 2016). These events enact breach of spatial limitations (Connell, 2017; Schoch-Spana et al., 2017; Wolicki et al., 2016) (3) Human-instigated fast-tracking of evolution (Jackson et al., 2001; Koblentz, 2017; Millett and Snyder-Beattie, 2017; Schoch-Spana et al., 2017) which constitutes breach of evolutionary limitations. To these three classes of events, existing, benign microbiotes becoming opportunistic pathogens due to defects of immunity and/or induction of dysbiosis may be added as a fourth one, instead of being incorporated into class (1).

The human-instigated thread of evolution can be further divided to products of biowarfare/bioterrorism/biocrime projects, namely, "engineered agents" and to by-products of the biotech industry, considered and classified mostly as "accidental events" (Frinking et al., 2016). The concept of "by-products" includes both novel, engineered strains accidentally released and "genetic garbage" from the genetic manipulation/engineering industry that flourishes nowadays with applications pertaining all life forms. Such genetic material may function as exogenome and find its way to microbiote hosts, thus fueling unpredictable bursts of microevolution, especially—but not exclusively—through bacterial transformation (Anonymous, 2004a,b; Schoch-Spana et al., 2017).

The spawn of the (1) and (2) classes of events constitute the "neo-pathogens" that perplex diagnosis and treatment (Frinking et al., 2016). The "metapathogens" (3), defined as "microbiotes engineered to increase their virulence", push such concerns to new highs. But the net impact of both meta- and neo-pathogens will be much more ominous: the core issue is the increase of the microbiotic base susceptible to, and available for, evolution into aggressive, virulent, and resistant forms, due to their participation in horizontal/lateral DNA exchange implemented by plasmids, transposons, viruses, and by any transferable/transposable element in general (Stokes and Gillings, 2011).

The selective pressure caused by abuse of antibiotics on such a wide and diverse substrate will probably result in explosive microbiote evolution due to enhanced natural selection process, which will eventually lead to an exponential emergence of superpathogens, an event sometimes referred to as "microbial apocalypse" (Wiles, 2015). The superpathogens now include multidrug resistant

(MDR), extensively drug resistant (EDR), or pandrug-resistant (PDR) agents, colloquially referred to as "Superbugs." But they may be worth of a wider definition than the one used today. So, instead of "pathogens displaying resistance to multiple classes of antibiotics" (Magiorakos et al., 2012), the definition may become "pathogens incorporating, by artificial modification or by natural recombination, virulence factors which neutralize immune response, perplex medical treatment or both."

The wider definition intends to expand to other microbiotes and to other forms of survivability against additional antimicrobial factors, beyond drugs. The new superviruses engineered to switch off immune responses (Jackson et al., 2001; Magiorakos et al., 2012) are most prominent examples. A superpathogen may result naturally or by engineering, and thus, in the second case, be classified as a metapathogen. But neither all metapathogens are superpathogens nor *vice versa*: the two categories overlap partially.

The sheer number and variability of superpathogens, irrespective of their origin/creation, might well outpace the capability of the human *technida* (defined as "the entirety of artificial objects and the procedures allowed or implemented by their use") to respond and adapt to novel pathogens (Laxminarayan et al., 2013), thus nullifying current diagnostic tests and therapeutic regimens. The more so, as the technida is encumbered less by inherent limitations and more by comparative profit considerations (Carlson and Wallstrom, 2017; Laxminarayan et al., 2013; Gould and Bal, 2013).

The different aspects of superpathogen/neo-pathogen emergence might display considerable, and perhaps hitherto unappreciated, impact onto health, economy, and mass psychology (Frinking et al., 2016). All the aforementioned objects of such impact offer prospect for illicit and malicious applications, namely, bioterrorism—ecoterrorism—agroterrorism, biocrime, and biological warfare (Koblentz, 2017; Madad, 2014).

Similarly, accidental events in biotechnological facilities, indifferent whether perpetrated or spontaneous, may release an array of unknown, experimental, or patented microbiotes, possibly pathogenic and definitely unidentifiable. Their evolution through the abovementioned mechanisms might produce microbiotes modified unpredictably in terms of content and extend of modification (Barbeschi, 2017). In such a context, a malicious intent may result in easily achievable biowarfare/bioterrorism events, accompanied by ecoterrorism and agroterrorism offshoots (Beale et al., 2002; Khalil and Shinwari, 2014; Madad, 2014) with or without further engineering efforts to achieve any kind of "optimization" (Madad, 2014).

The breaching of host-species barrier in infectious agents can be implemented through two very different strategies or "courses of events": one is maximalistic and the other minimalistic. The latter is seen routinely in viruses. The paradigm there is the HIV infection moving from primates to human (Brown, 1997; Morens and Fauci, 2013) and also the animal viruses causing hemorrhagic fevers to humans, as is the Venezuelan equestrian virus (Taylor et al., 2017). The event occurs through probabilistic steps, the two most important being (1) similarity in cellular receptor moieties combined with susceptibility of the respective cells to the particular viral infectious mechanism and (2) the transportation from the original to the novel host environment.

The adaptation is passive and rather wanton, and the agent does not have to evolve. The fewer the elements of the agent, the higher the probability to find novel, compatible hosts. This kind of novel hosting can materialize in hosts of extremely different phylogeny (Hart and Beeching, 2002) as the limiting/restraining factors are chance and possible incompatibilities; the latter are exemplified by the inability of human and animal viruses to insert their genome to a bacterium, due to the cell wall of the latter.

The maximalistic course of events for breaching the host-species barrier, on the other hand, is much more active and deterministic. It is routinely implemented by more evolved and complex pathogens, the fungi being a very good paradigm (Gräser et al., 1999). An evolved organism can implement actual evolutionary or adaptation measures, based on availability and selectivity, in order to secure additional hosting options. Evolving preferentially some of its mechanisms, such as digestive enzymes and dispersion strategies, instead of others, such as selectivity (Cullen et al., 2005; Meneau and Sanglard, 2005), is a way to expand its hosting base. For such strategies, spacious, complex genomes and feedback evolutionary mechanisms are needed, and thus eukaryotes are better poised for such events, although this is not an omnipotent rule.

Parasites of proto- and metazoan origin, especially obligate ones, offer some good counter-paradigm, as they evolved with their particular hosts, achieving remarkable single-host adaptation at the cost of adaptability to even the most permissive host-deprived environments (Cullen et al., 2005). Sacrificing the most basic autonomous survival functions tends to render them dependent on minimal rehosting events, as their evolutionary potential is dedicated almost exclusively to keeping pace with their original host; so much so that they cannot adapt to conditions (Costello et al., 2009). The example here is the vertical transmission (Howard et al., 2014), which limits rehosting to other members of the same host species- and relative members, to be precise.

The mechanism through which a pathogen enters a new host-species is no different from standard infection mechanisms. The carriage through carriers and vectors, inhalation, and the access through the food chain, mainly in fresh/unprocessed products (Hart and Beeching, 2002; Madad, 2014; Primmerman, 2000), are by far the most opportune infection routes for breaking the host-species barrier; the more so for microbiotes becoming virulent in an opportunistic manner, the most obvious such opportunity being the degradation of the immune system of the new hosts.

INTO THE FUTURE: PROJECTING A RESPONSIVE STRATEGY AND DEFINING OPERATIONAL PROCEDURES

Thus to gain time so as to stabilize, not expand, the humanome bubble within the biosphere (Vitousek et al., 1997), better use of resources and rational planning (Morens and Fauci, 2013) should be preferred to a—mostly obsessive—pursue of profitability based on optimization and competition (Fuchs, 2014; Woolf and Aron, 2013). Breach of barriers due to de-evolution of the human species at large can be tackled with joint, multisectorial health upkeep operations and joint drug development, which will remedy errors and failures in infection-control strategies.

The breach of host-species barrier is exemplified by paradigmatic plant pathogens, as are the fungal genera *Fusarium* and *Aspergillus*, which have eventually turned to extremely lethal human pathogens (Balajee et al., 2009; Guarro and Gené, 1995; Meneau and Sanglard, 2005) and *Claviceps purpurea*, which used to empoison entire communities (Miedaner and Geiger, 2015). Thus plant pathogens present both septic and toxic effect to humans and animals; both modes of action employ ingestion and inhalation as points of entry, although to different degrees; ingestion of mycotoxins is very morbid, while fungal cells and spores are more threatening when inhaled (Anderson et al., 1996; Švábenská, 2012; Velegraki-Abel, 1986).

The issue is not lost to animal pathogens wreaking havoc to humans, with the paradigm being the HIV and SARS infections (Morens and Fauci, 2013). Still, most important is the case with zoonoses (Fan and Moon, 2017); a considerable number of recognized bioterrorism agents are basically animal, not human, pathogens: *Bacillus anthracis*, *Burkholderia pseudomallei*, equine encephalitis virus are only some of the really long list (Ryan, 2008). Thus, wholescale iterations of current OMICS are expected under the prefix "agro-," with *agrigenomics* (or *agrogenomics*) already emerging to prominence (Misra and Panda, 2013).

The latter sector engages in both product and produce improvement, and in agrosecurity/agrosafety. Prompt, selective detection and differential identification of microbial phytopathogens in economically important crops by decentralized plant protection facilities may be crucial in cases of nonspecific signs and symptoms for timely intervention by stakeholders, thus suppressing disruption, famine, or massive loss of national income. In this way, safer and more sustainable farming is achieved, as the end product is less burdened by unnecessary use of biocides, free of opportunistically anthropopathogenic microbiotes, and unspoiled and thus of full market value.

The key for the above is adaptable, low cost yet reliable, and rapid diagnostics, specifications which practically spell molecular biology methods, preferably of increased sensitivity, but still basic and cost-effective enough to be implemented in decentralized agrosurveillance schemes integrated in multidomain biosecurity/biosafety pan-surveillance contexts (Kambouris et al., 2018c).

To this end, a multipronged approach is envisaged, including restructuring primary surveillance locations so as to merge agriculture, veterinary, and medical aspects of biosecurity and surveillance (Fig. 15.1). These primary outposts will be equipped with massively and competitively purchased

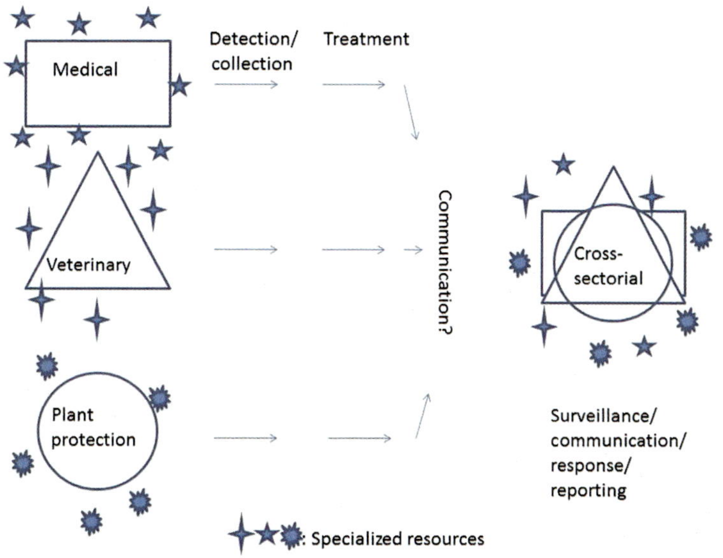

FIGURE 15.1

A functional merger of the three biosurveillance sectors to a joint format increases both coverage (in terms of density and range of the respective surveillance assets positioning) and data volume influx to a common processing pipeline.

basic equipment and manned with mixed scientists–technicians teams, to perform routine surveillance. For exceptional cases they will request assistance from network-connected centralized secondary structures operating state-of-the-art equipment and manned by experts and able to assist by dispatching field teams and first responders.

SURVEILLANCE–VIGILANCE

Conceptual and organizational issues

Cross-sectorial approaches in health-monitoring and promotion facilities for all three health risk sectors (medicinal, veterinary, phytopathological) allow better communication, cooperation, and coordination (Madad, 2014; Nuzzo and Shearer, 2017; Schoch-Spana et al., 2017). They can be better achieved through conceptual approaches focusing on the agent rather than on the host; consequently, infectomics and microbiomics will be at the spotlight—the latter including, among others, mycetomics and viromics. Consolidating three core health institutions, with respective boards, heads of departments, accounting offices, and policymakers to one cross-sectorial (Fig. 15.1), will release funds for "distributed pan-biosurveillance." The latter refers to a network of surveillance outposts throughout the realm (Doggett et al., 2016; Balajee et al., 2016) which detects and tracks biothreats irrespective of their possible, probable, or intended targets, hence the "pan-" prefix.

This operating scheme proposes the establishment of dispersed, primary biosurveillance installations, either unmanned platforms, with fixed surveyance equipment for both sensing and sampling (Kauchak, 2006; Ludovici et al., 2015), or manned outposts with standard crews of scientists and technicians. The stationed scientists will be receiving and processing both the actual samples and the sensor data influx and will be equipped mostly with rather inexpensive portable hardware, preferably point-of-need and/or energy independent (Mayboroda et al., 2016; Petrovick et al., 2007). The outposts will be linked to higher echelon hubs (Beale et al., 2002) staffed with specialists, endowed with the state-of-the-art hardware and poised to quick reaction. These hubs will be able to detail additional surveillance resources or to authorize targeted sampling/reconnaissance missions (Balajee et al., 2016).

Depending on the results of the detailed investigation, the end decision might be deferred from the hubs to higher echelons. In routine, or clear-cut or urgent situations, response and containment teams and assets may be dispatched by the hub proper, skipping higher authorization.

The manned outposts will focus on both syndromic surveillance and collection of samples (Madad, 2014) while engaging in limited *in situ* processing of samples. This approach is intended not only to avert saturation of the hubs during crisis (Madad, 2014; Ramage et al., 2016) but also to curtail the cost, delay, and risk inherent in the transportation of biohazardous samples (Balajee et al., 2016).

The network of manned outposts and fixed unmanned platforms of sensors may be enhanced or partially supplanted by autonomous, loitering platforms carrying stand-off (active from a distance) or stand-in (active within physical contact) sensing and sampling equipment, for assaying air but also ground and water masses (Table 15.1). Airborne, unmanned assets (unmanned aerial vehicles—UAV) are faster and more flexible in operation, as they may reach areas inaccessible or denied to dispatched teams (Halasz et al., 2002; Kambouris, 2018), especially if endowed with vertical motion so as to hover and sample, image, or deliver (Fig. 15.2). Still, manned airborne

Table 15.1 Main Detection Approaches and Their Compatibility With Different Platform Types.

Installation/Platform	Nucleic Acid	Structural (Ag-Ab, Aptomer)	UV-LIF	Biosampler
Fixed, unattended	?	?	+	+
Mobile	+	+	+	+
Mobile, unmanned	?	?	+	+
Airborne	?	?	+	+
Airborne, unmanned	?	?	+	+
Portable	+	+	−	+

+, *Developed;* ?, *possible;* −, *incompatible under current conditions and technology;* UV-LIF, *ultraviolet-laser induced fluorescence.*

FIGURE 15.2

The Naucicrate multicopter is an excellent example for increased accessibility granted by novel platforms for sampling, intervention and scanning.

Photo: UcanDrone PC, with permission.

platforms and waterborne or land platforms are not excluded and the acquisition choices will depend on operational environment, priorities, specification and cost (Grivas et al., 2008; Ludovici et al., 2015).

The use of UAVs is most common to agrosector and environmental applications, as it is possible to acquire by imaging and/or multispectral analysis the early sign macrodata not only on flora health but also on macrofauna, covering vast areas. Divergent behavior in the case of packs and flocks might raise an alarm for diseases vectored by such breeds so as to isolate and quarantine. Similarly, spectral analysis of the actual color of the greenery may point to early stages of a plant disease.

Medicine and public health sectors are belated in the use of UAVs. Thus multisectorial health-monitoring operations would allow exploitation of the plug-and-play provisions of modern, civilian-market UAVs used in agrosector so that they could shift mission from prairie, forest, and plantation—or animal stock—surveillance to air sample collection and incoming aerosol cloud classification in biosecurity and global catastrophic biorisks (GCBR) contexts, preferably but not exclusively in urban and rural environments.

The use of samplers (Kambouris and Velegraki, 2001) in such a context is rather straightforward. Especially in an airborne format, it allows extended coverage in three dimensions and in expedited timeframes, furnishing live, actual samples for definite identification by analytical instrumentation (identifiers) in a cost-effective manner. Moreover, it addresses safety concerns for the deployed personnel of dispatched investigation teams.

On the other hand, particle-sizers and imaging sensors such as ultraviolet (UV)-laser induced fluorescence (UV-LIF) laser exciters/imagers (Huffman et al., 2010; Joshi et al., 2013; Primmerman, 2000) can be integrated on moving or fixed platforms to provide early warning for aerosol clouds. This long-range sensing lacks the accurate readings needed for identification but is more than enough to initiate immediate protective measures (detect-to-protect) or to inform a chain of command for decision-making (detect-to-warn) (Kambouris et al., 2018a; Ludovici et al., 2015; Petrovick et al., 2007; Primmerman, 2000). Detectors lack the discrimination power to identify agents: in order to reject clutter and background noise, they perform a rudimentary classification of the detected signature (Kambouris et al., 2018a; Ludovici et al., 2015; Petrovick et al., 2007; Primmerman, 2000), usually by signal processing (Joshi et al., 2013; Primmerman, 2000), but the performance of such processes in these instruments is inadequate for identification, which demands more evolved and sophisticated capabilities—the actual definition of identifiers.

Still, analytical assays for agent identification, using spectroscopy (Ludovici et al., 2015; Madad, 2014), immunoassays, and other structure-inquiry approaches (Bartholomew et al., 2017; Madad, 2014; Pak, 2008; Petrovick et al., 2007; Sapsford et al., 2008), or nucleic acid analysis approaches (Christensen et al., 2006; Madad, 2014; Dedkov et al., 2018), are up to now reserved for larger, manned platforms at the very best. The state-of-the-art of such instrumentation may be issued to the personnel of the intervention teams dispatched from the hubs for immediate, *in situ* analyses at the deployment sites (Deshpande et al., 2016; Doggett et al., 2016; Mayboroda et al., 2016; Sapsford et al., 2008; Thavaselvam and Vijayaraghavan, 2010).

The state-of-the-art of the identifiers, irrespective of their operating principle, are portable and/or capable of resolving multiple agents in one sample, so as to discriminate them from benign flora (Ema et al., 2015; Mayboroda et al., 2016; Pak, 2008). But they remain selective, meaning they have to search for known signatures (Bartholomew et al., 2017; Euler et al., 2013; Ozanich et al., 2017; Pak, 2008). It is the quest of our time to develop reliable agnostic identifiers, compact or distributed, for warning on entirely novel, engineered threats, and the quest next of it, to achieve the fusion in multiple agnostic agents resolution (Kambouris et al., 2018a). An actual example of suitable, high-tech hardware is the 89 g MinION portable sequencing devise by Oxford Nanopore Technologies (Oxford Nanopore Technologies, 2018).

Technological solutions for detection and identification and their restrictions

In the field of surveillance technologies, structure-interrogating sensors/detectors will expand their ubiquity by improving performance in point-of-need, hardware-independent assays in expendable formats (such as strip tests). But the focus will be onto integrating detectors, samplers, and agnostic identifiers, as the last can detect novel agents and inherently offer co-infection resolution.

This new breed of sensors does not have a preconceived panel of signatures with which to compare the acquired signal. Instead, they register a sample signature in order to have it processed by much more elaborate equipment which includes agent-calling infrastructure. The latter may be remotely located and connected, or rather linked, to respective databases, so as to score on a next-of-kin basis or on any other decision-making concepts. Metagenomics formats in nucleic acid analysis approaches are the most probable paradigm to follow upon (Jakupciak and Colwell, 2009; Kambouris et al., 2018a).

Stand-off spectroscopy, defined as hardware which does not intake/inhale samples, to the tune of UV-LIF (Joshi et al., 2013), may implement a quantum leap, allowing some degree of identification potential, conditional on a signal-processing revolution. This could be combined with Big Data applications, similar to metagenomics (Thomas et al., 2012). The concept is tentatively referred to as spectroinformatics—or metaspectroscopy—and is likely to become the state-of-the-art in detector sensing. A similar course of events, called NCTR (noncooperative target recognition), has endowed military radars with the ability to identify fighter aircraft from the differences in the radiowave returns caused by the different shapes and features that creaet the uniqueness of the aerodynamic configuration of every type (Lopez-Rodriguez et al., 2014).

In terms of identifier sensors, things seem more or less straightforward. Despite the advent of high-tech methods, such as new-generation identification spectroscopy, exemplified by MALDI-TOF (Wieser et al., 2012), metagenomics (Nakamura et al., 2008), and new-generation sequencing (Voelkerding et al., 2009), there are two mainstream approaches to detect and identify a pathogen: serological detection, also known as Immunoassays (IA) and nucleic acids amplification tests (NAATs), represented mostly by PCR (Broussard, 2001).

Serology is sensitive, robust in austere conditions and applicable without any energy-dependent hardware: antibodies fixed on strips furnish expendable stand-alone assays, a useful feature in both field applications and in urban environments under failed infrastructures (Hjelle et al., 1995).

The bioagents were eventually declared a capital threat for the United States by the highest authority (The White House, 2004). The Research and Development initiatives following this deliberation resulted, among other concepts, in some interesting ideas designed specifically for surveillance and detection/classification purposes. In 2005 the annual audit to Congress reports on the CB-47 project of the US Department of Defense [DoD (Department of Defense), 2004a]; it was an enhanced serodiagnosis-based platform able to perform detection of accessible cellular and viral effector molecules plus different categories of cell-free effector molecules, such as toxins, prions, and viroids (Ding, 2010). The end result would have been a very promising and flexible tool, adaptable to new pathogens emerging at any given pace so as to cover different contingencies—and that in austere environments.

Still, serology has a number of drawbacks. The most important is that tampering with any level of the expression mechanism of a biological entity permits deletion or alteration of noneffector-related epitopes, without any loss of pathogenicity. Engineered pathogens are most likely to incorporate any measure of epitope concealment or passive deactivation of paratopes.

On the other hand, extremely high throughputs may make feasible conclusive antibody microarrays (see Chapter 6) which would compile a number of positive or even differential loci scores to synthesize a combined multiepitopic identity for each pathogen. This approach possesses inherent flexibility and adaptability, is agnostic-capable, and draws upon the ability to create microarrays with of up to 10^5 loci per assay (Palacios et al., 2007).

However, PCR allows DNA analysis; it requires more steps and far more resources than residual serology, but since both effector categories (cells of pathogens and toxic molecules) depend on

DNA-encoded protein sequences to function or to be produced (prions and virions excepted), the end result is a more robust approach (Nakamura et al., 2008). Multistep approaches exemplified by, but not restricted to, PCR-restriction fragment length polymorphism (RFLP) (which is relatively sensitive and very adaptable in detection and broad identification applications) were considered for field use before selecting single-channel chemistries that can produce results straight from the tube, as in the case with real-time PCR platforms.

The inherent degradation of the genetic code may allow a degree of concealment of the genetic signature of a modified pathogen through alternate coding to bypass sequence detection protocols, but this and other suitable applications of genetic/genomic engineering entail high costs, and high expenditure of material, temporal, and intellectual resources. At the same time, countermeasures existed already and included any option or combination of random priming, real-time applications, end-product prognosis and multilocus interrogation.

The CB-64 project [DoD (Department of Defense), 2006], developed practically in parallel with CB-47, was a genomic-oriented platform, destined to tackle the extremely wide field of genetically engineered/manipulated pathogens by using the then popular microarray concept (Kambouris, 2016; Lin et al., 2006; Wang et al., 2006; Heller, 2002) in its most advanced version: it enhanced the typical conclusive microarray breed (where the output is formed by the hierarchical combination of differential signals of positive loci—see Chapter 6) with *in situ* amplification and sequencing. Microarrays have extremely high throughput and offer an advantageous approach in cases of microbiotes where either fastidious requirements or the need for high-level biocontainment confound cultivation (Palacios et al., 2007). Moreover, in 2007 there were two documented case reports of multiplexed arrays detecting an infectious agent when traditional methods of investigation had failed, thus presenting a resounding proof-of-concept (Liu et al., 2009).

The CB-64 platform was supposed to detect, classify, and prioritize even bioengineered pathogens. The number of loci available on the microarray and their representativeness of actual sequences were limitations of low consequence at the time, as the state-of-the-art of the envisaged threat was the assembled genomes and the mainstream threat was the rudimentary altered ones (Kambouris et al., 2018a). The effort might have produced at least some applicable results: some platforms of mid-2000s (Lin et al., 2006; Wang et al., 2006; Malanoski et al., 2006) and even later (Gardner et al., 2010) might have been the spawn of CB-64.

Moreover, by the mid- to late 2000s, DNA microarrays were degenerate by design, with multiple different receptors per target, so as to reduce false-positives and allow better precision (Liu et al., 2009). Furthermore, the reporter oligonucleotides themselves were tolerant to a degree of mismatch, allowing hybridization between no perfect matches, so as to be able to detect slightly varying sequences, as is the case of related but unknown species/taxa and next-of-kin classification (Gardner et al., 2010; Palacios et al., 2007).

Once genomes designed and produced with detailed synthesis became available to mainstream research and development establishments, the above concept became obsolete. Only full-genome sequencing, in metagenomic formats and preferably on third-generation platforms, which dispense off amplification steps and require less depth and a fraction of the computational power (Heather and Chain, 2016), can tackle fully artificial microbes, at least so as to inform simpler techniques on the new signal identity, in order to embed it into more conventional, signal-seeking diagnostics.

Still, there is a drawback in NAAT: they are utterly inapplicable in the case of excreted, residual, or prepared toxins, one of the most prominent public health dangers and a most probable and handy bioterrorism agent (Anderson, 2012). This is the reason the US DoD envisaged the combined use of both the aforementioned platforms (CB-47 and CB-64) instead of downselecting one of them or expanding to additional ones.

INTERVENTION–CONTAINMENT–MANAGEMENT

Generic suppression

In massive incidents, broad-spectrum intervention is warranted to neutralize the sources of focal dispersions in both accidents and bioattacks. Such intervention is indiscriminate and microbicidal, based usually not only on standard decontamination with hot water and soap, broad-spectrum antibiotics, powerful biocidal, and/or oxidative chemicals (Delcomyn et al., 2006; Hart and Beeching, 2002; Ijaz et al., 2016; Primmerman, 2000; Wood et al., 2016) but also on different physical amenities such as high temperature, conflagration, shockwaves, irradiation (Kambouris et al., 2018b). The application of such amenities can be massive or individualized; in the second case, it may remain generic in terms of targeting. The massive option, for large volumes and/or many victims concentrated in limited space, is applied by respective generators carried on specialized vehicles or by loaded airborne assets dispersing said agent or producing said amenity.

The use of novel amenities is bound to allow a quantum leap in massive suppression and decontamination. The use of dynamic, electric, and other conceivable fields, in particular, can be scalable to accommodate crowds and large areas, thus lowering the burden, risk, and damage of powerful chemical decontaminants. Similarly, biological approaches as are natural enemies, predators, symbionts, and antagonists may also prove advantageous in cost, environmental impact, and efficiency. But their discovery and/or development presuppose heavy investment into applied OMICS and systems biology technological solutions, such as high-volume comparative culturomics, fast-track metagenomics, and also to improved and/or novel distribution methods, including massive use of drones (Fig. 15.3).

Selective treatment

Its purpose is to disinfect/decontaminate (decon)/sanitize contaminated sites, objects, and subjects from a biohazardous material but in a targeted manner, without exterminating benign natural microfauna/flora. Highly specific antibiotics come to mind, but well-tuned physical amenities [pressure waves, ultrasounds, and electromagnetic (EM) fields] as mentioned above may be applicable. If so, they are preferable for large volumes, both of open and enclosed spaces or cavities (Kambouris et al., 2018a,b).

The advent of *personalized medicine*, a concept of evidence-based medicine leaning toward treating the patient rather than the ailment as suggested by Hippocrates once upon a time in ancient Greece (5th century BCE), seems to revolutionize healthcare (Squassina et al., 2010). The concept favors treatment rather than prevention, through accurate diagnostics and specific, individualized therapeutic regimens (Squassina et al., 2010). But in massive syndromes/epidemics, personalized medicine is *not* applicable: it is slow, cumbersome, intermittent, expensive, and excessive (Gao and Keinan, 2016).

Instead, another concept is proposed, the *selective medicine*, still evidence-based but focusing on the ailment/pathogen more than the patient; it seems more suitable for combating infectious

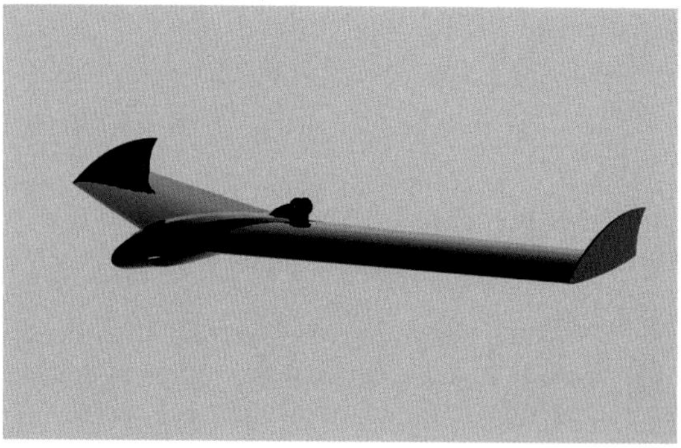

FIGURE 15.3

The agricultural and environmental sector is ahead in using UAV services. Moreover, modern drones have compact dimensions, straightforward operation syllabus, relatively permissive weight, and thus can be deployed on-call rather easily. Their modular, payload-agnostic bays, interfaced for quite a number of contingencies, allow cross-sectorial use with simple changes of payloads. Platforms similar to the illustrated Blackbird are mostly used for persistent multispectral monitoring of large areas and approach, physically or virtually, upon observing alarming signs and symptoms, for closer look. Though, their operational flexibility would make them suitable choice for targeted intervention by dispersing different kinds of antimicrobial agents. *UAV*, Unmanned aerial vehicles.

Photo: UcanDrone PC, with permission.

epidemics. Selective medicine is based on infectomics in conceptual terms rather than on human translational genomics and epidemiology, and partially overlaps with chem/bio/radio-hazardous material (HAZMAT) monitoring and management (Ludovici et al., 2015; Anonymous, 2017). As a result, it applies much more in prevention settings than in treatment, to exert effects before the pathogen harbors within the host and cause symptoms and health casualties, possibly variable ones due to the physiology, predisposition, and environment/ethology of each individual host. Most importantly, at this stage the pathogen is vulnerable and exposed to more drastic antimicrobial amenities than when it is within the tissues of the host. Such amenities include, as already mentioned, without being restricted to, different energy fields, oxidative/corrosive chemicals—especially acids—irradiation and high temperatures, either combined with pressure waves or not (Kambouris et al., 2018a,b).

But the most intriguing part of selective medicine is that it allows optimal use of purely biological solutions, nowadays called "Biologicals" or "biopharmaceuticals" (Declerck, 2012), such as the use of bacteriophages (Adhya et al., 2014; Golkar et al., 2014; Krylov, 2014), virophages (Desnues et al., 2012; Taylor et al., 2014), and hyperparasites (Parratt and Laine, 2016). All the above were originally delegated to microbial ecology and are now poised to be the highlights of the therapeutic microbiomics. As in both perpetrated and naturally occurring events, one or very few agents will be the major risk factors per incident, such selective and highly effective amenities suddenly become a preferred option as they are specific for their target organism and cause minimal environmental burden and collateral microbiota killing. Moreover, due to their mechanism of action, they are amenable themselves to genetic engineering and other translational genomics interventions so as to remain effective against evolving or new targets, increasing their net sustainability manifold.

After-infection intervention, known as medical countermeasures (MCM) in antibioterrorism communities (Ravi and Adalja, 2017), currently focuses on developing massively available antidotes, vaccines/antisera, and drugs, which are easy to stockpile, transport and administer. Such qualities would command minimal need for packaging, environmental control, handling, and distribution infrastructure and expenditure.

The dissolving microneedle patch vaccines (Sullivan et al., 2010) are excellent examples, as they can be administered to all age, sex, and race groups without any need for active cooperation (as in swallowing), irrespective of the condition of the skin, the respiratory, or the digestive tracts. In this respect they hold an advantage over pill vaccines and antidotes (Miles et al., 2018). In both these cases, any need to sterilize or keep sterile, to stockpile, and to transport administering equipment is abolished.

Physical amenities are also excellently suited: they do require energy, but in theory, one of them may substitute many different chemical compounds, each of the latter requiring careful stockpiling and distribution under supervision due to interactions and possible side effects (Kambouris et al., 2017). Instrumentation for physical amenities is usually tunable, allowing targeted and selective applications, as most come with bandwidth generators and/or tuning switches as standard; see Chapter 14.

Testing physical amenities to cope with emerging microbiotes is easier, compared to (bio)chemical agents, as it entails culturing in standard media, treating, and observing. Procedures for optimizing settings in newly detected pathogens or in the event of emergence of resistance should be developed as well, but they are likely to be nowhere near the cost and time required for new drugs. The object samples are easy to acquire, handle, and test conforming to the operating sequence "sample-assay-determine-apply," and descriptive culturomics approaches (see Chapter 8) may standardize and streamline the procedure (Kambouris et al., 2018b).

FRESH FROM THE PAST: ADAPTING OUR COGNITIVE DIMENSION TO AN EVOLVING UNIVERSE

In both civil and military surveillance applications concerning health issues (might those be of crops, livestock, or humans), the importance of threat definition, classification, and projection is paramount. Alert algorithms and respective procedures in both "detect-to-protect" and "detect-to-warn" applications are based onto a definition of what constitutes a threat (Kambouris et al., 2018a; Doggett et al., 2016; Wolicki et al., 2016).

The definition of the threat is a two-step process: in the first step a detected/collected entity/agent or object ("contact" or "datum") is provisionally included in or definitively excluded from a threat list/library according to its score on a range of criteria, qualitative and/or quantitative. The defining criteria may be accessed by the sensor system in an integrated process or applied separately, in a staged, hierarchical process. The second approach allows in turn either local processing by software, hosted within the sensor system, or distributed processing of the raw data at different on-line facilities, which would have access to higher computing efficiencies and diverse database resources (Doggett et al., 2016).

The second step uses the said threat list as a basis for the comparison of the detailed signature(s) of the contact, as apperceived by highly discriminative sensor system(s)—identifiers, to similar signatures of the threat library entities, in order to determine if the former constitute a threat (Deshpande et al., 2016; Petrovick et al., 2007).

TRAITS OF THE THREATS/COMPILATION OF A THREAT LIBRARY

Starting backwards, such threat libraries are compiled by stacking entries representing different entities, each one linked at the very least to its singular signature and its possible impact. In this tune, current such drafts might be used as a basis, with the best example being the regularly updated US Select Agents and Toxins list; the "Toxins" part might be left out of consideration, as when used independently, it verges to chemical warfare/chemoterrorism/chem-hazard. Strictly speaking, the biotoxins are biochemical and not biological agents (Joshi et al., 2013; Velegraki-Abel, 1986).

Thus the entries in Select Agents Lists are instrumental for the ability of an *in promptu* surveillance system to evaluate a detected entity/"contact." The evaluation is performed by comparing the acquired signal against threat libraries or sets of criteria so as to determine whether there is a valid threat and, if so, to implement risk assessment and lead to correct and timely decision-making. The latter might be reactive or proactive, as both options should be available and selection between them should depend on the crisis' pace and scale of development.

Although Select Agents List(s) is/are a good starting point, they remain exactly that: a starting point. Serious editing and expansion is warranted. A characteristic issue, exemplifying and underlining the need for review, is the practical exclusion from the US Government Select Agents List of fungi in particular, and of eukaryotes in general (Casadevall, 2017; Casadevall and Pirofski, 2006; Fierer and Kirkland, 2002). This issue is dwelled upon more thoroughly herein as a case study; other similar issues may be identified by careful scrutiny of the abovementioned and similar list(s).

Heated discussion is caused not only by the fact of this exclusion *per se*; even more alarming have been the questionable, if not outright weak, arguments used to support such a verdict. Currently, this specific list is used for deliberations and decision-making in many levels, despite remaining restricted to bacteria and viruses only, as far as the high-end, biosafety level (BSL)-4, live biothreats are concerned. Other views and opinions suggest that fungi should be counted within the bioagents of concern (Primmerman, 2000), but to no avail as they are not included to the respective lists and tables!

The fungus *Coccidioides*, being virulent to healthy individuals (*primary pathogen*) was originally the sole fungal entry (Casadevall, 2017; Casadevall and Pirofski, 2006; Dixon, 2001; Anonymous, 2005), obviously as an acknowledged issue by the US military due to its spontaneous virulence, its morbidity, and its endemic status in Continental US soil. Thus it featured as a handy organism for collection, cultivation, manipulation, and weaponization, conceivably by US services proper as well, to evade international attention. In the opposite sense, infiltrating malefactors might acquire, weaponize, and use it locally (Madad, 2014), striving to evade border controls and point-of-entry screening for bioagents. The latter attribute remains rather underestimated in respective literature.

Though, later drafts of the list contained no fungi at all (Anonymous, 2017; Casadevall, 2017). The reason for this deselection, which applies also to the briefly listed *Cryptococcus neoformans* yeast was arguably the perceived as rather low mortality of spontaneous fungal infections (Casadevall and Pirofski, 2006). Of course, such a verdict is debatable: mortality being *the* focal point is a medical perspective rather short-sighted and inflexible compared to more operationally focused standpoints, which would co-examine a larger number of more flexible and inclusive variables. A basic—or exemplary—such set of variables might well be (1) effectiveness, (2) logistics, and (3) upgrade/evolutionary potential.

Effectiveness

Although immunoefficient human hosts tend to fight off mycoses, the fungal virulence attributes might still inflict massive death toll in immunodeficient hosts, and this is true for many a fungal species. Besides, the tendencies observed recently within the genus *Cryptococcus*, which has become a major cause of death in sub-Saharan Africa and, more so, the emergence of *Candida auris*, an MDR eukaryote, introduce a next level of fungal threat compared to the naïve current beliefs mentioned above. At the same time, global warming might induce more fungal species to adapt to higher temperatures and thus become likely to negate the one of the two evolutionary shields the human species possess against systemic mycoses: the high temperature of human physiology, which is active even in people deprived of the other, the adaptive immunity (Casadevall, 2017).

In any case, this point is rather irrelevant in both tactical and strategic contexts of public health and biosecurity concerns. The main negative socioeconomic result of naturally occurring epidemics is the total disruptive effect in both physical and functional levels (Barbeschi, 2017; Cameron, 2017; Morens and Fauci, 2013; Schoch-Spana et al., 2017; Yassif, 2017). Understandably, this is the usual criterion of the projected effectiveness of any microbial candidate for weaponization (Croddy and Wirtz, 2005; Riedel, 2004).

Similarly, mycoses generally entail a prolonged gestation period (Casadevall and Pirofski, 2006). Far from constituting a reason to underestimate their threat status, this extended gestation denies early diagnosis and timely, correct administration of MCM (Madad, 2014). Similarly, security measures cannot be initiated in time so as to apprehend the perpetrators (Joshi et al., 2013), providing an opportunity to escape. The net result projects credible deterrence, thus becoming an attractive choice for perpetrators, compared to faster acting agents (Madad, 2014; Primmerman, 2000).

The low contagiousness of fungi, far from being another disadvantage for a propagator, should be viewed under the same light. Highly contagious pathogens entail high risk to backfire and affect own subjects/assets (Casadevall and Pirofski, 2006; Riedel, 2004) as allegedly happened with the Soviet implementation of biological warfare in Stalingrad (Weber, 1999). This leaves only long-distance strikes as viable operational options entailing low risk for self-inflicted casualties, and the more so when the operatives are ill-trained and poorly equipped, as was the case with Aum Shinrikyo terror group (Takahashi et al., 2004).

Pathogens of low contagiousness offer a more precise and permissive tool, as such choice provides multiple degrees of freedom for perpetrators and eases the mission planning: (1) less stringent standards apply in the handling sequence of a given pathogen; (2) storage, transportation, and deployment of the agent are simplified; and (3) the same applies to the selection of the human operators and their training (Croddy and Wirtz, 2005).

Logistics

Moreover, fungal agents are more prone to weaponization than other microbiota, especially bacteria, due to producibility and deployability factors (Casadevall and Pirofski, 2006). In many cases, bacteria considered as effective bioagents (Primmerman, 2000; Jakupciak and Colwell, 2009) might not be suitable for weaponization. The paradigm is *B. anthracis*: after meticulous selection of a suitable strain or its development from more benign ones by relatively simple genetic manipulations, as is plasmid insertion with conjugation or other techniques, a rather complicated process is required to optimize the cultivate for military-grade delivery and stockpiling (Inglesby et al., 2002). The case of Iraq is enlightening; the scientists of Saddam Hussein, in a stroke of genius, developed a better way to refine anthrax powder than conventional wisdom suggested (Shoham and Jacobsen, 2007). There are still doubts about the actual origin of the method, but the point remains: the bacterium required a degree of dedicated and specialized know-how to be weaponized.

Fungi, on the contrary, are in many cases almost self-developed biowarfare agents (Casadevall and Pirofski, 2006; Furcolow, 1961) and as such they had been used in simulations for biosecurity protocols (Cole, 1988; Riedel, 2004). Their ability to spontaneously form huge numbers of spores, sometimes able to travel great distances (O'Gorman, 2011; Pringle et al., 2005), amounts to almost half of the basic requirements for a bioweapon.

Moreover, they can be dispersed in many ways, they are difficult to diagnose, may cause significant levels of health casualties, and are susceptible to rather few drugs, many of which bear ill effects to human hosts (Casadevall and Pirofski, 2006). They also break vast host-species barriers rather easily (Esnakula et al., 2013; Ferrigo et al., 2016). Thus unknown or cryptic strains might arrive through trade and/or travel from intercontinental distances, rehost and multiply in environmental hosts, and present a sudden epidemic totally unknown to local medical authorities (Pringle et al., 2005).

Upgrade/evolutionary potential

But in the era of genetic manipulation, there is another issue: fungi are a class of organisms that can gain access to the skin through small abrasions and thus reach the subcutaneous tissue or exploit the follicular openings to infiltrate to it (O'Gorman, 2011; Mahajan, 2014). In this way, they may conceivably cause life-threatening, disseminated septic infections. The virulence characteristics make such a development something of a rarity, but the potential is there, and if some anti-virulence characteristics of the human species, such as thermal buffering, change, the impact may well multiply (Casadevall, 2017).

If the toxinogenicity of the kingdom as a whole is taken into account, the wide picture worsens. Extremely venomous mycotoxins causing acute effects are not readily produced by the more infectious species (Aleksic et al., 2017; O'Gorman, 2011; Velegraki-Abel, 1986). This can change though, and rather readily, by using simple gene transplantation techniques (Gibson et al., 2010), enhanced with overexpression cis regulatory elements, a technology mastered in plant biotechnology and other biotech industries (Biłas et al., 2016). The combination may result in very virulent, noncontagious fungal species able to inflict severe health casualties to all kinds of intended targets.

One should focus on the projected prospects to directly affect humans and animals by causing disease and destroy crops. Alternatively, more indirect results include empoisoning of primary sector's produce and thus cause food poisonings in a massive scale, or simply incur food shortage and social friction (Khalil and Shinwari, 2014; Velegraki-Abel, 1986).

Thus a novel threat emerges from a taxon up to now considered almost benign, at least in bioweaponeering terms. A biohazardous agent, suitable readily for disruptive use and encompassing both evolutionary and upgrade potential- which allow a much more sinister use, conditional on proper, but limited, manipulation.

"MEASURED, WEIGHED AND FOUND... THREATENING". ASSESSING THE THREAT FACTOR OF AN AGENT

To avoid such pitfalls in the threat definition of the future, quantifiable scoring systems should be developed and qualified by simulation in order to enable accurate -more or less- deduction of the biorisk posed by any given—or discovered, or engineered—microbiota. A pioneering concept, although rudimentary, has been previously proposed: the weapon potential (WP) of a microorganism (Casadevall and Pirofski, 2006). This has been a quantitative formula instead of semiquantitative empiric ones [Primmerman, 2000; DoD (Department of Defense), 2004a,b] for risk-assessment purposes. The latter are the basis for processing surveillance results so as to achieve actual intelligence.

The WP was defined as proportional to incidence, morbidity, contagiousness, and stability of the agent and inversely proportional to minimum morbid dose and gestation period: $WP = FsSC/IT$, where Fs is the incidence (the symptomatic fraction of the exposed population), I is the infectious dose (inoculum used for a given incapacitating result in terms of number and severity of cases), T is the gestation time (time elapsing between the assault and the onset of incapacitating symptoms), C is the contagiousness (the propensity of the agent to infect new hosts), and S is the stability of the agent.

These metrics need a bit of refinement, especially considering their microbiological dimension. The robustness of a microbe is more than its stability (S). Stability is related to the applicable production, stockpiling, and dispersion techniques. The latter may be spontaneous, instantaneous, overt, covert, pneumatic, or pyrotechnic (Pappas et al., 2009). This stability factor pertains the simplicity, accessibility, and affordability of the production facilities, the self-life of the processed agent, the infrastructure available for its conservation to a virulent but inert stage (refrigeration, oxygen deprivation) and its ruggedness in withstanding mechanical, thermal, or chemical shock associated with the dispersion method (Casadevall and Pirofski, 2006).

Robustness includes additional qualities. A very important one is the ability of the pathogen to remain aggressive through repeated infectious cycles in spreading contagious events among homogenous and heterogenous host populations. Another, possibly even more important, is the field persistence. It mostly depends on the agent's innate resistance to ambient and residual antimicrobial factors, such as is UV light, temperature, starvation, and low doses of antiseptics (Frinking et al., 2016). Last but not least is the resistance of the agent to specific MCM.

The availability of therapies is taken into consideration for the selection of a pathogen for weaponization. This parameter is also factored into the most basic classification, the BSLs: the definition of BSL-4, or rather, of the associated Risk Group 4 pathogens, explicitly identifies *the absence of available treatment or vaccination* as the basic inclusive criterion (Anonymous, 2004c). But in WP formulas, more exact parameters should be considered. Evolution indices, such as the frequency of stable conjugation and plasmid turnover, or the propensity for transposition and transformation are factors that (might) affect the selection process of a biowarfare/bioterrorism agent by the perpetrator.

Apart from the proposed elaboration of some parameters (as are the concepts of stability and robustness), there are three main areas to improve such formulas in terms of suitability, applicability, and accuracy. The first refers to the metrics proper. Any proposed formula must accommodate the need for different weighting of the parameters for significantly divergent scenarios. In addition, as already noted, more (micro)biologic parameters should be added, to better describe the malicious potential of any, natural or engineered, agent. And, finally, microbiomic parameters should be factored in as well, to supplement medical and microbiological ones with a wider and farther reaching perspective, where an agent potentially poses multiple threats and might have "accomplices," (not exclusively but) mainly microbial in nature, and, perhaps, enemies and antagonists of similar ilk just as well.

Regarding the used metrics and their specific weights, it should be remembered that the possible applications of bioassault differ significantly (Madad, 2014; Primmerman, 2000). For example, some microbes are better suited for offensive scenarios, disrupting targeted areas in precise spatiotemporal windows and with defined results, while others are applicable only for defensive use, with unpredictability and persistence used to discourage access to or through an area. Others still may be used in strategic context, with high secondary infectivity/contagiousness and some others for clandestine applications (Alibek and Handelman, 1999; Barbeschi, 2017; Madad, 2014; Schoch-Spana et al., 2017; Yassif, 2017).

All the aforementioned different applications require different gestation times for optimized performance. The same goes for lethality, pathogenicity, infectivity, virulence, and contagiousness: they are differently weighted in different mission profiles, and the More is not always the Better: high contagiousness sometimes is a reason for *not* using a pathogen (Primmerman, 2000).

The initial set of quantifiable parameters of WP, as proposed by the original publication, may be considered incomplete and medicine-oriented, although it is a great improvement over previous attempts with qualitative scoring (Primmerman, 2000). It includes no parameters outside standard epidemiology and infection aspects—and these incomplete. Mortality for example, an essential indicator of the severity of the expected health casualties and the degree of expected incapacitation is absent as a factor. But the key parameter in any malevolent scenario is stealthiness or, in simpler terms, undetectability (Madad, 2014), and it is absent altogether as well. The same applies to the suitability of the agent for genetic tampering/engineering and the agent's innate likelihood for misidentification/misdiagnosis (Madad, 2014).

To utilize prediction formulas (tentatively termed biothreat potential) in a wider context than immediate bioterrorism issues, other properties, of wider, microbiological nature, must be taken into account as well (Table 15.2) and be factored into improved expressions of said formulas. Such properties include but are not restricted to the following:

- Septic or toxic effect—the former involves adversely affecting tissue in physical continuity to the agent's cell, the latter implying adverse effect to tissues remote from the cellular agent or even residual adverse effects in its absence and/or while an agent is necrosed, decaying, or consumed.

Table 15.2 Basic Classification of Hazardous Bioagents and Selected Qualitative Risk Factor Scores.

Class of Agent	Pathogenic Mode	Nature	Currently Synthetic	Environmental Dormancy	Current Human Virulence
Bacteria	Septic/toxic	Prokaryote	+	Conditional	+
Fungi	Septic	Eukaryote	−	Conditional	+
Parasites	Septic	Eukaryote	−	Conditional	+
Viruses	Septic	Akaryote	+	−	+
Viroids	Septic	Akaryote	?	−	−
Prions	Toxic	Akaryote	−	−	−

Conditional, *Depending on species;* ?, *technically possible, but not acknowledged.*

- Cellular or acellular (virus, prion) form (Sapsford et al., 2008; Thavaselvam and Vijayaraghavan, 2010) of septic biological agents, which is an index of robustness, as cells are more vulnerable to some biochemical decontaminants and environmental conditions, exemplified by but not restricted to starvation.
- Ubiquity, which defines the probability of finding host and carrier populations (O'Gorman, 2011; Pringle et al., 2005).
- Infectivity/colonization of hosts by inhalation, which implies the presence of background signal (noise) due to pollen and dust in detection efforts; or by ingestion, which implies noise generated from other biological ingredients within samples (Madad, 2014; Petrovick et al., 2007).
- Ability for direct infection by penetrating intact, healthy skin (Hart and Beeching, 2002) or through hair follicles (Mahajan, 2014). It constitutes a very aggressive mode of infection, but not necessarily morbid and undeniably rare.
- Reproduction rate and mode, dissemination, and transmission mechanisms (Madad, 2014; O'Gorman, 2011), which affect the dispersion rate and impact heavily on prophylaxis, general precautions, decontamination protocols, and applicability of protective measures.
- Adaptability to adverse environment. Adaptation might be active, by developing resistance to antibiotics and toxins, or passive, by dormancy; in both modes, it increases the persistence and the robustness of the agent as it amplifies its resistance to decontamination efforts (Wood et al., 2013). Dormancy in particular may also hamper direct microbiological and serological diagnosis as metabolism and epitope presentation are suspended or altered (Aimanianda et al., 2009; Kambouris and Velegraki, 2001).

CONCLUSION—IS IT A DREAM OR A NIGHTMARE?

The emergence of the threat of pathogens reshuffled in a dual basis (novelty and frequency), combined with massive population movements of permanent or recurring nature, creates a most challenging environment for healthcare surveillance and monitoring. The number of possible public health hazards is multiplying without real prospect of attaining any effective tool of exclusion in the near future, which would assist diagnostic efforts. Thus high-specificity assays, absorbing huge resources of all kinds will become increasingly unaffordable and unsustainable. Military amenities for diagnostics on the other hand are nowadays fast and ruggedized to tackle time-sensitivity of operations while in austere, varied and usually adverse field conditions. Still, though, in their current format, they may cope with preciously few agents. Parallel formats for massive, simultaneous, and varied or, even better, agnostic interrogation of samples [DoD (Department of Defense), 2004b] are the future of diagnostics in both civil and military sectors.

A diagnostic revolution may be at hand, with fast turnovers, low developmental costs, and increased sustainability, conditional on two prospects. The first is the maturation of such concepts so as to show the necessary adaptability by retasking existing inquiring moieties/loci. Alternatively, the maturation might be approached by encompassing growth potential through integration of technologies enabling fast, customized and fast-track development of new assays (expedient prototyping) for new and/or unknown pathogens (agnostic formats).

Common logistic footprint for consumables and infrastructure, massive purchase for multiple, cross-sectorial tests and assays, timely implementation improving the prognosis and low developmental costs lead to an updated, more auspicious list of specifications for a new breed of diagnostics, focusing on affordability, accessibility, and flexibility in operational terms.

Such specifications updates should prove compatible with an extended surveillance scope (WHO, 2015c), encompassing more, most, or all sectors of biosecurity. This is a focal requirement, since the barriers among hosts have always been less formidable than acknowledged and continue to become even less so by the day as the pathogens themselves evolve and transit (Finlay and McFadden, 2006). The beefing up of such diagnostics with updated and adapting cognitive background, expressed mainly but not exclusively in decision-making and implemented mainly but not exclusively by software tools, might combine further with innovative intervention strategies to tackle ever intensifying and diversifying GCBRs (Kambouris et al., 2018d).

Despite different approaches and views, the airborne agent is the most feared case of perpetrated biothreat (Švábenská, 2012) and will continue being so, possibly in accidental and natural events as well. It is the most difficult to detect and contain, it disseminates fast and large, and affects extremely vital organs upon first contact. All the above constitute factors that enhance a massive disruptive effect (Barbeschi, 2017; Cameron, 2017; Connell, 2017; Schoch-Spana et al., 2017; Yassif, 2017), in times disproportionate to the actual destructive one (Madad, 2014; Primmerman, 2000).

The biothreat agents shall evolve. In the 21st century, it is rather restrictive to seek to simply augment a microbe's inherent pathogenicity by genetic manipulation—which usually implies the addition of virulence factors or of resistance genes, typically vectored by plasmids. This approach defines the second generation of bioagents, the best example being specific strains of *B. anthracis*. The most probable horizon of events predicts that once evolved and efficient genomic engineering protocols are fully developed and mastered, the septic constituent of the pathogen will be deemed obsolete as slow and suboptimal. The paradigm here is the CRISPR-Cas9 system, standing for "Clustered Regularly Interspaced Short Palindromic Repeats-CRISPR-associated gene-9" (Ran et al., 2013). In such a context the microbe will not be the weapon, but the vector, as is already the case in some natural instances, such as toxin-infested fungal spores (Aleksic et al., 2017).

This approach defines the third generation of bioagents, basically the "Metapathogens," where a robust, "optimized" microbe, producing an allogenic toxin (Patocka and Streda, 2006) or bioregulator upon germination or other environmental trigger, could wreak havoc while escaping medical response (diagnosis, prophylaxis, and treatment). The latter is so because the toxicosis' signs would be incompatible with the clinical picture of the septic phase/infection (Broussard, 2001) and diagnostic algorithms would be jammed. The desired, or dreaded, combination would consist of (1) the eukaryotic expression efficiency; (2) the ample genetic space available for insertion of genes or, rather, expression complexes in some eukaryotic taxa; and (3) a highly effective allogenic eukaryotic toxin or, for more precise and tunable results, an allogenic bioregulator (Anonymous, 2004b).

However, ultra-evolved bioagents, although within reach and clearly within the level of available technology, are not easy to acquire or develop. Less ambitious concepts include the microbiomic, not the microbiologic, aspects of a microbiot release (intentional or accidental). The new microbiote might be exterminated within a given, established microbiome, either environmental or hosted. But it may reshape and disturb a commensal microbiome, or it may transform some of its members to more aggressive and/or pathogenic varieties by horizontal gene transfer instead.

REFERENCES

Adhya, S., Merril, C.R., Biswas, B., 2014. Therapeutic and prophylactic applications of bacteriophage components in modern medicine. Cold Spring Harb. Perspect. Med. 4 (1), a012518.

Aimanianda, V., Bayry, J., Bozza, S., Kniemeyer, O., Perruccio, K., Elluru, S.R., et al., 2009. Surface hydrophobin prevents immune recognition of airborne fungal spores. Nature 460, 1117–1121.

Aleksic, B., Draghi, M., Ritoux, S., Bailly, S., Lacroix, M., Oswald, I.P., et al., 2017. Aerosolization of mycotoxins after growth of toxinogenic fungi on wallpaper. Appl. Environ. Microbiol. 83 (16), e01001–e01017.

Alibek, K., Handelman, S., 1999. Biohazard: The Chilling True Story of the Largest Covert Biological Weapons Program in the World, Told From the Inside by the Man Who Ran It. Random House.

Anderson, P.D., 2012. Bioterrorism: toxins as weapons. J. Pharm. Pract. 25 (2), 121–129.

Anderson, K., Morris, G., Kennedy, H., Croall, J., Michie, J., Richardson, M.D., et al., 1996. Aspergillosis in immunocompromised paediatric patients: associations with building hygiene, design, and indoor air. Thorax 51, 256–261.

Anonymous, 2005. HHS and USDA Select Agents and Toxins List. Available from: <https://jcesom.marshall.edu/media/36256/Select-Agent-List.pdf> (accessed 29.01.18.).

Anonymous, 2004a. WHO. Laboratory Biosafety Manual, third ed. WHO Report, vol. 178, Geneva, p. 7. Available from: <http://www.who.int/csr/resources/publications/biosafety/Biosafety7.pdf> (accessed 29.01.18.).

Anonymous, 2004b. WHO. Laboratory Biosafety Manual, third ed. WHO Report, vol. 178, Geneva, pp. 101–103. Available from: <http://www.who.int/csr/resources/publications/biosafety/Biosafety7.pdf> (accessed 29.01.18.).

Anonymous, 2004c. WHO. Laboratory Biosafety Manual, third ed. WHO Report, vol. 178, Geneva, pp. 1–2. Available from: <http://www.who.int/csr/resources/publications/biosafety/Biosafety7.pdf> (accessed 29.01.18.).

Anonymous, 2017. Select Agents and Toxins List. Available from: <https://www.selectagents.gov/selectagentsandtoxinslist.html> (accessed 29.01.18.).

Balajee, S.A., Borman, A.M., Brandt, M.E., Cano, J., Cuenca-Estrella, M., Dannaoui, E., et al., 2009. Sequence-based identification of *Aspergillus*, *Fusarium*, and *Mucorales* species in the clinical mycology laboratory: where are we and where should we go from here? J. Clin. Microbiol. 47 (4), 877–884.

Balajee, S.A., Arthur, R., Mounts, A.W., 2016. Global Health Security: building capacities for early event detection, epidemiologic workforce, and laboratory response. Health Secur. 14 (6), 424–432.

Barbeschi, M., 2017. A global catastrophic biological risk is not just about biology. Health Secur. 15, 349–350.

Bartholomew, R.A., Ozanich, R.M., Arce, J.S., Engelmann, H.E., Heredia-Langner, A., Hofstad, B.A., et al., 2017. Evaluation of immunoassays and general biological indicator tests for field screening of *Bacillus anthracis* and ricin. Health Sec. 15 (1), 81–96.

Beale, J., Brown, W., Cline, M., Cook, J., Desjardins, A., Fletcher, J., et al, 2002. Crop Biosecurity and Countering Agricultural Bioterrorism: Responses of the American Phytopathological Society. White Paper. Available from: <http://www.apsnet.org/publications/apsnetfeatures/Pages/Bioterrorism.aspx> (accessed 19.11.18.).

Biłas, R., Szafran, K., Hnatuszko-Konka, K., Kononowicz, A., 2016. *Cis*-Regulatory elements used to control gene expression in plants. Plant Cell Tissue Organ Cult. (PCTOC) 127, 269–287.

Broussard, L.A., 2001. Biological agents: weapons of warfare and bioterrorism. Mol. Diagn. 6, 323–333.

Brown, D.W., 1997. Threat to humans from virus infections of non-human primates. Rev. Med. Virol. 7, 239–246.

Cameron, E.E., 2017. Emerging and converging global catastrophic biological risks. Health Secur. 15 (4), 337–338.

Carlson, J., Wallstrom, S., 2017. Availability of Antibiotics. The Public Health Agency of Sweden 01229-2017-1. Solna & Stockhom. Available from: <https://www.folkhalsomyndigheten.se/contentassets/ab79066a80c7477ca5611ccfb211828e/availability-antibiotics-01229-2017-1.pdf>.

Casadevall, A., 2017. Don't forget the fungi when considering global catastrophic biorisks. Health Secur. 15, 341–342.

Casadevall, A., Pirofski, L.A., 2006. The weapon potential of human pathogenic fungi. Med. Mycol. 44 (8), 689–696.

Christensen, D.R., Hartman, L.J., Loveless, B.M., Frye, M.S., Shipley, M.A., Bridge, D.L., et al., 2006. Detection of biological threat agents by real-time PCR: comparison of assay performance on the R.A.P.I.D., the LightCycler, and the smart cycler platforms. Clin. Chem. 52, 141–145.

Cole, L.A., 1988. Clouds of Secrecy: The Army's Germ Warfare Tests over Populated Areas. Rowman & Littlefield.

Connell, N.D., 2017. The challenge of global catastrophic biological risks. Health Secur. 15, 345–346.

Costello, E.K., Lauber, C.L., Hamady, M., Fierer, N., Gordon, J.I., Knight, R., 2009. Bacterial community variation in human body habitats across space and time. Science 326, 1694–1697.

Croddy, E., Wirtz, J.J., 2005. Weapons of Mass Destruction: An Encyclopedia of Worldwide Policy, Technology, and History - ABC-CLIO. ABC-CLIO, Santa Barbara, CA. Available from: <https://www.abc-clio.com/ABC-CLIOCorporate/product.aspx?pc = A1440C> (accessed 20.01.18.).

Cullen, D.W., Toth, I.K., Pitkin, Y., Boonham, N., Walsh, K., Barker, I., et al., 2005. Use of quantitative molecular diagnostic assays to investigate *Fusarium* dry rot in potato stocks and soil. Phytopathology 95, 1462–1471.

Declerck, P.J., 2012. Biologicals and biosimilars: a review of the science and its implications. GaBI J. 1 (1), 13–16.

Dedkov, V.G., Magassouba, N.F., Safonova, M.V., Bodnev, S.A., Pyankov, O.V., Camara, J., et al., 2018. Sensitive multiplex real-time RT-qPCR assay for the detection of filoviruses. Health Secur. 16, 14–21.

Delcomyn, C.A., Bushway, K.E., Henley, M.V., 2006. Inactivation of biological agents using neutral oxone-chloride solutions. Environ. Sci. Technol. 40 (8), 2759–2764.

Deshpande, A., McMahon, B., Daughton, A.R., Abeyta, E.L., Hodge, D., Anderson, K., et al., 2016. Surveillance for emerging diseases with multiplexed point-of-care diagnostics. Health Secur. 14, 111–121.

Desnues, C., Boyer, M., Raoult, D., 2012. Sputnik, a virophage infecting the viral domain of life. Adv. Virus Res. 82, 63–89.

Ding, B., 2010. Viroids: self-replicating, mobile, and fast-evolving noncoding regulatory RNAs. Wiley Interdiscip. Rev. RNA 1 (3), 362–375.

Dixon, D.M., 2001. *Coccidioides immitis* as a select agent of bioterrorism. J. Appl. Microbiol. 91, 602–605.

DoD (Department of Defense), 2004a. CBRN Defense Program FY 2003-5 Performance Plan, p. 95. Available from: <http://www.acq.osd.mil/cp> (accessed 10.09.15.).

DoD (Department of Defense), 2004b. CBRN Defense Program FY 2003-5 Performance Plan, p. 57. Available from: <http://www.acq.osd.mil/cp> (accessed 10.09.15.).

DoD (Department of Defense), 2006. Chemical and Biological Defense Program Annual Report to Congress. F33. Available from: <https://fas.org/irp/threat/cbdp2006.pdf> (accessed 18.11.15.).

Doganay, M., Demiraslan, H., 2016. Refugees of the Syrian civil war: impact on reemerging infections, health services, and biosecurity in Turkey. Health Secur. 14 (4), 220–225.

Doggett, N.A., Mukundan, H., Lefkowitz, E.J., Slezak, T.R., Chain, P.S., Morse, S., et al., 2016. Culture-independent diagnostics for health security. Health Secur. 14, 122–142.

Ema, S.G., Mwadwo, K.-S., Paul, B.P., 2015. World News of Natural Sciences WNOFNS. Scientific Publishing House "DARWIN." Available from: <http://psjd.icm.edu.pl/psjd/element/bwmeta1.element.psjd-36da2d98-77d7-443c-8956-9dcd198565b3> (accessed 24.01.18.).

Esnakula, A.K., Summers, I., Naab, T.J., 2013. Fatal disseminated *Fusarium* infection in a human immunodeficiency virus positive patient. Case Rep. Infect. Dis. 2013, 379320.

Euler, M., Wang, Y., Heidenreich, D., Patel, P., Strohmeier, O., Hakenberg, S., et al., 2013. Development of a panel of recombinase polymerase amplification assays for detection of biothreat agents. J. Clin. Microbiol. 51, 1110–1117.

Fan, Y., Moon, J.J., 2017. Particulate delivery systems for vaccination against bioterrorism agents and emerging infectious pathogens. Wiley Interdiscip. Rev. Nanomed. Nanobiotechnol. 9 (1), 1403.

Ferrigo, D., Raiola, A., Causin, R., 2016. *Fusarium* toxins in cereals: occurrence, legislation, factors promoting the appearance and their management. Molecules 21, 627.

Fierer, J., Kirkland, T., 2002. Questioning CDC's "select agent" criteria. Science 295, 43.

Finlay, B.B., McFadden, G., 2006. Anti-immunology: evasion of the host immune system by bacterial and viral pathogens. Cell 124 (4), 767–782.

Frinking, E., Sweijs, T., Sinning, P., Bontje, E., Della Frattina, C., Abdalla, M., 2016. The Increasing Threat of Biological Weapons: Handle with Sufficient and Proportionate Care. Security 37, The Hague Center for Strategic Studies, The Hague.

Fuchs, V.R., July 23, 2014. Why Other Rich Nations Spend so Much Less on Healthcare. The Atlantic. Available from: <https://www.theatlantic.com/business/archive/2014/07/why-do-other-rich-nations-spend-so-much-less-on-healthcare/374576/> (accessed 30.10.18.).

Furcolow, M.L., 1961. Airborne histoplasmosis. Bacteriol. Rev. 25, 301–309.

Gao, F., Keinan, A., 2016. Explosive genetic evidence for explosive human population growth. Curr. Opin. Genet. Dev. 41, 130–139.

Gardner, S.N., Jaing, C.J., McLoughlin, K.S., Slezak, T.R., 2010. A microbial detection array (MDA) for viral and bacterial detection. BMC Genomics 11 (1), 668–689.

Gibson, D.G., Glass, J.I., Lartigue, C., et al., 2010. Creation of a bacterial cell controlled by a chemically synthesized genome. Science 329 (5987), 52–56.

Golkar, Z., Bagasra, O., Pace, D.G., 2014. Bacteriophage therapy: a potential solution for the antibiotic resistance crisis. J. Infect. Dev. Ctries. 8 (2), 129–136.

Gould, I.M., Bal, A.M., 2013. New antibiotic agents in the pipeline and how they can help overcome microbial resistance. Virulence 4, 185–191.

Gräser, Y., Kühnisch, J., Presber, W., 1999. Molecular markers reveal exclusively clonal reproduction in *Trichophyton rubrum*. J. Clin. Microbiol. 37, 3713–3717.

Grivas, K., Velegraki, A., Kambouris, M.E., 2008. Mid-term deployability and geointegration concerns in biodefence sampling and detection hardware design and procedures. Defence Pacis 22, 111–116.

Guarro, J., Gene, J., 1995. Opportunistic fusarial infections in humans. Eur. J. Clin. Microbiol. Infect. Dis. 14 (9), 741–754.

Halasz, L., Pinter, I., Solymar-Szocs, A., 2002. Remote sensing in the biological and chemical reconnaissance. AARMS 1, 39–56.

Hart, C.A., Beeching, N.J., 2002. A spotlight on anthrax. Clin. Dermatol. 20, 365–375.

Howard, E.J., Xiong, X., Carlier, Y., Sosa-Estani, S., Buekens, P., 2014. Frequency of the congenital transmission of *Trypanosoma cruzi*: a systematic review and meta-analysis. BJOG 121, 22–33.

Heather, J.M., Chain, B., 2016. The sequence of sequencers: the history of sequencing DNA. Genomics 107, 1–8.

Heller, M.J., 2002. DNA microarray technology: devices, systems, and applications. Annu. Rev. Biomed. Eng. 4, 129–153.

Hjelle, B., Jenison, S., Torrez-Martinez, N., Herring, B., Quan, S., Polito, A., 1995. Rapid and specific detection of Sin Nombre virus antibodies in patients with hantavirus pulmonary syndrome by a strip immunoblot assay suitable for field diagnosis. J. Clin. Microbiol. 35 (3), 600–608.

Huffman, J.A., Treutlein, B., Pöschl, U., 2010. Fluorescent biological aerosol particle concentrations and size distributions measured with an ultraviolet aerodynamic particle sizer (UV-APS) in Central Europe. Atmos. Chem. Phys. 10, 3215–3233.

Ijaz, M.K., Zargar, B., Wright, K.E., Rubino, J.R., Sattar, S.A., 2016. Generic aspects of the airborne spread of human pathogens indoors and emerging air decontamination technologies. Am. J. Infect. Control 44, S109–S120.

Inglesby, T.V., Henderson, D.A., Bartlett, J.G., Ascher, M.S., Eitzen, E., Friedlander, A.M., 2002. Anthrax as a biological weapon, 2002: updated recommendations for management. JAMA 287, 2236–2252.

Jackson, R.J., Ramsay, A.J., Christensen, C.D., Beaton, S., Hall, D.F., Ramshaw, I.A., 2001. Expression of mouse interleukin-4 by a recombinant *Ectromelia virus* suppresses cytolytic lymphocyte responses and overcomes genetic resistance to mousepox. J. Virol. 75 (3), 1205–1210.

Jakupciak, J.P., Colwell, R.R., 2009. Biological agent detection technologies. Mol. Ecol. Res. 9, 51–57.

Joshi, D., Kumar, D., Maini, A.K., Sharma, R.C., 2013. Detection of biological warfare agents using ultra violet-laser induced fluorescence LIDAR. Spectrochim. Acta A: Mol. Biomol. Spectrosc. 112, 446–456.

Kambouris, M.E., 2016. Population screening for hemoglobinopathy profiling: is the development of a microarray worth the while? Hemoglobin 40 (4), 240–246.

Kambouris, M.E., 2018. Mobile stand-off and stand-in surveillance against biowarfare and bioterrorism agents. In: Karampelas, P., Bourlai, T. (Eds.), Surveillance in Action. Advanced Sciences and Technologies for Security Applications., 2018. Springer, Cham, pp. 241–255.

Kambouris, M.E., Velegraki, A., 2001. *Aspergillus fumigatus*, *A. flavus* and *A. niger*: aerodynamic, immunological and metabolic virulence determinants. Arch. Hell Med. 18, 20–34.

Kambouris, M.E., Manoussopoulos, Y., Kantzanou, M., Velegraki, A., Gaitanis, G., Arabatzis, M., et al., 2018a. Rebooting bioresilience: a multi-OMICS approach to tackle global catastrophic biological risks and next-generation biothreats. OMICS 22 (1), 1–18.

Kambouris, M.E., Pavlidis, C., Skoufas, E., Arabatzis, M., Kantzanou, M., Velegraki, A., et al., 2018b. Culturomics: a new kid on the block of OMICS to enable personalized medicine. OMICS 22 (2), 108–118.

Kambouris, M.E., Manoussopoulos, Y., Kritikou, S., Milioni, A., Mantzoukas, S., Velegraki, A., et al., 2018c. Towards decentralized agrigenomics surveillance? A PCR-RFLP approach for adaptable and rapid detection of user-defined fungal pathogens in potato crops. OMICS 22 (4), 264–273.

Kambouris, M.E., Gaitanis, G., Manousopoulos, Y., Arabatzis, M., Kantzanou, M., Kostis, K., et al., 2018d. Humanome versus microbiome: games of dominance and pan-biosurveillance in the omics universe. OMICS 22 (8), 528–538.

Kambouris, M.E., Markogiannakis, A., Arabatzis, M., Manousopoulos, Y., Kantzanou, M., Velegraki, A., 2017. Wireless electrostimulation: a new approach in combating infection? Fut. Microbiol. 12, 255–265.

Kauchak, M., 2006. Letting Sensors Do the Work. Mil Med Technol. Available from: <http://www.military-medical-technology.com/mmt-home/149-mmt-2006-volume-10-issue-5/1310-letting-sensors-do-the-work.html> (accessed 29.06.16.).

Khalil, A.T., Shinwari, Z.K., 2014. Threats of agricultural bioterrorism to an agro dependent economy; what should be done? J. Bioterror. Biodef. 5 (1), 127–134.

Koblentz, G.D., 2017. The de novo synthesis of horsepox virus: implications for biosecurity and recommendations for preventing the reemergence of smallpox. Health Secur. 15, 620–628.

Krylov, V.N., 2014. Bacteriophages of *Pseudomonas aeruginosa*: long-term prospects for use in phage therapy. Adv. Vir. Res. 88, 227–278.

Laxminarayan, R., Duse, A., Wattal, C., Zaidi, A.K., Wertheim, H.F., Sumpradit, N., et al., 2013. Antibiotic resistance—the need for global solutions. Lancet Infect. Dis. 13, 1057–1098.

Lin, B., Wang, Z., Vora, G.J., Thornton, J.A., Schnur, J.M., Thach, D.C., et al., 2006. Broad-spectrum respiratory tract pathogen identification using resequencing DNA microarrays. Genome Res. 16, 527–535.

Liu, Y., Sam, L., Li, J., Lussier, Y.A., 2009. Robust methods for accurate diagnosis using pan-microbiological oligonucleotide microarrays. BMC Bioinformatics 10 (Suppl. 2), S11–S17.

Lopez-Rodriguez, P., Escot-Bocanegra, D., Fernandez-Recio, R., Bravo, I., 2014. Non-cooperative target recognition by means of singular value decomposition applied to radar high resolution range profiles. Sensors (Basel) 15 (1), 422–439.

Ludovici, G.M., Gabbarini, V., Cenciarelli, O., Malizia, A., Tamburrini, A., Pietropaoli, S., 2015. A review of techniques for the detection of biological warfare agents. Def. S&T Tech. Bull. 8 (1), 17–26.

Madad, S.S., 2014. Bioterrorism: an emerging global health threat. J. Bioterror. Biodef. 1, 5.

Magiorakos, A.P., Srinivasan, A., Carey, R.B., Carmeli, Y., Falagas, M.E., Giske, C.G., et al., 2012. Multidrug-resistant, extensively drug-resistant and pandrug-resistant bacteria: an international expert proposal for interim standard definitions for acquired resistance. Clin. Microbiol. Infect. 18, 268–281.

Mahajan, V.K., 2014. Sporotrichosis: an overview and therapeutic options. Dermatol. Res. Pract. 2014, 272376.

Malanoski, A.P., Lin, B., Wang, Z., Schnur, J.M., Stenger, D.A., 2006. Automated identification of multiple micro-organisms from resequencing DNA microarrays. Nucleic Acids Res. 34 (18), 5300–5311.

Mayboroda, O., Benito, A.G., Del Rio, J.S., Svobodova, M., Julich, S., Tomaso, H., et al., 2016. Isothermal solid-phase amplification system for detection of *Yersinia pestis*. Anal. Bioanal. Chem. 408 (3), 671–676.

Meneau, I., Sanglard, D., 2005. Azole and fungicide resistance in clinical and environmental *Aspergillus fumigatus* isolates. Med. Mycol. 43, S307–S311.

Miedaner, T., Geiger, H.H., 2015. Biology, genetics, and management of ergot (*Claviceps* spp.) in rye, sorghum, and pearl millet. Toxins (Basel) 7 (3), 659–678.

Miles, J.J., Tan, M.P., Dolton, G., Edwards, E.S., Galloway, S.A., Laugel, B., et al., 2018. Peptide mimic for influenza vaccination using nonnatural combinatorial chemistry. J. Clin. Invest. (Epub ahead of print) . Available from: <https://www.jcu.edu.au/news/releases/2018/march/taking-the-jab-and-the-chill-out-of-vaccination>.

Millett, P., Snyder-Beattie, A., 2017. Human agency and global catastrophic biorisks. Health Secur. 15, 335–336.

Misra, N., Panda, P.K., 2013. In search of actionable targets for agrigenomics and microalgal biofuel production: sequence-structural diversity studies on algal and higher plants with a focus on GPAT protein. OMICS 17, 173–186.

Morens, D.M., Fauci, A.S., 2013. Emerging infectious diseases: threats to human health and global stability. PLoS Pathog. 9 (7), e1003467.

Nakamura, S., Maeda, N., Miron, I.M., Yoh, M., Izutsu, K., Kataoka, C., et al., 2008. Metagenomic diagnosis of bacterial infections. Emerg. Infect. Dis. 14 (11), 1784–1786.

Nuzzo, J., Shearer, M., 2017. International engagement is critical to fighting epidemics. Health Secur. 15, 33–35.

O'Gorman, C.M., 2011. Airborne *Aspergillus fumigatus* conidia: a risk factor for aspergillosis. Fungal. Biol. Rev. 25, 151–157.

Oxford Nanopore Technologies, 2018. <www.nanoporetech.com> (accessed 30.10.18.).

Ozanich, R.M., Colburn, H.A., Victry, K.D., Bartholomew, R.A., Arce, J.S., Heredia-Langner, A., et al., 2017. Evaluation of PCR systems for field screening of *Bacillus anthracis*. Health Secur. 15 (1), 70–80.

Pak, T., 2008. A wireless remote biosensor for the detection of biological agents. In: Serving Nanoscale Science, Engineering & Technology, National Nanotechnology Infrastructure Network, pp. 24–25.

Palacios, G., Quan, P.L., Jabado, O.J., Conlan, S., Hirschberg, D.L., Liu, Y., et al., 2007. Panmicrobial oligonucleotide array for diagnosis of infectious diseases. Emerg. Infect. Dis. 13 (1), 73–81.

Pappas, G., Panagopoulou, P., Akritidis, N., 2009. Reclassifying bioterrorism risk: are we preparing for the proper pathogens? J. Inf. Pub. Health 2 (2), 55–61.

Parratt, S., Laine, A.-L., 2016. The role of hyperparasitism in microbial pathogen ecology and evolution. ISME J. 10, 1815–1822.

Patocka, J., Streda, L., 2006. Protein biotoxins of military significance. Acta Med. (Hradec. Kralove) 49 (1), 3–11.

Petrovick, M.S., Harper, J.D., Nargi, F.E., Schwoebel, E.D., Hennessy, M.C., Rider, T.H., et al., 2007. Rapid sensors for biological-agent identification. Lincoln Lab. J. 17 (1), 63–84.

Primmerman, C.A., 2000. Detection of biological agents. Lincoln Lab. J. 12 (1), 3–31.

Pringle, A., Baker, D.M., Platt, J.L., Wares, J.P., Latge, J.P., Taylor, J.W., 2005. Cryptic speciation in the cosmopolitan and clonal human pathogenic fungus *Aspergillus fumigatus*. Evolution 59 (9), 1886–1899.

Ramage, J.G., Prentice, K.W., DePalma, L., Venkateswaran, K.S., Chivukula, S., Chapman, C., et al., 2016. Comprehensive laboratory evaluation of a highly specific lateral flow assay for the presumptive identification of *Bacillus anthracis* spores in suspicious white powders and environmental samples. Health Secur. 14 (5), 351–365.

Ran, F.A., Hsu, P.D., Wright, J., Agarwala, V., Scott, D.A., Zhang, F., 2013. Genome engineering using the CRISPR-Cas9 system. Nat. Protocols 8 (11), 2281–2308.

Ravi, S., Adalja, A.A., 2017. Strengthening the US Medical Countermeasure Enterprise for biological threats. Health Secur. 15 (1), 12–14.

Riedel, S., 2004. Biological warfare and bioterrorism: a historical review. Proc. Baylor Univ. Med. Center 17, 400–406.

Ryan, C.P., 2008. Zoonoses likely to be used in bioterrorism. Public Health Rep. 123 (3), 276–281.

Sapsford, K.E., Bradburne, C., Delehanty, J., Medintz, I., 2008. Sensors for detecting biological agents. Mater. Today 11 (3), 38–49.

Schoch-Spana, M., Cicero, A., Adalja, A., Gronvall, G., Kirk Sell, T., Meyer, D., et al., 2017. Global catastrophic biological risks: toward a working definition. Health Secur. 15 (4), 323–328.

Shoham, D., Jacobsen, S.M., 2007. Technical intelligence in retrospect: the 2001 anthrax letters powder. Int. J. Intel. Counter. 20 (1), 79–105.

Squassina, A., Manchia, M., Manolopoulos, V.G., Artac, M., Lappa-Manakou, C., Karkabouna, S., et al., 2010. Realities and expectations of pharmacogenomics and personalized medicine: impact of translating genetic knowledge into clinical practice. Pharmacogenomics 11 (8), 1149–1167.

Stokes, H.W., Gillings, M.R., 2011. Gene flow, mobile genetic elements and the recruitment of antibiotic resistance genes into Gram-negative pathogens. FEMS Microbiol. Rev. 35 (5), 790–819.

Sullivan, S.P., Koutsonanos, D.G., del Pilar Martin, M., Lee, J.W., Zarnitsyn, V., Choi, S.O., et al., 2010. Dissolving polymer microneedle patches for influenza vaccination. Nat. Med. 16 (8), 915–920.

Švábenská, E., 2012. Systems for detection and identification of biological aerosols. Def. Sci. J. 62 (6), 404–411.

Takahashi, H., Keim, P., Kaufmann, A.F., Keys, C., Smith, K.L., Taniguchi, K., et al., 2004. *Bacillus anthracis* bioterrorism incident, Kameido, Tokyo, 1993. Emerg. Infect. Dis. 10 (1), 117–120.

Taylor, B., Cortez, M., Weitz, J., 2014. The virus of my virus is my friend: ecological effects of virophage with alternative modes of coinfection. J. Theor. Biol. 354, 124–136.

Taylor, K., Kolokoltsova, O., Ronca, S.E., Estes, M., Paessler, S., 2017. Live, attenuated Venezuelan equine encephalitis virus vaccine (TC83) causes persistent brain infection in mice with non-functional αβ T-cells. Front. Microbiol. 8, 81.

Thavaselvam, D., Vijayaraghavan, R., 2010. Biological warfare agents. J. Pharm. Bioallied Sci. 2 (3), 179–188.

The White House, April 28, 2004. Office of Press Secretary.

Thomas, T., Gilbert, J., Meyer, F., 2012. Metagenomics - a guide from sampling to data analysis. Microb. Inform. Exp. 2 (1), 3.

Velegraki-Abel, A., 1986. Mycotoxins. Athens (in Greek).

Vitousek, P.M., Mooney, H., Lubchenco, J., Melillo, J., 1997. Human domination of earth's ecosystems. Science 277 (5325), 494–499.

Voelkerding, K.V., Dames, S.A., Durtschi, J.D., 2009. Next-generation sequencing: from basic research to diagnostics. Clin. Chem. 55 (4), 641–658.

Wang, Z., Daum, L.T., Vora, G.J., Metzgar, D., Walter, E.A., Canas, L.C., et al., 2006. Identifying influenza viruses with resequencing microarrays. Emerg. Infect. Dis. 12, 638–646.

Weber, M., 1999. Secrets of the Soviet disease warfare program. J. Hist. Rev. 18 (2), 29–31.

WHO, 2015a. Available from: <http://www.who.int/drugresistance/Microbes_and_Antimicrobials/en/> (accessed 09.09.15.).

WHO, 2015b. Available from: <http://www.who.int/mediacentre/factsheets/fs395/en/> (accessed 12.11.15.).

WHO, 2015c. Available from: <http://www.who.int/mediacentre/factsheets/antibiotic-resistance/en/> (accessed 12.11.15.).

Wieser, A., Schneider, L., Jung, J., Schubert, S., 2012. MALDI-TOF MS in microbiological diagnostics—identification of microorganisms and beyond. Appl. Microbiol. Biotechnol. 93 (3), 965–974.

Wiles, S., 2015. All models are wrong, but some are useful: averting the 'microbial apocalypse'. Virulence 6 (8), 730–732.

Wolicki, S.B., Nuzzo, J.B., Blazes, D.L., Pitts, D.L., Iskander, J.K., Tappero, J.W., 2016. Public health surveillance: at the core of the global health security agenda. Health Secur. 14, 185–188.

Wood, T., Knabel, S., Kwan, B., 2013. Bacterial persister cell formation and dormancy. Appl. Environ. Microbiol. 79, 7116–7121.

Wood, J.P., Calfee, M.W., Clayton, M., Griffin-Gatchalian, N., Touati, A., Ryan, S., et al., 2016. A simple decontamination approach using hydrogen peroxide vapour for *Bacillus anthracis* spore inactivation. J. Appl. Microbiol. 121 (6), 1603–1615.

Public health and medical care systems. In: Woolf, S.H., Aron, L., National Research Council (US), Institute of Medicine (US) (Eds.), US Health in International Perspective: Shorter Lives, Poorer Health. National Academies Press, Washington, DC. Available from: <https://www.ncbi.nlm.nih.gov/books/NBK154484/>.

Woolhouse, M.E., 2002. Population biology of emerging and re-emerging pathogens. Trend Microbiol. 10, S3–S7.

Yassif, J., 2017. Reducing global catastrophic biological risks. Health Secur. 15 (4), 1–2.

FURTHER READING

Ackman, D., 2002. Pay Madness at Enron. Forbes. Available from: <https://www.forbes.com/2002/03/22/0322enronpay.html#37edb60e7a6d> (accessed 30.10.18.).

Bogle, J.C., 2002. Reflections on CEO Compensation. Academy of Management Perspectives, pp. 21–25. Available from: <http://webuser.bus.umich.edu/jpwalsh/PDFs/Bogle-2008-Reflectionson CEO compensation -- AMP paper.pdf> (accessed 30.10.18.).

DeWitte, S., 2014. Mortality risk and survival in the aftermath of the medieval Black Death, PLoS One, 9. p. e96513.

Lipsitch, M., 2017. If a global catastrophic biological risk materializes, at what stage will we recognize it? Health Secur. 15, 331–334.

Mackenbach, J.P., 2007. Global environmental change and human health: a public health research agenda. J. Epidemiol. Commun. Health 61, 92–94.

McFarland, L.V., 2008. Antibiotic-associated diarrhea: epidemiology, trends and treatment. Future Microbiol. 3, 563–578.

Power, E., 2006. Impact of antibiotic restrictions: the pharmaceutical perspective. Clin. Microbiol. Infect. 12 (Suppl. 5), 25–34.

Rees, W., 2010. What's blocking sustainability? Human nature, cognition, and denial. Sustain: Sci. Pract. Policy 6 (2), 13–25.

Shanahan, F., 2011. The colonic microflora and probiotic therapy in health and disease. Curr. Opin. Gastroenterol. 27, 61–65.

Stuart-Harris, C., 1984. Prospects for the eradication of infectious diseases. Rev. Infect. Dis. 6 (3), 405–411.

Türker, L., 2016. Thermobaric and enhanced blast explosives (TBX and EBX). Defence Technol. 12 (6), 423–445.

CHAPTER

EPILOGUE

16

Manousos E. Kambouris
The Golden Helix Foundation, London, United Kingdom

No clear prospect of the microbiomics may be projected with any certainty; they emerged in a hazy horizon of events. The revolution of molecular biology in terms of synthesis, followed by wildly available knowledge and skill, have brought about prospects or, rather, fears that biotechnology would disseminate into the actual household or even personal level, as happened with computers and IT in general (Regalado Antonio, 2012; Hemme Philip, 2016). From small, more or less licensed neighborhood labs to in-house applicable formats, affordability of basic expendables and instrumentation could, and actually may still, allow such a dissemination, which is a potential nightmare in both biosafety and biosecurity contexts. The projected risk areas range from spoilage of homemade brews and allergies due to a misengineered home strain to a multitude of crude but novel pathogens, created by any number of radicals and disgruntled students of life sciences, appearing at a monthly basis around the globe and each of them able, although not likely, to bring about a Global Catastrophic Biological Risk (GCBR) (Cameron, 2017).

But this leap of diffusion never came to being, and possibly shall not. It may, but the possibility is low. The reason is that it competes for actually limited resources, especially intellectual, within the limitations of a household, with IT and silica amenities. Human societies seem to have favored conventional technology and not being excited about the—very real—potential of biotechnology. Geeks prefer toiling over different machine codes and programming languages rather than over almost single-source DNA codes; innovation and workmanship are evidenced by shaping stone, wood, glass, metal, and plastic and combine, plug, and connect copper or semiconductor-cored parts rather than mixing biomolecules in home incubators, kettles, and Tupperware.

It is a matter of convenience, since the former has been around for two centuries and developed an unprecedented degree of familiarization with each and every member of the contemporary societies. It is also an issue of profiteering: colossal, behemoth corporations score high profits and are not bound to recast an existential competition for the sake of a new technology, where different rules and constants may apply, even if sustainability is the ultimate prize. Actually, they are loathsome to do so, *especially* if sustainability is at stake, since sustainability is patently bad for business based on consumption cycles—that is, for "business as usual."

But there is a more mundane reason for this, seemingly irreversible, choice: contrary to urban myth, nature is *not* the ultimate engineer, and the creation from the perspective of systems science leaves much to be desired which is evident by the ongoing process of evolution. The perceived inability of the human immune system to cope with metazoan parasites while in the process allergy

Microbiomics. DOI: https://doi.org/10.1016/B978-0-12-816664-2.00016-5

occurred, as a byproduct of a crucial but still ineffectual process, is one such example (Fitzsimmons et al., 2014). Autoimmunity and cancer point to blatant failures of regulation (Ivanova and Orekhov, 2015); on the other hand, the adaptability of microbiota to threats, antibiotics, and different environments in general is an unqualified success (Woappi et al., 2016). There may be a lesson in here, as to the extent in complexity where nature is at the moment at its best in systems design, and this may be a strong argument for the advancement of microbiomics.

The hidden reason for the mediocre engineering quality of the nature (or of whatever can be held responsible for this coherent and massive design effort we perceive as life forms) is very rational. It could not employ the best materials available, even in the context of this planet. Metals, ceramics, synthetic materials/plastics, semiconductors, and currently nanomaterials would have produced much better systems. This is obvious in the Enhanced Humanity concept (Buchanan, 2009), which was actually practiced for millennia, but conceptualized currently. In this concept, "spares" manufactured from the best available technological materials (mainly, but not exclusively, metals for most of this era) substitute for failed or otherwise compromised (i.e., by suffering wounds) parts and are shown to perform better and, especially during the last half century, actually improve on the original (Aydin, 2017). Of course the ultimate step is the proactive replacement to provide the enhanced, or optimized, performance, whence the currency of the "Enhanced Humanity" motto. The reason for this choice of lower-quality ingredients is self-evident in temporal prospect: Nature needed sustainability for its system of systems, and the need for autonomous reproduction, or semi-autonomous when exclusive sex species are considered, required a flexibility which could be afforded by the versatile but not optimal carbon chains. Being the quintessence of life in Earth biosphere, the carbon chain is versatile enough to use for very effective and diverse structures, some of them highly dedicated, while also allowing for coding the engineering plans in four dimensions (ontogeny-age being the fourth) and implementing them. A titanium skeleton is much more versatile than the original, made by the organic–inorganic alloy known as the bone (Taylor et al., 2007), but it cannot grow into being; much less replicate and draw molecules from the environment. It must be manufactured *de novo*.

Still, there *is* a very distinct possibility that microbiomics may prosper. The reason is the inherent sustainability and the diversification of the processes and entities developed, which may find application in numerous fields—examples being, selectively and not exclusively, the probiotics, pharmabiotics, and biopharmaceuticals (Gasbarrini et al., 2016; Hager and Ghannoum, 2017; Lee et al., 2018; Kooijman et al., 2012); such applications may be determined by the analytical power inherent in Big Data analyses and tackled by fine-tuning achievable through synthetic biology, most probably by genomic engineering but also by other translational intervention approaches. Microbiomics actually merge the previous collective concepts pertaining microbiology (microflora and microecology) to traceable networks; they also merge countable populations, mixtures, and effects into an integrated and functional whole, thanks to Big Data. In effect, a multidimensional, descriptive, and predictive picture emerges—an important dimension of the new Holly Grail of IT, the Quantified Planet (Özdemir, 2018), and by means of understanding and manipulating it, a true version of the Third Wave (Toffler, 1981) may be realized. Nature succeeded patently in distributed formats by flocks, herds, forests, and swarms (Trewavas, 2017; Zahadat et al., 2015), while technology strives with little success to do so. This distribution actually pertains sustainability, and once resources become the limiting factor even in technology, sustainability does become of essence. When reliable data can be efficiently collected and processed in the scale of the microbiome, which is Big Data (Navas-Molina et al., 2017), new opportunity arises for exploitation. Microbiota are excellent effectors, and may be improved manifold with switches, sensors, and new (bio)active molecules (Ford and Silver, 2015) and are prone to transform to reliable biosensors as well (Ling, 2017; Reen et al., 2017). Technologies may be developed not only by copying or using them as products but also by wiring them to interfaces so as to take advantage of the processing they spontaneously carry out (Kim et al., 2015; Reen et al., 2017). It is easier to listen to a microbiome for detecting an anomaly and to stimulate it to restore balance, than to conceive, design, and manufacture purely artificial, technical hardware to do so—and the microbiomic approach remains sustainable. In this, both start-ups and behemoths of the economy seem to consent the following: the use of structured bioresources (instead of raw materials) in both natural and artificial contexts (forestry and agriculture being examples of the former; biosensors and metabolite production/xenobiotics of the latter) is much more promising than in urban context and holds great potential for innovation, might that fuel profiteering or sustainability.

A biological compound due to its self-regulation and reproduction functions is much more efficient per weight compared to any active compound and is easier to produce: its extremely detailed and functional design, honed by many generations of evolution, requires much less effort to go to the production process by its natural carrier, allowing small units, moderate investments, distributed installations, with low environmental footprint, decreased transportation resources and affordable sizes, while making away with the need of structured, resource-hungry administrative, and staffing pyramidal models of production.

In a more practical, even technical, note, the subordinate biomes, such as virome, bacteriome, and mycetobiome (see Chapter 3: Myc(et)obiome: The Big Uncle in the Family) are prone to be discrete entities for quite some time. This is due to practicality: for any genomic and postgenomic kind of study, their collective complexity is too great, a fact seen with the panmicrobial arrays (Gardner et al., 2010) at a time when few microbial genomes had been sequenced. In addition, there have been considerable resources allocated to consensus gene metagenomic approaches, for both bacteria and fungi, and the stakeholders would not have anything to do with trashing such investments in the name of obsoleteness or redundancy. Still, slowly, things *are* bound to change. In the genomic sector alone, the development of long-range sequencing (LRS) to more affordable and reliable formats, either incorporating unbiased priming or in unprimed formats, will eventually allow shotgun metagenomics and metatranscriptomics in environmental (and, of course, clinical) samples. Advances in IT will allow Big Data being processed in more compact and affordable platforms, enabling true microbiomic-scale analyses of postgenomics fields (transcriptomics, metabolomics, and infectiomics).

At the same time, as the current generation of specialized microbiologists (mycologists, virologists, etc.) retire, the field, in professional terms, will be reshuffled: the new methodological, conceptual, and technological approaches demand dedicated training and special skills and there are neither the resources nor the will to fan out today's status of specialization any further, expanding current differentiation and specialization with additional levels and fields. Thus the near - future breeds of microbiologists will come full circle to more generic basis (Fig. 16.1) so as to eliminate current division and tackle all microbes, not just some taxa-Kingdoms, while their specialty and differentiation will be either functional (metabolome, genetics/genomics) or in terms of methodology and of technology (spectrometry, nucleic acids, immunomethods, imaging/microscopy), advancing the actual impact of the first (subcellular) and third (methodological) dimensions of omics, at the expense of the second (Kambouris et al., 2018).

Current specialties	Bacteriologist				
	Mycologist				
	Virologist				
Future specialties—expanded	*Bacteriologist*	*Genomicist*	*Metabolomicist*	*Ecologist/ interactomist*	*Translationalist*
	Mycologist	*Genomicist*	*Metabolomicist*	*Ecologist/ interactomist*	*Translationalist*
	Virologist	*Genomicist*	*Metabolomicist*	*Ecologist/ interactomist*	*Translationalist*
Future specialties—reshuffled	Microbiologist	Genomicist	Metabolomist	Ecologist/ interactomist	Translationalist

FIGURE 16.1

Indicative reshuffle projections of microbiology specialties under microbiomic concept.

REFERENCES

Aydin, C., 2017. The posthuman as hollow idol: a nietzschean critique of human enhancement. J. Med. Philos. A: Forum Bioeth. Philos. Med. 42, 304–327.

Buchanan, A., 2009. Human nature and enhancement. Bioethics 23, 141–150.

Cameron, E.E., 2017. Emerging and converging global catastrophic biological risks. Health Secur. 15, 337–338.

Fitzsimmons, C.M., Falcone, F.H., Dunne, D.W., 2014. Helminth allergens, parasite-specific IgE, and its protective role in human immunity. Front. Immunol. 5, 61.

Ford, T.J., Silver, P.A., 2015. Synthetic biology expands chemical control of microorganisms. Curr. Opin. Chem. Biol. 28, 20–28.

Gardner, S.N., Jaing, C.J., McLoughlin, K.S., Slezak, T.R., 2010. A microbial detection array (MDA) for viral and bacterial detection. BMC Genomics 11, 668–689.

Gasbarrini, G., Bonvicini, F., Gramenzi, A., 2016. Probiotics history. J. Clin. Gastroenterol. 50, S116–S119.

Hager, C.L., Ghannoum, M.A., 2017. The mycobiome: role in health and disease, and as a potential probiotic target in gastrointestinal disease. Dig. Liver. Dis. 49, 1171–1176.

Hemme Philip, 2016. 3 Major impacts biotechnology could have by 2030. In: Falling Walls Fragments. Falling Walls Fragm. <https://www.fallingwallsfragments.com/2016/02/16/3-major-impacts-biotechnology-could-have-by-2030/> (accessed 05.10.19.).

Ivanova, E.A., Orekhov, A.N., 2015. T helper lymphocyte subsets and plasticity in autoimmunity and cancer: an overview. Biomed. Res. Int. 2015, 1–9.

Kambouris, M.E., Gaitanis, G., Manoussopoulos, Y., Arabatzis, M., Kantzanou, M., Kostis, G.D., et al., 2018. Humanome versus microbiome: games of dominance and pan-biosurveillance in the omics universe. OMICS 22, 528–538.

Kim, M., Lim, J.W., Kim, H.J., Lee, S.K., Lee, S.J., Kim, T., 2015. Chemostat-like microfluidic platform for highly sensitive detection of heavy metal ions using microbial biosensors. Biosens. Bioelectron. 65, 257–264.

Kooijman, M., van Meer, P.J., Moors, E.H., Schellekens, H., 2012. Thirty years of preclinical safety evaluation of biopharmaceuticals: did scientific progress lead to appropriate regulatory guidance? Expert. Opin. Drug. Saf. 11, 797–801.

Lee, E.-S., Song, E.-J., Nam, Y.-D., Lee, S.-Y., 2018. Probiotics in human health and disease: from nutribiotics to pharmabiotics. J. Microbiol. 56, 773–782.

Ling, F., 2017. Microbial communities as biosensors for monitoring urban environments. Microb. Biotechnol. 10, 1149–1151.

Navas-Molina, J.A., Hyde, E.R., Sanders, J.G., Knight, R., 2017. The microbiome and big data. Curr. Opin. Syst. Biol. 4, 92–96.

Özdemir, V., 2018. The dark side of the moon: the Internet of Things, Industry 4.0, and the quantified planet. OMICS 22, 637–641.

Reen, F.J., Gutiérrez-Barranquero, J.A., O'Gara, F., 2017. Mining microbial signals for enhanced biodiscovery of secondary metabolites. Methods Mol. Biol. (Clifton, N.J.). 1539, 287–300.

Regalado Antonio, 2012. Doing Biotech in My Bedroom - MIT Technology Review. MIT Technol. Rev. <https://www.technologyreview.com/s/426885/doing-biotech-in-my-bedroom/> (accessed 05.10.19.).

Taylor, D., Hazenberg, J.G., Lee, T.C., 2007. Living with cracks: damage and repair in human bone. Nat. Mater. 6, 263–268.

Toffler, A., 1981. The Third Wave. Bantam Books, New York.

Trewavas, A., 2017. The foundations of plant intelligence. Interface Focus 7, 20160098.

Woappi, Y., Gabani, P., Singh, A., Singh, O.V., 2016. Antibiotrophs: the complexity of antibiotic-subsisting and antibiotic-resistant microorganisms. Crit. Rev. Microbiol. 42, 17–30.

Zahadat, P., Hahshold, S., Thenius, R., Crailsheim, K., Thomas Schmickl, T., 2015. From honeybees to robots and back: division of labour based on partitioning social inhibition. Bioinspir. Biomim. 10, 066005.

Index

Note: Page numbers followed by "*f*" and "*t*" refer to figures and tables, respectively.

D

E

F

G

H

I

J

K

L

M

N

O

P

X

Z

Printed in the United States
By Bookmasters